The Corpus Callosum

Mehmet Turgut • R. Shane Tubbs
Ahmet Tuncay Turgut • Cuong C. J. Bui
Editors

The Corpus Callosum

Embryology, Neuroanatomy, Neurophysiology, Neuropathology, and Surgery

Springer

Editors
Mehmet Turgut
Department of Neurosurgery
Aydın Adnan Menderes University
Faculty of Medicine
Aydın, Turkey

Ahmet Tuncay Turgut
Department of Radiology
Ankara Medipol University
Faculty of Medicine
Ankara, Turkey

R. Shane Tubbs
Department of Neurosurgery
Tulane School of Medicine
New Orleans, LA, USA

Cuong C. J. Bui
Department of Neurosurgery
Ochsner Medical Center
New Orleans, LA, USA

ISBN 978-3-031-38113-3 ISBN 978-3-031-38114-0 (eBook)
https://doi.org/10.1007/978-3-031-38114-0

This Springer imprint is published by the registered company Springer Nature Switzerland AG
The registered company address is: Gewerbestrasse 11, 6330 Cham, Switzerland

Paper in this product is recyclable.

Preface

Our knowledge of the human neuroanatomy has expanded in recent years with, for example, improved methods of imaging the brain with greater resolution. Additionally, the clinical data gathered from many years of surgical procedures involving the brain has added to our understanding of its various functions. However, one area of the brain that has had less attention is the corpus callosum (CC). The importance of this largest of the commissural tracts has become clearer recently, but still many of its functions are unclear. Newer imaging modalities such as tractography and functional MRI have taught us much and have shown us that the CC is affected in more disease processes than once thought.

In this comprehensive textbook devoted to the CC, the editors have collected 41 chapters from experts from around the world. These offerings are divided into six parts: Embryology and Neuroanatomy of the Corpus Callosum, Neurophysiology of the Human Corpus Callosum, Congenital and Acquired Neuropathology of the Corpus Callosum, Surgery of the Corpus Callosum, Cognitive Neuroscience as it relates to the CC, and Other Features of the Corpus Callosum. Taken together, it is the hope of the editors that the reader will become better acquainted with this fascinating part of the human central nervous system by having a wider understanding of its anatomy, physiology, pathology, and treatment of diseases that affect it. Although there is still much to learn about the human CC, the knowledge contained in this textbook will highlight our current understanding.

Aydın, Turkey — Mehmet Turgut
New Orleans, LA, USA — R. Shane Tubbs
Ankara, Turkey — Ahmet Tuncay Turgut
New Orleans, LA, USA — Cuong C. J. Bui

Contributors

Niyazi Acer Department of Anatomy, Arel University School of Medicine, İstanbul, Turkey

Abad Cherif El Asri Department of Neurosurgery, Mohamed V Military Hospital, Rabat, Morocco

Müge Ayanoğlu Department of Pediatric Neurology, Aydın Adnan Menderes University Faculty of Medicine, Aydın, Turkey

Fayize Maden Bedel Department of Pediatric Genetics, Meram Medical Faculty, Necmettin Erbakan University, Konya, Turkey

Sanjay Behari Department of Neurosurgery, Sanjay Gandhi Postgraduate Institute of Medical Sciences, Lucknow, India

Mohammed Benzagmout Department of Neurosurgery, University Hospital of Fez, Fez, Morocco

Clinical Neurosciences Laboratory, Faculty of Medicine, Pharmacy and Dentistry, University Sidi Mohamed Ben Abdellah, Fez, Morocco

Kyle Biggins Tulane University School of Medicine, New Orleans, LA, USA

Okan Bilge Department of Anatomy, Ege University Faculty of Medicine, İzmir, Turkey

Nagehan Bilgeç Department of Pediatric Genetics, Meram Medical Faculty, Necmettin Erbakan University, Konya, Turkey

Nigel Blackwood Tulane University School of Medicine, New Orleans, LA, USA

Cuong C. J. Bui Department of Neurosurgery, Ochsner Medical Center, New Orleans, LA, USA

Burcu Çalışkan Necmettin Erbakan University Faculty of Medicine, Department of Pediatric Neurology, Necmettin Erbakan University, Konya, Turkey

Department of Pediatric Neurology, Faculty of Medicine, Necmettin Erbakan University, Konya, Turkey

Abdullah Canbal Necmettin Erbakan University Faculty of Medicine, Department of Pediatric Neurology, Necmettin Erbakan University, Konya, Turkey

Department of Pediatric Neurology, Faculty of Medicine, Necmettin Erbakan University, Konya, Turkey

Servet Çelik Department of Anatomy, Ege University Faculty of Medicine, İzmir, Turkey

Application and Research Center of Cord Blood Cell-Tissue, Ege University, İzmir, Turkey

Pınar Çeltikçi Department of Radiology, Ankara Bilkent City Hospital, Ankara, Turkey

Michael Cesarek Tulane University School of Medicine, New Orleans, LA, USA

Mitchell W. Couldwell Tulane University School of Medicine, New Orleans, LA, USA

Kuntal Kanti Das Department of Neurosurgery, Sanjay Gandhi Postgraduate Institute of Medical Sciences, Lucknow, India

Balachandar Deivasigamani Department of Neurosurgery, Sanjay Gandhi Postgraduate Institute of Medical Sciences, Lucknow, India

Okan Derin Department of Anatomy, Ege University Faculty of Medicine, İzmir, Turkey

Priyadarshi Dikshit Department of Neurosurgery, Sanjay Gandhi Postgraduate Institute of Medical Sciences, Lucknow, India

Pelin Dilsiz Anesthesiology and Reanimation Clinic, Menteşe State Hospital, Menteşe, Muğla, Turkey

Oreste de Divitiis Division of Neurosurgery, Department of Neurosciences and Reproductive and Odontostomatological Sciences, Università degli Studi di Napoli Federico II, Naples, Italy

Graham Dupont Department of Cellular and Structural Biology, Tulane School of Medicine, New Orleans, LA, USA

Vikas Dwivedi Department of Neurosurgery, Sanjay Gandhi Postgraduate Institute of Medical Sciences, Lucknow, India

Uduak-Obong I. Ekanem Department of Neurosurgery, Tulane University School of Medicine, New Orleans, LA, USA

Fedi Ercan Department of Obstetrics and Gynecology, Division of Perinatology, Aydin Adnan Menderes University School of Medicine, Efeler, Aydın, Turkey

Damla Balkan Erdoğan Kocaeli City Hospital, İzmit, Turkey

Huriye Berra Ertuğrul Department of Pediatrics, Aydın Adnan Menderes University Faculty of Medicine, Aydın, Turkey

Mengzhao Feng Department of Neurosurgery, First Affiliated Hospital of Zhengzhou University, Zhengzhou, Henan Province, China

International Joint Laboratory of Nervous System Malformations, First Affiliated Hospital of Zhengzhou University, Zhengzhou, Henan Province, China

Vaitsa Giannouli Aristotle University of Thessaloniki, Thessaloniki, Macedonia, Greece

Ryan Glynn Department of Neurosurgery, Tulane University Hospital, New Orleans, LA, USA

Viktoriya Grayson Department of Neurosurgery, Tulane University School of Medicine, New Orleans, LA, USA

Melisa Gülcan Department of Anatomy, Ege University Faculty of Medicine, İzmir, Turkey

Erkan Gümüş Department of Histology and Embryology, Aydın Adnan Menderes University Faculty of Medicine, Aydın, Turkey

Fuyou Guo Department of Neurosurgery, First Affiliated Hospital of Zhengzhou University, Zhengzhou, Henan Province, China

International Joint Laboratory of Nervous System Malformations, First Affiliated Hospital of Zhengzhou University, Zhengzhou, Henan Province, China

Nihal Gürlek Çelik Department of Anatomy, Amasya University School of Medicine, Amasya, Turkey

Sara Hanaei Universal Scientific Education and Research Network (USERN), Tehran, Iran

Department of Neurosurgery, Imam Khomeini Hospital Complex, Tehran University of Medical Sciences (TUMS), Tehran, Iran

Borderless Research, Advancement, and Innovation in Neuroscience Network (BRAINet), Tehran, Iran

Meryem Himmiche Department of Neurosurgery, University Hospital of Tangiers, Tangiers, Morocco

Ayhan Kanat Department of Neurosurgery, Medical Faculty, Recep Tayyip Erdogan University, Rize, Turkey

Mehmet Saim Kazan Department of Neurosurgery, Akdeniz University School of Medicine, Antalya, Turkey

Hayriye Nermin Keçeci Necmettin Erbakan University Faculty of Medicine, Department of Pediatric Genetics, Necmettin Erbakan University, Konya, Turkey

Department of Histology and Embryology, Aydın Adnan Menderes University Health Sciences Institute, Aydın, Turkey

Yaşar Türk Radiology Department, Şişli Kolan International Hospital, İstanbul, Turkey

Yigit Uyanıkgil Application and Research Center of Cord Blood Cell-Tissue, Ege University, İzmir, Turkey

Department of Histology and Embryology, Ege University School of Medicine, İzmir, Turkey

Osman Vırıt Department of Psychiatry, Muğla Sıtkı Koçman University School of Medicine, Muğla, Turkey

Aaron Yu Tulane University School of Medicine, New Orleans, LA, USA

Yuchao Zuo Department of Neurosurgery, First Affiliated Hospital of Zhengzhou University, Zhengzhou, Henan Province, China

International Joint Laboratory of Nervous System Malformations, First Affiliated Hospital of Zhengzhou University, Zhengzhou, Henan Province, China

Part I

Embryology and Neuroanatomy of the Corpus Callosum

Editor's Summary

Proper development of the corpus callosum (CC) is critical for optimized brain function. Derailment of this embryological process can lead to various pathological conditions. Therefore, knowledge of the neuroembryological and neuromolecular processes that form the CC is important. Moreover, a thorough understanding of the gross anatomy of the CC is important to not only anatomists but also clinicians who visualize this part of the brain such as radiologists and neurosurgeons who operate through or near this structure. This section of the textbook will outline the key embryological and neuroanatomical concepts relating to the CC.

History of the Corpus Callosum 1

Nikolaos Syrmos, Vaitsa Giannouli, Sotirios Kottas, and Mehmet Turgut

1.1 Introduction

The corpus callosum (CC) is a large, structurally and functionally important white matter tract that connects the two brain hemispheres. It enables humans to perceive depth and allows the two sides of the brain to communicate. "CC" is Latin for "tough body." It is the great white matter complex in the human brain in terms of both size and number of axonal connections between the two hemispheres. Precise knowledge of the mechanisms of the cerebral cortex is essential for describing and explaining mental processes, psychiatric functions, and the operative status of the whole human central nervous system. The study of the cerebral cortex remains a challenging aspect of neurosciences (neurology, neurosurgery, and psychiatry) and of related scientific fields such as neuroanatomy, neurophysiology, neuroradiology-neuroimaging, neurobehavioral studies, and neuropsychology [1–5].

1.2 Galen of Pergamos

The Hellenic Galen of Pergamos (131–201 CE) was the first to describe the CC, according to Franz Joseph Gall and various other authors. Galen considered it to be part of a complex arrangement for suspending the brain hemispheres from the skull. The leading physician, anatomist, surgeon, and philosopher of his era, he is considered one of the most accomplished medical researchers of the ancient world. He influenced various scientific disciplines (anatomy, physiology, pathology, surgery, neurology, neurosurgery pharmacology, logic, and philosophy). His eighteenth century depiction by Georg P. Busch is known worldwide [1, 6–12] (Fig. 1.1).

N. Syrmos (✉)
Aristotle University of Thessaloniki,
Thessaloniki, Macedonia, Greece

Human Performance and Health, Master Science Program, Aristotle University of Thessaloniki, Thessaloniki, Macedonia, Greece

V. Giannouli
Aristotle University of Thessaloniki,
Thessaloniki, Macedonia, Greece

S. Kottas
Papanikolou General Hospital,
Thessaloniki, Macedonia, Greece

M. Turgut
Department of Neurosurgery, Aydın Adnan Menderes University Faculty of Medicine, Aydın, Turkey

Department of Histology and Embryology, Aydın Adnan Menderes University Health Sciences Institute, Aydın, Turkey

M. Turgut et al. (eds.), *The Corpus Callosum*, https://doi.org/10.1007/978-3-031-38114-0_1

Fig. 1.1 Galen 131–201 BC

1.3 The Vesalius Studies

Andreas Vesalius (Latin version of Andries van Wezel) was born on December 31, 1514, in Brussels, which at that time was region of the Habsburg Netherlands. Vesalius is recognized as a famous sixteenth-century neuroanatomist and author of one of the most famous human anatomy books, *De Humani Corporis Fabrica Libri Septem*. He served as professor of human anatomy in Padova, Italy, between 1537 and 1542. Later, he got imperial physician at the court of Emperor Charles V. He died on the Hellenic island of Zakynthos (Zante) in the Ionian Sea in 1564. Vesalius is often considered the founder of modern human anatomy and morphology. During 1555, he performed anatomical and physiological studies on the CC. He described both its appearance and its gross position precisely. There

Fig. 1.2 Vesalius 1514–1564

is a famous portrait of him by Jan van Calcar, a German-born Italian artist [2, 9, 11–17] (Fig. 1.2).

1.4 Peyronie, Monro, and Gall

François Gigot de la Peyronie (January 15, 1678–April 25, 1747) was a surgeon from Montpellier, France. His name is related with a condition also called "Peyronie's disease" in medical literature. According to his anatomical and physiological studies, the CC is the seat of the human soul. The same opinion was expressed by Lanisi in 1713 [1, 9, 18–26] (Fig. 1.3).

Alexander Monro of Craiglockhart and Cockburn (May 22, 1733–October 2, 1817) was an anatomist, physiologist, and researcher from Scotland. Monro was from a distinguished family of physicians. He described the CC and the lymphatic system clearly, providing details of the musculoskeletal system and other parts of the human body. He is also recognized for the "Monro-Kellie doctrine" on intracranial pressure, a hypothesis developed by Monro and his student George Kellie, later a famous surgeon in the port of Leith [2, 9] (Fig. 1.4).

Fig. 1.3 Peyronie 1678–1747

Fig. 1.4 Monro 1733–1817

Franz Gall was a German anatomist, neuroscientist, physiologist, psychologist, anthropologist, and researcher in the field of brain and behavior. He was born on March 9, 1758, and he died on August 22, 1828. He believed that regional differences in the human cerebral cortex could affect the shape of the skull. He performed various cranioscopic, craniometric, and phrenological studies, both anatomical and physiological. He made clear, detailed, and lucid descriptions of the structure of the CC. In 1818, he published a five-volume work together with Johann Gaspar Spurzheim, a promoter of phrenology, about the anatomy and physiology of the nervous systems of humans and other animals. In this extensive work, there are observations and notes on the possibility of understanding the many moral and intellectual dispositions not only of humans but also of animals from the configurations of their heads [3, 9, 27–34] (Fig. 1.5).

Fig. 1.5 Gall 1758–1828

1.5 The Studies of Félix Vicq d'Azyr, Romberg, and Gowers

Félix Vicq d'Azyr (April 23, 1748–June 20, 1794) was a famous physician of his era and also a homologist, biologist, scientist, comparative anatomist, and physiologist. He was born in Valognes, Normandy, France. He studied medicine at the University of Paris, and after his graduation, he continued his studies at the Jardin du Roi (currently the Paris Museum of Natural History). Later, he was chosen member of the Académie des Sciences and afterward a lifetime member of the Académie Française and also a creator of an important medical society (Société Royale de Médecine). He was the last doctor of medicine of Queen Marie Antoinette. At the School of Alfort, he served as a professor of veterinary medicine and also as supervisor of epidemiological and other studies. He performed anatomical studies and managed to describe the CC, venturing an interpretation of its functions. He wrote his anatomical observations in an anonymous collection. He died during a tuberculosis pandemic. An anthology of some of his documents is held at the National Library of Medicine in Bethesda, Maryland, USA [9, 24–38] (Figs. 1.6 and 1.7).

Fig. 1.6 Vicq d'Azyr 1748–1794

Moritz Heinrich Romberg (November 11, 1795–June 16, 1873) was a German neurologist born in Meiningen, and he published his classic textbook in the 1840s. He performed various medical studies of optical paths and sensory ataxia. He studied the CC and the effect of various diseases on normal human brain and nervous system function. A truly innovative neurologist in Europe, he spent a 30-year period (1820–1850) on various clinical observations and deductions concerning a discipline that was then in its infancy. He is recognized by many authors as "the first clinical neurologist." He defined "Romberg's sign" in his original description of tabes dorsalis as an entity caused by syphilis involving the dorsal roots of the spinal nerve which emerge from the spinal cord [1, 9, 23–38] (Fig. 1.8).

Sir William Richard Gowers (March 20, 1845–May 4, 1915) was a British neuroscientist. According to MacDonald Critchley in 1949, he was "probably the greatest clinical neurologist of all time." For 40 years (1870–1910), he worked mainly in the National Hospital for the Paralysed and Epileptics, Queen Square (National Hospital for Neurology and Neurosurgery). He performed various clinical neurological studies to elucidate the function of the CC. He published an extensive book entitled *A Manual of Diseases of the*

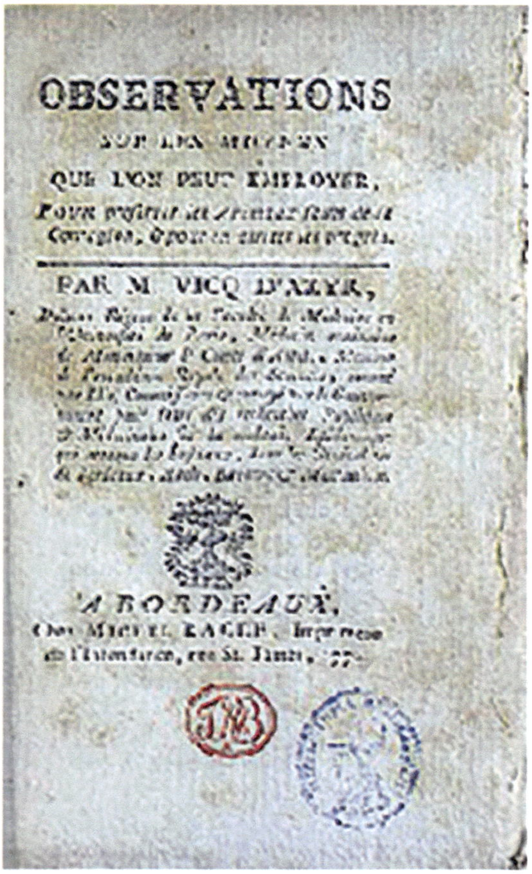
OBSERVATIONS

QUE L'ON PEUT EMPLOYER,

PAR M. VICQ D'AZYR,

A BORDEAUX,

Fig. 1.7 Vicq d'Azyr, observations (collections of papers)

Fig. 1.8 Romberg (1755–1873)

Nervous System, a masterpiece for its era, the so-called "bible of neurology" for many authors. Gowers gave his name to Gowers's sign, the Gowers' tract, Gowers' syndrome, and Gowers' round. He was one of the inventor members of the National Society for the Employment of Epileptics together with other scientists, Sir John Hughlings Jackson and David Ferrier [2, 9, 27–34, 36–38] (Fig. 1.9).

Fig. 1.9 Gowers 1845–1915

1.6 Reil and Santiago Ramón y Cajal

Johann Christian Reil (February 20, 1759–November 22, 1813) was an anatomist and psychiatrist from Germany. He managed to harden cadaveric brains in alcohol and dissected them from below, enabling him to see various structures including the CC more clearly. He studied the morphology and the anatomy of CC very carefully using the technology of his era. In 1812, he reported a case of agenesis of the CC. Various anatomical structures and medical situations, such as Reil's finger and the Island of Reil (insula), are named later. In 1809, he was the first physician to define the white fiber tract called "arcuate fasciculus." He also studied the locus coeruleus, which was firstly defined by Félix Vicq d'Azyr. Reil collaborated with Johann Friedrich Blumenbach, and he worked in a hospital in Halle (Saale), Germany. He managed to develop an innovative medical training program based mainly on Friedrich Schelling's *Naturphilosophie*. He authorized a well-known journal of physiology named *Archiv für die Physiologie*. In 1810, in Berlin, he got a lecturer of psychiatry in the university [3, 9, 14–20, 22, 27–34, 37–39] (Fig. 1.10).

Fig. 1.10 Guttmacher (1898–1966)

Santiago Ramón y Cajal (May 1, 1852–October 17, 1934) was a pioneering Spanish neuroscientist, neuroanatomist, neurophysiologist, pathologist, and histologist. He performed various studies on the morphology and functionality of the central nervous system in humans. He described and interpreted the connections of the CC. With Camillo Golgi, he won a Nobel Prize. Born in Petilla de Aragón in the Navarra Region, he attended the medical school of Zaragosa and then served as a captain in the military expedition to Cuba. After returning to Spain, he received his doctorate in medicine in Madrid. Later, he became the director of the Zaragoza Museum, and afterward, he served as professor of anatomy in the University of Valencia. In 1887, he moved to Barcelona, and in 1892, he became professor at Madrid. In 1899, he became director of the Instituto Spaniolo de Higiene and in 1922 founded the Laboratorio de Investigaciones Biológicas (Instituto Cajal). His studies in various disciplines influenced ways of thinking and acting throughout the European scientific community [1, 9, 13–21, 27–34, 37–39] (Fig. 1.11).

Fig. 1.11 Ramón y Cajal (1852–1934). Ramón y Cajal as a young captain in Cuba, 1874

1.7 Studies by Bean, Holloway, and Lacoste-Utamsing Regarding Differences in the Size of the Corpus Callosum

The first study regarding gender and race differences in the size of the CC was performed by Robert Bennett Bean (1874–1944), an anatomist, in 1906 [6, 13]. On the basis of his studies, he suggested that an exceptional CC size could mean exceptional intellectual activity, possibly indicating differences between genders and between races. In the following years, other researchers suggested that the CC is wider in women's than men's brains, resulting in greater interference between the brain hemispheres, likely the basis for women's insight [6, 13].

In 1982, on the basis of their studies, Ralph L. Holloway and Christine De Lacoste-Utamsing suggested that sex difference in morphology of human brain caused differences in perception. In contrast to their results, a meta-analysis of 49 studies revealed no sex difference in the size of the CC [6, 13]. In 2006, an MRI study showed no difference in thickness of the CC in relation to the size of the subject [34] (Fig. 1.12).

Fig. 1.12 Bean (1874–1944)

1.8 Redvers Ironside, Manfred Guttmacher, and the Comparative Anatomy of the Nervous Systems of Vertebrates, Including Man

Redvers Ironside was a Scottish scientist, physician, anatomist, and clinical neurologist who performed anatomical, physiological, and clinical studies of the function and properties of the CC. In his early studies, he attempted to correlate its normal and pathological appearances with morphological, anatomical, and physiological (functional) bases. During 1929, he and Manfred Guttmacher described the "syndrome of the CC" as a pathology with symptoms such as mental changes, apathy, apraxia, memory deficits, and drowsiness. Many years later, in 1961, he and

F.D. Bosanquet and W.H. McMenemy described a case of central demyelination of the CC, the so-called "Marchiafava-Bignami disease" [2, 8, 13–19, 27–34, 37–39].

He was born in Aberdeen. He studied first in grammar school and then he served in the Royal British Artillery. After a successful military career, he trained in neurology first at the Westminster and West London Hospitals, then at The National Hospital (Queen Square), and later at Guy's Hospital under the auspices of his teachers, Ashley Mackintosh, Sir James Purves Stewart, and George Riddoch. After his basic training, he became a successful consultant neurologist, serving the West London Hospital and the Hospital of St. John & St. Elizabeth. In Maida Vale Hospital, he was appointed first as a member of staff and later as dean and adviser on various research projects and physiological and clinical studies. He participated actively in the North African campaign during the Second World War. He received various honors as president of the Neurological Section of the Royal Society of Medicine, as master member of honor of the Society of Apothecaries and as member of the Order of St. John of Jerusalem. In his late career, he served as neurologist at Charing Cross Hospital [1, 9, 15–19, 27–34, 37–39].

Manfred Schanfarber Guttmacher (May 19, 1898–November 7, 1966) was a forensic psychiatrist and also a chief medical officer in the USA (Fig. 1.13). He studied the connection between psychiatry and criminal law. He wrote several books and studied the function of the CC in order to understand psychiatric diseases better [2, 8, 17, 18, 27–34, 36–38] (Fig. 1.13).

Another very important step in elucidating the function of CC was the publication in 1936 of the two volumes entitled *The Comparative Anatomy of the Nervous System of Vertebrates, Including Man* by Kappers, Huber, and Crosby. This book gave an exhaustive review of the comparative anatomy of vertebrate nervous systems. It attempted not only to accumulate anatomical data but also to investigate the phylogenetic development of the nervous system through the study of various structure-function correlations and principles of neurobiotaxis. The aim was to settle various arguments such as those concerning neuron distributions, the exact locations of nuclei, the function and the properties of the CC, the rise of new complex structures, etc. The text contained both theoretical discussions and informative illustrations [22].

Fig. 1.13 Reil the anatomist: a portrait from 1811

1.9 The Work of Hugo Karl Liepmann and Other Experimental Animal Studies

Hugo Karl Liepmann (April 9, 1863–May 6, 1925) was a German neurologist, psychiatrist, philosopher, and pioneer scientist. First, he worked closely with Carl Wernicke at Breslau. Then, he became head physician at Dalldorf and later he served as medical director (Städtische Irrenanstalt zu Lichtenberg, Herzberge). He pioneered various scientific studies of cerebral localizations of function and the role of the CC. He suggested that lesions of the CC could produce

Fig. 1.14 Liepmann (1863–1925)

Fig. 1.15 Baron Sándor Korányi von Tolcsva (1866–1944)

similar symptoms in experimental animals, with results contradicting theory [3, 9, 15–20, 27–34, 37–39] (Fig. 1.14).

From anatomical studies, he moved to physiological investigations of the parietal lobe of the dominant hemisphere in the brain. He described apraxia as a disorder of coordination of voluntary muscle movements. On the basis of his studies, he distinguished three types of apraxia: ideational, ideomotor, and kinetic [1, 9, 24–36, 39, 40].

Experimental studies, dissections, and surgical approaches on animals yielded controversial results regarding the functions of the CC:

- Baron Sándor Korányi de Tolcsva (6/18/1866–04/12/1944) in Hungary (Fig. 1.15).
- Hartmann and Trendelenburg, on monkeys with no obvious disturbances.
- The students of Pavlov, Anrep (1923), and Bykoff (1925), on dogs.
- Various other researchers and physicians, on primates.

In 1913, van Valkenburg reviewed the experimental and pathological-anatomical research on the CC up to that point [3, 9, 27–34, 36–38].

In June 1937, Olan R. Hydryman and Wilder Penfied reported a case of agenesis of the CC with the help of ventriculography study. Up to 1933, Baker and Graves encountered more than 80 reports describing this condition in the medical literature, which was firstly described in 1812, by Reil. During 1956, Bremer Brihaye and Andres Basilux performed various experimental studies on the CC [2, 9, 15–24, 27–34, 36–38].

1.10 The Important Contributions of Lashley and Sperry

Karl Spencer Lashley (June 7, 1890–August 7, 1958) was a psychologist and behaviorist from the USA, and his contributions to the study of normal neurological processes such as learning and memory are well known in the medical literature. He performed extensive dissections of the CC in many animals including primates. His

1.13 Conclusions

Many attempts have been made over the ages to understand the anatomical, physiological, morphological, structural, and behavioral aspects of the CC. All of these contributions demonstrate the need for reciprocal interaction to achieve optimal, helpful, and useful results in various medical disciplines. The confirmation of the neuron doctrine and the principle of cortical localization, together with advances in medical technology, were critical developments.

References

1. Aboitiz F, Montiel J. One hundred million years of interhemispheric communication: the history of the corpus callosum. Braz J Med Biol Res. 2003;36(4):409–20.
2. Aboitiz F, López J, Montiel J. Long distance communication in the human brain: timing constraints for inter-hemispheric synchrony and the origin of brain lateralization. Biol Res. 2003;36(1):89–99.
3. Aboitiz F. Brain connections: interhemispheric fiber systems and anatomical brain asymmetries in humans. Biol Res. 1992;25(2):51–61.
4. Bamiou DE, Sisodiya S, Musiek FE, Luxon LM. The role of the interhemispheric pathway in hearing. Brain Res Rev. 2007;56:170–82.
5. Baynes K, Eliassen JC, Lutsep HL, Gazzaniga MS. Modular organization of cognitive systems masked by interhemispheric integration. Science. 1998;280:902–90.
6. Bean WB. Robert Bennett Bean 1874-1944. Science. 1945;101(2623):346–8.
7. Beevor CE. Hamilton theory concerning the corpus cllosum. Brain. 1886;9:63–73.
8. Bradley L, Bryant PE. Categorizing sounds and learning to read—a causal connection. Nature. 1983;301:419–21.
9. Catiglioni A. Storia della Medicina. Milan: A. Mondadori; 1936.
10. Chicoine AJ, Proteau L, Lassonde M. Absence of interhemispheric transfer of unilateral visuomotor learning in young children and individuals with agenesis of the corpus callosum. Dev Neuropsychol. 2000;18:73–94.
11. Clarke JM, Lufkin RB, Zaidel E. Corpus callosum morphometry and dichotic listening performance: individual differences in functional interhemispheric inhibition? Neuropsychologia. 1993;31:547–57.
12. Crow TJ, Paez P, Chance SA. Callosal misconnectivity and the sex difference in psychosis. Int Rev Psychiatry. 2007;19:449–57.
13. De Lacoste-Utamsing C, Holloway RL. Sexual dimorphism in the human corpus callosum. Science. 1982;216:1431–2.
14. Faul F, Erdfelder E, Lang AG, Buchner A. G*Power 3: a flexible statistical power analysis program for the social, behavioral, and biomedical sciences. Behav Res Methods. 2007;39:175–219.
15. Fujimoto C, Ito K, Iwasaki S, Nakao K, Sugasawa M. Reversible impairment of auditory callosal pathway in 5-fluorouracil-induced leukoencephalopathy: parallel changes in function and imaging. Otol Neurotol. 2006;27:716–9.
16. Gadea M, Marti-Bonmati L, Arana E, Espert R, Casanova V, Pascual A. Dichotic listening and corpus callosum magnetic resonance imaging in relapsing-remitting multiple sclerosis with emphasis on sex differences. Neuropsychology. 2002;16:275–81.
17. Gazzaniga MS. Cerebral specialization and interhemispheric communication: does the corpus callosum enable the human condition? Brain. 2000;123(Pt 7):1293–326.
18. Hagelthorn KM, Brown WS, Amano S, Asarnow R. Normal development of bilateral field advantage and evoked potential interhemispheric transmission time. Dev Neuropsychol. 2000;18:11–31.
19. Hyndyman OR, Penfield W. Agenesis of corpus callosum. Its recognition by ventriculography. Arch Neurol Psychiatry. 1937;37(6):1251–70.
20. Ironside R, Guttmacher M. The Corpus callosum and its tumours. Brain. 1929;52(4):442–83.
21. Ironside R, Bosanquet FD, Mcmenemey. Central demyelination of the corpus callosum (Marchiafava-Bignami disease) with report of a second case in Great Britain. Brain. 1961;84:212–30.
22. Kappers CUA, Huber GC, Crosby EC. The comparative anatomy of the nervous system of vertebrates, including man (2 Vols). New York: Macmillan; 1936. p. 709.
23. Kimura D. Functional asymmetry of the brain in dichotic listening. Cortex. 1967;3:163–78.
24. Lenroot RK, Giedd JN. Brain development in children and adolescents: insights from anatomical magnetic resonance imaging. Neurosci Biobehav Rev. 2006;30:718–29.
25. Luders E, Narr KL, Bilder RM, Thompson PM, Szeszko PR, Hamilton L, Toga AW. Positive correlations between corpus callosum thickness and intelligence. NeuroImage. 2007;37:1457–64.
26. Martin GN, Carlson NR, Buckist W. Psychology. 3rd ed. London: Pearson Education Company; 2007.
27. Racle M. Observations sur les moyens que l'on peut employer, pour préserver les animaux sains de la contagion, et pour en arréter les progrès (in French), Bordeaux, 1774.
28. Peng SJ, Hsin YL. Functional connectivity of the corpus callosum in epilepsy patients with secondarily generalized seizures. Front Neurol. 2017;8:446.
29. Syrmos N, Ampatzidis G, Fachantidou A, Mouratidis A, Syrmos C. Historical back training in most impor-

tant points of neurosurgery. Ann General Psychiatry. 2010;9(Suppl 1):S89.
30. Syrmos N. Historical back training in most important points of neurosurgery. Master thesis, Aristotle University of Thessaloniki, Thessaloniki, vol. 9, 2009.
31. Syrmos N, Syrmos Chr. History of neurosurgery (in Greek), Thessaloniki, Greece, 2010.
32. Syrmos N. Endoscopic anatomy of the roof of the IV ventricle. PhD thesis, Aristotle University of Thessaloniki, Thessaloniki, 2009.
33. Syrmos N. Italian-Greek pediatric neurosurgery terminology. Post-PhD thesis, Aristotle University of Thessaloniki, Thessaloniki, 2022.
34. Westerhausen R, Woerner W, Kreuder F, Schweiger E, Hugdahl K, Wittling W. The role of the corpus callosum in dichotic listening: a combined morphological and diffusion tensor imaging study. Neuropsychology. 2006;20:272–9.
35. Myers RE, Sperry RW. Interhemispheric communication through the corpus callosum mnemonic carry-over between the hemisphere. AMA Arch Neurol Psychiatry. 1958;80(3):298–303.
36. Petersson KM, Silva C, Castro-Caldas A, Ingvar M, Reis A. Literacy: a cultural influence on functional left-right differences in the inferior parietal cortex. Eur J Neurosci. 2007;26:791–9.
37. Rauch RA, Jinkins JR. Analysis of cross-sectional area measurements of the corpus callosum adjusted for brain size in male and female subjects from childhood to adulthood. Behav Brain Res. 1994;64:65–78.
38. Reeves GA, Roberts DW. Advances in behavioral biology, ABBI, vol. 45. New York: Springer; 1995.
39. Pollmann S, Maertens M, von Cramon DY, Lepsien J, Hugdahl K. Dichotic listening in patients with splenial and nonsplenial callosal lesions. Neuropsychology. 2002;16:56–64.
40. Pujol J, Vendrell P, Junque C, Marti-Vilalta JL, Capdevila A. When does human brain development end? Evidence of corpus callosum growth up to adulthood. Ann Neurol. 1993;34:71–5.

Embryonic Development and Myelination of the Corpus Callosum

2

Aaron Yu and R. Shane Tubbs

2.1 Introduction

The corpus callosum (CC), consisting of approximately 200 million topographically organized axons of varying degrees of myelination and diameter, is the greatest of the three interhemispheric commissures (anterior commissure, hippocampal commissure, and CC) [1–3]. As the major pathway of association between the cerebral cortex of the left and right cerebral hemispheres, the CC facilitates complex cognitive integration and coordination between the two hemispheres [2]. Therefore, any injury to the structure (via stroke, trauma, tumor, or iatrogenic damage during invasive neurosurgical procedures) can incur profound cognitive and manual deficits due to a disturbance of interhemispheric connections [2]. Importantly, complete callosal agenesis or partial dysgenesis is the consequence of maldevelopment during crucial embryological steps from early midline telencephalic patterning to commissural axonal guidance [1]. Such developmental anomalies of the CC vary in degree of clinical presentation, occurring either as a rather asymptomatic isolated finding or, more frequently, accompanying other severe congenital brain abnormalities [1, 2, 4].

As such, an understanding of the normal callosal development is of particular clinical interest, both from a radiological and embryological perspective, in the recognition of related malformations as well as in the distinction between in utero developmental dysgenesis from destructive brain insults of the CC [4, 5]. Early indications of callosal maldevelopment thus have potential value in guiding safe pregnancy management and neonatal support [6]. Herein, we elucidate the normal developmental morphology of the CC, placing emphasis on its sequence of embryogenesis, timing of myelination, and diagnostic imaging, to position the reader to better appreciate the clinical implications of its embryological disruption in subsequent sections.

2.2 Corpus Callosum Embryogenesis

In placental mammals, the embryogenesis of the CC is a phylogenetic innovation that involves the complex process of interhemispheric midline fusion with migration of specialized glial cells guiding callosal axons toward their respective contralateral hemispheres [7]. This process ultimately results in the development of an independent CC, crossing the midline separately from the anterior and hippocampal commissures [7, 8].

A. Yu (✉)
Tulane University School of Medicine,
New Orleans, LA, USA
e-mail: ayu2@tulane.edu

R. S. Tubbs
Department of Neurosurgery, Tulane School of Medicine, New Orleans, LA, USA
e-mail: rtubbs@tulane.edu

M. Turgut et al. (eds.), *The Corpus Callosum*, https://doi.org/10.1007/978-3-031-38114-0_2

Early in development, after the closure of the neural tube at week 4, the region of the rostral wall of the telencephalon anterior to the chiasmatic plate rapidly grows in its dorsal portion near the paraphysis around 6–8 weeks of gestation [4, 5]. The lamina reuniens of His represents this dorsal portion and is a proliferative cellular zone, a median pediment from which the hemispheric vesicles, ganglionic eminences, and olfactory buds arise [7, 8]. The lamina terminalis, the anterior aspect closer to the chiasmatic ridge, does not demonstrate the same increase in thickness and instead remains thin [5].

The ventral aspect of the lamina reuniens, located near the lamina terminalis, becomes the area precommissurals, developing into the future septal area containing the anterior commissure [5, 7]. At 8–9 weeks, fiber bundles grow medially from the ventrolateral wall of each hemispheric vesicle; and around a week later, the bundles cross at midline to form the anterior commissure within the lamina reuniens [4]. The dorsal portion of the lamina reuniens is where the hippocampal commissure will develop [5, 7]. During the eighth week of gestation, the dorsal lamina reuniens folds into the median groove, a sulcus also referred to as the "sulcus medianus telencephali medii," with the bilateral expansion of the hemispheric vesicles around the prospective hippocampus [4, 5, 7].

At about 10 weeks, the subsequent fusion of the 2 superior lips of the sulcus is immediate and forms a substantial cell mass, the so-called massa commissuralis, which serves as the bed for the crossing commissural fibers of the CC [4, 7, 8]. Around this time, it was observed that glial and neuronal cells in mouse models migrate toward the midline to invade the primitive meninge, forming the massa commissuralis [7]. This same structure has been referred by some authors as a "glial sling," a necessary interhemispheric junction that permits the crossing of pioneer callosal fibers [7].

By 11–12 weeks of gestation, the earliest callosal fibers are evident in penetrating the massa commissuralis, bridging the pocket of the anterior interhemispheric fissure formed between the banks of the median groove [4, 5]. This pocket is eventually closed by the rostrum and is known as the cavum septi pellucidi, dorsal to the septal area and anterior commissure [7, 8]. Around 12–13 weeks, callosal fibers continue to grow in number and form a definitive callosal commissural plate, a distinct structure from the anterior commissure [8]. By 18–20 weeks, the CC resembles that of its adult form and position; however, it remains immature in respect to its rostrocaudal extent and thickness [8].

From front to back, the CC is structurally divided into four main parts: the rostrum, genu, body, and splenium. The isthmus marks the focal narrowing, the fusion between two separate segments: the anterior portion of the CC, comprising of the rostrum, genu, and body, and the posterior portion, comprising of the splenium [2]. The fusion of the anterior portion, derived from the glial sling, with the posterior portion, associated with the hippocampal commissure, occurs just anterior to the hippocampal commissure [7]. The posterior displacement of the hippocampal commissure along with the callosal splenium is the result of the marked expansion of the frontal cortex early in development, accompanying the expansion of the anterior CC [1]. The dorsal displacement of the splenium by the anterior growth of the frontal lobe and the growth of the other callosal segments leads to an apparent backward progression of the CC [7]. The two-locus origin of the CC, from the anterior commissure and the hippocampal commissure, has significant implications when considering partial agenesis of the CC [1, 9].

Callosal neurons stem from layers II/III, V, and VI of the neocortex. Observed in both human and mouse models, midline crossing of neocortical neuros is preceded by the crossing of cingulate cortex associated axons [1]. At approximately 11–12 weeks of gestation, the earliest fibers of the human CC begin to rapidly develop with callosal fibers crossing midline by the 12th week [5, 9, 10]. The rudimentary CC is rather short, stretching from the anterior commissure to the hippocampal commissure [7]. By 14–15 weeks of gestation, the lamina rostralis, genu, and body (fused with the splenium) are readily recognizable [9]. Then, by weeks 18–19, the splenium assumes a more prominent, complete form. At this stage, all segments of the CC can be visualized on ultrasonography [11]. By week 20, the CC essentially assumes its final shape, though, with a still immature cross-sectional area (5% of mature size) [9]. By the time of birth, both the splenium and genu display uniform cross-sectional area [9].

Throughout postnatal development, the CC enlarges with the corresponding growth of the cortex with cortical plates extending association axons to cross to a contralateral hemisphere via the CC [4, 9]. Axons of the CC continue to display exuberant growth in volume with an initial addition of fibers [7, 9]. Subsequent pruning of axonal fibers is compensated by myelination [7]. The CC has been demonstrated to reach its target volume and adult levels of myelinization by the age of 6–9 years; however, some studies have indicated that callosal growth may continue well into late adolescence or early adulthood [2, 12]. Hence, structural maturation of the callosal fibers continues throughout postnatal development even though the number of callosal fibers are rather fixed at birth [1].

2.2.1 Anterior-to-Posterior Theory of Callosal Development

Two main contending theories concerning the sequence of callosal development exist in neuroradiological literature to date. The previously prevailing theory of callosal development in neuroradiologic literature maintained that the callosal axons initially cross the midline proceeding in an anterior-to-posterior direction with the splenium following the formation of the genu and body, the rostrum being the final structure to develop [5, 7, 9]. Kier and Truwit [5], in their rebuttal against such postulation, suggest that the anterior-to-posterior theory may have originated in an article published in 1967 [5]. The article introduces the notion that the most anterior segment of the CC begins development first as a bud formed from the lamina terminalis, eventually becoming the rostrum [5]. Development then proceeds upward and posteriorly accompanied by the formation of the genu, followed by the body, and ending with the splenium. Over time, several revisions of this original theory were made to suggest that the rostrum or even the posterior genu, not the splenium, was the final segment to develop [5].

In a retrospective review of MR imaging studies, Barkovich and Norman [4] also concluded that callosal development proceeded in an anterior-to-posterior direction. Between 11 and 12 weeks of gestation, 50- to 60-mm crown-rump length (CRL), fibers enter the massa commisuralis; and in the following week, these pioneer callosal fibers form a definite CC near the commissural plate to eventually become the genu [4]. In the next 5–7 weeks, Barkovich and Norman [4] found that growth continued caudally, corresponding to the caudal growth of the cerebral hemispheres. After the growth of the genu, body, and splenium, there is an exception to this directional development in that the rostrum forms last at 18–20 weeks of gestational age [4].

2.2.2 Bidirectional Theory of Callosal Development

On the other hand, neuroradiological evidence from both early and more recent investigations in the literature has led to the proposal of the development of the CC progressing bidirectionally, stemming from an evolutionary and radiological standpoint [9]. Of these investigations, notably, in an MR imaging study of 1800 patients with normal callosal development and 113 patients with callosal abnormalities, Kier and Truwit [5] concluded that the CC first begins its development more centrally, in the future anterior body (not the genu), subsequently growing bidirectionally. According to this postulation, the anterior body of the CC develops prior to the definitive genu [5]. For Kier and Truwit [5], the unidirectional theory was not consistent, since if the genu were the first callosal structure to form, then the callosal body could not be present without the genu [5]. It was thus found that the normal human genu always projects anterior to the so-called

MAC line, a reference landmark drawn from the mamillary body through the anterior commissure and CC [5].

Kier and Truwit [5] reached two conclusions in favor of a bidirectional callosal development: firstly, that no cases in the study were discovered to consist of only the genu in front of the MAC line with the absence of all other segments, and secondly, that an abnormal CC would be entirely behind or at the point of the MAC line. Further, when the genu is present, it is always present in association with the presence of the anterior body [5]. In the 113 cases of callosal maldevelopment in their study, Kier and Truwit [5] confirmed that the region of the body of the CC is the first callosal segment to develop. As such, the foundation of the CC forms near and superior to the lamina terminalis, precluding a unidirectional development, instead of bidirectional with more prominent anterior growth as the callosal primordium develops before the genu and splenium [5, 9]. It has now been quite established that the hippocampal commissure comes first, followed by the anterior callosum, and then the commissural plate from the lamina rostralis to the splenium [7].

The previous notion that the rostrum was perhaps the last to develop was a point of contention as Kier and Truwit [13] further found that the fetal lamina rostralis segment, a thin fibrous layer which tethers the beaked rostral segment, is present well before the development of the genu and splenium [13]. At 14 weeks of gestation, the lamina rostralis is already present anchoring the rudimentary CC anteriorly [13]. It is hypothesized that the shape of the posteriorly pointing beaked rostral segment results from the tethering effect of the lamina rostralis. In the absence of a normal genu, the rudimentary beaked segment of the rostrum may also be present. This beaked rostral segment was found to develop concurrently with genu maturation [13]. Therefore, Kier and Truwit [13] conclude that the lamina rostralis segment is present even at the earliest developmental stages of the rudimentary CC near the region of the future callosal body well before the genu and splenium form [13].

2.3 Myelination

While the number of callosal fibers reaches its target volume already at the time of birth, the CC continues to undergo structural changes during postnatal development due to axonal pruning, redirection, and myelination [14]. Myelination in the CNS describes a vital sequential process of brain maturation by which oligodendrocytes, specialized glial cells derived from the neuroepithelium, produce sheaths of myelin that insulates axonal fibers [15–17]. Primarily found in white matter with negligible amounts in gray matter, axonal myelination is essential for adequate growth and maturation of the axon, permitting the rapid transmission of electric action potentials down the neuronal axonal fiber [16]. The integrity of the axon relies on the myelinating cell body for support, likely due to the role of myelin on the regulation of ion composition and fluid volume surrounding the axon [16].

Assessment of myelination is performed qualitatively, primarily using magnetic resonance imaging (MRI) since no imaging techniques are currently in use to directly view the myelin lipid bilayer [16, 17]. Given that the process of myelination is an orderly, sequential process with a generally predictable course, MRI provides a crucial means to study the progression and extent of myelination [15]. Estimation of this progression is of particular importance in pediatric neuroimaging as normal milestones of clinical development correlate with the extent of myelination of the infant brain [15, 18]. Importantly, the appearance on imaging of the normal progression in a myelinating infant brain contrasts starkly with the appearance of an adult brain and is thus not to be mistaken for white matter abnormalities [15, 16].

Various pathological processes can result in delayed progression of myelination, including ischemic events, congenital malformations, chromosomal disorders, inborn errors of metabolism, and congenital and postnatal infections [17]. Therefore, recognition of the normal process of myelination, namely, of the CC, is important when considering early detection of aberrant brain development due to various leukodystrophies and hypomyelinating conditions [16].

2.3.1 Normal Sequence of Myelination

Myelination of the brain begins at the fifth fetal month, commencing with the cranial nerves and continuing its progression throughout life [15, 16]. According to Barkovich [15], generally, myelination of the brain progresses from the center to the periphery (deep to superficial), bottom to top (caudocranial), and back to front (posterior to anterior) [15]. Adhering to these rules, the following processes of myelination, therefore, should make sense. Proceeding caudocranially, the brain stem myelinates before the cerebellum and basal ganglia [15]. The brain stem and cerebellum together commence myelination prior to the cerebral hemispheres [15]. The occipital lobe myelinates before the frontal lobe [17]. The basal ganglia and thalamus then begin myelination before the white matter [16]. Additionally, the posterior limb of the internal capsule myelinates before the anterior limb and the central corona radiata before the subcortical regions [16].

Another general postulation by Barkovich [15] is that the dorsal regions of any part of the brain tend to myelinate first [2]. Further, myelination of the brain progresses much more readily in functional systems that are utilized early on in child development [15]. The process of myelination rapidly progresses until approximately 2 years of age, slowing down significantly after this point [15]. Regardless, myelination of the brain continues into later adulthood around the third and fourth decades of life [15, 19].

2.3.2 Radiological Imaging of Myelination

Postnatal conventional anatomical MR imaging is routinely utilized in clinical practice for visualizing the continuous process of myelination at various stages of development [15, 17]. It is important to note that current standard MR modalities do not specifically provide an actual quantification of myelin; rather, they provide an assessment of apparent myelination, depending on the different field strength of scanning or pulse sequences utilized [17]. Hence, such techniques examine changes in axonal density, membrane integrity, and water content that would allow us to infer the progression of myelination [16]. Alterations in MR imaging parameters, such as a reduction of T1 and T2 relaxation times, reduced water diffusion, increased diffusion anisotropy, and increased magnetization transfer, are associated with brain maturation, namely, resulting from myelination [20].

Myelination of the white matter is assessed using MRI sequences that reflect differences in tissue water, specifically T1-weighted (T1W) and T2-weighted (T2W) images with additional implementation of diffusion-weighted imaging (DWI) and diffusion tensor imaging (DTI) [15–17]. Cerebrospinal fluid appears dark on T1W images, while fatty tissues appear bright. Myelinated white matter appears hyperintense when compared to gray matter on T1W images, whereas myelinated white matter appears hypointense on T2W images [15]. T1W hyperintensity results from the high lipid content of myelin (particularly galactocerebroside and cholesterol) with T2W hyperintensity being due to the hydrophobic nature of myelin and tightening of the myelin sheath around the axon, leading to reduced water content relative to gray matter [15, 16]. The CC, due to its composition of tightly packed fiber bundles, appears even brighter on T1W since most free water is extruded from their myelin fibers.

For the first 2 years after birth, brain maturation is assessed by signal changes on MR that occur secondary to myelination that varies in timing and rate of progression. During the first 6–8 months of postnatal life, T1W images are considered ideal in the visualization of white matter maturation [15]. At the time of birth, the medulla, dorsal portion of the pons, midbrain, posterior limb of the internal capsule, and perirolandic cortex display myelination on T1W images [17].

9. Paul LK. Developmental malformation of the corpus callosum: a review of typical callosal development and examples of developmental disorders with callosal involvement. J Neurodev Disorder. 2011;3:3–27.
10. Malinger G, Zakut H. The corpus callosum: normal fetal development as shown by transvaginal sonography. AJR Am J Roentgenol. 1993;161(5):1041–3.
11. Kirthika CP, Siva T, Rajeswaran R, Kalpana R, Francis YM. Morphology and morphometry of foetal corpus callosum using MRI – a retrospective study. Biomed Pharmacol J. 2021;14(2):663.
12. Allin M, Nosarti C, Narberhaus A, et al. Growth of the corpus callosum in adolescents born preterm. Arch Pediatr Adolesc Med. 2007;161(12):1183–9.
13. Kier EL, Truwit CL. The lamina rostralis: modification of concepts concerning the anatomy, embryology, and MR appearance of the rostrum of the corpus callosum. AJNR Am J Neuroradiol. 1997;18(4):715–22.
14. Luders E, Thompson PM, Toga AW. The development of the corpus callosum in the healthy human brain. J Neurosci. 2010;30(33):10985–90.
15. Barkovich AJ. Magnetic resonance techniques in the assessment of myelin and myelination. J Inherit Metab Dis. 2005;28(3):311–43.
16. Branson HM. Normal myelination: a practical pictorial review. Neuroimaging Clin N Am. 2013;23(2):183–95.
17. Prabhu SM, Yadav V, Prakash D, Sudhakar S, Mani S. Myelination clock: a simplified step-by-step approach to normal myelination. Hong Kong J Radiol. 2016;19:208–16.
18. Adam A, Dixon AK, Gillard H, Schaefer-Prokop C. Grainger & Allison's diagnostic radiology: a textbook of medical imaging. 7th ed. New York: Elsevier; 2021.
19. Buyanova IS, Arsalidou M. Cerebral white matter myelination and relations to age, gender, and cognition: a selective review. Front Hum Neurosci. 2021;15:662031. Published 2021 Jul 6.
20. Barkovich AJ. Concepts of myelin and myelination in neuroradiology. AJNR Am J Neuroradiol. 2000;21(6):1099–109.
21. Ballesteros MC, Hansen PE, Soila K. MR imaging of the developing human brain. Part 2. Postnatal development. Radiographics. 1993;13(3):611–22.
22. Welker KM, Patton A. Assessment of normal myelination with magnetic resonance imaging. Semin Neurol. 2012;32(1):15–28.
23. Barkovich AJ, Kjos BO. Normal postnatal development of the corpus callosum as demonstrated by MR imaging. AJNR Am J Neuroradiol. 1988;9(3):487–91.
24. Knyazeva MG. Splenium of corpus callosum: patterns of interhemispheric interaction in children and adults. Neural Plast. 2013;2013:639430.
25. Murakami JW, Weinberger E, Shaw DW. Normal myelination of the pediatric brain imaged with fluid-attenuated inversion-recovery (FLAIR) MR imaging. AJNR Am J Neuroradiol. 1999;20(8):1406–11.
26. Provenzale JM, Isaacson J, Chen S. Progression of corpus callosum diffusion-tensor imaging values during a period of signal changes consistent with myelination. AJR Am J Roentgenol. 2012;198(6):1403–8.
27. Aung WY, Mar S, Benzinger TL. Diffusion tensor MRI as a biomarker in axonal and myelin damage. Imaging Med. 2013;5(5):427–40.
28. Kühne F, Neumann WJ, Hofmann P, et al. Assessment of myelination in infants and young children by T1 relaxation time measurements using the magnetization-prepared 2 rapid acquisition gradient echoes sequence. Pediatr Radiol. 2021;51:2058–68.
29. Menkes JH, Sarnat H, Maria BL. Child neurology. 7th ed. Philadelphia, PA: Lippincott Williams & Wilkins; 2006. p. 279–80.
30. Sarnat H, Curatolo P. Malformations of the nervous system (volume 87 of the handbook of clinical neurology). 1st ed. New York, NY: Elsevier; 2007. p. 69–70.

3 Histogenesis and Developmental Disorders of the Corpus Callosum

Canberk Tomruk, Cansin Şirin, Kubilay Doğan Kılıç, Okan Derin, Servet Çelik, Ali Çağlar Turgut, and Yigit Uyanıkgil

3.1 Definition and Etymology

The corpus callosum (CC) was described by Andreas Vesalius in 1543. Etymologically, its name derives from the Latin words corpus (body) and callosum (hard skinned). The CC accounts for the largest volume in the substantia alba tract that connects the two cerebral hemispheres. Homologous structures in the central nervous system (CNS) are also connected by points called commissures, which is a version of the Latin-derived word "commissura" meaning unite or connect [1].

3.2 Corpus Callosum Histology

The CC is one of the five main commissures in the brain. The others are the anterior, posterior, hippocampal, and habenular commissures. The CC, which is the largest, occurs only in placental mammals; the hippocampal and anterior commissures can be found in mammalia. The interhemispheric coordination of motor, sensory, and cognitive functions is realized by information exchange through some 200 million axons in the main tract, the CC [2, 3].

The CC is a transverse bundle of nerve fibers. The number of fibers it contains is evident at birth, though their structural and functional maturation continues postnatally. The presence of different fiber types in different parts of the CC suggests functional differences in interhemispheric communication. Although most callosal fibers are homotopic linkages that connect analogous regions in the two hemispheres, there are also heterotypic links that connect different areas.

Model primate studies have shown that the number of axons is maximal in the neonatal CC; axon elimination during the postnatal period and myelination starting from the third postnatal month are critical in forming the final morphology of the structure.

C. Tomruk · C. Şirin · K. D. Kılıç
Department of Histology and Embryology, Ege University Faculty of Medicine, İzmir, Turkey

O. Derin
Department of Anatomy, Ege University Faculty of Medicine, İzmir, Turkey

S. Çelik
Department of Anatomy, Ege University Faculty of Medicine, İzmir, Turkey

Application and Research Center of Cord Blood Cell-Tissue, Ege University, İzmir, Turkey

A. C. Turgut
Department of Radiology, Ege University School of Medicine, İzmir, Turkey

Department of Histology and Embryology, Aydın Adnan Menderes University Health Sciences Institute, Aydın, Turkey

Y. Uyanıkgil (✉)
Application and Research Center of Cord Blood Cell-Tissue, Ege University, İzmir, Turkey

Department of Histology and Embryology, Ege University School of Medicine, İzmir, Turkey

M. Turgut et al. (eds.), *The Corpus Callosum*, https://doi.org/10.1007/978-3-031-38114-0_3

Only 30–40% of CC fibers contain myelin. Thanks to the compact structure of the substantia alba tracts consisting of dense fibers, they are relatively resistant against conditions such as interstitial edema, infection, and tumor infiltration [4].

In humans, the CC is divided into five regions. From front to back, these are rostrum, genu, corpus (truncus), isthmus, and splenium. In mice, there are three regions: genu, trunk, and splenium. Micrographs of hematoxylin and eosin-stained preparations of cadaver-derived tissues from these five regions in humans are shown in Fig. 3.1. The histological structures of the samples taken from five different areas of the CC are shown in Fig. 3.2. The connection of the premotor and frontal cerebral cortex regions via the rostrum and genu links the motor, somatosensory, and parietal regions. Fibers passing through the genu carry neuronal signals necessary for motor function, while fibers forming the posterior corpuscle carry somatosensory inputs. Axons in the isthmus are responsible for transmitting auditory signals and visual information through the splenium. The most posterior part of the CC is the bulb-shaped splenium [5]. Although studies show that the splenium occupies a larger area in females, the reason for this and how gender differences affect cognitive functions have not been clarified [6]. The splenium fibers are temporal, parietal, and occipital cortex projections. Hippocampal commissure fibers and splenial fibers connect the inferior temporal, occipital, and posterior parietal cortices of the two hemispheres. Along with the genu, the middle and anterior splenium contains the highest volume of thin axonal fibers connecting the medial temporal and parietal lobes, while thicker fibers are found in the posterior part of the splenium, which fuses the semi-representation of visual space [5].

Via fibers in these regions, the CC connects the limbic and insular lobes, the occipital, frontal, parietal, and temporal lobes, and also the basal ganglia. It is an inhomogeneous structure in which the anterior cortical areas are connected by the anterior CC and the posterior cortical areas by fibers with different structures forming the posterior CC. The dorsal part connects the medial region, and the ventral part connects the lateral cortical regions. The fibers connecting the temporoparietal and prefrontal association areas are mostly weakly myelinated, small-diameter,

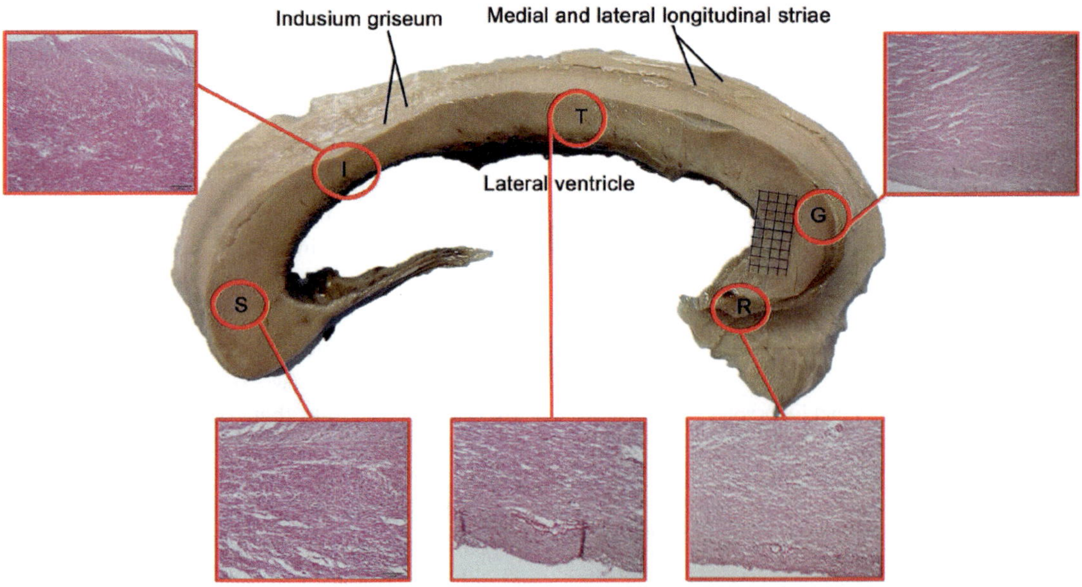

Fig. 3.1 Macroscopic view of the human CC structure used for histological sampling. Areas where sections were taken are indicated

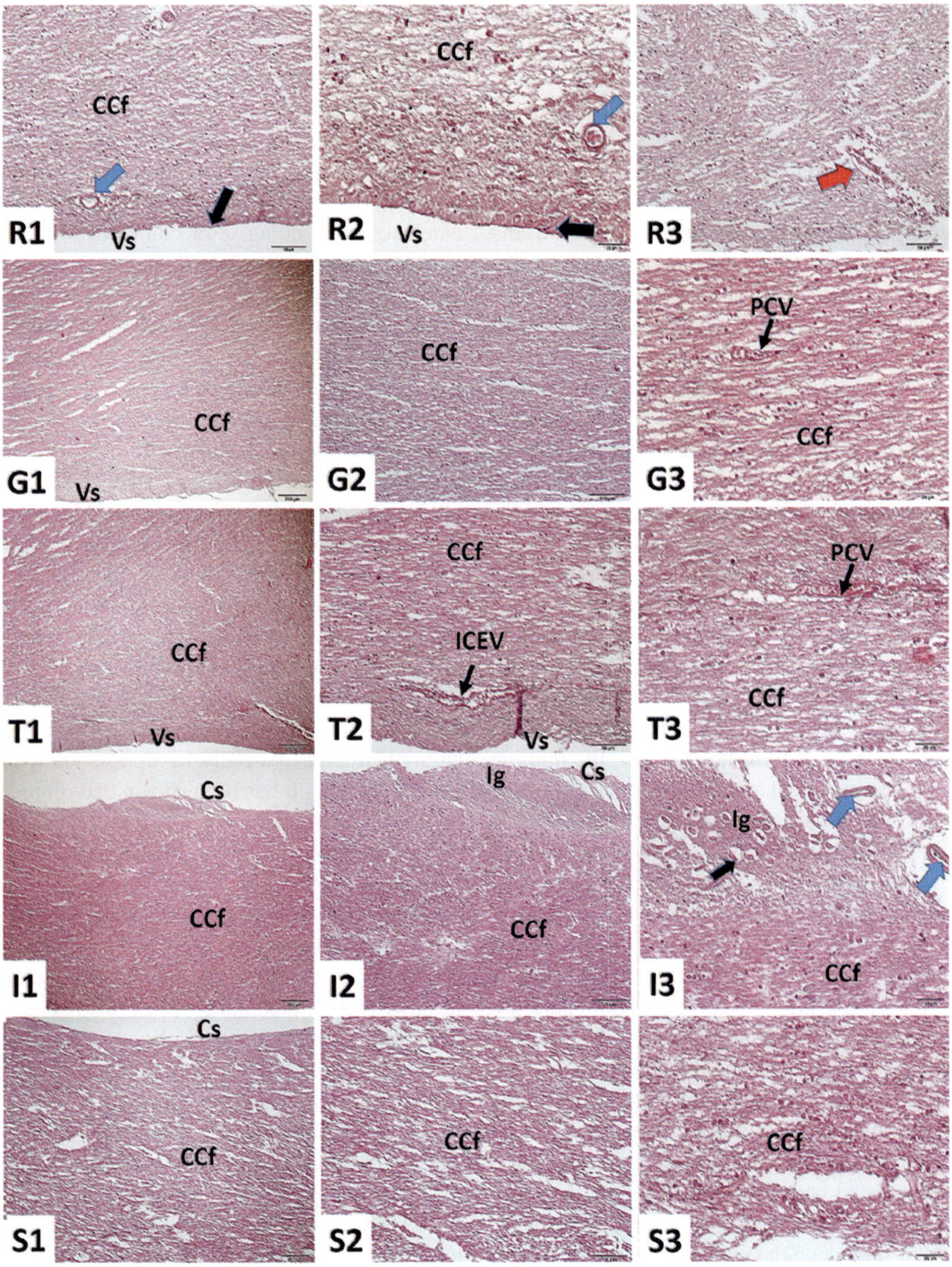

Fig. 3.2 Photomicrographs of different regions of the CC, hematoxylin and eosin stain. **R1-R2-R3**: Rostrum, ×200, ×400, and ×200 magnification, respectively. **Black arrow**; ependymal cells. **Red arrow**; the callosal vein leading to the subependymal vein. **Blue arrow**; branches of the pial plexus. **G1-G2-G3:** Genu, ×40, ×100, and ×400 magnification, respectively. **Vs**: Ventricular surface. **CCf**: CC fibers. **PCV**: Pericallosal vein. **T1-T2-T3:** Truncus, ×40, ×200, and ×400 magnification, respectively. **Vs**: Ventricular surface. **CCf**: CC fibers. **ICEV**: Infracommissural ependymal vein. **PCV**: Pericallosal vein**. I1-I2-I3:** Isthmus, ×40, ×100, and ×400 magnification, respectively. **Cs**: Cortical surface. **CCf**: CC fibers. **Ig**: Indusium griseum. **Blue arrow**; indusial artery. **Black arrow**; neuron. **S1-S2-S3**: Splenium, ×40, ×100, and ×400 magnification, respectively**. Cs**: Cortical surface. **CCf**: CC fibers

slow-conducting fibers, while those connecting the primary and secondary sensorimotor areas are fast and strongly myelinated and have diameters of more than 3 μm. More than half of all callosal axons start from layer II and III neurons; less than a quarter come from neurons in layer V [7]. These neurons called callosal projection neurons, protrude within the telencephalon and have only one projection to the target of attachment in the contralateral cortex and one of the striata and a double projection to the ipsilateral frontal cortex [8].

Deep layer cortical projection neurons can show bilateral axonal branching, an ipsilateral branch projecting into the internal capsule as well as to the commissural branch. These neurons commonly make long-distance connections both ipsilateral and contralateral to the striatum or the anterior or sensorimotor cortex, both homotopically and heterotopically.

In line with the development of the cortical layer, callosal projection neurons in the fifth and sixth layers are formed earlier and pass the midline in the embryonic period, while neuron axons in the second and third layers reach the midline during the postnatal period. They pass through the midline to the target area in the contralateral hemisphere, where callosal axons reach from the substantia alba to the cortical plate, branching and forming synaptic connections [7].

Interstitial neurons located deep in the CC are mostly involved in directing the callosal axons in the midline during CC development. There are very few of them in adulthood [9].

Midline glia-associated neurons include immature glutamatergic cells derived from the later-developing GABAergic cells and the pallial ventricular region. Glutamatergic neurons, which are associated with astroglia before the first callosal axons arrive, are found in the superficial, middle, and deep layers of the midline. As a result of this organization, which is completed by the migration of GABAergic neurons to the midline prior to the cingulate fibers start to pass through them, the cingulate and frontal callosal axons extend between the superficial and middle layers, and the parietal callosal axons move between the middle and deep layers [7].

Axons extend and negotiate the neural parenchyma by reacting to axon pathfinding molecules. At the end of each axon is a growth cone with microtubules at its center. Axons elongate or retract by adjoining or removing tubulin units. Surrounding the growth cone are actin-rich lamellipodia and their filopodia with cytoplasmic extensions that elongate or contract in response to molecular signals. Axons can be steered heterotopically (decussation) or homotopically (commissure) to local or distant ipsilateral targets or to a contralateral target, thanks to molecular signals from cellular scaffolds and the specific axonal membrane receptors responsible for their responses [10].

Oligodendrocytes in the substantia alba mediate the feeding of axonal activity. Metabolites and signaling molecules travel through channels in compacted myelin from the oligodendrocyte soma to the innermost ends surrounding the axon. Lactate is an energy metabolite delivered by oligodendrocytes. Monocarboxylate transporter 1 (MCT1) expressed by oligodendrocytes releases lactate into the periaxonal space and is thought to strengthen axons partially by uptake via axonal MCT2 [11, 12]. It has been shown that oligodendrocyte NMDA receptors are also important in metabolic cooperation between axons and oligodendrocytes.

In the CC, the connexin isoforms Cx30 and Cx43 are expressed by astrocytes, while Cx32 and Cx47 are expressed by oligodendrocytes, and these cells show panglial coupling. Cx47 ablation disrupts the attachment of oligodendrocytes to astrocytes, resulting in smaller-than-normal networks consisting of oligodendrocytes [11].

CC development can be understood in three stages: a cellular mass develops between the two telencephalic vesicles during commissure formation. The primitive lamina terminalis corresponds to the closure point of the anterior neuropore. The dorsal part enlarges during the sixth to eighth intrauterine weeks and forms a lamina fusion. These lamina fusions form the praecommissuralis, the source of the anterior commissure, the primordium hippocampi, and the massa commissuralis, from ventral to dorsal. Fibers from the developing hemispheres pass through this primitive CC. The growth of the CC is followed by enlargement of the hemispheres by rostrocaudal and then dorsoventral circular motion. The last part of the CC to be formed is the rostrum. Maturation occurs after birth and corresponds to axonal elimination and myelination, gradually changing the callosal connection pattern of the newborn and infant to the adult pattern [13].

The CC develops between the 12th and 16th and 20th days of pregnancy. The traditional notion that development begins in the genus and progresses backward when the rostrum has last appeared has been dispelled in contemporary studies by magnetic resonance (MR) tractography. Instead, it is suggested that the anterior trunk develops first and then continues bidirectionally, the anterior parts (genu) developing earlier/more distinctly than the posterior parts (splenium). However, this has not yet been universally accepted. The CC becomes myelinated in the opposite direction, forward from the splenium. It should be noted that the CC develops together with the relevant areas of the brain and continues to increase in volume after birth [14].

CC development, animal models, and humans with CC anomalies have been studied, revealing that the CC begins to develop in the embryo and continues for a long period after birth. Therefore, it takes longer to develop than other structures. First, after fusion of the hemispheres, the midline structures form a pathway for interhemispheric axons. Subsequently, the neurons of the cingulate cortex are the first to extend their axons and pass along this path following certain guiding cues. These leading axons serve to guide the subsequent axons of the cortical neuron stack. Although some developmental callosal projections are preserved, a significant number are retracted. Finally, with increased myelination, the contralateral projections invade the cortical plate branch and are further remodeled. Both axonal thinning and myelination are directed in an activity-dependent manner, suggesting that CC development is highly plastic [15].

The width and thickness of the CC increase between the 19th and 21st weeks. CC development is not prescribed and the process does not seem linear. Myelination affects morphometric development. The CC begins to myelinate at about 4 months of age and continues throughout adolescence, as confirmed by magnetic resonance imaging (MRI) and histology. Studies of children and adults have shown that myelination affects the size of the CC [16, 17].

Sex could be an important factor in CC development, as determined by the presence or absence of the Y chromosome and gonadal hormone production. Although fetal testosterone levels contribute to the asymmetric development of the CC, no correlation has been reported between CC size and fetal testosterone. However, those results do not agree with previous studies. For example, some researchers have claimed that the splenium area of the CC is significantly larger in females than males, while others have reported such a gender difference for the isthmus. These differences could be due to the age range of the participants studied. It is important to consider age when examining changes in the CC [18].

When we examine development at the molecular level, we see that many transcription factors and signals work together, as in every stage of neurulation. Transcription factors that are effective in prenatal development continue to be effective postnatally [19].

The transcription factor neurogenin 2 elongates neurons to form callosal axons.

3.3 Developmental Anomalies Associated with the Corpus Callosum

3.3.1 Holoprosencephaly

Holoprosencephaly is incompleteness of a portion of the forebrain with three classic grades, alobar, semi-lobar, and lobar, in order of decreasing severity. This classical definition seems clear but meets the following problems: less prominent ends of the olfactory pathways and bulbar agenesis of the CC occur at less severe ends of the phenotypic spectrum, including hypopituitarism and single upper central incisor. Serious CNS malformation entails a continuum between the brain and other structures, accentuated by com-

Structural changes have been observed in the substantia alba regions of patients with schizophrenia, and recent studies have shown that the levels of specific lipid species in the substantia alba are changed. It is hypothesized that deterioration of the high myelin content of the CC as a result of abnormal lipid homeostasis could help to elucidate the changes seen in schizophrenia [33].

Kim Peek, who inspired the movie Rain Man, was born with a defect in the CC, affecting communication between the two lobes of the brain. Kim Peek, who has enhanced memory owing to a physical difference that allowed the right lobe of the brain to evolve, was also quite adept at piano playing and arithmetic. Thanks to his enhanced memory, he could read and memorize two pages at a time. Kim Peek has almost 9000 books in his memory. The film "Rain Man," which he inspired, was a turning point for him too. Peek, who had previously withdrawn from society and locked himself in his room after returning from work, gained confidence after the film, both from the communication he established with the producers and the celebrations following the success of the film.

References

1. Pearce JMS. Corpus callosum. Eur Neurol. 2007;57(4):249–50.
2. Shah A, Jhawar S, Goel A, Goel A. Corpus callosum and its connections: a fiber dissection study. World Neurosurg. 2021;151:e1024–35.
3. Splittgerber R. Snell's clinical neuroanatomy. Philadelphia: Lippincott Williams & Wilkins; 2018.
4. Shen S, Feng S, Liu H, Jiang J, Yu X. Associations of histological and molecular alterations with invasion of the corpus callosum in gliomas. Acta Neurochir. 2020;162(7):1691–9.
5. Blaauw J, Meiners LC. The splenium of the corpus callosum: embryology, anatomy, function and imaging with pathophysiological hypothesis. Neuroradiology. 2020;62(5):563–85.
6. Hwang SJ, Ji EK, Lee EK, Kim YM, Shin DY, Cheon YH, Rhyu IJ. Gender differences in the corpus callosum of neonates. Neuroreport. 2004;15(6):1029–32.
7. Ku R, Torii M. New molecular players in the development of callosal projections. Cell. 2020;10(1):29.
8. Aboitiz F, Montiel J. One hundred million years of interhemispheric communication: the history of the corpus callosum. Braz J Med Biol Res. 2003;36(4):409–20.
9. Jovanov-Milošević N, Petanjek Z, Petrović D, Judaš M, Kostović I. Morphology, molecular phenotypes and distribution of neurons in developing human corpus callosum. Eur J Neurosci. 2010;32(9):1423–32.
10. Raybaud C. Corpus callosum. Neuroimaging Clin N Am. 2019;29(3):445–59.
11. Meyer N, Richter N, Fan Z, Siemonsmeier G, Pivneva T, Jordan P, Steinhäuser C, Semtner M, Nolte C, Kettenmann H. Oligodendrocytes in the mouse corpus callosum maintain axonal function by delivery of glucose. Cell Rep. 2018;22(9):2383–94.
12. Saab AS, Tzvetanova ID, Nave KA. The role of myelin and oligodendrocytes in axonal energy metabolism. Curr Opin Neurobiol. 2013;23(6):1065–72.
13. de León Reyes NS, Bragg-Gonzalo L, Nieto M. Development and plasticity of the corpus callosum. Development. 2020;147(18):dev189738.
14. Mostafa SN, Akhter A, Sultana T, Islam MN. Corpus callosal lipoma with agenesis of corpus callosum - a case report. Faridpur Med Coll J. 2018;13(1):47–9.
15. Brown WS, Paul LK. The neuropsychological syndrome of agenesis of the corpus callosum. Int Neuropsychol Soc. 2019;25(3):324–30.
16. Huang HC, Chou HC, Tsao PN, Chen CY. Linear growth of corpus callosum and cerebellar vermis in very-low-birth-weight preterm infants. Formos Med Assoc. 2020;119(8):1292–8.
17. Tanaka-Arakawa MM, Matsui M, Tanaka C, Uematsu A, Uda S, Miura K, Sakai T, Noguchi K. Developmental changes in the corpus callosum from infancy to early adulthood: a structural magnetic resonance imaging study. PLoS One. 2015;10(3):e0118760.
18. Allouh MZ, al Barbarawi MM, Ali HA, Mustafa AG, Alomari SO. Morphometric analysis of the corpus callosum according to age and sex in middle eastern Arabs: racial comparisons and clinical correlations to autism spectrum disorder. Front Syst Neurosci. 2020;14:30.
19. Takase H, Hamanaka G, Ohtomo R, Park JH, Chung KK, Gelman IH, Kim KW, Lok J, Lo EH, Arai K. Roles of A-kinase Anchor Protein 12 in astrocyte and oligodendrocyte precursor cell in postnatal corpus callosum. Stem Cell Rev Rep. 2021;17(4):1446–55.
20. Dubourg C, Bendavid C, Pasquier L, Henry C, Odent S, David V. Holoprosencephaly. Orphanet J Rare Dis. 2007;2(1):8.
21. Kasprian G, Brugger PC, Schöpf V, Mitter C, Weber M, Hainfellner JA, Prayer D. Assessing prenatal white matter connectivity in commissural agenesis. Brain. 2013;136(1):168–79.
22. Tang PH, Bartha AI, Norton ME, Barkovich AJ, Sherr EH, Glenn OA. Agenesis of the corpus callosum: an MR imaging analysis of associated abnormalities in the fetus. AJNR Am J Neuroradiol. 2009;30(2):257–63.
23. Gebarski SS, Gebarski KS, Bowerman RA, Silver TM. Agenesis of the corpus callosum: sonographic features. Radiology. 1984;151(2):443–8.

24. Szczupak D, KossmannFerraz M, Gemal L, Oliveira-Szejnfeld PS, Monteiro M, Bramati I, Vargas FR, Lent R, Silva AC, Tovar-Moll F. Corpus callosum dysgenesis causes novel patterns of structural and functional brain connectivity. Brain Commun. 2021;3(2):fcab057.
25. Richards L, Plachez C, Ren T. Mechanisms regulating the development of the corpus callosum and its agenesis in mouse and human. Clin Genet. 2004;66(4):276–89.
26. Oba H, Barkovich AJ. Holoprosencephaly: an analysis of callosal formation and its relation to development of the interhemispheric fissure. AJNR. 1995;16(3):453–60.
27. Paul LK, Brown WS, Adolphs R, Tyszka JM, Richards LJ, Mukherjee P, Sherr EH. Agenesis of the corpus callosum: genetic, developmental and functional aspects of connectivity. Nat Rev Neurosci. 2007;8(4):287–99.
28. Ghi T, Carletti A, Contro E, Cera E, Falco P, Tagliavini G, Michelacci L, Tani G, Youssef A, Bonasoni P, Rizzo N, Pelusi G, Pilu G. Prenatal diagnosis and outcome of partial agenesis and hypoplasia of the corpus callosum. Ultrasound Obstet Gynecol. 2010;35(1):35–41.
29. Lerman-Sagie T, Ben-Sira L, Achiron R, Schreiber L, Hermann G, Lev D, Kidron D, Malinger G. Thick fetal corpus callosum: an ominous sign? Ultrasound Obstet Gynecol. 2009;34(1):55–61.
30. Wang P, Wang J, Tang Q, Alvarez TL, Wang Z, Kung YC, Lin CP, Chen H, Meng C, Biswal BB. Structural and functional connectivity mapping of the human corpus callosum organization with white-matter functional networks. Neuroimage. 2021;227:117642.
31. Hategan A, Bourgeois JA, Cheng T, Young J. Aging with neurodevelopmental disorders: intellectual disability and autism spectrum disorder. In: Geriatric psychiatry study guide. Berlin: Springer International Publishing; 2018. p. 333–49.
32. Ganeshan D, Das K. Congenital malformations of the brain. Br J Hosp Med. 2008;69(11):640–6.
33. Shimamoto-Mitsuyama C, Nakaya A, Esaki K, Balan S, Iwayama Y, Ohnishi T, Maekawa M, Toyota T, Dean B, Yoshikawa T. Lipid pathology of the corpus callosum in Schizophrenia and the potential role of abnormal gene regulatory networks with reduced microglial marker expression. Cereb Cortex. 2021;31(1):448–62.

4 Morphological Anatomy of the Corpus Callosum

Servet Çelik, Okan Bilge, Okan Derin, Melisa Gülcan, Canberk Tomruk, and Ali Çağlar Turgut

4.1 Introduction

The corpus callosum (CC) is the largest commissural pathway that connects functional cortical areas in the cerebral hemispheres. It is a white matter structure composed of dense myelinated fibers. Most of these fibers are axons of pyramidal cells that cross the midline. Approximately 180–200 million are heavily myelinated. The CC is located in the middle of the brain and at the base of the longitudinal cerebral fissure. When the right and left cerebral hemispheres are separated from the longitudinal cerebral fissure, the outer or peripheral surface of the CC is visible in the midline [1].

One of the primary functions of the CC is integration and transformation of the information from both cerebral hemispheres to process sensory, motor, and higher-level cognitive signals. Detailed information about CC functions has been obtained from studies of callosotomy and of patients with separate brain syndrome. In topographical organization, the anterior parts of the CC are involved in transmitting perceptual information, and the posterior parts are involved with visual, auditory, and somatosensory information [2]. The anterior callosal fibers have roles in transferring motor signals between the frontal lobes, while the posterior ones have roles in processing of somatosensory (posterior midbody), auditory (isthmus), and visual (splenium) information [3].

4.2 Gross Parts of the Corpus Callosum

Histological accounts in classical anatomy books divide the CC into four parts in sagittal sections of the brain. From front to back, these parts are *rostrum*, *genu*, *body*, and *splenium* (Fig. 4.1a). They connect different functional cortical areas between the hemispheres. In addition, the narrow component between the poste-

S. Çelik (✉)
Department of Anatomy, Ege University Faculty of Medicine, İzmir, Turkey

Application and Research Center of Cord Blood Cell-Tissue, Ege University, İzmir, Turkey

Interventional Anatomy and Plastination Application and Research Center, Ege University, İzmir, Turkey

O. Bilge
Department of Anatomy, Ege University Faculty of Medicine, İzmir, Turkey

Interventional Anatomy and Plastination Application and Research Center, Ege University, İzmir, Turkey

O. Derin · M. Gülcan
Department of Anatomy, Ege University Faculty of Medicine, İzmir, Turkey

C. Tomruk
Department of Histology and Embryology, Ege University Faculty of Medicine, İzmir, Turkey

A. C. Turgut
Department of Radiology, Ege University Faculty of Medicine, İzmir, Turkey

Department of Histology and Embryology, Aydın Adnan Menderes University Health Sciences Institute, Efeler, Aydın, Turkey

M. Turgut et al. (eds.), *The Corpus Callosum*, https://doi.org/10.1007/978-3-031-38114-0_4

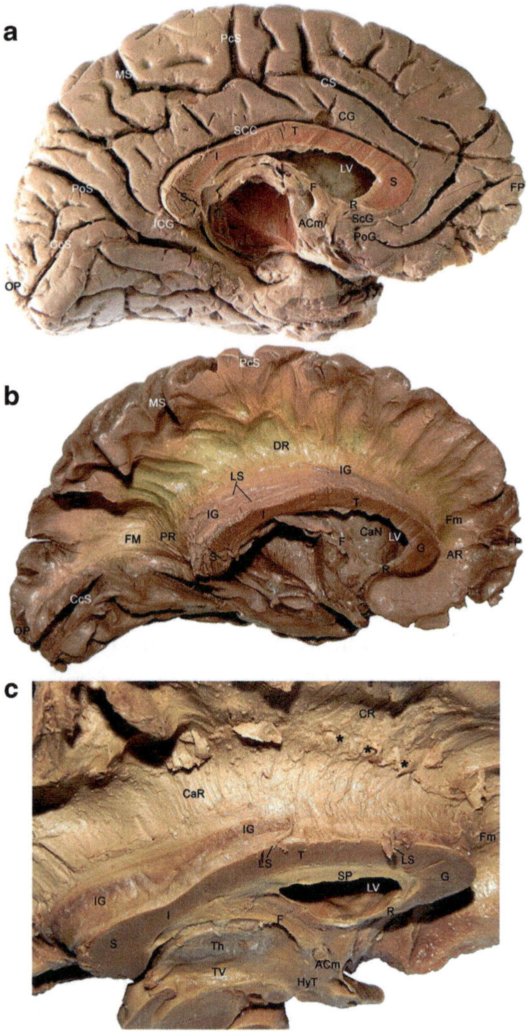

Fig. 4.1 (**a**) Corpus callosum (CC) in medial surface aspect of hemisphere. (**b**) Demonstration of callosal radiations by fiber dissection. (**c**) Posterior-superior view of CC radiations; part of the indusium griseum above the body is dissected and the ends of the longitudinal striae are deviated medially. (*With permission from the Ege University School of Medicine Department of Anatomy).* (*: Intersection of callosal radiation with corona radiata fibers; ACm, anterior commissure; AR, anterior radiation of CC; CaN, caudate nucleus; CaR, callosal radiations; CcS, calcarine sulcus; CR, corona radiata; CcS, calcarine sulcus; CS, cingulate sulcus; DR, dorsal radiation of CC; FM, forceps major; Fm, forceps minor; FP, frontal pole; G, genu; HyT, hypothalamus; I, isthmus; ICG, isthmus of cingulate gyrus; IG, indusium griseum; LS, longitudinal stria; LV, lateral ventricle; MS, marginal sulcus; OP, occipital pole; PcS, paracentral sulcus; PoG, paraolfactory gyrus; PoS, parietooccipital sulcus; PR, posterior radiation of CC; R, rostrum; S, splenium; SCC, sulcus of CC; ScG, subcallosal gyrus; SP, septum pellucidum; T, truncus (body); Th, Thalamus; and TV, third ventricle)

rior part of the body and the splenium is called the *isthmus* [1].

The rostrum is a leaf-shaped part of the CC, located at the anterior lower side and tapering downward. It continues anteriorly with the genu and posteriorly and downward with the lamina terminalis. The rostrum is located on the anterior wall of the third ventricle. The genu is the anteriorly thick knee-shaped part of the CC, its most anterior end in sagittal section, and it continues downward posteriorly with the rostrum. The genu is located in the anterior upper part of the septum pellucidum. The body is the longest part of the CC, extending from the genu anteriorly to the splenium posteriorly. The splenium is the posterior and thickest part of the CC (Fig. 4.1a), inserted posteriorly into the quadrigeminal cistern. It can adopt two shapes, bulbous and tubular; the bulbous shape is more common than tubular one [4]. Some authors report that both types have similar frequencies (52% tubular, 47% bulbous); the tubular shape is more common in men and the bulbous shape in women [5].

4.2.1 Dimensions of the Corpus Callosum

The lengths and thicknesses of the CC and its segments have been measured in midsagittal sections of the cadaveric brain, fresh brains from autopsy, and living magnetic resonance imaging (MRI) images.

In the midsagittal section, the average length of the CC from the rostrum to the splenium is 7.8 cm. Its average length from the tip of the genu to the most posterior aspect of the splenium is 7.5 cm. The mean rostrum width is 3.2 cm, genu width 3.4 cm, anterior body width 3.7 cm, posterior body width 4.1 cm, and splenium width 4.5 cm. The transverse width of the dorsal callosal radiations from one hemisphere to the other is 8.1 cm [6].

In 107 female and male cadaver brains, the average thickness of the body from the midpoint was 5.46 mm, the mean rostrum thickness was 5.91 mm, and the average splenium thickness was 8.4 mm. The length of the line passing from the lowest points of the rostrum and splenium to the highest point of the body, that is, the CC height, was 23.37 mm on average [7].

Differences in CC size and shape were detected in cadaver brains. Embalming postmortem brains with formalin can cause such differences [8]. In some studies, while the measurements of cadavers embalmed with formalin and live MRI images were similar, the dimensions were larger in autopsy brains [3]. In a comparison of formalin-embalmed brains with live MRI images, the CCs proved slightly smaller in the formalin brains [7].

There is much variation in the shape and size of the CC among individuals, regardless of gender and age. It is not possible to predict gender or age from the shape or size of any CC [4]. In a total of 44 (22 women and 22 men) formalin-embalmed cadavers, the distance between the most prominent anterior point of the CC and the frontal pole was 3.54 cm, the distance between the most prominent posterior point and the occipital pole was 5.63 cm, and the shortest distance from the body to the cortex surface averaged 3.59 cm [3]. The distance from the genu to the fornix was 3.15 cm, and the distance to the anterior commissure was 3.35 cm [9]. These measurements can be important for callosotomy approaches.

The mean CC cross-sectional area on live MRI images was 5.5 cm^2 in adults: 5.8 cm^2 in men and 5.3 cm^2 in women [10]. This is twice the cross-sectional area of all other commissural tracts [11].

MRI measurements of a total of 301 women and men aged 11–84 years gave a mean CC volume of 47.657 mm^3. The volume was greater in males than females. In addition, the left CC volume proved greater than the right in both sexes, apart from the body [12]. More details of CC dimensions will be presented in other parts of this book.

4.2.2 Morphometric Subunits of the Corpus Callosum

Thanks to measurements and functional studies, the sections of the CC are becoming more clearly defined, and subsections have been distinguished [13]. Some authors suggest different classifications and subdivide the CC further [14, 15]. According to the Witelson classification,

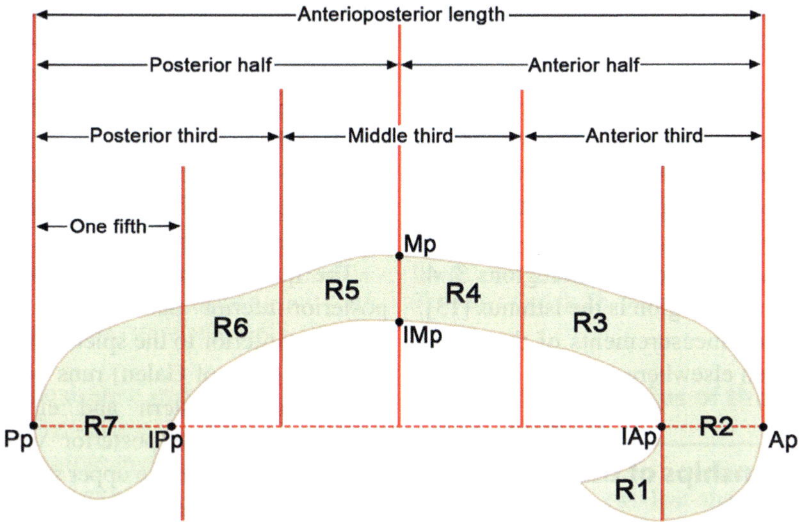

Fig. 4.2 Morphometric subunits of the CC. (Ap, anterior-most prominent point of CC; IAp, the most anterior point of the inner surface of CC; IMp, midpoint of inner surface of CC; IPp, the most posterior point of the inner surface of CC; Mp, midpoint of outer surface of CC; Pp, posterior-most prominent point of CC; R1, region 1, rostrum; R2, region 2, genu; R3, region 3, rostral body; R4, region 4, anterior midbody; R5, region 5, posterior midbody; R6, region 6, isthmus; and R7, region 7, splenium)

part of the inner surface of the body. The inner surface of the splenium is adjacent to the commissure of the fornix, which gradually expands posteriorly (Figs. 4.1c and 4.3a). The lateral parts of the inner surface of the body form the roof of the anterior horn and the central part of the lateral ventricles. The inner surface of the genu forms the anterior wall of the anterior horn of the lateral ventricle, while the inner wall of the rostrum forms its base [1, 17].

4.4 Radiation of Corpus Callosum

The extensions of the tightly packed CC fibers in the midline to the cerebral hemispheres are called *the radiations of the CC.* In fiber dissections, radiations of the CC form the fourth (central) group of white matter fibers. Thanks to the callosal radiations, almost all regions of both hemispheres are directly or indirectly connected [21].

4.4.1 Directions of Corpus Callosum Fibers

In the midsagittal section of the brain, the anteroposterior length of the CC is much less than the length of the hemispheres. This size difference compresses the commissural fibers coming from the frontal lobe anteriorly and from the occipital lobe posteriorly, leading to the anterior and posterior ends of the CC, respectively, and passing from there to the opposite hemisphere. Consequently, the genu and splenium through which these fibers pass contain more fibers than the body and appear thicker [1, 6].

The difference in length between the CC and the hemisphere in the midsagittal section also causes the callosal radiation fibers extending from different CC sections to follow different courses toward the hemispheres. While fibers passing through the body follow a transverse course in the white matter, those passing through the genu and splenium cannot do so (Fig. 4.1c). In transverse sections of the brain, the fibers passing through the genu connecting the two frontal lobes bend forward on the sides and extend into the frontal lobe. This is called the *forceps frontalis (or forceps minor)* because it resembles a pincer in horizontal sections (Fig. 4.1b). Similarly, the fibers passing through the splenium and connecting the two occipital lobes curve backward to enter those lobes. These curled-back fibers are called *forceps occipitalis (or forceps major)* because they too resemble pincers (Fig. 4.1b). Some fibers passing through the body and splenium curve downward to reach the lower parts of the temporal and occipital lobes. They pass through the upper and outer posterior horn and the outer side of the inferior horn of the lateral ventricle. The fibers that cover the lateral wall of the lateral ventricle in this section are called *the tapetum.* The fibers of the tapetum separate the temporal horn of the lateral ventricle from the fibers of the optic radiation. Fibers leaving the body extend on the sides as rays and cross the projection and association fibers as they travel to the cerebral cortex [1, 21].

The fibers of the CC have been distinguished by fibril dissection. They are classified according to their direction as *superior or dorsal radiation, inferior or ventral radiation, anterior radiation,* and *posterior radiation* (Fig. 4.1b, c) [6].

4.4.1.1 The Superior or Dorsal Callosal Radiation

Dorsal callosal radiations originating from the body first extend laterally in the hemisphere, passing under the cingulum and superior longitudinal fasciculus. These fibers are then directed upward, pass through the lateral aspect of the superior longitudinal fasciculus, and merge with the fibers of the corona radiata. In our callosal fibril dissection experience, it is difficult to separate the callosal fibers from the corona radiata fibers with a wooden stick at this intersection (Fig. 4.1c). The width of the dorsal callosal radiations in both hemispheres is 8.1 cm on average [6].

4.4.1.2 The Inferior or Ventral Callosal Radiation

In general, the ventral callosal radiations extend downward in the cerebral hemisphere after exiting the CC. They consist of *anterior* and *posterior components* [6].

The anterior component fibers extend laterally and downward just anterior to the genu region. They then run backward and terminate around the caudate nucleus. The anterior component of the ventral callosal radiations is near the fibers of the inferior fronto-occipital fascicle. The average length of the anterior component of the ventral callosal radiations is 4.0 cm [6].

The posterior component fibers of the ventral radiations of the CC extend laterally to form the roof of the lateral ventricular body, atrium, and occipital horn. They then orient downward to form tapetum fibers. They average 5.5 cm in length and separate the lateral ventricular wall from the optic radiation fibers [6].

4.4.1.3 The Anterior Callosal Radiation

The anterior callosal radiations are generated by fibers originating from the genu and rostrum.

Most of them extend anteriorly to the medial orbitofrontal gyrus, forming forceps minor fibers located on the inner wall of the frontal horn of the lateral ventricle (Fig. 4.2b). The forceps minor fibers extending most anteriorly from the genu have an average length of 3.8 cm. Some of the anterior callosal fibers emerging from the rostrum pass through the medial side of the uncinate fasciculus and travel laterally to the insular region [6].

4.4.1.4 The Posterior Callosal Radiation

Posterior callosal radiation fibers originating from the splenium extend posteriorly to reach the occipital lobe and form the forceps major on the inside of the occipital horn of the lateral ventricle (Fig. 4.2b). They form the atrium in the wall of the lateral ventricle. The extreme fibers of the forceps major are 6.5 cm from the splenium [6].

4.4.2 Cortical Regions Connected by Radiations of the Corpus Callosum

The CC fibers are classified as *homotopic* and *heterotopic* according to the hemisphere regions to which they attach. Homotopic fibers constitute the majority and connect homologous cortical fields between the two hemispheres. Heterotopic fibers connect different functional cortical areas in the hemispheres asymmetrically [22]. The CC fibers connect the frontal, parietal, temporal, occipital, and the *insular* cortex (also *insula,* island of Reil, and *insular lobe*) as well as limbic lobe and the basal ganglia in both hemispheres [6].

According to the Witelson classification, fibers connecting the caudal or orbital prefrontal and inferior premotor cortexes of the hemispheres pass through the rostrum. Fibers passing through the genu connect the prefrontal cortices; those passing through the rostral body connect the premotor cortices and supplementary motor areas. Fibers passing through the anterior midbody connect the motor cortex, and those passing through the posterior midbody connect the somaesthetic, posterior parietal cortex. The CC fibers in the isthmus connect the superior temporal and posterior parietal cortices between the hemispheres. Fibers in the splenium connect the occipital and inferior temporal cortices [13].

Fibril dissection studies have also revealed the regions bound by CC radiations. The dorsal callosal radiations connect the motor and supplementary motor areas. The anterior truncal fibers contributing to the dorsal callosal radiation connect the medial surfaces of the superior frontal gyrus. Fibers emerging from the posterior part of the body connect the posterior portion of the medial superior frontal gyrus, called supplementary motor area, precentral and postcentral gyri, or the paracentral lobule. The ventral parts of callosal radiations connect the temporal lobes and the basal ganglia of the two hemispheres to each other. The forceps minor fibers formed by anterior callosal radiations connect the medial orbitofrontal gyri in both hemispheres. A portion of the anterior callosal fibers emerging from the rostrum connects the two insular regions. The forceps major, formed by posterior callosal radiation fibers, connects the occipital regions [6].

4.5 Vascularization of the Corpus Callosum

While the arterial supply to the CC is from the outer (peripheral) side of the CC, most of the venous drainage is from the inner (central) side. There is less venous drainage from the outer surface; this connects with the superficial veins of the brain [23, 24].

4.5.1 Arterial Supply of the Corpus Callosum

While the rostrum, genu, and body of the CC are supplied from *the internal carotid system*, the splenium is supplied from *the vertebrobasilar system*. The anterior cerebral artery (ACA) is the terminal branch of the internal carotid artery. Most the CC is supplied by the PcA branch of the ACA. The PcA runs posteriorly in the SCC and gives perforating branches that enter vertically to feed the rostrum, genu, and body. Rarely, the median artery originating from the anterior communicating artery (ACmA) participates in supplying the anterior parts of the CC. The posterior cerebral artery (PCA) is the terminal branch of the basilar artery. The posterior pericallosal artery (PPcA) branch of the PCA supplies the posterior parts of the CC via perforating branches [25, 26]. The details of the arteries supplying the CC are as follows:

4.5.1.1 Subcallosal Artery

The subcallosal artery (ScA) arises from the caudo-dorsal aspect of the ACmA as a single artery, 1.46 mm from the right ACA and 1.13 mm from the left. It extends upward to the rostrum and genu in front of the lamina terminalis, ending at the most protruding point of the genu [27]. The ScA has a frequency of 50–91% in cadaveric dissections and is 0.5 mm in diameter. It gives proximal and distal branches. The distal branches average 0.25 mm in diameter and supply the medial parts of the genu and rostrum [26, 27]. The ScA gives blood to the rostrum in 84%, the genu in 79%, and the body in 37% of cases [28].

4.5.1.2 Median Callosal Artery

The median callosal artery (McA) is also considered *the third pericallosal artery*. It arises from the ACmA and is 0.9 mm in diameter. In cadaver studies, it was detected with a frequency of 9–30%, often as a single artery in the midline [26, 27]. In 5% of cases, it arises as a separate single artery and in 25% as a common stump with the hypothalamic artery [26]. The course of the McA is similar to that of the ScA but extends more posteriorly. It generally supplies the rostrum and genu but also the body and splenium when it is long. The McA is rarely double [29].

4.5.1.3 Pericallosal Artery

Some authors also refer to the PcA as *the anterior pericallosal artery* [30]. It originates from the A2 segment of the ACA and has an average diameter of 2.6 mm on the right and 2.5 mm on the left. In 65% of cadavers, it courses backward via the SCC in the pericallosal cistern (Fig. 4.3a) [31].

There are variations in the emergence and course of the PcA. In approximately one-third of hemispheres, one of its distal segments runs within the cingulate sulcus (CS). Rarely, all of its distal segments are in the CS and do not contribute to CC nutrition; it can even pass to the opposite hemisphere and supply blood to the area. A PcA passing to the opposite hemisphere is also called *the bihemispheric pericallosal artery*. In 67% of hemispheres, the PcA extends over the splenium in a thin coiled fashion like the posterior pericallosal artery. Its terminal end, running in the CS, anastomoses with the splenial branches of the PCA over the posterior part of the CC and forms *the pericallosal pial plexus (PcPP)*. Branches from the PcPP often attach to the tela choroidea of the third ventricle and anastomose with the medial posterior choroidal arteries. The anterior and posterior pericallosal arteries anastomose with each other and form *the perisplenial arterial ring* around the splenium [31].

4.5.1.4 Pericallosal Pial Plexus

Most of the PcPP is formed by the long callosal arteries and cingulocallosal arteries arising from the PcA. Branches of the ScA, McA, and posterior PPcA also participate in forming it. The PcPP is most prominent on the splenium, where the callosal arteries also join. While the branches participating in forming the PcPP are 0.6–0.9 mm in diameter at the origin, they have a uniform diameter of 0.3–0.7 mm subsequently. Twigs arising from the PcPP form connections among themselves before they enter the brain tissue. Branches that emerge from the PcPP and perforate to supply the CC extend in the anteroposterior direction [32]. Branches from the PcPP provide blood to parts of the CC and callosal radiations as well as the CG [26].

4.5.1.5 Short Callosal Artery (Perforating Artery)

The perforating branches of the arteries around the CC that enter and supply it are called *callosal arteries* (Fig. 4.3b). They come directly from the artery adjacent to the CC, which they enter without branching. In almost all (96.6%) hemispheres, these perforating branches take the form of short callosal arteries [33]. They also arise from the PcPP created by the callosal arteries. The short callosal arteries arise as 2 or 4 clusters in a hemisphere, and there can be up to 20 such arteries in a hemisphere. Perforating arteries supply the CC throughout the midline [26, 31].

4.5.1.6 Long Callosal Artery

The long callosal arteries emerge from the PCA and course in parallel with it in the SCC. There are 0–3 of them in 55% of hemispheres and their diameter is 0.62 mm on average. They end at the body or splenium giving many perforating branches. The branches of the long callosal arteries are the cingulocallosal arteries and the short callosal perforating branches. These supply blood to the CG and the lateral parts of the CC and callosal radiations. The long callosal arteries give off branches that participate in forming the PcPP. Most of them extend posteriorly and anastomose with the PPcA [25, 26]. Some authors have shown that long callosal arteries are long non-branching forms of short callosal arteries. Arteries of this type arise from the PcA, run parallel to it a few centimeters, and then enter the CC [31].

4.5.1.7 Cingulocallosal Artery

The cingulocallosal arteries arise directly from the inferolateral aspect of the PcA and run outward in the SCC. In cadaver dissection, they are present in all hemispheres [33]. Their number is in the range 2–23 and their diameter averages 0.25 mm. Cingulocallosal arteries give perforating branches to the CG, the lateral sides of the CC, and callosal radiations. They participate in forming the PcPP by anastomoses with ScA branches, McA branches, long callosal arteries, and other cingulocallosal arteries [26, 33].

4.5.1.8 Recurrent Cingulocallosal Artery

These arteries originate from the cortical branches of the PcA, pass down the medial surface of the CG, and course toward the CC. In cadaver dissections, 1–7 recurrent cingulocallosal arteries are seen in 45% of hemispheres, and their diameter is 0.25 mm on average [26].

4.5.1.9 Posterior Pericallosal Artery

The PPcA is also called *the splenial artery*. It arises from the PCA or more often from its cortical branches. In cadaveric dissections, it occurs in 88% of hemispheres [30] and averages 0.7 mm in diameter. *The proximal type of PPcA* that emerges proximally from the PCA rises behind the splenium and upward to its outer surface. *The distal originated type* of PpcA is located behind the splenium. Rarely, a *double type PPcA* arises from both proximal and distal parts of the PCA or its cortical branches. The PPcA divides into *superior* and *inferior branches*. The superior branch courses within the SCC and participates in forming the PcPP above the splenium, often by anastomosis with the PCA or more rarely with the McA or long callosal arteries. The inferior branch supplies the third or the lateral ventricle. The PPcA provides blood particularly to the splenium [26].

4.5.1.10 Industrial Artery

Industrial arteries are longitudinally extending arterial branches in the medial longitudinal striae. In cadaver studies, they were detected in 9 of 22 brains. These branches participated in forming the PcPP [32].

4.5.1.11 Accessory Posterior Pericallosal Artery

In 25% of hemispheres, another posterior pericallosal artery arises from the precuneal branch of the parietooccipital, hippocampal, medial posterior choroidal, or lateral posterior choroidal artery. It has an average diameter of 0.3 mm and is called *the accessory posterior pericallosal artery*. It contributes to the nutrition of the splenium [26]. The PCA, which arises from the basilar artery, gives terminal and choroidal branches that nourish the splenium [34]. It has been stated that *the subependymal arteries*, which are terminal branches arising from the lateral posterior choroidal and anterior choroidal arteries, nourish the white matter fibers of the tapetum as it supplies the walls of the lateral ventricle [35].

4.5.2 Venous Drainage of the Corpus Callosum

Venous drainage of the CC involves two groups of veins, *the first group of veins* arising from the inner (central) face toward the ventricle and *the second group of veins* arising from the outer (peripheral) face toward the cortex. The main venous drainage of the CC is toward the deep cerebral veins through veins that arise from its inner surface and are located in the walls of the lateral ventricle. The veins that exit on the outer face are also known as the *accessory venous drainage* circulation [16].

4.5.2.1 Central Venous Drainage of the Corpus Callosum

Central venous drainage of the CC is via *the callosal veins (CV)* and *callosal cingulate veins (CCV)* arising from the inner surface of the CC. Another part of the central venous drainage comprises *the deep medullary veins (DMV)* passing between the callosal radiation fibers to reach the *subependymal veins (SeV)*, which ultimately drain the blood inside to *the internal cerebral veins (ICV)* of the deep venous system of the brain [16, 23].

Callosal Veins and Callosocingulate Veins

The postcapillary venules and *the callosal venules* between the CC fibers form the CVs, which extend perpendicular to the CC fibers toward the ventricle and collect the callosal venules. These veins are defined as *short*, *medium*, and *long CVs* according to the distance they course in the CC. A long CV spans the entire thickness of the CC and opens into the SeV in the ventricular wall. The short CVs are often found in the rostrum and splenium and travel a shorter distance in the CC [16]. The CCVs originate from the CG and enter the CC from the more lateral parts of its peripheral surface, span its entire thickness, and then join the SeV. Veins of this kind that travel the entire distance of the CC are also called *transcallosal veins* [16].

Deep Medullary Veins

The DMVs collect blood from the white matter, including the callosal radiations. They pass through four aggregation sites in the white matter and go toward the lateral ventricle [24, 36]. They appear as a fan from the lateral ventricle to the cortex and form *a caput medusa image* at the edge of the lateral ventricle. These veins receive blood as it passes between the callosal fibers and deliver it to the SeVs [24, 37]. The subcortical part of the white matter drains through the superficial medullary veins and subcortical veins to *the pail venous plexus* over the cerebral cortex [24, 36, 38].

Subependymal Veins

The SeVs are located in the ventricle walls just below the ependyma of the ventricles. They collect in the corners of the lateral ventricle and form the *anterior septal vein (anterior vein of the septum pellucidum)*, the *posterior septal vein (posterior vein of the septum pellucidum)*, and the *medial* and *lateral atrial veins* running on the medial or lateral ventricle walls. *The lateral group* of these veins drains blood from the lateral ventricle walls and joins the ICV via *the superior thalamostriate vein*; *the medial group* drains CC blood directly into the ICV. Some veins emerging from the CC run in the lateral ventricular wall below the CC before opening into the SeV. These are called *infracomissural veins* [16, 24]. The anterior vein of the septum pellucidum drains blood from the genu and empties into the ICV. The posterior vein of the septum pellucidum receives blood from the body and also opens into the ICV. Blood coming from the posterior parts of the CC and tapetum flows into the medial (atrial) and lateral (atrial) veins of the lateral ventricle. These veins drain into the ICV or greater cerebral vein [18].

4.5.2.2 Peripheral (Accessory) Venous Drainage of the Corpus Callosum

The veins emerging from the outer surface of the CC are *the anterior vein of the* CC and *the posterior vein of the* CC [39, 40].

Anterior Vein of Corpus Callosum

This is a pair of veins, also called *the anterior pericallosal vein (APcV)*. The mean diameter of these vessels is 0.8 mm [18]. The APcV is seen in 90% of cases. It runs anteriorly in the pericallosal cistern above the CC and empties into the anterior cerebral vein, which transmits blood from the CC to the basal vein. The APcV sometimes drains into the lower end of the inferior sagittal sinus. It drains blood from the genu and rostrum of the CC [40].

Posterior (Dorsal) Vein of the Corpus Callosum

This is a pair of veins, also called *the posterior pericallosal vein (PPcV)* or *splenial vein*. The mean diameter of these vessels is 1.1 mm [18]. The PPcV is detected in 60% of brain dissections. It starts from the posterior third of the outer surface of the CC and drains the blood there. It runs posteriorly in the pericallosal cistern, close to the CG, extends downward through the upper surface of the splenium, and often (66%) ends by connecting to the greater cerebral vein. In fewer cases, The PPcV empties into the ICV (16%) or occipital vein (16%). It drains blood from the posterior parts of the CC [39, 40].

4.6 Conclusion

The CC is located in the center of the brain and is the largest commissural tract. It transfers information from both cerebral hemispheres. From front to back, its parts are the rostrum, genu, body, isthmus, and splenium. The boundaries and subunits of these parts are distinguished in morphometric studies. Callosal fibers extending to the hemispheres form the superior or dorsal, inferior or ventral, and anterior and posterior radiations. Although the arterial blood supply to the CC comes from its outer surface, venous drainage is mainly from the inner surface. There is also a connection with the superficial veins of the brain. It is necessary to know the CC segments, sizes, radiations, and vessels in order to manage various CC lesions and for surgical approaches.

References

1. Catani M, Zilles K. Chapter 32: Cerebral hemispheres. In: Standring S, editor. Gray's anatomy: the anatomical basis of clinical practice. 42nd ed. Beijing: Elsevier Health Sciences; 2021. p. 512–39.
2. Tzourio-Mazoyer N. Intra-and inter-hemispheric connectivity supporting hemispheric specialization. In: Kennedy H, Van Essen DC, Christen Y, editors. Micro-, meso-and macro-connectomics of the brain. Basel: Springer; 2016. p. 129–46.
3. Gupta T, Singh B, Kapoor K, Gupta MM, Kochar S. Corpus callosum morphometry: comparison of fresh brain, preserved brain and magnetic resonance imaging values. Anat Sci Int. 2008;83:162–8.
4. Byne W, Bleier R, Houston L. Variations in human corpus callosum do not predict gender: a study using magnetic resonance imaging. Behav Neurosci. 1988;102(2):222–7.
5. Pervin S, Nurunnabi ASM, Ara S, Haroon K. Morphometry of the splenium of the corpus callosum—a study on 60 cadaveric brains. Bang J Neurosurg. 2021;11(1):30–5.
6. Shah A, Jhawar S, Goel A, Goel A. Corpus callosum and its connections: a fiber dissection study. World Neurosurg. 2021;151:e1024–35.

Morphometry of the Corpus Callosum

5

Niyazi Acer and Nihal Gürlek Çelik

5.1 Introduction

The corpus callosum (CC) has the feature of being the largest and most important commissural pathway that integrates the activities of the two cerebral hemispheres by providing information transfer [1, 2]. There are approximately 300 million axons within the CC, and these contribute to processesing memory, attention, language, intelligence, and high-level cognitive information [3, 4]. These fibers, distributed in the substantia alba, are divided into homotopic and heterotopic [1]. While homotopic fibers form a mirror image by connecting the same regions of both hemispheres, heterotopic fibers form an asymmetrical image by connecting different regions [5].

We can see the shape, dimensions, and sections of the CC on sagittal sections of the brain [6]. Anatomically, the CC is divided into four parts. From front to back, these segments are the rostrum, genu, body, and splenium [1, 7]. Each of these parts connects with different areas of the cerebral cortex [8]. The beak-shaped anterior part of the CC is the rostrum. Connected with the frontal lobe, it extends posteriorly and downwardly to the lamina terminalis [1]. The genu is the thick part of the CC found anteriorly [1]. The fibers passing through the genu travel inside the frontal lobe, and these result in the formation of the forceps minor. The fibers passing through the splenium posteriorly and traveling inside the occipital lobe form the forceps major. The fibers passing through the body form the corona radiata by moving transversely along the cerebral cortex [1, 6]. In addition, the CC has connections with the lateral ventricle and fornix [6].

It has been reported in the literature that many factors such as growth development, age, diseases, and environmental factors can all cause morphological changes in the structure of the brain. One of the structures affected by these factors is the CC [9–11]. Many studies have shown that age, sex, prenatal and perinatal trauma, hypoxic injuries, variation, genetic factors, environmental factors, ethnicity, demyelinating disorders, and neurological and some psychiatric diseases such as Alzheimer's disease, Williams syndrome, Down syndrome, and bipolar disorder are associated with changes in the shape or size of the CC [9, 12–18].

Many studies have been conducted on the association of variations of the CC with age and sex. Due to the anatomical localization of the CC, it is possible that it is affected by changes in the cortical and subcortical parts of the brain. Any deterioration in the structure of the CC may be a sign of a pathological condition. In cases where medical treatment is insufficient, surgical divi-

N. Acer (✉)
Department of Anatomy, Arel University School of Medicine, İstanbul, Turkey

N. Gürlek Çelik
Department of Anatomy, Amasya University School of Medicine, Amasya, Turkey

M. Turgut et al. (eds.), *The Corpus Callosum*, https://doi.org/10.1007/978-3-031-38114-0_5

sion of the CC for patients with severe epilepsy can be performed.

Changes in the morphology of the CC caused by brain atrophy have been detected in Alzheimer's disease, multiple sclerosis, and Huntington's disease [19–23]. Also, it has been stated that CC morphology in neuropsychiatric diseases such as schizophrenia, bipolar disorder, and depression is altered [24–26].

When looking at some longitudinal studies on the shape and size of the CC [27, 28]. Splenium atrophy was observed in the early stages of Alzheimer's disease. It has been reported that changes in some parts of the CC are age related, and the total area decreases with increasing age [29, 30]. The size of the CC has been reported to decrease with aging in men [31].

In this section, data on the differences in morphological and biometric characteristics of the CC due to age, sex, disease, and environmental factors are discussed.

5.2 Different Imaging Methods for Evaluating the Corpus Callosum

Several different imaging modalities can be used in the assessment of the CC. The CC morphology has been studied on Magnetic Resonance Imaging (MRI), Ultrasonography (USG), and Computed Tomography (CT).

5.2.1 Utrasonography (USG)

USG has advantages such as low cost, easy access, easy repetition, no radiation, and being performed at the patient's bedside [32, 33]. USG of the CC can be used for the purpose of scanning in the prenatal period and only in the early postnatal period (unil the time that the anterior fontanelle is closed) [33, 34].

5.2.2 Computed Tomography (CT)

CT was first used in 1972 [35]. CT can be helpful in detecting bleeding, edema, mass effect, calcification, and necrosis [36]. However, it is not best at evaluating the CC due to less soft tissue contrast and exposure to high radiation [36, 37].

5.2.3 Magnetic Resonance Imaging (MRI)

MRI has advantages such as providing sufficient soft tissue contrast and the absence of radiation [37]. MRI demonstrates the development, length, and detection of vascular lesions of the CC [38–40]. For measurement of the CC, MR images in the midsagittal plane are best [9, 41] (Fig. 5.1).

5.2.4 Morphometric Measurement of the Corpus Callosum

Usual measurements of the CC include its length, thickness, height, and area [9]. Midsagittal sections are important for measurements of the CC as the septum pellucidum and interthalamic adhesions as well as the anterior and posterior commissures are clear [42].

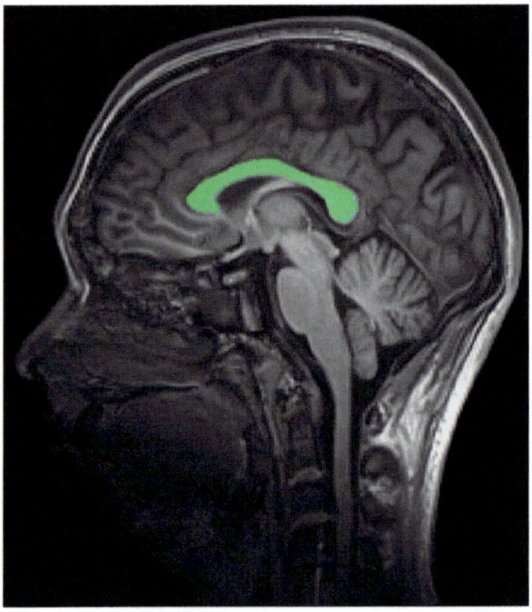

Fig. 5.1 Area of the CC on T1-weighted MR images

5.2.5 Length of the Corpus Callosum

The length of the CC is measured as the curvilinear line that runs through the middle of the CC starting from the rostrum to the splenium [9] (Fig. 5.2).

5.2.6 Anterior-Posterior Diameter of the Corpus Callosum

The anterior-posterior diameter (APD) of the CC is the measurement of the distance between the anterior surface of the genu and the posterior surface of the splenium [9] (Fig. 5.2).

5.2.7 Corpus Callosum Thickness

The thickness of each of the parts of the CC is made as shown in Fig. 5.2. The measurement of the thickness of the isthmus is made from the level where it joins the fornix [9] (Fig. 5.2).

5.2.8 Corpus Callosum Height

The height of the CC is made at the midpoint of the distance between a line tangent to the lower boundary of the genu and the splenium and the lines tangent to the body of the CC [43].

5.2.9 Evaluation of the Splenium Position

First, a straight line is drawn along the dorsal surface of the brain stem. Then, a second line is drawn parallel to the first line from the posterior end of the splenium. The splenium/tegmentum distance between these two lines is measured to coincide with the level of the roof of the fourth ventricle [9].

5.2.10 Corpus Callosum Area

The area of the CC can be calculated using many different softwares such as ImageJ and

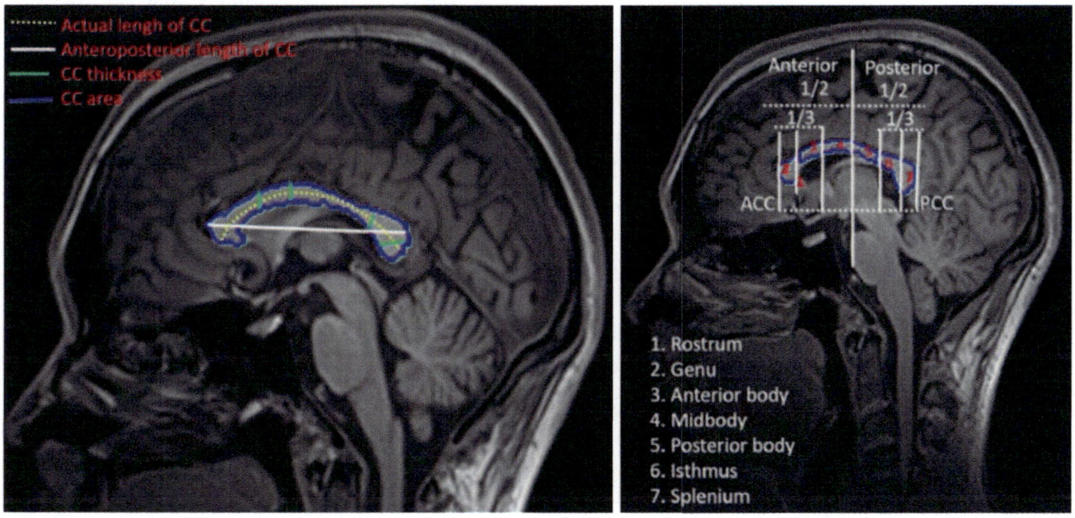

Fig. 5.2 Measurement of the thickness, area, and length of the CC

PACS. Witelson divided the CC into seven subsegments according to a proportional segmentation scheme [31]. These segments included 1/3, 1/2, 2/3, and 4/5 of the CC total length from anterior to posterior in the midsagittal plane. First, the length of the anterior-posterior distance of the CC was drawn, and then, a vertical line was drawn in the middle of this line. The vertical line divides the CC into two equal segments. The segment in the front was called the anterior CC, and the segment at the back was called the posterior CC. The CC is then divided into three parts as anterior, middle, and posterior segments. Another vertical line was drawn from the kinking part of the CC, and the anterior segment was divided into four parts: the first part as the rostrum in the anterior and tip, the second part as the genu, and then the other two parts. Finally, the CC was divided into five parts for the surface area of the splenium. Thus, the posterior segment was divided into two parts, and in total, the CC was divided into seven segments [31] (Fig. 5.2).

5.3 Discussion

In Witelson's study of human cadavers, it was found that there was no difference between the sexes, but the size of the CC was larger in males. In the same study, it was found that while the size of the CC decreased in men with aging, this was not the case in women [31]. However, some have determined that while there was a decrease in CC size in men with aging, there was no difference in women [44] (Table 5.1). Many studies have reported that the CC size is greater in males [31, 50, 51], and some stated that the isthmus portion of the CC was larger in right-handed women [31].

Some authors have measured the area of the CC in the midsagittal section of the cadaveric brain [44, 52]. Some authors measured the midsagittal callosal area, thickness, and length of the CC on MRI [53, 54] and detected a weak but statistically significant negative correlation between age and the thicknesses of the genu [53] (Table 5.1). Weis et al. found significant age-related changes in the genu and anterior portion of the CC. It was also found that the total area of the CC decreased with increasing age [30].

Previous studies have investigated callosal size in obsessive-compulsive disorder, schizophrenia, bipolar disorder, and major depressive disorder [55–59] (Table 5.2).

Piras et al. found that patients with obsessive-compulsive disorder and patients with schizophrenia had a thinner body of the CC than controls [59] (Table 5.2).

Others have studied the morphology of the CC on the MRIs of different diseases such as dyslexic children, Down and Williams syndromes, frontotemporal dementia, progressive supranuclear palsy, and Alzheimer's disease [60–62]. Hynd et al. reported that the genu was significantly smaller in dyslexic children [60]. Wang et al. found that the width of the rostrum of the CC decreases in patients with Down syndrome; Yamauchi et al. reported CC atrophy in frontotemporal dementia, progressive supranuclear palsy, and Alzheimer's disease [61, 62].

Ahmadvand et al. found that the length and area of the CC were shorter and smaller in patients with schizophrenia compared to controls. Greater differences were identified in males [41].

Table 5.1 Measurement parameters and results for the CC

Study	Number	Methods	Measurement parameters	Results
Junle et al. (2008) [45]	286 healthy adults, including 127 males and 159 females, and ranging in age from 20 to 81 years	MRI	The total longitudinal dimension, maximum height, widths of the genu, central region, and splenium	Larger in females
Koşar et al. (2012) [46]	42 male and 48 female three group as 6–9, 10–13, and 14–17 years old	MRI	Anteroposterior diameter of the CC	Differences between age group results were found to be statistically significant
Abdolmaleki et al. (2016) [47]	68 females and 60 males	MR T2 Pms DViewer	Area of the rostrum and genu to the area of the CC, area of the anterior midbody to ACC, area of the posterior midbody to the ACC, area of the isthmus to the ACC, and area of the splenium to the ACC	All segments of the CC are larger in men compared to women
Habibi et al. (2021) [33]	160 boys and 160 girls	MR midsagittal T1-weighted images	Area of the CC, supracallosal area, length of the CC, genu thickness, body thickness, splenium thickness	All parameter values increased with age
Choudhury et al. (2020) [48]	37 brains, 24 male, 13 female	Cadaveric	Different diameters of the brain and CC	Positive linear correlation between length of the CC and the length and vertical diameters of the brain
Pervin et al. (2022) [49]	60 brains 36 male and 24 female	Cadaveric	Length and width of the splenium were measured	The mean length of the splenium larger in male in comparison to female, the mean width of the splenium greater in female than male
Poleneni et al. (2021) [43]	40 formalin-fixed, full brain specimens	Cadaveric	Longitudinal dimension of the CC	Positive correlation between CC length and longitudinal dimension of the brain

26. Walterfang M, Yücel M, Barton S, Reutens DC, Wood AG, Chen J, Lorenzetti V, Velakoulis D, Pantelis C, Allen NB. Corpus callosum size and shape in individuals with current and past depression. J Affect Disord. 2009;115(3):411–20. https://doi.org/10.1016/j.jad.2008.10.010.
27. Hampel H, Teipel SJ, Alexander GE, Pogarell O, Rapoport SI, Möller HJ. In vivo imaging of region and cell type specific neocortical neurodegeneration in Alzheimer's disease. Perspectives of MRI derived corpus callosum measurement for mapping disease progression and effects of therapy. Evidence from studies with MRI, EEG and PET. J Neural Transm. 2002;109:837–55.
28. Marcus DS, Fotenos AF, Csernansky JG, Morris JC, Buckner RL. Open access series of imaging studies (OASIS): longitudinal MRI data in nondemented and demented older adults. J Cogn Neurosci. 2010;22(12):2677–84. https://doi.org/10.1162/jocn.2009.21407.
29. Lyoo IK, Satlin A, Lee CK, Renshaw PF. Regional atrophy of the corpus callosum in subjects with Alzheimer's disease and multi-infarct dementia. Psychiatry Res. 1997;74(2):63–72. https://doi.org/10.1016/s0925-4927(97)00009-7.
30. Weis S, Kimbacher M, Wenger E, Neuhold A. Morphometric analysis of the corpus callosum using MR: correlation of measurements with aging in healthy individuals. AJNR Am J Neuroradiol. 1993;14:637–45.
31. Witelson SF. Hand and sex differences in the isthmus and genu of the human corpus callosum. Brain. 1989;112:799–835.
32. Anderson NG, Laurent I, Woodward LJ, Inder TE. Detection of impaired growth of the corpus callosum in premature infants. Pediatrics. 2006;118(3):951–60. https://doi.org/10.1542/peds.2006-0553.
33. Habibi HA, Gül OB, Öztürk M, Çalışkan E. Morphometric analysis of corpus callosum with magnetic resonance imaging in children; correlation with age and gender. İzmir Dr Behçet Uz Çocuk Dergisi. 2021;11(3):277–85. https://doi.org/10.5222/buchd.2021.93709.
34. Hwang SJ, Ji EK, Lee EK, Kim YM, Shin DY, Cheon YH, Rhyu IJ. Gender differences in the corpus callosum of neonates. Neuroreport. 2004;15(6):1029–32.
35. Matteson SR, Deahl ST, Alder ME, Nummikoski PV. Advanced İmagıng methods. Crit Rev Oral Biol Med. 1996;7(4):346–95.
36. Ho ML, Moonis G, Ginat DT, Eisenberg RL. Lesions of the corpus callosum. AJR 2013;200:W1-W16 0361–803X/13/2001–W1. https://doi.org/10.2214/AJR.11.8080.
37. Tavlayan E, Olacak N, Yavuz A. Beyin tümörlerinin radyoterapi planlamasında, BT ve MRG görüntü eşleştirilmesinin hedef volüm belirlenmesine ve doz dağılımına etkisi. Türk Onkoloji Dergisi. 2011;26(2):67–75. https://doi.org/10.5505/tjoncol.2011.188.
38. Chrysikopoulos H, Andreou J, Roussakis A, Pappas J. Infarction of the corpus callosum: computed tomography and magnetic resonance imaging. Eur J Radiol. 1997;25:2–8.
39. Grassi W, Filippucci E. A brief history of ultrasound in rheumatology: where we were. Clin Exp Rheumatol. 2014;32(Suppl.80):S3–6.
40. Liu F, Cao SK, Liu JR, Du ZF, Guo ZM, Ren CJ. Ultrasound measurement of the corpus callosum and neural development of premature infants. Neural Regen Res. 2013;8(26):2432–40. https://doi.org/10.3969/j.issn.1673-5374.2013.26.004.
41. Ahmadvand A, Shahidi SB, Talari H, Ghoreishi FS, Mousavi GA. Morphology of the corpus callosum and schizophrenia: a case-control study in Kashan, Iran. Electron Physician. 2017;9(10):5478–86.
42. Allouh MZ, Barbarawi MMA, Ali HA, Mustafa AG, Alomari SO. Morphometric analysis of the corpus callosum according to age and sex in middle eastern arabs: racial comparisons and clinical correlations to autism spectrum disorder. Front Syst Neurosci. 2020;14:30. https://doi.org/10.3389/fnsys.2020.00030.
43. Poleneni SR, Jakka LD, Chandrupatla M, Vinodini L, Ariyanachi K. Morphometry of corpus callosum in south indian population. Ann Neurosci. 2021;28(3–4):150–5. https://doi.org/10.1177/09727531211059892.
44. Holloway RL, de Lacoste MC. Sexual dimorphism in the human corpus callosum: an extension and replication study. Hum Neurobiol. 1986;5:87–91.
45. Junle Y, Youmin G, Yanjun G, Mingyue M, Qiujuan Z, Min X. A MRI quantitative study of corpus callosum in normal adults. J Med Coll PLA. 2008;23(6):346–51. https://doi.org/10.1016/s1000-1948(09)60005-8.
46. Kosar MI, Erdil FH, Sabanciogullari V, Karacan K, Cimen M, Atalar M. Morphometry of corpus callosum related with gender and age: magnetic resonance imaging study. Pak J Med Sci. 2012;28(3):408–12.
47. Abdolmaleki A, Mastery Farahani R, Ghoreishi SK, Shaerzadeh F, Aliaghaei A, Mirjavadi SH, Abbaszadeh HA. Magnetic Resonance Imaging-based morphometric assessment of sexual dimorphism of corpus callosum. Anatom Sci. 2016;13(2):117–24.
48. Choudhury PR, Choudhury P, Baruah P. Morphometric assessment of human corpus callosum on cadaveric brain specimens. J Evol Med Dent Sci. 2020;9(7):388–93. https://doi.org/10.14260/jemds/2020/89.
49. Pervin S, Nurunnabi ASM, Ara S, Haroon K. Morphometry of the splenium of the corpus callosum – a study on 60 cadaveric brains. Bangladesh J Neurosurg. 2022;11(1):30–5. https://doi.org/10.3329/bjns.v11i1.57990.
50. Demeter S, Ringo JL, Doty RW. Morphometric analysis of the human corpus callosum and anterior commissure. Hum Neurobiol. 1988;6(4):219–26.
51. Weber G, Weis S. Morphometric analysis of the human corpus callosum fails to reveal sex- related differences. J Hirnforsch. 1986;27(2):237–40.

52. De Lacoste-Utamsing C, Holloway RL. Sexual dimorphism in the human corpus callosum. Science. 1982;216:1431–2. https://doi.org/10.1182/blood-2009-06-229658.
53. Arda KN, Akay S. The relationship between corpus callosum morphometric measurements and age/gender characteristics: a comprehensive MR imaging study. J Clin Imaging Sci. 2019;9:33. https://doi.org/10.25259/JCIS-13-2019.
54. Ardekani BA, Figarsky K, Sidtis JJ. Sexual dimorphism in the human corpus callosum: an MRI study using the OASIS brain database. Cereb Cortex. 2013;23(10):2514–20. https://doi.org/10.1093/cercor/bhs253.
55. Downhill JE, Buchsbaum MS, Wei T, Spiegel-Cohen J, Hazlett EA, Haznedar MM, Silverman J, Siever LJ. Shape and size of the corpus callosum in schizophrenia and schizotypal personality disorder. Schizophr Res. 2000;42(3):193–208.
56. Brambilla P, Nicoletti MA, Sassi RB, Mallinger AG, Frank E, Kupfer DJ, Keshavan MS, Soares JC. Magnetic resonance imaging study of corpus callosum abnormalities in patients with bipolar disorder. Biol Psychiatry. 2003;54(11):1294–7.
57. Ballmaier M, Kumar A, Elderkin-Thompson V, Narr KL, Luders E, Thompson PM, Hojatkashani C, Pham D, Heinz A, Toga AW. Mapping callosal morphology in early- and late-onset elderly depression: an index of distinct changes in cortical connectivity. Neuropsychopharmacology. 2008;33(7):1528–36. https://doi.org/10.1038/sj.npp.1301538.
58. Di Paola M, Luders E, Rubino IA, Siracusano A, Manfredi G, Girardi P, Martinotti G, Thompson PM, Chou Y-Y, Toga AW, Caltagirone C, Spalletta G. The structure of the corpus callosum in obsessive compulsive disorder. Eur Psychiatry. 2013;28(8):499–506. https://doi.org/10.1016/j.eurpsy.2012.07.001.
59. Piras F, Vecchio D, Kurth F, Piras F, Banaj N, Ciullo V, Luders E, Spalletta G. Corpus callosum morphology in major mental disorders: a magnetic resonance imaging study. Brain Commun. 2021;3(2):fcab100. https://doi.org/10.1093/braincomms/fcab100.
60. Hynd GW, Hall J, Novey ES, Eliopulos D, Black K, Gonzalez JJ, Edmonds JE, Riccio C, Cohen M. Dyslexia and corpus callosum morphology. Arch Neurol. 1995;52:32–8.
61. Wang PP, Doherty S, Hesselink JR, Bellugi U. Callosal morphology concurs with neurobehavioral and neuropathological findings in two neurodevelopmental disorders. Arch Neurol. 1992;49:407–11.
62. Yamauchi H, Fukuyama H, Nagahama Y, Katsumi Y, Hayashi T, Oyanagi C, Konishi J, Shio H. Comparison of the pattern of atrophy of the corpus callosum in frontotemporal dementia, progressive supranuclear palsy, and Alzheimer's disease. J Neurol Neurosurg Psychiatry. 2000;69:623–9.

Sex- and Age-Related Differences in the Corpus Callosum

6

Michael Cesarek and R. Shane Tubbs

6.1 Introduction

The brain changes morphologically and microstructurally across a lifetime, as well as differs in its morphology and connectivity, between sexes [1–3]. These variations have been of particular interest for decades in medicine as they present the possibility of explaining the etiology of many different neurological and psychiatric disorders [4]. Rightly so, with the corpus callosum (CC) being the largest commissural tract in the human brain, containing over 300 million axons and having both homotopic and heterotopic interhemispheric connections, it has been of particular interest in attempting to define morphological and microstructural age and sexual differences [5, 6].

Fetal callosal fibers can be first identified at 8 weeks of gestation and from then on, much like the rest of the cortices that it connects, develops in a rostral-to-caudal direction and continues well into adulthood [1, 7, 8]. This directional co-development between the cortices and the CC ultimately results in a topographical distribution of subregions based on the cortical and subcortical regions in which it connects [5, 7]. This makes the CC an ideal candidate to study differential age effects, as it anatomically represents the interhemispheric connection of numerous sensory, motor, and cognitive cortical areas and allows for the analysis of axonal and white matter development and age-related changes [5, 7]. Thus, much of the current literature centers around the goal of characterizing the differential development as well as the age-related changes of the CC and relating functional and anatomical callosal changes with that of age-related neurological and psychological disease states.

In addition to age, sexual differences in the CC have been of clinical interest ever since DeLacoste-Utamsing in 1982 suggested that the female splenium was "both more bulbous and larger than the male counterpart" [9]. From then on, callosal structural and functional differences between sexes have become a prevalent topic of investigation. The prevalence and symptomology of many different neurologic and psychiatric conditions, as well as cognitive abilities, are linked to callosal structure and often vary with sex [4, 10–12]. As such, the goal of teasing out the neurological factors related to the biological sex variable has been an important goal of clinical research, as made evident by *The Journal of Neuroscience* recognizing that "sex fundamentally influences the brain" and requiring the "consideration of sex as a biological variable" [13]. This notion of sexual differences is no doubt reflected in neurological research with differences in neuroanatomy at both the global and regional brain levels [3, 14]. However, there is

M. Cesarek (✉)
Tulane University School of Medicine,
New Orleans, LA, USA
e-mail: mcesarek@tulane.edu

R. S. Tubbs
Department of Neurosurgery, Tulane School of Medicine, New Orleans, LA, USA

M. Turgut et al. (eds.), *The Corpus Callosum*, https://doi.org/10.1007/978-3-031-38114-0_6

still debate within the neuroscience community as to the exact sexual differences, macro- and microstructurally, observed in the CC [6, 15].

Despite the large abundance of literature on both topics of age and sex differences in the CC, there are inherent challenges that have led to a dispute over methodology, validity, analysis, and interpretation of the age- and sex-related differences found. In particular, the sexual differences found in the CC have presented topics of debate in multiple aspects of the literature such as the sampling population, particularly whether the data is generalizable and comparable between, and within, postmortem or in vivo studies [16]. Additionally, there is debate over the methodology of imaging, mainly in relation to magnetic resonance imaging (MRI), as there are numerous protocols in which researchers use different slice thicknesses, pixel sizes, signal-to-noise ratios, as well as differences in the plane of orientation [16, 17]. Further, debate has developed on the partitioning of the CC, as there is a lack of callosal gross anatomical landmarks leading to challenges in repeatably and reliably sectioning or subsectioning the CC and the best modality to do as such [5, 6, 16]. In addition, there is controversy over the proper statistical analysis utilized, from issues of controlling for brain size [15, 18] to controlling for confounding variables like handedness, socioeconomic status, intelligence quotient, as well as genetic factors [4, 6, 16, 19].

As a result, the goal of this chapter will be to review and explore the current ideas presented in the current literature regarding both the age-related and sex-related differences in the CC. Emphasis will be placed on exploring the changes observed macroscopically and microscopically in relation to development and aging and the inherent issues in characterizing these changes presented within the literature. As for sex-related differences, emphasis will be on the variety of different arguments within the current literature with respect to the presence or absence of sexual differences, as well as on the current challenges and topics of debate within the literature. In order to report on this topic, multiple database searches were conducted utilizing key terms such as "age-related differences," "age related," "sexual differences," "sex-related differences," and "sexual dimorphisms" all in register with "CC" and with a date range of 2000 to present. The searches resulted in approximately 967 articles, of which 94 were selected based on relevance to human subjects and the generalizability to a healthy, normally aging population.

6.2 Age-Related Differences in the Corpus Callosum

As early as 8 weeks of gestation, fibers will start to form in the area that will eventually become the genu, and by 17–20 weeks, the CC forms into its characteristic structure [8]. The CC will then continue to grow with age until it plateaus in growth in the third decade of life and eventually will decline in size around the sixth decade of life [20]. These morphological and microstructural changes that the CC undergoes across a lifetime, as demonstrated by Aboitiz et al. and Frederiksen, have functional manifestations. Thus, studying the development and aging process of the CC provides significant clinical relevance from characterizing normal fetal brain growth to aid in accurate determination of gestational age to better understanding functional and cognitive decline with aging [21, 22].

Fetal development of the CC occurs in a rostral-to-caudal and a ventral to dorsal direction, in which the fibers are roughly distributed in a topographical manner representing the cortical structures in which they connect [5, 7, 8, 23]. Growth, in terms of thickness, length, and width, has been proven to be relatively linear from week seventeen on with maximal growth occurring in the second trimester, followed by a plateau in the third trimester [22, 24]. Volumetrically, as demonstrated by Machado-Rivas et al., the CC increases in volume by 10.4% per week of gestation [25]. These findings present clinical significance as CC length has shown to have significant predictive power for accurately determining gestational age, aiding in the delivery of correct obstetric care, providing normative measures on which to base prenatal diagnosis and prognosis of potential neurological and psychological diseases, as well as provide potential explanatory value for morphometric differences later in life [22, 24, 25].

As discussed, the CC develops prenatally with a relatively linear relationship and grows in relation to the whole brain [8]. However, postnatally, the growth of the CC becomes non-linear and changes in an over-proportionalized fashion, which is seen as rapid growth in childhood and adolescence that is not proportional to changes in forebrain size [20, 26]. Postnatally and into early adulthood, CC development, both morphologically and microstructurally, is non-linear with respect to total CC measurements as well as with respect to subregional development [7, 20, 26, 27]. The gross morphological development of the CC rapidly grows until late childhood, at which point the rate of growth slows and will start to plateau into the third decade of life [20, 26, 28]. As for the growth of the CC subsections, the rostrum is characterized by a gradual decrease in thickness from its peak thickness, into young adulthood and continuing into old age. The genu, truncus, isthmus, as well as splenium are shown to grow into young adulthood, with the relative growth of the splenium being much greater than the anterior subsections—genu, truncus, and isthmus [20]. This observed accentuated posterior regional development is hypothesized to be related to callosal microstructural, both axonal and white mater, development and growth [7, 20, 27]. This hypothesis is substantiated by the presence of regional differences in peak white matter content being not only more in the posterior regions, compared to the anterior regions, but also peaking at a later age, with splenial white matter peaking somewhere in the sixth decade of life and the genu peaking around the fourth decade of life [7, 29]. Further, this anterior-to-posterior microstructural development results in differential callosal subregional aging characteristics.

After the plateau of growth in the third decade of life, the CC is reported to stabilize for a period of time followed by a point, in the fifth to sixth decade of life, where there is an inflection point and the CC starts to undergo age-related decline [20, 28, 30]. Multiple morphometric modalities including histological, MR, and diffusion imaging not only support this age-related decline observed relative to the entire CC [20, 28] but have additionally shown that the callosal subregions are differentially affected, with the effects of aging being more pronounced in the anterior regions of the CC as opposed to the posterior regions, forming an anterior-to-posterior gradient of age-related decline [7, 20, 31–41]. However, there is still speculation as to the exact mechanisms behind this anterior-to-posterior gradient, with authors referencing numerous causes. Corticocentric causes of the observed anterior-to-posterior gradient include cortical Wallerian degeneration [34, 35], and age-related cortical white matter and/or gray matter changes [21, 32, 34, 37–41]. While the reasoning that more focused on the CC, included aberations in microstructural white matter [20, 21, 31, 32, 42, 43], callosal axonal aberrations [7, 20, 30, 31, 33], as well as attributing the anterior-to-posterior gradient to callosal function with the observed preferential maintenance of the splenium being related to its importance in basic perception and visual processing [20, 34]. Additionally, age-related changes in the CC are thought to be in part, attributable to the decline in intrahemispheric gray matter efficiency in cognitive and motor function, thus increasing the need for interhemispheric transmission and decreased lateralization for a given task, as seen in aging [33, 34, 40]. Interestingly, while the accentuated aging effects found in the genu were supported, a study conducted by Bennett et al., investigating a population of 94 "oldest-old" subjects ranging from 90–103 years of age, found significant aging effects in the splenium of the CC, as opposed to the genu, in this age range. This shift in aging pattern was thought to be associated with characteristic "oldest-old" adult neuropathologies such as neurofibrillary tangles, neuritic plaques, lacunes, and amyloid angiopathy [31].

In summary, it is generally accepted, with some debate on exact age and timing, that the CC develops in a rostral-to-caudal fashion in utero. From then on, the CC continues to grow in a non-linear fashion until it plateaus in adulthood, afterwhich the CC begins to degenerate morphometrically in a typical anterior-to-posterior gradient. Thus, taken together, it forms an inverted "U"-shaped curve across a lifetime. Given that the function of the CC is the interhemispheric connection between

Table 6.1 (continued)

Author (year)	Title	Type of Report	*n*	Modality	Finding
Shiino et al. (2017) [59]	Sex-related difference in human white matter volumes studied: inspection of the CC and other white matter by VBM	Secondary	37	MRI	In the CC, the genu was the subregion showing significant sex-associated difference, being larger in women when analyzed by permutation test
Genc et al. (2018) [27]	Age, sex, and puberty related development of the CC: a multi-technique diffusion MRI study	Secondary	370	MRI/DTI/DWI/NODDI	Observed higher AFD in females compared with males across the CC, most considerably in the anterior splenium and posterior genu
Wang et al. (2019) [30]	Application of super-resolution track-density technique: earlier detection of aging-related subtle alterations than morphological changes in CC from normal population?	Primary	131	MRI/TDI	Only CC1 showed a larger volume in female subjects. CC2, CC3, CC4, and CC5 of male and female subjects showed no significant differences in volume, FA, or TDI
Xin et al. (2019) [60]	Brain differences between men and women: evidence from deep learning	Secondary	1065	MRI	Results confirmed that there is likely a morphological difference at the genu of CC between man and women, with FA being higher in women
Male favored					
Sullivan et al. (2001) [61]	Sex differences in CC size: relationship to age and intracranial size	Primary	92	MRI	With the exception of one ratio measure, the different statistical adjustments for the contribution of sex differences in brain size to CC size all indicated that men had larger corpora callosa than women for their brain size
Suganthy et al. (2003) [62]	Gender- and age-related differences in the morphology of the CC	Primary	100	MRI	The CC was longer in males and the discriminant score to differentiate gender was determined with an accuracy of 66%
Tuncer et al. (2005) [63]	Sexual dimorphism and handedness in the human CC based on magnetic resonance imaging	Primary	80	MRI	Right-handed males had larger rostrums and isthmuses than right-handed females
Luders et al. (2006) [64, 65]	Parasagittal asymmetries of the CC	Secondary	60	MRI	Results showed significant rightward asymmetries of callosal thickness predominantly in the anterior body and anterior third of the callosum, suggesting a more diffuse functional organization of callosal projections in the right hemisphere. Asymmetries were increased in men, supporting the assumption of a sexually dimorphic organization of male and female brains

Silveri et al. (2006) [45]	Sex differences in the relationship between white matter microstructure and impulsivity in adolescents	Primary	21	DTI	In the present study, males displayed significantly higher FA values in the left forward-projecting WM region relative to females
Menzler et al. (2011) [66]	Men and women are different: diffusion tensor imaging reveals sexual dimorphism in the microstructure of the thalamus, CC and cingulum	Primary	50	DTI	FA was increased in male as compared to female subjects in parts of the CC and cingulum as well as in the capsula extrema and midbrain
Westerhausen et al. (2011) [67]	A critical re-examination of sexual dimorphism in the CC microstructure	Primary	75	DTI	Only the sex difference located in the anterior genu subregions could be unequivocally interpreted. This was the only cluster where the diffusion parameters did not correlate with regional callosal size. The present results indicate a stronger inter-hemispheric connectivity between the frontal lobes in males than females, which might be related to sex differences in hemispheric asymmetry and brain size
Spasojević et al. (2015) [68]	Sex differences of human CC revealed by polar coordinate system: magnetic resonance imaging study	Secondary	30	MRI	Found significantly larger CCA, C3 segment, and the TCC in males
Jung et al. (2019) [69]	Sex differences in white matter pathways related to language ability	Primary	67	MRI/DWI/DTI	Males showed higher fractional anisotropy (FA) in the left anterior thalamic radiations (ATR), right cingulum-angular bundle, right corticospinal tract, bilateral superior longitudinal fasciculus temporal terminations, bilateral uncinate fasciculus (UNC), and CC-forceps minor when compared with the female group
Schmied et al. (2020) [70]	Sex differences associated with CC development in human infants: a longitudinal multimodal imaging study	Secondary	104	MRI/DWI	Found that male and female human infants were similar in CC area and thickness at age 6 months, followed by diverging trajectories into toddlerhood characterized by a higher growth rate for males

(continued)

Table 6.1 (continued)

Author (year)	Title	Type of Report	*n*	Modality	Finding
Tétreault et al. (2020) [71]	Diffusion time dependency along the human CC and exploration of age and sex differences as assessed by oscillating gradient spin-echo diffusion tensor imaging	Primary	20	DTI	The greater OGSE-PGSE difference in the genu of males could reflect larger axon diameters than females
Machado-Rivas et al. (2021) [25]	Normal growth, sexual dimorphism, and lateral asymmetries at fetal brain MRI	Primary	98	MRI	On average, male fetuses had higher volumes in all structures where significant sexual dimorphism was found (intracranial volume, CSF, lateral ventricles, brain parenchyma, hemispheric white matter, proliferative compartment, ventricular zone, CC, deep gray matter, lentiform nucleus, thalamus, brainstem, and cerebellar hemisphere). The largest sex-related differences were 22.1% for the lateral ventricles, 12.9% for the CC, and 12.4% for the cerebellar hemispheres
No significant difference					
Luders et al. (2006) [64, 65]	Gender effects on callosal thickness in scaled and unsealed space	Secondary	60	MRI	When callosal thickness data were compared in scaled space (achieved through correcting images for brain dimensions), we did not observe any significant differences between men and women
Ota et al. (2006) [37]	Age-related degeneration of CC measured with diffusion tensor imaging	Primary	42	DTI	We assessed age-related changes in fractional anisotropy and mean diffusivity in the subregions of the CC using diffusion tensor imaging. There were no significant interactions of sex x region
Leonard et al. (2008) [72]	Size matters: cerebral volume influences sex differences in neuroanatomy	Primary	200	MRI	The major findings of the present study are that women and men have very reliable differences in cerebral volume that account for much of the variance in brain structure size that at first glance might appear to be attributable to sex. Sex did not contribute unique variance to the relationship between relative CC size and cerebral volume. Both women and men with lower cerebral volumes have relatively larger callosa

Luders and Toga (2010) [73]	Sex differences in brain anatomy	Review	–	–	When applying a novel technique which circumvents pitfalls associated with traditional callosal parcellation by automatically estimating callosal thickness at 100 equidistant surface points, it was observed that sex differences were completely absent when data were adjusted for individual brain size
Tepest et al. (2010) [74]	CC size in adults with high-functioning autism and the relevance of gender	Primary	58	MRI	Show that the difference of the ratio of CC to TBV can be convincingly explained as a function of brain size per se, but not of gender, thus rejecting in the studied population the sexual dimorphism hypothesis of the CC
Bruner et al. (2012) [75]	Gender-based differences in the shape of the human CC are associated with allometric variations	Secondary	102	MRI	Provide three major conclusions regarding the morphological variation of the CC. First, it is very variable because of scarce integration among its parts, due to intrinsic or extrinsic factors. Second, there is a moderate allometric component showing that larger brains have a relatively smaller CC cross-section and relatively smaller splenium. Third, differences between males and females are secondary consequences of this allometric pattern, the former having mean larger size than the latter
Luders et al. (2014) [76]	Why size matters: differences in brain volume account for apparent sex differences in callosal anatomy: the sexual dimorphism of the CC	Secondary	48	MRI	Findings suggest that individual differences in brain size account for apparent sex differences in the anatomy of the CC
Kurth et al. (2018) [77]	Sex differences in associations between spatial ability and CC morphology	Secondary	38	MRI	Male and female corpora callosa seem to have similar dimensions when accounting for brain size
Guz et al. (2019) [28]	Analysis of CC size depending on age and sex	Secondary	1108	MRI	There are no statistically significant differences between women and men in the measurements of CC

(continued)

Table 6.1 (continued)

Author (year)	Title	Type of Report	*n*	Modality	Finding
Eliot et al. (2021) [11]	Dump the "dimorphism": comprehensive synthesis of human brain studies reveals few male-female differences beyond size	Review	–	–	Found no sex difference in CC area or volume
Variable					
Dubb et al. (2003) [78]	Characterization of sexual dimorphism in the human CC	Secondary	189	MRI	There is a statistically significant area of expansion in the splenium in females with respect to males and in the genu of males with respect to females
Hwang et al. (2004) [53]	Gender differences in the CC of neonates	Primary	200	Ultrasonography	The male CC had significantly more height than that of the female whereas there was no difference in width of the CC. The estimated average thickness of CC was greater in the female splenium
Liu et al. (2010) [36]	Sex differences in the human CC microstructure: a combined T2 myelin-water and diffusion tensor magnetic resonance imaging study	Primary	22	DTI	Results of this study demonstrate regionally dependent sex differences in microstructural composition and organization of the CC. The MWF in males was significantly greater than females for the rostral body and posterior midbody, whereas the splenium MWF in males was significantly less than females. The DTI analysis revealed significantly increased FA in males compared with females within the genu of the CC
Björnholm et al. (2017) [79]	Structural properties of the human CC: multimodal assessment and sex differences	Secondary	835	MRI/MTR/DTI	Sex differences in mean MTR were similar across the CC (males > females) while these in FA differed by the callosal segment (body: M > F; splenium: F > M)
López-Vicente et al. (2021) [1]	White matter microstructure correlates of age, sex, handedness and motor ability in a population-based sample of 3031 school-age children	Secondary	3031	DTI	Girls showed higher FA in the bilateral ILF and in the FMa, whereas boys showed higher FA in the CB, CST, FMi, and the UF. Girls and boys also differed in MD, with girls showing lower MD in all tracts, except the FMi, the right CST, and the UF. AD was lower in girls in most of the tracts. In girls, RD was lower in FMa, ILF, and SLF tracts than in boys, and higher in the left CB, and FMi

6.4 Challenges with Characterizing Sex- and Age-Related Differences

Concerning both age- and sex-related differences, determining the morphological changes and differences in the CC has certainly led to significant insights but has also resulted in inconsistent findings and much debate on the "proper" way in which to discern these changes and differences. Given that our brain's neuronal structure is influenced by genetic, epigenetic, and social-environmental factors, the differences that are either strictly related to gender or age may, however, be either too small or simply obscured by the numerous confounding variables that influence our brain's structure [11]. In addition, despite the advancements in imaging and data analysis that have allowed for tremendous insight and characterization of the human brain, the lack of a standard methodological and analytical approach to characterizing the CC has only compounded these inconsistencies [11, 12, 16, 18, 78, 80]. Thus, the following discussion will present some of these challenges, to include methodological inconsistencies like imaging, measurements, and analytical discrepancies as well as discuss the influence of genetic factors, disease states, and social-environmental impacts and their potential confounding effects on the structure of the CC.

6.4.1 Imaging Variability

Despite the lack of callosal anatomical landmarks on which to clearly delineate functional sections, attempts to characterize the CC date back to 1836 and were mostly limited to gross anatomical and histological analysis in postmortem samples [83]. The analysis subsequently evolved with improved imaging technologies and now is dominated by different MRI modalities, such as diffusion tensor imaging (DTI) [31, 52], quantitative MRI [29], multimodal MRI [79], diffusion tensor tractography [10], hybrid diffusion imaging (HYDI) [41], neurite orientation dispersion and density imaging (NODDI) [14], and track-density imaging (TDI) [30] to name a few. Still, yet, others utilize different methods such as ultrasonography [24, 53] to analyze fetal CC, magnetic resonance spectroscopy (MRS) [42, 43] to analyze metabolic changes within the CC, and magnetization transfer (MT) imaging [47] to investigate myelin integrity. All this variability has allowed for tremendous discovery but has conversely potentially led to a lack of comparability, as factors such as image quality, determination of regions of interest (ROIs), and callosal measurements all vary between these different modalities. Even for MRI studies, which is the dominant imaging modality, issues with exact imaging parameters and image analysis present potential variation [16]. The first point of MRI variability is that of brain orientation and subsequent determination of ROIs. Many MRI studies utilize a midsagittal section; however, their methodology of controlling for head rotation and determining the true midsectional plane in which to base ROIs varies. Attempts to mitigate this variability by using anatomical landmarks, both extracranial and intracerebral, are predicated on the assumption that those landmarks are not inherently variable themselves [16, 35, 56]. Additionally, the second point of variability is that of MR imaging parameters such as the signal-to-noise ratio, pixel size, and slice thickness which can lead to variability in tissue discrimination, image resolution, and result in a partial volume effect [16]. Additionally, image postprocessing adjustments such as spatial smoothing and its utilization have been called into question [10, 35]. Finally, in the issue of repeatability, some of these imaging parameters are arbitrary or not discussed in the study, thus making it difficult to corroborate and repeat the findings [56].

6.4.2 Measurement and Normalization Variability

The next methodological issue is that of the variability present in callosal measurement and subregional partitioning. In order to quantify morphological changes, many researchers turn to the use of either metrics of volume, thickness, and area or utilize a partitioning method in which to characterize the volumes/areas of arbitrarily defined callosal subsections [16, 78].

As for the use of callosal thickness, cross-sectional area, or volume, two issues have potentially led to conflicting results. The first is the use of the measurements being in relation to a mid-sagittal section of the CC. As previously mentioned, this presents the possibility of imaging variability—orientation, imaging parameters, quality, resolution, and determination of a true mid-callosal section—limiting the comparability of the results obtained. The second is the issue of normalization of the obtained callosal measurements. The determination of the relative size of the CC has been a topic of much debate [15–18, 80]. Researchers when not adjusting for brain size often find that the male CC is larger; however, when brain size is adjusted for often, the findings flip and support larger female callosal values [17, 18, 56, 57]. However, depending on the normalization technique, overcorrection is possible leading to falsely inflated measures in smaller brains [15]. When methods of normalization for brain size are applied, it is usually in one of three ways, the stereotaxic method (normalizing by scaling MRI into a standard space), the ratio method, and the covariate method that both utilize one of the many different indices of controlling for brain size including, brain weight, forebrain volume, intracranial volume, or midsagittal cross-sectional cerebral area [17]. Unfortunately, these correctional measures do not always completely adjust for brain size, as the denominator often used in the ratio method is often itself dimorphic [16], and there is still debate as to whether or not these variables are related to CC size [17, 18]. This issue is highlighted in the work done by Bermudez and Zattorre, in which they compared the utilization of the three different methods and found that the different normalization approaches are not equivalent or interchangeable [17]. In attempts to get away from these normalization methods, some have tried to control for brain size by comparing brain-size matched male and female subjects [57, 59, 61, 76]. This method of brain-size matching has shown some validity but unfortunately is limited in effect size as finding a larger than average brain-size female subject and a smaller than average brain-size male subject in which to compare is difficult [15].

As it pertains to the utilization of callosal partitioning methods, there, unfortunately, is no gold standard by which to divide the CC into subregions. Given that there is a lack of callosal anatomical landmarks on which to clearly delineate functional sections, multiple different methods of partitioning the CC have been used. By in large, the most common method utilized is the Witelsons partitioning scheme, in which the CC is subdivided into five subregions using vertical sections that are perpendicular to a line connecting the longest line possible across a midsagittal section of the CC, in an effort to delineate the cortical topographical interhemispheric connections that travel through the CC [34, 80]. However, as demonstrated by Huang et al., the utilization of DTI tractography resulted in the conclusion that the tracts that travel through the CC are highly variable, with only the most anterior and posterior portions, containing frontal and parietal lobe tracts, respectively, being usable as reliable landmarks for structure-function-based partitioning [5]. Other methods include various straight-line methods, similar to the Witelsons partitioning scheme in which perpendicular lines vertically section out the CC, radial partitioning methods, as well as bent and curved line partitioning methods [6, 16]. The utilization of these methods is sometimes arbitrary or is dependent on application objectives; regardless, not only are the partitioning methods subject to the inherent variability based on the image on which they are based, but they are also difficult to compare across studies [6]. Thus, it is not beyond reason that both methods in which individual researchers utilize to measure the CC and normalize the measurements obtained provide concerns with interchangeability and comparability, as well as significantly increase the potential for variable results.

6.4.3 Confounding Variables

In addition to the numerous methodological variations resulting in conflicting and ambiguous results, multiple confounding variables have only compounded this ambiguity as they have largely been uncontrolled for in investigative studies. Much of the research regarding morphological measures recognizes and controls for two confounding variables, age and handedness. Both age, as discussed previously, and handedness have been shown to tremendously influence callosal morphology [1, 63, 84]. However, other variables including genetic factors, disease states, as well as environmental factors have been shown to influence the morphology of the CC.

First, a large part of the goal in characterizing age- and sex-related differences in the CC is to explain potential correlates with psychological and neurological disease states. However, caution must be taken when studying morphological differences in the CC as many neuropsychological disorders including attention-deficit hyperactivity disorder (ADHD), autism, Down syndrome, dyslexia, Tourette's syndrome, Alzheimer's disease and dementia, schizophrenia, as well as fetal alcohol syndrome, and prenatal antidepressant exposure have been linked to aberrations in callosal microstructure and morphology and must be controlled for in studies of the CC [6, 85–87].

Secondly, as demonstrated by Woldehawariat et al., the size and morphology of the CC substructures are highly heritable and influenced by multiple genetic factors [88]. For example, both ethnicity [62, 89, 90] and prenatal [58, 91] and adolescent [92] sex hormone exposure have been shown to variably influence white matter and structural development of the CC. Interestingly, genetic mutations such as a BDNF val66met polymorphism have been linked to a predisposition of widespread white matter deterioration, especially in the splenium of the CC [93]. In addition, sex-related X and Y chromosome aneuploidies have shown an association between callosal morphometry with sex-wise dosages of X and Y chromosomes [94]. Lastly, it has long been accepted that our experiences and environment influence the global structure and microstructure of our brains. This idea has also been shown to apply to the size and microstructure of the CC. As demonstrated by Takeuchi et al., family socioeconomic status (SES) was found to have differential effects on callosal anisotropy and diffusivity in relation to sex, and further, Ohashi et al. found sex-specific and time-specific effects of childhood maltreatment on callosal anatomy and microstructure [19, 82]. Even differential cognitive abilities, influenced by experience and learning, have been reported to affect the structure of the rostrum and anterior midbody of the CC [81].

Genetic factors, disease states, as well as environmental factors have significant potential impacts on the morphology of the CC and often go uncontrolled for in callosal morphometric research. Thus, it stands to reason that the conflicting results and continued debate, present in both age-related and even more so in sex-related differences in the CC, have only been exacerbated by variables that have been previously not thought of to control for, or in some cases nearly impossible to control for. Eliot et al. could not have expressed this same idea any better than when they stated: "Given abundant evidence that experience alters neuronal structure and function, as well as growing knowledge about epigenetic influences on central nervous system development, it is impossible to discern the degree to which group-level differences between human males and females are attributable to inborn sex factors versus social-environmental gender learning, acting through lifelong neuroplasticity" [11].

6.5 Conclusion

The CC is the largest commissural tract in the human brain, containing over 300 million axons with both homotopic and heterotopic interhemispheric connections [5]. It has largely been investigated for its relation of structure to function, particularly in relation to sex- and age-related changes. However, for the past four decades, there has been debate over the exact morphological differences with respect to when these changes occur across a human lifetime, to what extent these differences manifest in male vs. female corpora callosa, and the reasoning behind the observed or lack of observed differences. The goal of this chapter was not to argue for one side or the other and ultimately decide on a conclusion; rather, it was to present a review of the current literature on both age- and sex-related differences in the CC and speculate as to why there is still debate and conflicting results, even after the many years of investigation into the two topics. In summary, a large portion of the literature regarding age-related differences agrees that the CC develops in a rostral-to-caudal fashion in utero; from then on, it continues to grow in a non-linear fashion until it plateaus in adulthood, afterwhich the CC begins to degenerate morphometrically in a typical anterior-to-posterior gradient. Thus, forming an inverted "U"-shaped curve across an individual's lifetime. However, there is still debate about the exact timing of these events, as well as the potential causes of these events. In relation to sex-related differences, much less is agreed upon. Rather, there are large bodies of evidence supporting varying degrees of male-favored or female-favored callosal sizes, densities, thicknesses, as well as microstructural compositions.

As to the causes of these debates, it has been speculated that much of it is due to a lack of standard methodological procedures as well as many confounding variables that impact brain morphometry. Many authors have recommended different modalities in which to standardize, analyze, and control for confounding variables, including novel highly repeatable morphometric methods utilizing anatomical landmarks [56], factor-based analysis [55], and a 3D convolutional neural network (CNN) methods, as well as advised methods to better statistically analyze the differences found [18, 80] and even control for the brain-size effect utilizing brain-size matched subjects [57, 59, 61, 76].

However, even with these proposed standards, some researchers feel that defining morphological differences in the CC is somewhat of a futile pursuit due to the numerous confounding variables that affect the CC, as discussed, the difficulty in defining the functional relevance behind structural brain differences [12, 18], and the challenges associated with interpreting the relevance of callosal morphometry when no brain region acts in isolation and thus ignoring the sheer complexity of the human brain [6, 18]. Despite the abundance of controversy and the inconclusive results, characterizing the differences in the CC in relation to age and sex, and defining the functional ramifications of such differences, is of important clinical relevance, as doing so has the potential to elucidate the etiology of numerous population-specific neurological and psychological disorders and potentially explain numerous different behavioral and cognitive processes. However, in order to obtain some sort of conclusive evidence on which to base these functional ramifications, a repeatable standardized methodology of measurement, analysis, and controls must be established and subsequently followed.

References

1. López-Vicente M, Lamballais S, Louwen S, Hillegers M, Tiemeier H, Muetzel RL, White T. White matter microstructure correlates of age, sex, handedness and motor ability in a population-based sample of 3031 school-age children. NeuroImage. 2021;227:117643. https://doi.org/10.1016/j.neuroimage.2020.117643.
2. Potvin O, Dieumegarde L, Duchesne S. Normative morphometric data for cerebral cortical areas over the lifetime of the adult human brain. NeuroImage. 2017;156:315–39. https://doi.org/10.1016/j.neuroimage.2017.05.019.
3. Ruigrok AN, Salimi-Khorshidi G, Lai MC, Baron-Cohen S, Lombardo MV, Tait RJ, Suckling J. A meta-analysis of sex differences in human brain structure. Neurosci Biobehav Rev. 2014;39(100):34–50. https://doi.org/10.1016/j.neubiorev.2013.12.004.
4. Jahanshad N, Thompson PM. Multimodal neuroimaging of male and female brain structure in health and disease across the life span. J Neurosci Res. 2017;95(1–2):371–9. https://doi.org/10.1002/jnr.23919.
5. Huang H, Zhang J, Jiang H, Wakana S, Poetscher L, Miller MI, van Zijl PC, Hillis AE, Wytik R, Mori S. DTI tractography based parcellation of white matter: application to the mid-sagittal morphology of corpus callosum. NeuroImage. 2005;26(1):195–205. https://doi.org/10.1016/j.neuroimage.2005.01.019.
6. Thompson PM, Narr KL, Blanton RE, Toga AW. Mapping structural alterations of the corpus callosum during brain development and degeneration. In: The parallel brain: the cognitive neuroscience of the corpus callosum. Issues in clinical and cognitive. Cambridge, MA: MIT Press; 2003. p. 93–130.
7. Lynn JD, Anand C, Arshad M, Homayouni R, Rosenberg DR, Ofen N, Raz N, Stanley JA. Microstructure of human corpus callosum across the lifespan: regional variations in axon caliber, density, and

myelin content. Cereb Cortex. 2021;31(2):1032–45. https://doi.org/10.1093/cercor/bhaa272.
8. Rakic P, Yakovlev PI. Development of the corpus callosum and cavum septi in man. J Comp Neurol. 1968;132(1):45–72. https://doi.org/10.1002/cne.901320103.
9. DeLacoste-Utamsing C, Holloway RL. Sexual dimorphism in the human corpus callosum. Science. 1982;216(4553):1431–2. https://doi.org/10.1126/science.7089533.
10. Choi CH, Lee JM, Koo BB, Park JS, Kim DS, Kwon JS, Kim IY. Sex differences in the temporal lobe white matter and the corpus callosum: a diffusion tensor tractography study. Neuroreport. 2010;21(1):73–7. https://doi.org/10.1097/WNR.0b013e3283345eb0.
11. Eliot L, Ahmed A, Khan H, Patel J. Dump the "dimorphism": comprehensive synthesis of human brain studies reveals few male-female differences beyond size. Neurosci Biobehav Rev. 2021;125:667–97. https://doi.org/10.1016/j.neubiorev.2021.02.026.
12. Fjell AM, Westlye LT, Amlien I, Espeseth T, Reinvang I, Raz N, Agartz I, Salat DH, Greve DN, Fischl B, Dale AM, Walhovd KB. Minute effects of sex on the aging brain: a multisample magnetic resonance imaging study of healthy aging and Alzheimer's disease. J Neurosci. 2009;29(27):8774–83. https://doi.org/10.1523/jneurosci.0115-09.2009.
13. Prager EM. Addressing sex as a biological variable. J Neurosci Res. 2017;95(1–2):11. https://doi.org/10.1002/jnr.23979.
14. Ritchie SJ, Cox SR, Shen X, Lombardo MV, Reus LM, Alloza C, Harris MA, Alderson HL, Hunter S, Neilson E, Liewald DCM, Auyeung B, Whalley HC, Lawrie SM, Gale CR, Bastin ME, McIntosh AM, Deary IJ. Sex differences in the adult human brain: evidence from 5216 UK biobank participants. Cereb Cortex. 2018;28(8):2959–75. https://doi.org/10.1093/cercor/bhy109.
15. Luders E, Kurth F. Structural differences between male and female brains. Handb Clin Neurol. 2020;175:3–11. https://doi.org/10.1016/b978-0-444-64123-6.00001-1.
16. Dorion AA, Capron C, Duyme M. Measurement of the corpus callosum using magnetic resonance imaging: analyses of methods and techniques. Percept Mot Skills. 2001;92(3 Pt 2):1075–94. https://doi.org/10.2466/pms.2001.92.3c.1075.
17. Bermudez P, Zatorre RJ. Sexual dimorphism in the corpus callosum: methodological considerations in MRI morphometry. NeuroImage. 2001;13(6 Pt 1):1121–30. https://doi.org/10.1006/nimg.2001.0772.
18. Smith R. Interpretation of ratios in research on sexual dimorphism in the human corpus callosum. Relative size versus controlling for size. Curr Anthropol. 2005;46(2):249–73. https://doi.org/10.1086/427117.
19. Takeuchi H, Taki Y, Nouchi R, Yokoyama R, Kotozaki Y, Nakagawa S, Sekiguchi A, Iizuka K, Yamamoto Y, Hanawa S, Araki T, Miyauchi CM, Sakaki K, Nozawa T, Ikeda S, Yokota S, Magistro D, Sassa Y, Kawashima R. The effects of family socioeconomic status on psychological and neural mechanisms as well as their sex differences. Front Hum Neurosci. 2018;12:543. https://doi.org/10.3389/fnhum.2018.00543.
20. Danielsen VM, Vidal-Piñeiro D, Mowinckel AM, Sederevicius D, Fjell AM, Walhovd KB, Westerhausen R. Lifespan trajectories of relative corpus callosum thickness: regional differences and cognitive relevance. Cortex. 2020;130:127–41. https://doi.org/10.1016/j.cortex.2020.05.020.
21. Liu Y, Hsu CH, Huang CC, Zhang Y, Zhao J, Tsai SJ, Chen LK, Lin CP, Lo CZ. Connectivity-based topographical changes of the corpus callosum during aging. Front Aging Neurosci. 2021;13:753236. https://doi.org/10.3389/fnagi.2021.753236.
22. Shi Y, Xue Y, Chen C, Lin K, Zhou Z. Association of gestational age with MRI-based biometrics of brain development in fetuses. BMC Med Imaging. 2020;20(1):125. https://doi.org/10.1186/s12880-020-00525-9.
23. Barkovich AJ, Norman D. Anomalies of the corpus callosum: correlation with further anomalies of the brain. AJR Am J Roentgenol. 1988;151(1):171–9. https://doi.org/10.2214/ajr.151.1.171.
24. Achiron R, Lipitz S, Achiron A. Sex-related differences in the development of the human fetal corpus callosum: in utero ultrasonographic study. Prenat Diagn. 2001;21(2):116–20.
25. Machado-Rivas F, Gandhi J, Choi JJ, Velasco-Annis C, Afacan O, Warfield SK, Gholipour A, Jaimes C. Normal growth, sexual dimorphism, and lateral asymmetries at fetal brain MRI. Radiology. 2021;303(1):162–70. https://doi.org/10.1148/radiol.211222.
26. Tanaka-Arakawa MM, Matsui M, Tanaka C, Uematsu A, Uda S, Miura K, Sakai T, Noguchi K. Developmental changes in the corpus callosum from infancy to early adulthood: a structural magnetic resonance imaging study. PLoS One. 2015;10(3):e0118760. https://doi.org/10.1371/journal.pone.0118760.
27. Genc S, Malpas CB, Ball G, Silk TJ, Seal ML. Age, sex, and puberty related development of the corpus callosum: a multi-technique diffusion MRI study. Brain Struct Funct. 2018;223(6):2753–65. https://doi.org/10.1007/s00429-018-1658-5.
28. Guz W, Pazdan D, Stachyra S, Świętoń F, Poręba P, Bednarz M, Lis A, Kazańska A, Krukowska J, Klęba J, Urbanik A. Analysis of corpus callosum size depending on age and sex. Folia Morphol (Warsz). 2019;78(1):24–32. https://doi.org/10.5603/FM.a2018.0061.
29. Berman S, West KL, Does MD, Yeatman JD, Mezer AA. Evaluating g-ratio weighted changes in the corpus callosum as a function of age and sex. NeuroImage. 2018;182:304–13. https://doi.org/10.1016/j.neuroimage.2017.06.076.
30. Wang D, Chen YJ, Li YH. Application of super-resolution track-density technique: earlier detection of aging-related subtle alterations than morphological changes in corpus callosum from normal population?

autism and the relevance of gender. Psychiatry Res. 2010;183(1):38–43. https://doi.org/10.1016/j.pscychresns.2010.04.007.
75. Bruner E, de la Cuétara JM, Colom R, Martin-Loeches M. Gender-based differences in the shape of the human corpus callosum are associated with allometric variations. J Anat. 2012;220(4):417–21. https://doi.org/10.1111/j.1469-7580.2012.01476.x.
76. Luders E, Toga AW, Thompson PM. Why size matters: differences in brain volume account for apparent sex differences in callosal anatomy: the sexual dimorphism of the corpus callosum. NeuroImage. 2014;84:820–4. https://doi.org/10.1016/j.neuroimage.2013.09.040.
77. Kurth F, Spencer D, Hines M, Luders E. Sex differences in associations between spatial ability and corpus callosum morphology. J Neurosci Res. 2018;96(8):1380–7. https://doi.org/10.1002/jnr.24260.
78. Dubb A, Gur R, Avants B, Gee J. Characterization of sexual dimorphism in the human corpus callosum. NeuroImage. 2003;20(1):512–9. https://doi.org/10.1016/s1053-8119(03)00313-6.
79. Björnholm L, Nikkinen J, Kiviniemi V, Nordström T, Niemelä S, Drakesmith M, Evans JC, Pike GB, Veijola J, Paus T. Structural properties of the human corpus callosum: multimodal assessment and sex differences. NeuroImage. 2017;152:108–18. https://doi.org/10.1016/j.neuroimage.2017.02.056.
80. Pettey DJ, Gee JC. Sexual dimorphism in the corpus callosum: a characterization of local size variations and a classification driven approach to morphometry. NeuroImage. 2002;17(3):1504–11. https://doi.org/10.1006/nimg.2002.1130.
81. Newman SD. Differences in cognitive ability and apparent sex differences in corpus callosum size. Psychol Res. 2016;80(5):853–9. https://doi.org/10.1007/s00426-015-0688-3.
82. Ohashi K, Anderson CM, Khan A, Rohan ML, Bolger EA, McGreenery CE, Teicher MH. Sex and sensitive period differences in potential effects of maltreatment on axial versus radial diffusivity in the corpus callosum. Neuropsychopharmacology. 2022;47(4):953–64. https://doi.org/10.1038/s41386-021-01260-7.
83. Aboitiz F, Montiel J. One hundred million years of interhemispheric communication: the history of the corpus callosum. Braz J Med Biol Res. 2003;36(4):409–20. https://doi.org/10.1590/s0100-879x2003000400002.
84. Welcome SE, Chiarello C, Towler S, Halderman LK, Otto R, Leonard CM. Behavioral correlates of corpus callosum size: anatomical/behavioral relationships vary across sex/handedness groups. Neuropsychologia. 2009;47(12):2427–35. https://doi.org/10.1016/j.neuropsychologia.2009.04.008.
85. Campbell KSJ, Williams LJ, Bjornson BH, Weik E, Brain U, Grunau RE, Miller SP, Oberlander TF. Prenatal antidepressant exposure and sex differences in neonatal corpus callosum microstructure. Dev Psychobiol. 2021;63(6):e22125. https://doi.org/10.1002/dev.22125.
86. Panizzon MS, Hoff AL, Nordahl TE, Kremen WS, Reisman B, Wieneke M, Harris D, Goodman C, Espinoza S, Liu W, Lim K. Sex differences in the corpus callosum of patients with schizophrenia. Schizophr Res. 2003;62(1–2):115–22. https://doi.org/10.1016/s0920-9964(02)00432-2.
87. Tønnesen S, Kaufmann T, Doan NT, Alnæs D, Córdova-Palomera A, Meer DV, Rokicki J, Moberget T, Gurholt TP, Haukvik UK, Ueland T, Lagerberg TV, Agartz I, Andreassen OA, Westlye LT. White matter aberrations and age-related trajectories in patients with schizophrenia and bipolar disorder revealed by diffusion tensor imaging. Sci Rep. 2018;8(1):14129. https://doi.org/10.1038/s41598-018-32355-9.
88. Woldehawariat G, Martinez PE, Hauser P, Hoover DM, Drevets WW, McMahon FJ. Corpus callosum size is highly heritable in humans, and may reflect distinct genetic influences on ventral and rostral regions. PLoS One. 2014;9(6):e99980. https://doi.org/10.1371/journal.pone.0099980.
89. Allouh MZ, Al Barbarawi MM, Ali HA, Mustafa AG, Alomari SO. Morphometric analysis of the corpus callosum according to age and sex in middle eastern Arabs: racial comparisons and clinical correlations to autism Spectrum disorder. Front Syst Neurosci. 2020;14:30. https://doi.org/10.3389/fnsys.2020.00030.
90. Nouri Hosseini H, Mohammadi MR, Aarabi M, Mohammadi N, Golalipour MJ. Ethnicity influences corpus callosum dimensions. Neurol Res Int. 2018;2018:8916035. https://doi.org/10.1155/2018/8916035.
91. Chura LR, Lombardo MV, Ashwin E, Auyeung B, Chakrabarti B, Bullmore ET, Baron-Cohen S. Organizational effects of fetal testosterone on human corpus callosum size and asymmetry. Psychoneuroendocrinology. 2010;35(1):122–32. https://doi.org/10.1016/j.psyneuen.2009.09.009.
92. Ho TC, Colich NL, Sisk LM, Oskirko K, Jo B, Gotlib IH. Sex differences in the effects of gonadal hormones on white matter microstructure development in adolescence. Dev Cogn Neurosci. 2020;42:100773. https://doi.org/10.1016/j.dcn.2020.100773.
93. Kennedy KM, Rodrigue KM, Land SJ, Raz N. BDNF Val66Met polymorphism influences age differences in microstructure of the Corpus Callosum. Front Hum Neurosci. 2009;3:19. https://doi.org/10.3389/neuro.09.019.2009.
94. Wade BS, Joshi SH, Reuter M, Blumenthal JD, Toga AW, Thompson PM, Giedd JN. Effects of sex chromosome dosage on corpus callosum morphology in supernumerary sex chromosome aneuploidies. Biol Sex Differ. 2014;5:16. https://doi.org/10.1186/s13293-014-0016-4.

7 Animal Studies Related with the Corpus Callosum

Erkan Gümüş

7.1 Introduction

The corpus callosum (CC) is the major white matter commissure connecting the left and right hemispheres of the brain in all placental mammals and is the biggest white matter structure of the human brain [1–4]. The CC begins to develop around the eighth week after conception and is fully formed at 18–20 weeks of gestation [5, 6]. CC development is continuous until adolescence due to neural development, such as myelination of axons. Therefore, CC thickness increases during childhood and adolescence [5–7]. The main function of the CC is the integration and processing of motor, sensory, and cognitive information by connecting neurons in the two brain hemispheres, thus facilitating the higher-order functions of the cerebral cortex [1, 3, 8]. Over the years, more detailed information about the development and function of CC has been obtained from high-resolution imaging studies of patients and from experimental animal studies. The structural defects that occur during the development of the CC affect the individual throughout his/her life due to changes in brain wiring during early development. Therefore, it is difficult to predict exactly how structural defects in the CC will affect an individual's life and behavior. Most of the animal studies on the CC have consisted of experiments on neurodevelopmental disorders and fiber composition. Experimental animal models are very important in terms of providing a complementary contribution to better describe these complex disorders and understand their underlying causes.

Animal studies related to the CC will provide a theoretical basis for understanding in detail both the formation and development of CC and related neurodevelopmental disorders such as schizophrenia and autism spectrum disorder (ASD). In this chapter, a brief overview of animal studies of the CC is provided.

7.2 Animal Studies Related with the CC

7.2.1 Animal Studies Related with the Development of CC

Mice and rats are the ideal experimental animal model due to their genetic and physiological similarity to humans, their small size, short life cycles, ease of use and maintenance and cost-effectiveness, and abundant genetic resources. In addition, these animals are also preferred because they can be easily manipulated genetically and transgenic models can be generated [9]. Mice and rats are the most widespread laboratory animal models for studies of brain formation, development, and function. Experiments using these models constitute the majority of studies on brain

E. Gümüş (✉)
Department of Histology and Embryology, Aydın Adnan Menderes University Faculty of Medicine, Aydın, Turkey

M. Turgut et al. (eds.), *The Corpus Callosum*, https://doi.org/10.1007/978-3-031-38114-0_7

and neural function [10]. Therefore, the CC-related animal studies reviewed for this chapter mainly consisted of rodents.

The partial or complete absence of the CC causes a rare congenital disorder known as agenesis of the CC (ACC) [5]. Malformations of the CC may be partial, hypoplasia throughout the entire structure, or complete agenesis, and ACC is associated with more than 50 different human congenital anomalies [11]. The incidence of ACC is about 1 in 4000 live births and around 2–4 per 100 in individuals with developmental disabilities [12]. 30–45% of ACC cases are genetic, of which about 10% have chromosomal abnormalities, while the remaining 20–35% are correlated with a single-gene mutation [13]. ACC causes disabilities that range from mild to severe, depending on the associated brain anomalies. Such disabilities include mental retardation and learning difficulties, malnutrition and swallowing difficulties, language and social communication disorders, and vision and hearing disorders [11, 13]. Magnetic resonance imaging (MRI), high-resolution ultrasound, and other imaging modalities are used to diagnose ACC [5]. However, ACC can be diagnosed at the earliest in the second trimester (18–20 weeks of gestation) of pregnancy by using imaging techniques [14]. The CC develops in a complex and well-orchestrated process of neuronal proliferation and migration and axonal growth and/or guidance. Therefore, disruptions in these processes may cause developmental malformations of the CC [13]. These malformations that result in ACC are due to mutations in some genes responsible for certain factors such as axonal guidance (both attractive and repulsive), transcription factors, and growth factors [15]. The knowledge on the development of the CC has been obtained through animal studies and detailed structural neuroimaging studies [13]. Animal studies related to the CC remain popular as an important tool in determining the cellular and molecular mechanisms involved in the development of the CC and the genes that coordinate this process. However, the molecular mechanisms underlying these processes have not yet been fully explored, and it is not always possible to relate these mechanisms to human syndromes, depending on whether these mechanisms are conserved across species. In fact, it is thought that many single-gene mouse models that have been developed result in death during embryonic or fetal development, which may explain the absence of these genes in human populations. Indeed, studies in animal models associated with ACC have identified critical processes involved in callosal development, and these data have been confirmed by analysis of human fetal brains [13]. Although the molecular mechanisms underlying the development of CC are not clearly known, the axons in the CC arise from neocortical neurons in layers II/III and V [2, 13]. It is known that disruptions in the mechanisms regulating the development and migration of these neurons and involved in the growth and orientation of axons cause different brain malformations as well as the development of CC. Indeed, it has been reported that disruptions in the expression of some axon guide molecules affect glial development and cause ACC [13, 15]. The current knowledge of the development of the CC at the cellular and molecular levels is primarily derived from studies using animal models. A summary of animal model studies related with the development of CC are given as below in detail (Table 7.1).

In one of these model studies, the molecular specification of commissural plates was demonstrated in mice [16]. Data on the molecular organization of the human commissural plate are still limited. In the study by Moldrich et al., the authors explored the molecular areas of the commissural plate in human [16]. Nuclear factor I (NFI) genes are expressed in embryonic mouse brain, and the mutations in the Nfia, one of the four NFI genes, cause ACC and hydrocephalus [26]. In their study, Moldrich et al. demonstrated the molecular analysis and identification of subdomains of CC by using specific transcription factors such as Nfia, Emx1, Six3, and Zic2. The results of this study showed that the mouse commissural plate consists of four domains, each of them associated with a specific commissural projection. Furthermore, the dorsal domain disruptions have been reported in strains with commissural defects such as Nfia-, Satb2-, and

Table 7.1 Animal studies on CC development

Author(s)/ref. #	Year	Species/strain	Age	Analysis/techniques	Conclusion
Moldrich et al. [16]	2010	Mouse/Nfia, Emx, Satb knockout	Ed 14 to 17	DTMRI Histology/IHC and ISH	ACC ($Emx2^{-/-}$ and $Satb2^{-/-}$ mice) Abnormal CC ($Nfia^{-/-}$ mice) Commissural plate consists of four domains in mouse
Edwards et al. [17]	2020	Mouse/BTBR N2	Pd 15 to 20 Pd 80 to 82	Histology dMRI, tractography	Characterized for the first time the brain connectivity of a mouse model of complete and partial absence of CC
Orioli et al. [18]	1996	Mouse/EphB2, Eph B3 knockout	Ed 14 to 16.5 Pd0	Molecular Histology/IHC and ISH	The mutation of EphB3 and EphB2 receptors causes defects in pathfinding of axons and CC formation A single mutation of EphB3 causes CC defects and a double mutation of EphB3 and EphB2 causes ACC
Mendes et al. [19]	2006	Mouse/Eph receptors (B1, B2, B3, and A4) knockout B-class ephrins (B1, B2, and B3) knockout	Ed 15 to 18 Pd 0	Genetic Histology/LM, IHC, and ISH Cell culture	All mutant mice have CC defects ranging from mild hypoplasia to complete agenesis The CC defects are most prominent in the double-knockout mouse mutants
Fothergill et al. [20]	2014	Mouse/NTN1, DCC knockout	Ed 15.5 to 17 Pd 0	Molecular Histology/IHC and ISH Cell culture	NTN1-DCC signaling facilitates and coordinates the migration of pioneering callosal axons NTN1 and DCC mutant mice both display ACC and formed ipsilateral Probst bundles
Morcom et al. [21]	2021	Mouse/NTN1, DCC knockout	Ed 12 to 17 Pd 0	Molecular Histology/IHC and ISH Cell culture	NTN1 and DCC genes are required for interhemispheric fissure remodeling DCC plays role in regulating astroglial morphology and migration
Qiu et al. [22]	1996	Mouse/Emx-1 knockout	10 week old	Histology/LM	Homozygous mutant mice lack most or all their CC Heterozygotes mutant mice show partial absence of CC
Dattani et al. [23]	1998	Mouse/Hesx1 knockout	Ed 9.5 to 12.5 Pd8	Molecular Histology/ISH	Hesx1 mutation causes various abnormalities in the CC including agenesis, anterior and hippocampal commissures, and septum pellucidum

(continued)

Table 7.1 (continued)

Author(s)/ref. #	Year	Species/strain	Age	Analysis/techniques	Conclusion
Dahme et al. [24]	1997	Mouse/LICAM knockout	2 to 7 month old	Histology/LM and EM Tissue culture	L1CAM mutation causes aberrant pyramidal decussations, reduced axonal association with nonmyelinating Schwann cells, enlarged lateral ventricles, and impaired neurite outgrowth
Tapanes-Castillo et al. [25]	2010	Mouse/LICAM knockout	Pd 21 to39 3 month old	Molecular Phenotypic	L1-6D mutation causes hydrocephalus with thinner deep cortical layers (layers V, VI), ACC, and abnormal thalamocortical tract
Yoo et al. [15]	2017	Mouse/Nogo receptors (NgR1, NgR2, and NgR3) knockout	12 week old 1 year old	Histology/IHC and immunoblotting Behavior	Defects in Nogo receptor expression change in the orientation of axons and cause complete ACC

Ed embryonic day, *Pd* postnatal day, *DTMRI* diffusion tensor magnetic resonance imaging, *dMRI* diffusion magnetic resonance imaging, *ACC* agenesis of the corpus callosum, *IHC* immunohistochemistry, *ISH* in situ hybridization, *LM* light microscopy, *EM* electronmicroscopy

Emx2-knockout mice. This study demonstrated that the Emx- and Satb2-knockout mice had ACC, while Nfia-knockout mice had abnormal CC. In addition, it has been shown that Fgf8, a morphogen, expression plays an important role in the initial modeling of the forebrain and in the formation of the commissural plate [16].

In recent years, a new mouse model investigating the presence, plasticity, and function of changes in structural connectivity across humans and mice was reported. In this study, variable degrees of callosal malformations were characterized in a mouse model. To this end, complete ACC BTBR T+ tf/J (BTBR) mice were backcrossed twice with wild-type C57Bl/6J mice to produce littermates with either full CC, complete, or partial ACC. Ultimately, a mouse model was created that predicted the features of structural brain connections in humans with ACC. In this study, the mathematical modeling of axonal connections was performed with high-resolution ex vivo diffusion MRI (dMRI) and tractography, and these findings were confirmed by intrauterine electroporation and immunohistochemistry [17]. This study demonstrated that the whole brain network features of BTBR N2 mouse with ACC summarize the organization of the brain structures observed in humans with ACC. Considering the complexity of white matter networks in the human brain, the availability of safe and effective models is critical. Therefore, it has been demonstrated with this study that the BTBR N2 mouse may be a model in which developmental and functional analysis can be performed to understand the mechanisms underlying axonal plasticity in the ACC brain [17]. For normal CC development, the growth and orientation of the axons that cross from one cerebral hemisphere to the other are crucial. Indeed, some molecules associated with axon and/or axon guide that result in an ACC phenotype when mutated in various animal models have been reported [18, 22–25].

In fact, it has been shown in various animal studies that Eph receptors and ligands have important functions in commissural axon guidance in CC development [18, 19]. First, the expression of the receptors B-class Eph (EphB2 and EphB3) and EphA in the CC fibers of mice suggested that these molecules might be important in the development of CC. Orioli et al. demonstrated the roles of EphB3 and EphB2 receptors, which bind to a group of cells involved in the guidance and fasciculation of axons in mouse. In this study, the mutation of EphB3 and EphB2 receptors were investigated during brain development by molecular and histological analysis in E16.5 embryo and newborn of chimeric mice model. According to this study, the mutation of EphB3 and EphB2 receptors was shown to cause defects in pathfinding of axons and CC formation. In addition, it has been shown that a single mutation of EphB3 causes CC defects and a double mutation of EphB3 and EphB2 causes ACC. In the same study, it was claimed that other Eph receptors and ligands may also be associated with ACC in the human brain due to similar roles, given that EphB3 and EphB2 are members of the Eph-associated receptor protein tyrosine kinase family [18]. The roles of Eph receptors and ligands in mediating the midline pathfinding process are highly complex. In 2006, Mendes et al. conducted a study to determine which types of Eph receptors and ligands are functional in the development of CC. In their study, Eph receptors (B1, B2, B3, and A4) and B-class ephrins (B1, B2, and B3) were investigated by histological and genetic analyses and cell culture techniques in mutant embryos and newborns. The results of this study showed that CC defects in mutant mice ranged from mild hypoplasia to complete agenesis and Probst bundles formation and the defects were most pronounced in the double-knockout mutants [19].

As it is known, the left and right hemispheres of the brain communicate with each other via commissural axons that cross the midline during development, and various molecules that are protected between species mediate the guidance of these axons. The consequences that may be caused by mutations or deficiencies in the expression of these molecules during brain development are investigated, particularly in knockout animal models. Deleted in colorectal carcinoma (DCC) and netrin 1 (NTN1) have been well-studied molecules and have been shown to play

Table 7.2 Animal studies related with the CC and autism spectrum disorder (ASD)

Author(s)/ref. #	Year	Species/strain	Age	Analysis/techniques	Conclusion
Peça et al. [29]	2011	Mouse/SHANK3 knockout	Pd 7 to 21	Histology/LM, EM, ISH MRI Behavior Electrophysiology	SHANK3 mutation causes histological brain abnormality, various synaptic defects, and ASD-related behavioral abnormalities
Malara et al. [30]	2022	Mouse/SHANK3 knockout	Pd 15 to 20 Pd 140	Histology/LM, EM, IHC MRI Organoid/cell culture	SHANK3 is expressed in oligodendrocytes and Schwann cells. SHANK3 mutation causes loss of myelin basic protein in the CNS and zero loss of myelin protein in the PNS SHANK3 mutation causes reduction of the white matter and of the CC volume
Uccelli et al. [31]	2021	Rat/the valproic acid (VPA) model	Pd 15 to 36	Histology/LM, EM, IHC FDG-PET imaging Behavior	Cellular disorganization, decreased number of myelinated fibers, and less compact myelin sheath in the CC of VPA animals Hypomyelination and/or abnormal myelination of the CC may underlie the white matter changes observed in the CC of individuals with ASD
Tabuchi et al. [32]	2007	Mouse/NL3R451C knock-in (NL3 KI)	8 week old 2 to 4 month old	Molecular Biochemical Histology/LM, EM, IHC, and immunoblotting Electrophysiological behavior	NL3-KI mouse model could be a model system for the development of new treatments for patients with ASD It could be possible to improve ASD-related behavioral anomalies using decrease of inhibitory synaptic transmission
Ellegood et al. [33]	2011	Mouse/NL3 KI	Pd108	MRI	In the brain of mutant mice, both gray matter structures (hippocampus and thalamus) and white matter structures (CC) are small compared to healthy ones Volumetric change of CC is probably due to fewer axons or less mature axons
Peñagarikano et al. [34]	2011	Mouse/CNTNAP2 knockout	2 to 8 month old	Molecular Histology/IHC and ISH Biochemical electrophysiological Behavior	CNTNAP2 mutant mice display three core behavioral characteristics of ASD CNTNAP2 mutation causes abnormal migration of neurons The presence of ectopic neurons was detected in the CC of CNTNAP2 mutant mice

Fairless et al. [35]	2008	Mouse/BALB/cJ inbred	Pd30	Histology Behavior	There is a striking variability in CC development in BALB/cJ mice, with a significant positive correlation between sociability and the size of the CC
Fairless et al. [36]	2012	Mouse/BALB/cJ inbred	Pd19–70	Histology Behavior	Sociability increased with age in BALB/cJ mice There was no correlation between CC size and sociability
Stephenson et al. [37]	2011	Mouse/BTBR	8 to 10 week old	Histology/LM, IHC, ISH, and stereology	Complete ACC

ASD autism spectrum disorder, *CNS* central nervous system, *PNS* peripheral nervous system, *Tg* transgenic, *Pd* postnatal day, *MRI* magnetic resonance imaging, *ACC* agenesis of the corpus callosum, *IHC* immunohistochemistry, *ISH* in situ hybridization, *LM* light microscopy, *EM* electron microscopy

Table 7.3 Animal studies related with the CC and schizophrenia (SCZ)

Author(s)/ref. #	Year	Species/strain	Age	Analysis/techniques	Conclusion
Chin et al. [38]	2011	Rat/methylazoxymethanol acetate (MAM) SCZ model	Pd 45 to 65	Histology/LM MRI	Results were consistent with the findings observed in SCZ Decreased CC Volumetric change of CC is probably due to the occurrence of demyelination in these white matter fiber tracts
Xiu et al. [39]	2015	Mouse/MK-801 SCZ model	8 week old	Histology/LM, EM, IHC MRI Organoid/cell culture	Abnormal CC with the splitting lamellae of myelin sheaths, the demyelination, and a reduction in the CC size
Roy et al. [40]	2007	Mouse/DN-erbB4 Tg	Pd 112	Histology/LM, EM, IHC, and ISH Behavior	Loss of erbB signal resulted in changes in the number and morphology of oligodendrocyte (OL), reduction in myelin thickness, and slower conduction velocity in axons of central nervous system neurons The normal density of myelinated and unmyelinated axons and decreased thickness of the myelin sheath in the CC of transgenic mice erbB signaling plays an important role in regulating the white matter structure
Shen et al. [41]	2008	Mouse/DISC1 Tg	Ed 17.5 2 month old	Molecular Histology/LM, IHC, and ISH Behavior Cell culture	In Disc1 transgenic mice, enlarged lateral ventricles, partial ACC, reduced cerebral cortex, and neural proliferation Behavior patterns compatible with SCZ Fewer and shorter neurites in transgenic neurons

SCZ schizophrenia, *Ed* embryonic day, *Pd* postnatal day, *Tg* transgenic, *MRI* magnetic resonance imaging, *ACC* agenesis of the corpus callosum, *IHC* immunohistochemistry, *ISH* in situ hybridization, *LM* light microscopy, *EM* electron microscopy ration of the PDZ domain. SHANK3-KO mice were backcrossed five generations with wild-type C57Bl/6 mice. In this study, the disruptions in the SHANK3 gene were assessed by histological, biochemical, molecular, electrophysiological, and behavioral analyses in young and old mutant mice [29]. In this study, SHANK3 mutant mice were shown to exhibit ASD-related behavioral abnormalities, including various synaptic defects and self-injurious repetitive grooming and impairments in social interaction. In addition, it has been shown that SHANK3 play a crucial role in the normal development of neuronal connectivity. Thus, a strong correlation has been shown between a disruption in the SHANK3 gene and the occurrence of autistic-like behaviors in mice [29]. Based on imaging studies, marked structural changes in white matter have been demonstrated in the brain of ASD patients. As mentioned before, besides the reduction in the volume of CC [48–50], some other reports have demonstrated a reduction in functional connectivity between hemispheres with decreased fractional anisotropy and axial diffusivity in the CC of patients with ASD [52–54]. In fact, defects in axonal diameter and myelin ultrastructure and a decrease in the number of myelinated axons are demonstrated in both animals' and humans' brains with ASD [31]. Although changes in CC in ASD patients have been demonstrated, the mechanisms underlying these changes are not well known. Thus, studies in animal models of ASD are helpful in this regard. A recent study on the SHANK3-knockout mouse model examined the effects of the SHANK3 gene on myelination of axonal fibers and the molecular changes in white matter in SHANK3 deficiency. In this study, disruptions in the SHANK3 gene were evaluated by histological, ultrastructural, and molecular examination in young and old mutant mice. Malara et al. demonstrated for the first time that SHANK3 is expressed in oligodendrocytes and Schwann cells. Moreover, it has been shown that SHANK3 deficiency causes loss of myelin basic proteins in the central nervous system (CNS) which are important myelin proteins and zero loss of myelin protein in the peripheral nervous system (PNS). Therefore, it has been suggested that loss of SHANK3 in neurons may also affect the differentiation of oligodendrocytes and Schwann cells. In addition, they concluded that dysregulation of MBP expression observed in the SHANK3 mutant mice might give a possible explanation for myelin alterations in neurodevelopmental disorders [30]. Another recent study evaluated the composition and structure of CC on the valproic acid (VPA) rat model, which is a nongenetic experimental animal model of ASD by behavioral, histopathological, and neuroimaging techniques [31]. In this study, Uccelli et al. showed cellular disorganization, decreased number of myelinated fibers, and less compact myelin sheath in the CC of VPA animals. In fact, the decrease in the number of myelinated axons without changes in the total number of axons suggested that VPA treatment did not affect the development of CC. Therefore, it has been emphasized that hypomyelination and/or abnormal myelination of the CC may underlie the white matter changes observed in the CC of individuals with ASD [31]. The neuroligin and neurexin are synaptic cell adhesion genes, and disruptions of these genes have been associated with neurodevelopmental disorders, including ASD. In 2003, the R451C mutation in NLGN3 gene, which encodes the neuroligin-3 protein, was detected for the first time in two brothers diagnosed with autism [55]. The autism-related $NL3^{R451C}$-knock-in (NL3-KI) mouse model was generated to study this mutation [32]. Tabuchi et al. created NL3-KI mice by gene targeting the R451C substitution to the endogenous neuroligin-3 gene. Furthermore, they also analyzed neuroligin-3-knockout (KO) mice to determine the function of the R451C substitution. These mutant mice were assessed by histological, biochemical, molecular, electrophysiological, and behavioral analyses. In this study, it was shown that an enhancement in inhibitory synaptic transmission without changing excitatory synaptic transmission was observed in NL3-KI mice brain. In addition, they found that the mutant mice had disturbed social behaviors. In summary, this study claimed that the NL3-KI mouse model could be a model system for the development of new treatments for patients with ASD and that it would be possible to improve ASD-related behavioral disorders using attenuation of inhibitory synaptic transmission [32]. Although conflicting results have been reported, NL3-KI mouse model is one of the well-characterized autism-related genetic animal models for ASD research [33]. In a study examining brain abnormalities using magnetic resonance imaging (MRI) in the NL3-KI mouse model, both gray matter structures such as the thalamus and hippocampus and white matter structures such as the CC had been reported to be smaller than healthy mouse brains. In this study, NL3-KI mice were backcrossed three to nine generations with wild-type C57Bl/6 mice, and mature mice were investigated. In the same study, it had been reported that this volumetric change of CC is due to fewer axons or less mature axons rather than an overall deterioration in the microstructure of the tissue [33]. Peñagarikano et al. characterized another knockout mouse model to study CNTNAP2 function the pathophysiology of ASD. CNTNAP2 is the member of the neurexin family and play a role in axonal differentiation and neural development [34]. CNTNAP2 mutation was evaluated by histological, biochemical, molecular, electrophysiological, and behavioral analyses. Mice with CNTNAP2 mutation display three core behavioral characteristics of ASD [34, 56]. In this study, deficiency of CNTNAP2 in the mouse has been shown to resemble many of the behavioral and cognitive profiles of patients with idiopathic autism and to cause abnormal migration of neurons. In addition, the presence of ectopic neurons was detected in the CC of CNTNAP2 mutant mice [34].

However, a detailed meta-analysis summarizing the volumetric changes of different brain structures in people with autism reported a reduction in the CC size [57]. This reduction in CC size in patients with ASD suggested that the impairment in social interactions observed in these patients may be related to CC size or callosal disconnection [35, 36]. Evaluation of sociability in the animal model is critical for determining the ASD. The most effective method used for validation of ASD animal models is behavioral assays sensitive to social deficits. One of the animal models to examine the interactions between decreased sociability and CC underdevelopment is the BALB/cJ inbred mouse strain. Fairless et al. investigated this relation in 30-day-old BALB/cJ inbred mice. In this study, it has been shown that there is a striking variability in CC development in BALB/cJ mice, with a significant positive correlation between sociability and the size of the CC relative to brain weight [35]. In another study, the same group examined the effect of CC underdevelopment on social behavior at different ages (19, 23, 31, 42, or 70 days of age) with larger sample sizes. In this study, BALB/cJ mice were tested for sociability and measured brain weights and CC areas. They reported that sociability increased with age in BALB/cJ mice and, unlike their previous studies, there was no correlation between CC size and sociability [36]. BTBR T+Itpr3tf/J (BTBR) inbred mouse strain is another well-defined model for investigating the behavioral characteristics found in ASD. BTBR mice display complete agenesis of CC, and several studies have demonstrated reduced social behaviors in this strain by evaluating the differences in social behavior [37].

7.2.4 Animal Studies Related with the CC and Schizophrenia (SCZ)

Schizophrenia (SCZ) is a neurodevelopmental and neurodegenerative disease characterized by profound disruptions in cognition and emotion. It is associated with an impaired white matter connectivity. Postmortem studies have shown that CC plays a critical role in the neurobiology of SCZ, with abnormalities in CC size and reductions in callosal fiber density in these cases [58, 59]. The possible causes of the reductions in the CC and white matter in SCZ are being investigated in animal models. A summary of animal model studies related with the CC and SCZ are given as below in detail (Table 7.3).

One of these models is the methylazoxymethanol acetate (MAM), the best-known model of SCZ. MAM is a methylating agent that causes impaired neurogenesis and abnormalities in brain development when administered prenatally. In 2011, Chin et al. conducted a study to investigate the reductions of white matter volume and the possible reason in rat MAM model at different ages (puberty and adulthood) [38]. In this study, it was determined that the results were consistent with the findings observed in SCZ and the diffusion fractional anisotropy retrieved from the CC and cingulum was significantly decreased. In conclusion, the authors suggested that this reduction was probably due to the occurrence of demyelination in these white matter fiber tracts [38]. The animal model of SCZ induced by MK-801, NMDA receptor antagonists, is another widely studied animal model to investigate the pathogenesis of SCZ. Xiu et al. investigated the possible causes of SCZ-related white matter abnormalities in MK-801-induced adult mice by behavioral and histomorphometric analyses. The results of this study reported the abnormalities of CC with the splitting lamellae of myelin sheaths, the demyelination, the loss of the myelinated fibers length, and a reduction in the CC size. Therefore, this study suggested that CC may play a crucial role in the pathophysiology of SCZ and may also provide a basis for further studies investigating the treatment of SCZ by remyelination of myelinated fibers [39]. White matter disturbances in psychiatric disorders such as SCZ have been thought to be responsible for the development of psychotic symptoms. The assumed role of white matter abnormalities in SCZ is being studied in knockout animal models [60].

One of the candidate "risk" genes in the development of SCZ is neuregulin-1 (NRG1) and its receptor ERBB4. NRG1-mediated erbB signaling plays critical roles in neural and glial development as well as in neuronal migration and in regulating the myelination process during development and functioning of the nervous system. Although the effects of NRG1-mediated erbB signaling in the development of SCZ are known, the mechanisms by which they exert their effects on this disease are still unknown [60].

In a transgenic (Tg) mice model that express a dominant-negative erbB receptor (DN-erbB4) completely blocks erbB2, erbB3, and erbB4 receptor signaling were generated to test whether erbB signaling effects to SCZ by regulating the white matter structure. In this study, it was determined that the results were consistent with the behavioral changes observed in SCZ and that loss of erbB signal resulted in changes in the number and morphology of oligodendrocyte (OL), reduction in myelin thickness, and slower conduction velocity in axons of central nervous system neurons. In addition, it was shown by electron microscopy that the density of myelinated and unmyelinated axons in the CC of transgenic mice was normal, but the thickness of the myelin sheath was significantly reduced. Therefore, this study demonstrated that erbB signaling plays a critical role in regulating the white matter structure and alterations in white matter structure cause defects in the dopaminergic system and result in psychiatric disorders [40].

Disrupted-in-schizophrenia 1 (DISC1) is a gene involved in processes that regulate neural development and brain maturation and is known to be a risk factor for diseases associated with dopamine disorders such as SCZ, depression, and bipolar disorders. Mutations in DISC1 gene located at 1q42 have been reported to be related with ACC [41, 61]. How the mutation in DISC1 leads to these psychiatric disorders is still not fully elucidated [41]. Shen et al. generated a transgenic mice model to explore the role of Disc1 gene in brain development [41]. The abnormalities observed in the Disc1 transgenic mouse model were found to be like those in severe SCZ. These transgenic mice were investigated by histological, molecular, and behavioral analyses and cell culture techniques. In Disc1 transgenic mice, enlarged lateral ventricles, partial ACC, reduced cerebral cortex, and neural proliferation were detected. Furthermore, it has been shown in primary neuronal cell culture experiments that transgenic neurons grow fewer and shorter neurites. In addition to these findings, it was also shown that transgenic mice exhibited behavioral patterns compatible with SCZ [41]. This study provided strong data that the partial ACC observed in transgenic mice may be due to decrease cortical neuron number and/or reduce neurite outgrowth.

44. Varghese M, Keshav N, Jacot-Descombes S, Warda T, Wicinski B, Dickstein DL, Harony-Nicolas H, De Rubeis S, Drapeau E, Buxbaum JD, Hof PR. Autism spectrum disorder: neuropathology and animal models. Acta Neuropathol. 2017;134(4):537–66.
45. Barnea-Goraly N, Frazier TW, Piacenza L, Minshew NJ, Keshavan MS, Reiss AL, Hardan AY. A preliminary longitudinal volumetric MRI study of amygdala and hippocampal volumes in autism. Prog Neuropsychopharmacol Biol Psychiatry. 2014;48:124–8.
46. Ha S, Sohn IJ, Kim N, Sim HJ, Cheon KA. Characteristics of brains in autism Spectrum disorder: structure, function and connectivity across the lifespan. Exp Neurobiol. 2015;24(4):273–84.
47. Schumann CM, Hamstra J, Goodlin-Jones BL, Lotspeich LJ, Kwon H, Buonocore MH, Lammers CR, Reiss AL, Amaral DG. The amygdala is enlarged in children but not adolescents with autism; the hippocampus is enlarged at all ages. J Neurosci. 2004;24(28):6392–401.
48. Egaas B, Courchesne E, Saitoh O. Reduced size of corpus callosum in autism. Arch Neurol. 1995;52(8):794–801.
49. Frazier TW, Hardan AY. A meta-analysis of the corpus callosum in autism. Biol Psychiatry. 2009;66(10):935–41.
50. Frazier TW, Keshavan MS, Minshew NJ, Hardan AY. A two-year longitudinal MRI study of the corpus callosum in autism. J Autism Dev Disord. 2012;42(11):2312–22.
51. Wang W, Li C, Chen Q, van der Goes MS, Hawrot J, Yao AY, Gao X, Lu C, Zang Y, Zhang Q, Lyman K, Wang D, Guo B, Wu S, Gerfen CR, Fu Z, Feng G. Striatopallidal dysfunction underlies repetitive behavior in Shank3-deficient model of autism. J Clin Invest. 2017;127(5):1978–90.
52. Aoki Y, Abe O, Nippashi Y, Yamasue H. Comparison of white matter integrity between autism spectrum disorder subjects and typically developing individuals: a meta-analysis of diffusion tensor imaging tractography studies. Mol Autism. 2013;4(1):25.
53. Just MA, Cherkassky VL, Keller TA, Kana RK, Minshew NJ. Functional and anatomical cortical underconnectivity in autism: evidence from an FMRI study of an executive function task and corpus callosum morphometry. Cereb Cortex. 2007;17(4):951–61.
54. Vogan VM, Morgan BR, Leung RC, Anagnostou E, Doyle-Thomas K, Taylor MJ. Widespread white matter differences in children and adolescents with autism spectrum disorder. J Autism Dev Disord. 2016;46(6):2138–47.
55. Jamain S, Quach H, Betancur C, Råstam M, Colineaux C, Gillberg IC, Soderstrom H, Giros B, Leboyer M, Gillberg C, Bourgeron T, Paris Autism Research International Sibpair Study. Mutations of the X-linked genes encoding neuroligins NLGN3 and NLGN4 are associated with autism. Nat Genet. 2003;34(1):27–9.
56. Anderson GR, Galfin T, Xu W, Aoto J, Malenka RC, Südhof TC. Candidate autism gene screen identifies critical role for cell-adhesion molecule CASPR2 in dendritic arborization and spine development. Proc Natl Acad Sci U S A. 2012;109(44):18120–5.
57. Stanfield AC, McIntosh AM, Spencer MD, Philip R, Gaur S, Lawrie SM. Towards a neuroanatomy of autism: a systematic review and meta-analysis of structural magnetic resonance imaging studies. Eur Psychiatry. 2008;23(4):289–99.
58. Patel S, Mahon K, Wellington R, Zhang J, Chaplin W, Szeszko PR. A meta-analysis of diffusion tensor imaging studies of the corpus callosum in schizophrenia. Schizophr Res. 2011;129(2–3):149–55.
59. Zhuo C, Liu M, Wang L, Tian H, Tang J. Diffusion tensor MR imaging evaluation of callosal abnormalities in schizophrenia: a meta-analysis. PLoS One. 2016;11(8):e0161406.
60. Xu H, Li XM. White matter abnormalities and animal models examining a putative role of altered white matter in schizophrenia. Schizophr Res Treatment. 2011;2011:826976.
61. Blackwood DH, Fordyce A, Walker MT, St Clair DM, Porteous DJ, Muir WJ. Schizophrenia and affective disorders--cosegregation with a translocation at chromosome 1q42 that directly disrupts brain-expressed genes: clinical and P300 findings in a family. Am J Hum Genet. 2001;69(2):428–33.

8 Microsurgical Anatomy of the Corpus Callosum

Genevieve Korst, Cuong C. J. Bui, and R. Shane Tubbs

8.1 Introduction

The human brain has multiple regions of interhemispheric connections known as commissures. These regions serve to allow for connection and integration of complex signals, allowing for coordination of information from bilateral hemispheres [1]. The corpus callosum (CC) (Figs. 8.1, 8.2, 8.3, 8.4,8.5, 8.6, and 8.7) is the largest commissure or junction between the hemispheres of the brain, and it connects the cerebral hemispheres at the interhemispheric fissure. White fiber tracts within the CC serve to connect bilateral frontal, temporal, occipital, insular, and limbic lobes in addition to the basal ganglia [2, 3].

There are four portions of the CC: the rostrum, which serves as the floor of the frontal horn; the genu, forming the anterior wall of the frontal horn; the body, which forms the roof of the frontal horn, as well as the roof of the body of the lateral ventricle; the final and most posterior portion, the splenium, which forms the forceps major (Fig. 8.4) [4]. Additionally, the isthmus is the narrow portion of the CC, stretching between the body and the splenium [5].

Functionally, like all commissures, the CC allows transfer of information between hemispheres of the brain. This allows for interhemispheric integration of sensory, motor, and high-function signaling. Anterior fibers, connecting the frontal lobes, have been associated with higher-cognitive transfer and processing, in addition to motor signals and coordination between hemispheres [3, 6]. Fibers located more posteriorly are associated with integration of somatosensory, auditory, and visual cues via the posterior midbody, isthmus, and splenium, respectively [3]. These connections allow for the progressive refinement and coordination of movement and cognitive function as the brain continues to develop in early life [7]. Shah et al. described four types of radiations involved in the CC: superior, inferior, anterior, and posterior.

The superior radiations extend from the totality of the body of the CC transversely into the medial hemispheric surfaces, generally connecting the motor and supplementary motor regions [2]. Anterior superior fibers arising from the body of the CC connected the CC to the medial superior frontal gyrus. Posterior superior fibers, also from the body, serve to connect the posterior portion of the medial superior frontal gyrus, the precentral and postcentral gyri. Continuing

G. Korst
Department of Neurosurgery, West Virginia University, Morgantown, WV, USA

C. C. J. Bui
Department of Neurosurgery, Ochsner Medical Center, New Orleans, LA, USA
e-mail: cjbui@ochsner.org

R. S. Tubbs (✉)
Department of Neurosurgery, Tulane School of Medicine, New Orleans, LA, USA

M. Turgut et al. (eds.), *The Corpus Callosum*, https://doi.org/10.1007/978-3-031-38114-0_8

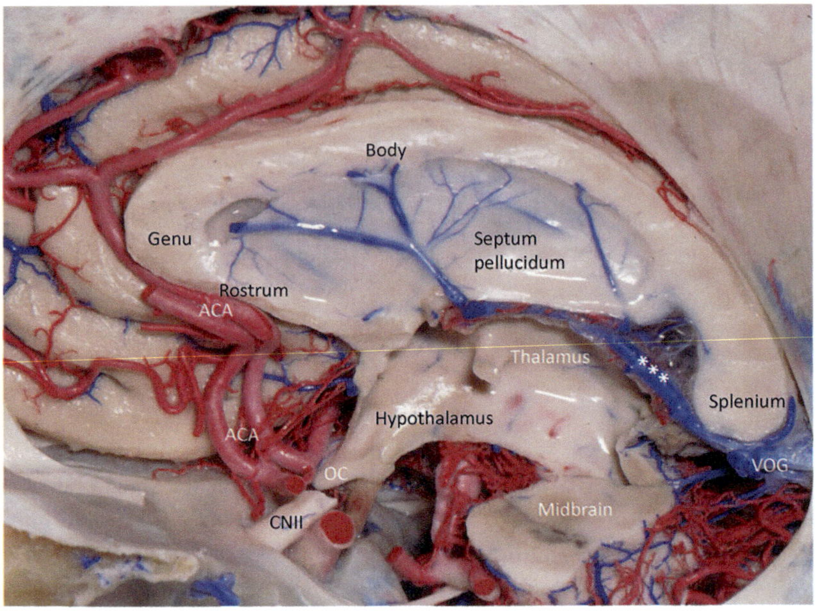

Fig. 8.1 Sagittal brain (injected) section noting the parts of the CC and related structures. (Courtesy Dr. Albert Rhoton)

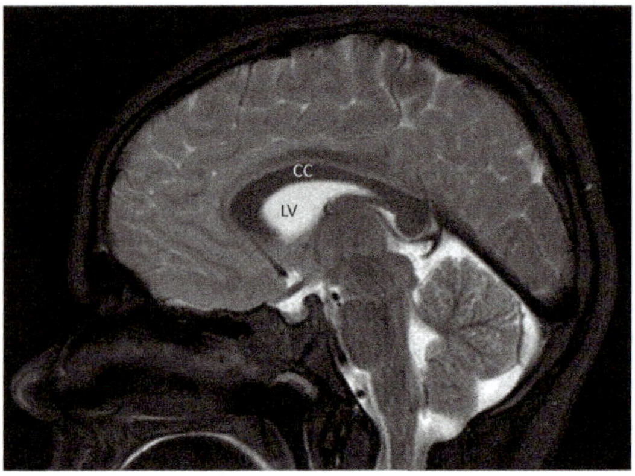

Fig. 8.2 T2-weighted MRI noting the CC and lateral ventricle (LV)

posteriorly, the superior radiations arising from the genu to just anterior to the splenium of the CC track to the precuneus [2, 3, 7].

The inferior radiations fall into anterior and posterior segments and serve to connect the temporal lobes as well as the basal ganglia [7]. The anterior portion traverses inferiorly to the inferior fronto-occipital fasciculus, while the posterior portion of these radiations travels laterally, serving to create the roof of portions of the lateral ventricle, including the body, atrium, and occipital horn. These posterior fibers then run inferiorly along the lateral ventricle, forming the inner portion of the lateral wall of the temporal and occipital horns [2].

The anterior callosal radiations rise from the genu and the rostrum of the CC and serve to connect bilateral insular regions. From the genu and the rostrum, these fibers travel anteriorly toward the medial orbitofrontal area of the brain, creating the forceps minor, or the medial wall of the frontal horn of the lateral ventricle. Finally, the posterior radiations, which arise from the splenium, serve to forms the forceps major and travel posteriorly to the medial occipital region of the brain. These radiations connect bilateral medial occipital lobes.

Given the extensive radiations and their connections with the CC, due consideration to these connections is important with surgical approaches. Additionally, understanding of the radiations of the CC may allow for further understanding of the spread of a corpus callosal gliomas [2].

From an operative standpoint, these complex radiations have been characterized into five zones: [8] Region I, the most anterior portion, contains fibers that project to the prefrontal cortex. Region II is the rest of the anterior CC and is made up of fibers connecting to the supplementary motor and premotor cortices. Region III,

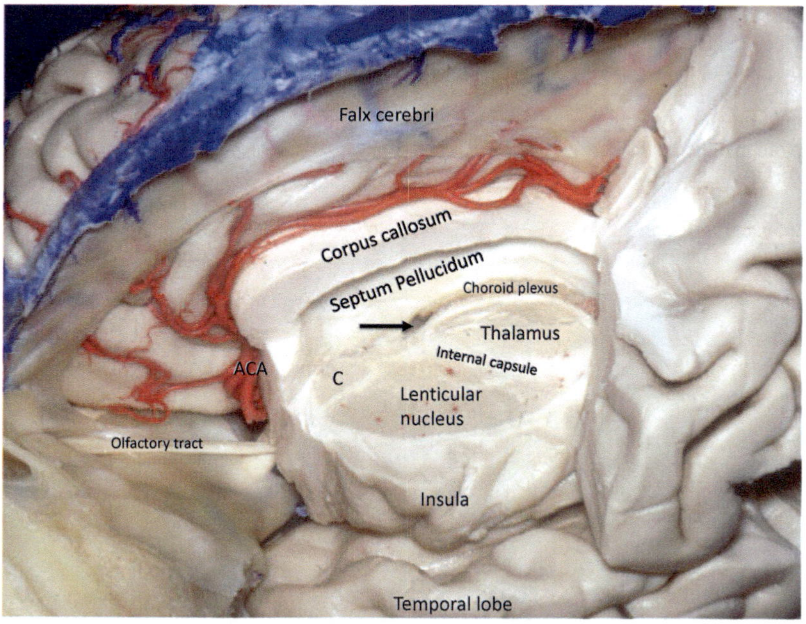

Fig. 8.3 Brain section (injected) noting the CC and related structures. The arrow marks the foramen of Monro. Also note the head of the caudate nucleus (C) and anterior cerebral arteries (ACA). (Courtesy Dr. Albert Rhoton)

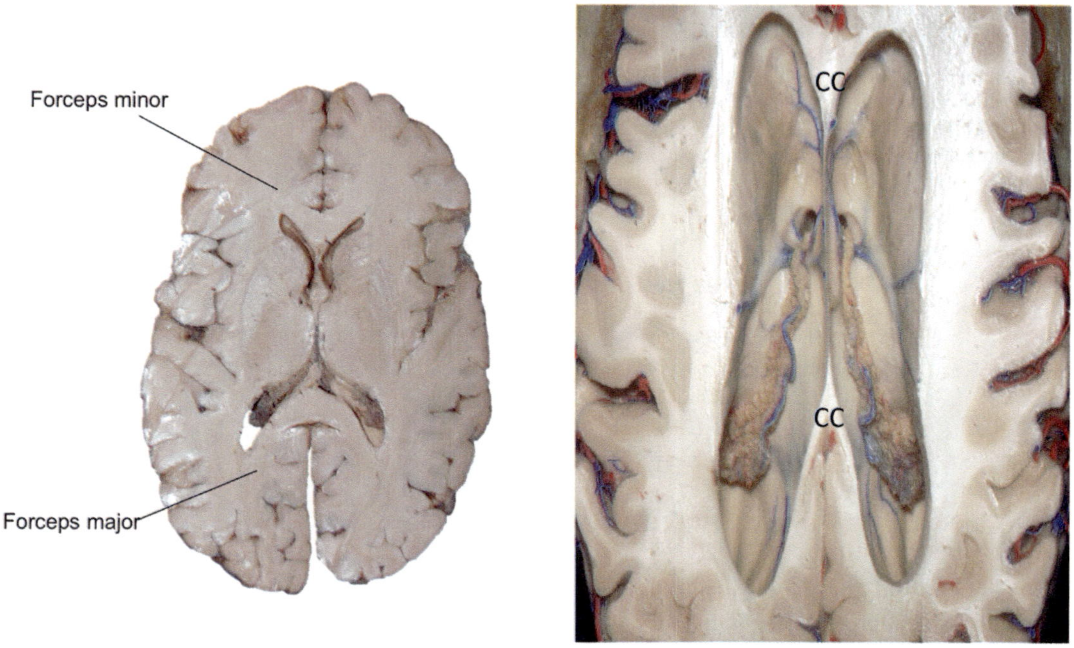

Fig. 8.4 Left, axial brain section noting the extensions of the CC, the forceps minor and forceps major. Right, axial section noting the CC related to the lateral ventricles. (Courtesy Dr. Albert Rhoton)

involving the posterior midbody, contains fibers of somatosensory projections. Region IV, the posterior third of the CC (excluding the final posterior one-fourth), is also composed of somatosensory fibers. The posterior most one-fourth of the CC is region V, which contains parietal, occipital, and temporal projections [8, 9].

The blood that supplies the CC arises from branches of the internal carotid artery. The anterior cerebral artery (ACA) (Fig. 8.8) gives rise

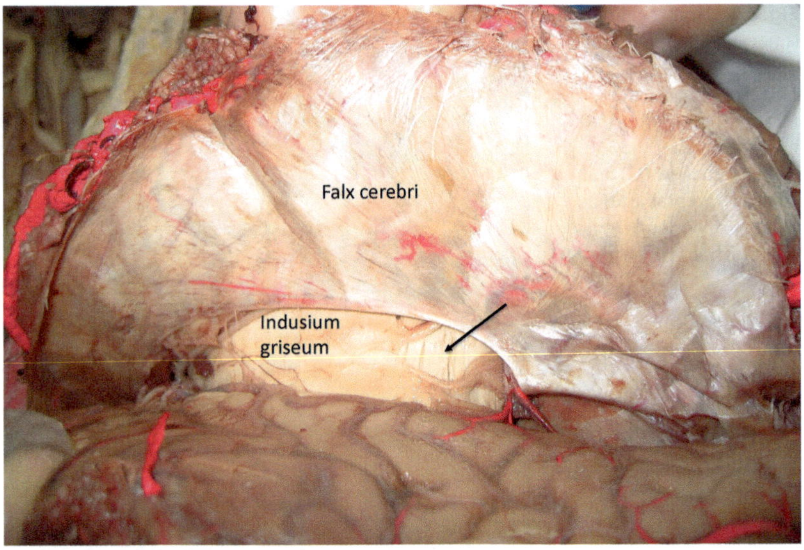

Fig. 8.5 Cadaveric view noting the relationship between the falx cerebri and underlying CC (arrow). Note that the dorsal surface of the CC is covered with the indusium griseum

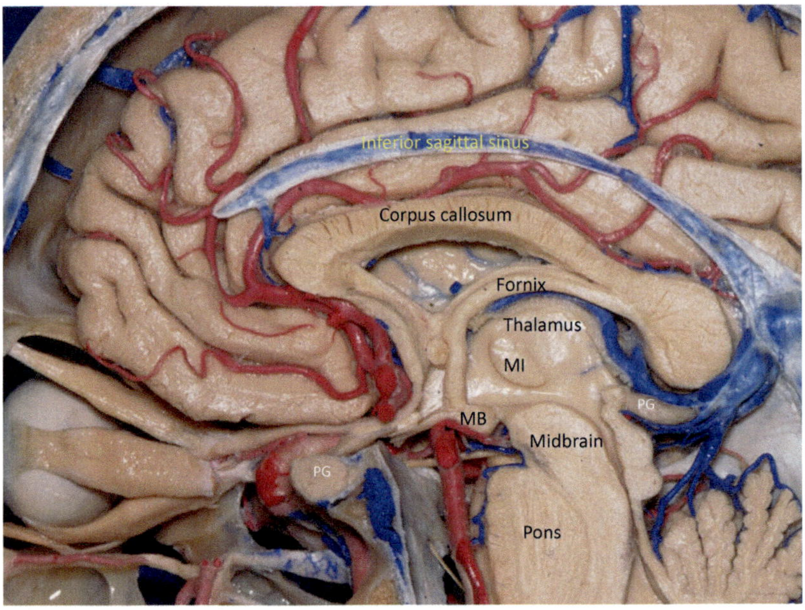

Fig. 8.6 Sagittal brain (injected) section noting the parts of the CC and related structures. Note the massa intermedia (MI), mammillary body (MB), and pituitary gland (PG). (Courtesy Dr. Albert Rhoton)

to the pericallosal artery typically from its third segment, A3, the pre-callosal segment, and runs to the post-callosal segment, or A5. A3 arises after the branch off of the callosomarginal artery [10, 11]. Occasionally, the pericallosal artery may refer to the ACA from just posterior to the origin of the anterior communicating artery, in A2. The pericallosal artery then extends to above the body of the CC, terminating posterior to the plane of the coronal suture. When the pericallosal artery arises from segment A2, the A2 portion is referred to as the proximal segment, A3 the middle, and segments A4 and A5 the distal segment [10–12]. However, the splenium's arterial supply is different than the rest of the CC, as it is supplied by the posterior pericallosal artery, which arises from the posterior cerebral artery [13].When operating near the CC, it is important to be aware of these arteries, in addition to the anterior communicating artery.

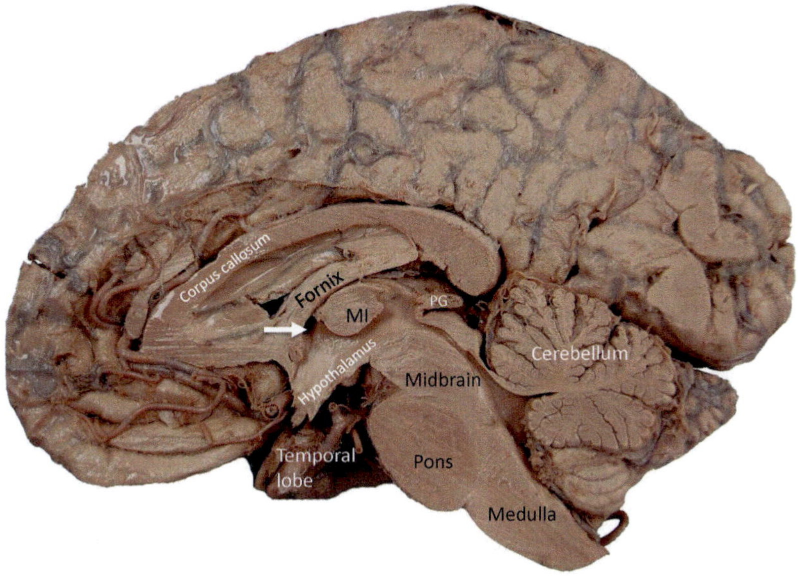

Fig. 8.7 Non-injected sagittal brain section noting the CC. Note the pineal gland (PG) and massa intermedia (MI). The right foramen of Monro is seen at the arrow

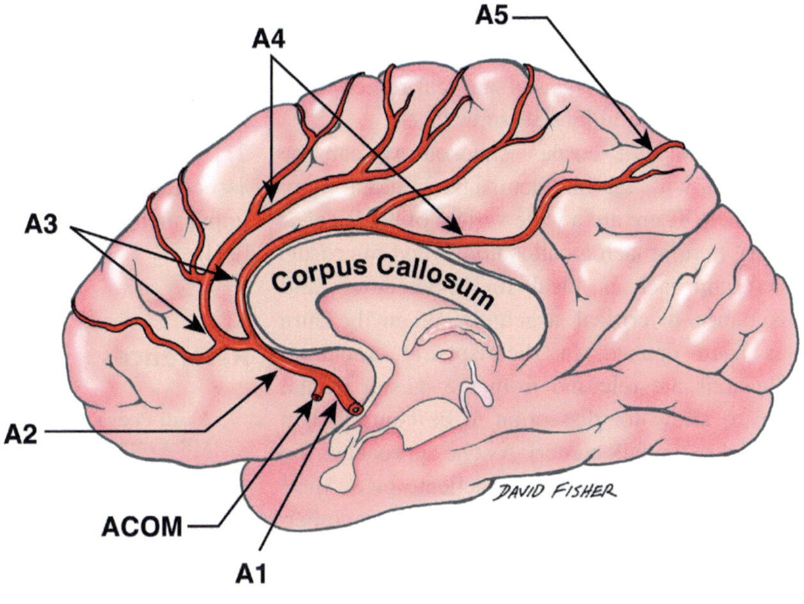

Fig. 8.8 Schematic drawing of the parts of the anterior cerebral artery (A1–A5) as they relate to the CC. Also note the position of the anterior communicating artery (ACOM)

There is a wide variety of anatomical variations in these arteries, so general awareness of their locations is highly important [9].

From a venous perspective, the CC is drained by pericallosal, callosal, and callosocingulate veins, which drain differently dependent upon the portion of the CC they affect. Some anterior callosal veins, responsible for the genu and the rostrum, drain into the anterior cerebral vein or directly into the anterior portion of the inferior sagittal sinus. The veins responsible for draining the central portion of the CC, the callosal and callosocingulate veins, travel inferiorly to join the internal cerebral veins, which then drain into the straight sinus. Posterior callosal veins pass medially around the splenium to empty into the vein of Galen or, in some variations, the straight sinus [14].

Within epilepsy, the CC serves to propagate seizure activity across hemispheres, expanding epileptic activity throughout the brain [15, 16].Callosotomy, either partial or total, has been found to aid in treatment of refractory epilepsy, particularly those affected by drop attacks [17]. The purpose of this procedure is to sever the connections between hemispheres, thus dampening the ability for widespread ictal propagation [1]. While partial callosotomies have been beneficial in some cases, greater extent of callosotomy in the pediatric population has been demonstrated in some cases to have a greater reduction in seizure activity [17]. Reasoning for a partial callosotomy, such as an anterior callosotomy, is to preserve some fibers responsible for sensorimotor integration, largely concentrated in the posterior fibers, as mentioned above. Partial callosotomies may range from the anterior one-half to the anterior 80% [1]. This is a palliative procedure for patients who may not be ideal candidates for a more focused surgical resection or unlocalized frontal lobe seizures [1, 9].

Practically, a corpus callosotomy can be achieved with an interhemispheric craniotomy approach. There are a wide variety of angles necessary to approach a callosotomy, and microsurgical approach is ideal for appreciation and avoidance of critical structures. When the dura has been opened, interhemispheric arachnoid adhesions are released from the mesial frontal lobe to allow for dissection to be continued in parallel to the falx cerebri. As one approaches the cingulate gyri, separation is delicate given their tendency toward adherence to one another. Separation allows for protection of the callosomarginal artery in the superior plane and the pericallosal artery inferior to the gyri. Once they have been sufficiently separated, the CC should be visible, bounded on lateral sides by the pericallosal arteries [9].

While the corpus callosotomy has demonstrable benefits in certain patient populations, it is not without risks. Anterior callosotomies have been attempted to allow for continued transfer of sensorimotor information but also to limit disconnection syndrome [1]. Disconnection syndrome is a wide-ranging term that encompasses syndromes associated with damage to the communicating fibers in the cerebrum but is importantly not associated with lesions to cortical brain matter [18]. Colloquially, the disconnection syndrome involving the CC is referred to as "split brain." Of note, the two regions continue to function independently as normal except in those areas where coordination and communication are necessary for function [19].

Diagnostic apraxia is one manifestation of a corpus callosal disconnection syndrome, associated with posterior truncal lesions. In addition, posterior truncal lesions are associated with crossed tactile anomia or the inability to recognize objects by touch as well as tactile disorientation [20]. Further posterior, lesions to the splenium of the CC have been associated with the inability to recognize written words in the affected field or alexia [21]. When both the splenium and the posterior truncus are involved, auditory extinction and crossed optic ataxia, the inability to reach for or point toward objects in the opposite visual field, have been noted [20, 21]. Anterior corpus callosal lesions have been associated with impaired prosody of speech or impairment of rhythm, intonation, and accent [21].

References

1. Asadi-Pooya AA, Sharan A, Nei M, Sperling MR. Corpus callosotomy. Epilep Behav. 2008;13:271–8. https://doi.org/10.1016/j.yebeh.2008.04.020.
2. Shah A, Jhawar S, Goel A, Goel A. Corpus callosum and its connections: a fiber dissection study. World Neurosurg. 2021;151:e1024–35. https://doi.org/10.1016/j.wneu.2021.05.047.
3. Goldstein A, Covington BP, Mahabadi N, Mesfin FB. Neuroanatomy, corpus callosum. StatPearls; 2022. https://www.ncbi.nlm.nih.gov/books/NBK448209/. Accessed 26 June 2022.
4. de Melo Mussi AC, da Luz de Oliveira EP. Ventricular anatomy. In: Comprehensive overview of modern surgical approaches to intrinsic brain tumors; 2019. p. 107–18. https://doi.org/10.1016/B978-0-12-811783-5.00005-7.
5. Sakai T, Mikami A, Suzuki J, et al. Developmental trajectory of the corpus callosum from infancy to the juvenile stage: comparative MRI between chimpanzees and humans. PLoS One. 2017;12(6):e0179624. https://doi.org/10.1371/JOURNAL.PONE.0179624.

6. Tzourio-Mazoyer N. Intra- and inter-hemispheric connectivity supporting hemispheric specialization. Res Perspect Neurosci. 2016;9783319277769:129–46. https://doi.org/10.1007/978-3-319-27777-6_9.
7. Musiek FE. Neuroanatomy, neurophysiology, and central auditory assessment. Part III: corpus callosum and efferent pathways. Ear Hear. 1986;7(6):349–58. https://doi.org/10.1097/00003446-198612000-00001.
8. Hofer S, Frahm J. Topography of the human corpus callosum revisited—comprehensive fiber tractography using diffusion tensor magnetic resonance imaging. NeuroImage. 2006;32(3):989–94. https://doi.org/10.1016/J.NEUROIMAGE.2006.05.044.
9. Haridas A, Sood S, Asano E, Rutka J, Cohen-Gadol A. Corpus callosotomy. In: Neurosurgical atlas. Neurosurgical Atlas, Inc.; 2020. https://doi.org/10.18791/nsatlas.v7.ch05.02.
10. Kakou M, Destrieux C, Velut S. Microanatomy of the pericallosal arterial complex. J Neurosurg. 2000;93(4):667–75. https://doi.org/10.3171/JNS.2000.93.4.0667.
11. Perlmutter D, Rhoton AL. Microsurgical anatomy of the distal anterior cerebral artery. J Neurosurg. 1978;49(2):204–28. https://doi.org/10.3171/JNS.1978.49.2.0204.
12. Lehecka M, Dashti R, Hernesniemi J, et al. Microneurosurgical management of aneurysms at the A2 segment of anterior cerebral artery (proximal pericallosal artery) and its frontobasal branches. Surg Neurol. 2008;70(3):232–46. https://doi.org/10.1016/J.SURNEU.2008.03.008.
13. Sparr SA, Bieri PL. Infarction of the splenium of the corpus callosum in the age of COVID-19: a snapshot in time. Stroke. 2020;51(9):e223–6. https://doi.org/10.1161/STROKEAHA.120.030434.
14. Wolfram-Gabel R, Maillot C. The venous vascularization of the corpus callosum in man. Surg Radiol Anat. 1992;14(1):17–21. https://doi.org/10.1007/BF01628038.
15. Aboitiz F, Montiel J. One hundred million years of interhemispheric communication: the history of the corpus callosum. Braz J Med Biol Res. 2003;36(4):409–20.
16. Küçükyürük B, Uzan M, Avyasov R, Tahmazo Glu B, Sanus GZ, Tanrıö Ver N. Evaluation of ideal extent of corpus callosotomy based on the location of intracallosal motor fibers. World Neurosurg. 2020;144:e568–75. https://doi.org/10.1016/j.wneu.2020.09.006.
17. Graham †david, Tisdall M, Gill †deepak. Corpus callosotomy outcomes in pediatric patients: a systematic review. Epilepsia. 2016;57(7):1053–68. https://doi.org/10.1111/epi.13408.
18. Kaufman DM, Geyer H, Milstein MJ, Rosengard J. Kaufman's clinical neurology for psychiatrists. 9th ed. Philadelphia: Elsevier; 2023.
19. Catani M, Mesulam M. What is a disconnection syndrome? Cortex. 2008;44(8):911–3. https://doi.org/10.1016/J.CORTEX.2008.05.001.
20. Shozawa H, Futamura A, Saito Y, et al. Diagonistic apraxia: a unique case of corpus callosal disconnection syndrome and Neuromyelitis optica spectrum disorder. Front Neurol. 2018;9:653. https://doi.org/10.3389/FNEUR.2018.00653/BIBTEX.
21. Klouda GV, Robin DA, Graff-Radford NR, Cooper WE. The role of callosal connections in speech prosody. Brain Lang. 1988;35(1):154–71. https://doi.org/10.1016/0093-934X(88)90106-X.

9 Structural Connectivity of the Corpus Callosum to Other Cortical Regions

Isabella G. McCormack and R. Shane Tubbs

9.1 Introduction

The term *corpus callosum* (CC) is derived from the Latin *corpus* meaning "body" and *callosum* meaning "tough." Composed of approximately 200 million axonal fibers, the CC is the largest commissure in the brain [1]. Due to its broad connectivity throughout the brain, the CC is implicated in a wide array of pathologies including congenital malformations, neurodegenerative disorders, and epileptic disorders, among many others [2, 3]. From anterior to posterior, the CC is divided into five sections: the rostrum, genu, body, isthmus, and splenium. The anatomical connections (Fig. 9.1) formed through these various sections have long been a subject of scientific study but have garnered increasingly more interest as neurosurgical procedures involving the CC have become more common. In recent times, Gamma Knife radiosurgery is being explored as an option for treating arteriovenous malformations of the CC as well as a potential alternative to microsurgical callosotomy in cases of medically refractory epilepsy [4–7]. For cases of glioblastomas involving the CC, including the so-called butterfly glioblastomas, recent studies are reevaluating the feasibility and advantages of gross total resection over traditional more conservative approaches [8–10]. Research utilizing newer technology such as diffusion tensor imaging has found significant alterations in the microstructure of the CC in Huntington's disease, lower limb amputation, and borderline personality disorder [11–13].

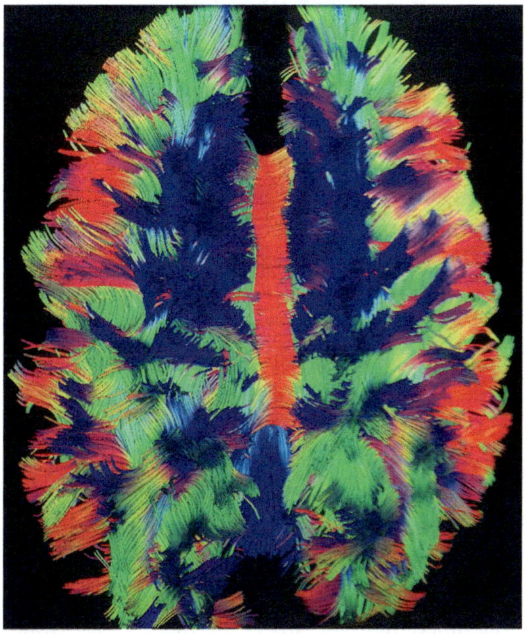

Fig. 9.1 Axial tractography of the brain. Note the dorsal aspect of the CC (orange band at center of the image)

The structural connectivity of the CC with other cortical regions has been extensively stud-

I. G. McCormack (✉) · R. S. Tubbs
Department of Neurosurgery, Tulane University School of Medicine, New Orleans, LA, USA
e-mail: imccormack@tulane.edu

M. Turgut et al. (eds.), *The Corpus Callosum*, https://doi.org/10.1007/978-3-031-38114-0_9

ied throughout history, but modern advancements in technology and surgical technique have sparked new interest in understanding the anatomy of the CC and its role in the disease. In this chapter, we will explore the historical scientific discourse regarding the structural connectivity of the CC, the most current research regarding its anatomy, and the clinical implications that have arisen from discoveries regarding its microstructure.

9.2 Historical Scientific Discourse Regarding the Structural Connectivity of the CC

Once regarded as the seat of the soul by the French surgeon La Peyronie (1678–1747) and Italian anatomist Lancisi (1654–1720), the CC has long been a subject of neuroanatomical interest [1]. However, it was not until the German physicians Franz Gall and Johann Spurzheim began experiments on the CC that its structural connectivity would be considered. Spurzheim wrote in *The Anatomy of the Brain* (1826), "Until Dr. Gall and I published, it was the custom to take merely mechanical views of these [cerebral commissures], without attempting to discover their relations with the other cerebral parts, their derivations, or the causes of their dissimilarity in different animals... [14]." Through their work, they concluded that the fibers of the CC are topographically organized, that fibers originating from the lower parts of the anterior brain form the genu of the CC, fibers from the upper parts of the brain form the body, and fibers from the lower parts of the posterior brain form the splenium.

In 1886, Hamilton asserted that the CC is not a commissure comprised of transcortical association fibers but is rather comprised of decussating descending fiber tracts that originate in the cortex, course through the internal capsule, and terminate in the contralateral thalamus and caudate nucleus [15]. This was contested by Bruce (1889) who analyzed a brain in which there was complete agenesis of the CC and found that the internal capsule was present bilaterally without evidence of degeneration [16]. Bruce further argued that studies of brain development also contradict Hamilton's conclusions, as in the human infant at 3 months the CC is non-myelinated, while the internal capsules are almost completely myelinated. Mott and Schaefer (1890) conducted experiments on monkeys in which they removed one hemisphere of the brain and then electrically stimulated the cut surface of the CC. They found that stimulation of CC fibers causes motor activity originating from the intact hemisphere in a general topographical manner – fibers in the anterior CC coordinate movement of the eyes and head and fibers in the posterior CC coordinate movement in the lower limbs. They concluded that the fibers of the CC are transcortical and terminate in several motor centers of the contralateral hemisphere rather than descend into the internal capsule [17]. This would be further contested until Mettler (1935) and Combs (1949) concluded that the fibers of the CC do not descend in the internal capsule [18, 19].

Further research conducted by numerous scientists such as Sherrington (1889), Cajal (1911), Van Valkenburg (1913), and Milch (1932) focused on the fiber connections made between the hemispheres via the CC and found that the fibers of the CC connect homologous and heterologous brain foci [20–22]. That is, the fibers of the CC connect a brain focus with its contralateral homologous focus as well as various contralateral association regions related to that focus. In 1940, Curtis's experiment using the oscillographic method in monkeys corroborated this hypothesis, though he noted his results showed no connection between area 17 in both hemispheres [23]. He postulated that there could be special connections between area 17 bilaterally that current methods could not detect that would have implications for understanding binocular vision.

Krieg (1954) conducted experiments specifically searching for evidence of non-heterologous connections through the CC but found none, while Pandya, Hallett, and Mukherjee (1969) found through their research that there are indeed non-homotopic connections as the caudal primary auditory cortex projects to ipsilateral association areas as well as the contralateral primary auditory cortex, frontal and parietal lobes, and cingulate gyrus [24, 25]. In 1963, Powell's study in rats showed that fibers from the cingulate gyrus cross the CC and terminate in the septal nuclei [26]. Further studies by Locke and Yakovlev (1965) in human brains found that there are two components in the CC, corticocortical fibers and transcallosal projection fibers from the cingulate gyrus. Their work showed that the fibers from the cingulum form dorsal and ventral condensations in the CC, and in between, these condensations lie the corticocortical commissural fibers [27].

9.3 Current Evidence Regarding the Structural Connectivity of the CC

In modern times, more advanced imaging modalities such as diffusion tensor imaging (DTI) and fiber tractography have clarified our understanding of the microstructure of the CC. In 1989, Witelson's study on nonhuman primate brains established the current widely used segmentation scheme for the CC, dividing the CC into the anterior one-third, the anterior and posterior midbody, the posterior one-third, and the posterior one-fifth [28].

These segments were explored by Hofer and Frahm (2006) using DTI in human brains, and they found several important differences from the Witelson's scheme in their human subjects [29]. Based on their tractography data, they concluded that the CC should be segmented into five regions as follows: Region 1 is the anterior one-sixth of the CC and contains fibers coursing between the prefrontal cortexes within the forceps minor, Region 2 is the remaining anterior one-half of the CC and contains fibers coursing between the premotor and supplementary motor cortices, Region 3 is the posterior one-half minus the posterior one-third of the CC and contains fibers coursing between the primary motor cortices, Region 4 is the posterior one-third minus the posterior one-fourth of the CC and contains fibers coursing between the primary somatosensory cortices, and Region 5 is the posterior one-fourth of the CC and contains fibers coursing between the parietal, temporal, and occipital cortices within the forceps major.

Measurements of fractional anisotropy (FA), which increases when fluid moves more uniformly in a single direction, indicated higher FA in Region 1 and 5 and lower FA in Region 3 and 4. The authors noted that this could potentially be explained by the microstructural differences in fiber diameter and myelination that are present within the CC, as Regions 3 and 4 are comprised of larger-diameter myelinated fibers, which allow more space for fluid movement within the individual axons and thus less tightly oriented fluid movements along a specific direction. Regions 1 and 5 have smaller fibers with light myelination, allowing for more uniform fluid movement along the axons and thus higher FA values. Notably, they found that Region 1 has lower FA than Region 5, which the authors propose could be due to the presence of more obliquely oriented transcallosal fiber tracts in the genu that connect heterotopic foci within the brain.

Studies using functional magnetic resonance imaging (fMRI) have explored the functional connectivity of the CC by analyzing the blood-oxygen-level-dependent (BOLD) signal in the CC in response to various stimuli. A study conducted by Fabri et al. (2011) found that gustation and tactile stimuli to the tongue evoke activation foci in the genu and anterior body of the CC, with sweet and bitter tastes also activating the posterior CC [30]. The anterior and posterior body showed activation foci in response to hand motor activity, and tactile stimuli to various body regions evoked activation in the isthmus. Their research also indicated that proximal body representations are carried through fibers in the posterior isthmus and anterior splenium. Tactile stimuli to the hand and foot evoked activation of the same regions of the CC that were activated with hand motor activity in the anterior and posterior body as well as a posterior focus near the splenium. Interestingly, these data were obtained in the absence of tasks requiring interhemispheric transfer, indicating that all information reaching the cortex is shared with the corresponding contralateral cortex to maintain an accurate global representation of the environment at all times.

Several studies examining BOLD signal activation in the CC during interhemispheric transfer tasks found that during the Poffenberger paradigm, in which light is flashed in one visual hemifield and the subject must either respond with the ipsilateral or contralateral hand, activation foci are elicited in the genu [31–33]. Following these results, another interhemispheric transfer task, the Sperry paradigm, began being employed in studies of CC connectivity. In the Sperry paradigm, crossed conditions are created when the subject is presented with a word in the left visual hemifield or a face in the right hemifield, as the left hemisphere is relatively specialized for word recognition and the right hemisphere is relatively specialized for facial recognition [34]. Using the Sperry paradigm, researchers consistently found activation of the CC in the posterior CC near the splenium [35–37].

Baird et al. (2005) sought to correlate the structural and functional connectivity of the CC using DTI combined with BOLD fMRI and found that faster reaction time on a task was associated with a lower BOLD signal in the cortical regions responsible for initiating the task and a higher FA in the fibers of the splenium, while slower reaction time was associated with a higher BOLD signal in the cortical regions of interest and higher FA in the fibers of the genu [38]. Recent work from Wang et al. (2021) has sought to explore the correlation between structure and function through independently mapping and analyzing the correspondence between the fiber tractography and fMRI data from the CC [39]. Their work showed there is a high degree of correspondence between the structural and functional pathways in the CC, indicating that neural function generally follows anatomical scaffolding, and raises important questions for future work. Further research combining DTI and BOLD fMRI is needed to further understand the relationships between structural and functional connectivity in the CC in health and disease.

9.4 Clinical Implications of the Structural Connectivity of the CC

Owing to its central role in brain processing, the CC is implicated in an extensive array of pathologies. Studies of the structural connectivity of the CC in disease have primarily used DTI and fiber tractography to analyze structural abnormalities. Accumulating evidence using DTI in patients with autism spectrum disorder (ASD) has shown that there is a reduction in the total volume of the CC in ASD patients, particularly in the anterior fibers in the genu, which is in line with previous research showing reduced frontal lobe volume in ASD [40–42]. Work done by Sadek et al. (2021) in female patients with borderline personality disorder (BPD) has also shown decreased FA in the genu of the CC, which was correlated with increased impulsivity, in line with decreased functional connectivity between the prefrontal cortices [13]. They also found increased mean diffusivity (MD) through the body of the CC which was positively correlated with motor impulsiveness, and they concluded that such structural alterations of the CC may underlie the emotional impulsivity and dysregulation observed in BPD.

Regarding neurodegenerative diseases, work in Huntington's disease (HD) patients has also shown distinct topographical alterations in the CC. Rosas et al. (2010) found significant reductions in FA in Region 2 of the CC in patients who were more than a decade before the expected onset of their HD [12]. Meanwhile, in patients who were closer to the onset or already experiencing HD, there was significant global reduction in FA in all regions of the CC. They also found increased radial diffusivity (RD), an index of myelin dysfunction, in Regions 2 and 5 of the CC in patients who were far from the onset of HD, and global increases in RD in all regions in patients who were closer to or already experiencing HD. Longitudinal research on CC atrophy in Alzheimer's disease (AD) has revealed that compared to normal aging patients, AD patients exhibit enhanced loss of fibers in the rostrum and splenium [3].

This correlates with the pattern of cortical atrophy seen in AD, namely, the frontal association cortices that supply fibers through Region 1 and pronounced atrophy of temporal and parietal association cortices that supply fibers through Region 5. Research in patients with relapsing-remitting multiple sclerosis (RR-MS) has also shown distinct regional structural abnormalities in the CC early in the disease course, as described by Zito et al. (2014). Using MRI and electroencephalographic recordings to correlate the structure of the CC with functional interhemispheric electrical activity, they found that patients with RR-MS who exhibit worse performance on an isometric handgrip task showed enhanced atrophy of the anterior midbody of the CC [43].

Numerous epilepsy studies have also found structural alterations in the CC correlated with disease severity and progression. Cao et al. (2017) found decreased FA in the region of the CC that connects bilateral primary sensorimotor cortices in patients with benign epilepsy with centrotemporal spikes (BECTS) with bilateral spiking patterns compared to BECTS patients with unilateral spikes [44]. The authors suggest this could be due to a more detrimental effect on the CC induced by the bilateral seizure activity compared to unilateral. O'Muircheartaigh et al. (2011) utilized DTI to study the CC in patients with juvenile myoclonic epilepsy (JME) and found reduced FA in the rostrum of the CC corresponding with fibers coursing to the supplementary motor area as well as in the splenium corresponding with fibers projecting to mesial temporal lobe and posterior parietal regions [45]. These results have implications for our understanding of idiopathic generalized epilepsies such as JME, which were canonically believed to lack discrete epileptogenic foci. Further work corroborated these findings and revealed a correlation between the reduction in FA in the genu and increased generalized tonic-clonic seizure activity in JME [46]. Future studies of the CC in various forms of epilepsy could allow for improved diagnosis, treatment, and overall quality of life for patients.

9.5 Conclusion

The structural connectivity of the CC has been the focus of scientific study since the early nineteenth century, and our understanding has become more refined as advanced imaging modalities have become available. Today, DTI tractography analysis and fMRI have allowed researchers to explore the structural and functional connections formed through the fibers of the CC, and current research is increasingly discovering the regional microstructural alterations that occur within the CC during the disease. There are myriad conditions in which the CC microanatomy has been implicated, and future studies will continue to define the correlation between CC structure and function, allowing for improved medical and surgical interventions for a wide host of pathologies.

References

1. Tomasch J. Size, distribution, and number of fibres in the human corpus callosum. Anat Rec. 1954;119:119–35. https://doi.org/10.1002/ar.1091190109.
2. Krupa K, Bekiesinska-Figatowska M. Congenital and acquired abnormalities of the corpus callosum: a pictorial essay. Biomed Res Int. 2013;2013:265619. https://doi.org/10.1155/2013/265619.
3. Teipel SJ, Bayer W, Alexander GE, Zebuhr Y, Teichberg D, Kulic L, Schapiro MB, Möller H-J, Rapoport SI, Hampel H. Progression of corpus callosum atrophy in Alzheimer disease. Arch Neurol. 2002;59:243–8. https://doi.org/10.1001/archneur.59.2.243.
4. Bodaghabadi M, Bitaraf MA, Aran S, Alikhani M, Ashrafian H, Zahiri A, Alahverdi M. Corpus callosotomy with gamma knife radiosurgery for a case of intractable generalised epilepsy. Epileptic Disord. 2011;13:202–8. https://doi.org/10.1684/epd.2011.0436.
5. Maruyama K, Shin M, Tago M, Kurita H, Kawamoto S, Morita A, Kirino T. Gamma knife surgery for arteriovenous malformations involving the corpus callosum. J Neurosurg. 2005;102:49–52. https://doi.org/10.3171/sup.2005.102.s_supplement.0049.
6. Mohanty CB, Devi BI, Somanna S, Bhat DI, Dawn R. Corpus callosum arteriovenous malformation with persistent trigeminal artery. Br J Neurosurg. 2011;25:736–40. https://doi.org/10.3109/02688697.2011.554583.
7. Sachdev S, Sita TL, Shlobin NA, Gopalakrishnan M, Sucholeiki R, Régis J, Bandt SK. Completion corpus callosotomy with stereotactic radiosurgery for

Neuroimaging Techniques for Investigation of the Corpus Callosum

10

Pınar Çeltikçi and Ahmet Tuncay Turgut

10.1 Introduction

The corpus callosum (CC) is the human brain's largest commissural structure, consisting of densely packed white matter tracts connecting homologous regions of the two cerebral hemispheres. It is divided into four main parts from anterior to posterior: rostrum, genu, body, and splenium (Fig. 10.1). The narrowed portion between the body and the splenium is termed the isthmus [1]. Because it is compact, the CC creates a relative barrier to interstitial edema and the spread of tumors, which can be helpful for identifying specific aggressive pathologies such as glioblastoma. On the other hand, tightly bundled white matter tracts can be susceptible to traumatic injury [2]. In this chapter, imaging modalities and techniques for investigating the CC are discussed, followed by a brief overview of the imaging features of the normal structure and its pathologies.

P. Çeltikçi (✉)
Department of Radiology, Ankara Bilkent City Hospital, Ankara, Turkey

A. T. Turgut
Department of Radiology, Ankara Medipol University Faculty of Medicine, Ankara, Turkey

M. Turgut et al. (eds.), *The Corpus Callosum*, https://doi.org/10.1007/978-3-031-38114-0_10

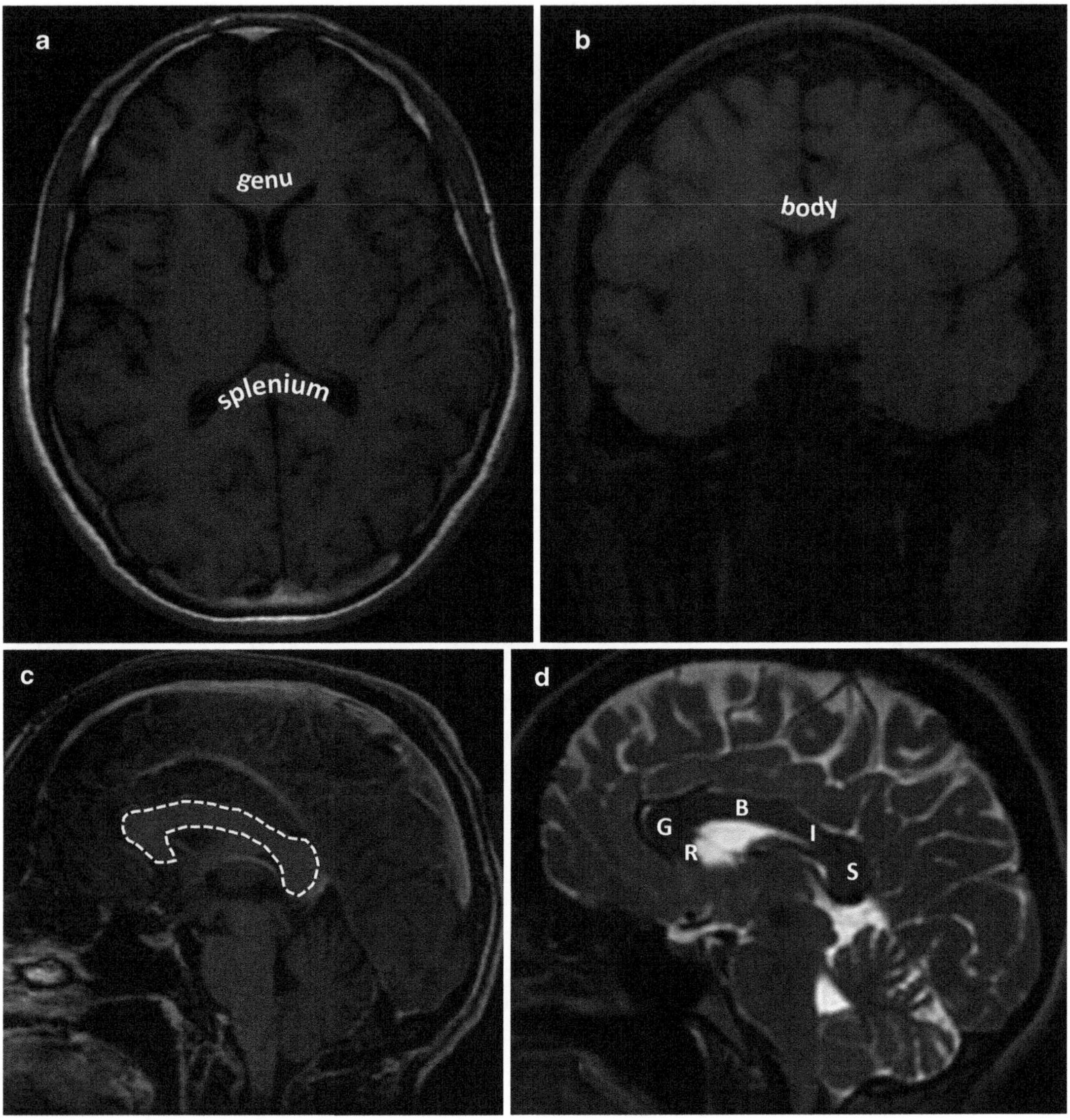

Fig. 10.1 Normal anatomy of the CC (marked with dotted line in **c**). Axial (**a**), coronal (**b**), and sagittal (**c**) T1-weighted images showing the genu, body, and splenium of the CC. Sagittal T2-weighted image (**d**) shows the rostrum (R), genu (G), body (B), isthmus (I), and splenium (S) segments

10.2 Imaging Modalities

10.2.1 Ultrasonography

Ultrasonography (US) is a noninvasive imaging modality that uses high-frequency sound waves to characterize tissues on the basis of their acoustical properties. US requires no ionizing radiation; it is safe, readily available for bedside evaluation, inexpensive, and repeatable. Best imaging results are achieved with soft tissue rather than with bones or air-containing organs (e.g., lungs, the lumen of the gastrointestinal system) owing to acoustic impedance differences [3]. This limits the value of US for examining brain parenchyma encased in calvarial bones and the CC alongside it. However, the neonatal skull with its open fontanelles allows the brain of the neonate to be evaluated by US [4]. The CC can be examined through both the anterior and posterior fontanelles. Congenital pathologies such as CC agenesis/dysgenesis and pericallosal lipoma, hemorrhage, or hypoxic-ischemic encephalopathy affecting the CC are among the lesions that can be diagnosed by US.

10.2.2 Computed Tomography

Computed tomography (CT) uses ionizing radiation to reconstruct cross-sectional images of the body part of interest based on the degree of attenuation of X-rays. Basically, CT scanners comprise one (or more) rotating X-ray tube(s) corresponding to a row of detectors positioned in a circular gantry. The X-ray tube rotates around the patient, while the detectors, placed opposite, measure the attenuation of the imaged body part at different angles. The acquired data are then processed by various algorithms into cross-sectional images. Bone, air, fat, and water can be identified on CT images, their different densities being quantified in Hounsfield units (HU). Intravenous iodinated radiocontrast material can be administered during scanning to obtain angiograms of the cerebral vessels or to evaluate tumors in cases where magnetic resonance imaging (MRI) is contraindicated. CT is a fast and efficient neuroimaging technique, especially for trauma patients. Although soft-tissue contrast resolution is inferior to that from MRI, it is reliable for detecting hemorrhage, edema, mass effect, calcifications, and bone-related pathologies [5]. These attributes can help in examining the CC for tumors, vascular lesions with calcification and/or hemorrhage, and traumatic lesions (e.g., hematomas, hemorrhagic diffuse axonal injuries).

10.2.3 Magnetic Resonance Imaging

MRI requires no ionizing radiation and relies on the differing responses of tissues to magnetic fields to obtain cross-sectional images. It provides superior soft-tissue contrast resolution in multiple imaging planes; therefore, it is the ideal modality for neuroimaging. Nevertheless, MRI is less sensitive than CT for detecting subarachnoid hemorrhages, calcification, and some bone-related pathologies. Also, it is contraindicated in patients with a cardiac implantable electronic device, non-MRI-compatible metallic implants or similar devices, and metallic fragments (e.g., bullets). MRI-compatible metallic materials such as dental braces or spinal fixation hardware would cause artifacts resulting in suboptimal image quality. Claustrophobia, inability to be motionless, and morbid obesity are also limiting factors for this modality [6].

Various MRI pulse sequences are developed for the further identification of tissues. Conventional MRI sequences for brain imaging usually comprise multiple plane (axial, coronal, and sagittal) T1-weighted, T2-weighted spin-echo (SE) images, which are developed to reflect various tissue relaxation effects. T1-weighted images (T1WI) are used to evaluate the overall anatomy and contrast enhancement following administration of gadolinium contrast agent. Water has a low signal and appears darker in T1WI, while fat, melanin, methemoglobin (in subacute hematoma), protein-rich fluids, and minerals such as manganese appear bright. Water and fat both have high signals in T2-weighted images (T2WI) and appear bright. Fat-suppression techniques can be applied to both T1WI and T2WI, causing fat tissue to appear darker. Fat-suppressed T1WI images are better for contrast enhancement, while fat-suppressed T2WI images are more sensitive to edema. Fluid-attenuated inversion recovery (FLAIR) sequence is indispensable in brain imaging for achieving T2WI with fluid suppression, where water (cerebrospinal fluid) appears dark. FLAIR sequence improves the visualization of T2 bright lesions of the brain parenchyma such as edema, tumors, ischemic-gliotic foci, and demyelinated plaques of multiple sclerosis. Gradient-echo images can be incorporated to create susceptibility-weighted imaging (SWI) to reveal substances that cause magnetic field inhomogeneities such as blood products, calcium, and other mineral depositions. Diffusion-weighed imaging (DWI) is essential for diagnosing cytotoxic edema and is evaluated alongside apparent diffusion coefficient (ADC) maps to determine diffusion restriction. Diffusion restriction is commonly secondary to ischemic stroke or cytotoxic lesions of the CC (CLOCCs).

Advanced neuroimaging techniques such as MR spectroscopy (MRS), perfusion MRI, tractography, and functional MRI should be considered when the above routine techniques fail to identify the pathology or for presurgical planning. MRS provides data regarding the presence and concentration of various metabolites in the area of interest. Depending on the lesion size, homogeneity, and location, this technique can be applied on single voxel or multi-voxel. A spectrum of metabolites is acquired for evaluation. MRS is most commonly used to identify tumors, leukodystrophies, demyelination, and infections affecting the CC.

Perfusion MRI can be acquired using three different techniques: arterial spin labeling (ASL), dynamic susceptibility contrast (DSC) perfusion MRI, and dynamic contrast-enhanced (DCE) perfusion MRI. ASL does not require the administration of gadolinium contrast medium and allows cerebral blood flow (CBF) to be assessed. DSC perfusion MRI relies on the decrease in signal caused by intravascular gadolinium on T2WI over a period of time. Multiple perfusion parameters such as relative cerebral blood flow (rCBV), relative cerebral blood volume (rCBF), mean transit time (MTT), and time to peak (TTP) can be analyzed on color maps. DCE perfusion MRI exploits the signal increase in blood and tissue due to gadolinium on T1WI during the image acquisition interval. Capillary permeability parameters such as the transfer constant (K^{trans}), fractional plasma volume (v_p), and fractional volume of the tissue extracellular space (v_e) can be determined. Perfusion MRI is useful for differentiating areas of penumbra (at risk for stroke) from areas already infarcted. Also, it is commonly employed for predicting tumor grade or providing a differential diagnosis.

Tractography is a DWI MRI technique that uses anisotropic diffusion of water molecules along white matter tracts to estimate axonal organization. Qualitative evaluation of white matter tracts is possible on color maps in which different directions are coded with different colors and three-dimensional tractography images (Fig. 10.2). Quantification is also provided through the analysis of diffusion parameters such as fractional anisotropy, mean diffusivity, axial diffusivity, and radial diffusivity.

Fig. 10.2 Magnetic resonance tractography image reveals the three-dimensional structure of the CC

The aim of functional MRI is to locate cortical brain areas corresponding to specific tasks such as motor function and speech. The activity of a particular brain region increases oxygen and glucose consumption, which changes regional blood flow. The blood-oxygen-level-dependent (BOLD) contrast method is the standard technique for functional MRI, relying on local alterations in the deoxyhemoglobin/oxyhemoglobin ratio triggered by a task, as each metabolite has different signal properties on T2WI. In current clinical practice, tractography and functional MRI are tools for the presurgical planning of lesions in eloquent brain areas. These techniques are also used widely in neuroscience research.

In summary, a first-line brain MRI study usually includes T1WI, T2WI, and FLAIR sequence images, at least one sequence acquired in all three planes, accompanied by DWI and SWI. Contrast administration or further advanced MRI techniques can be considered following the interpretation of these images. Protocols are adjusted depending on the preliminary diagnosis, patient factors, or the radiologist's preference.

10.3 Normal Imaging Properties and Lesions of the Corpus Callosum

10.3.1 Normal Imaging Properties

The CC is a white matter structure; therefore, it has similar echogenicity, density, and intensity on US, CT, and MRI images. The intensity of the CC is higher on T1WI and lower on T2WI than the gray matter (Fig. 10.1). There should be no diffusion restriction or altered perfusion.

10.3.2 Lesions

Lesions of the CC are summarized in Table 10.1. The mnemonic VINDICATE was proposed to memorize the wide variety of pathologies: vascular, infection, neoplastic, demyelinating, idiopathic, congenital or genetic, autoimmune, traumatic or toxic, and embolic [7] (Figs. 10.3, 10.4, and 10.5).

Table 10.1 Lesions of the CC (mnemonic: VINDICATE)

Vascular	Infarction
	Arteriovenous malformation
	Cerebral cavernous malformation (cavernoma)
	Aneurysm
	Gliosis
	Hypoxic-ischemic encephalopathy
	Enlarged perivascular spaces
Infection	Abscess
	HIV encephalitis
	Progressive multifocal leukoencephalopathy
	Tuberculosis
	Aspergillosis
	Ventriculitis
	Subacute sclerosing panencephalitis
Neoplastic	Glioblastoma (Fig. 10.3)
	Primary CNS lymphoma (Fig. 10.4)
	Other glial tumors
	Metastasis
	Meningioma
Demyelinating	Multiple sclerosis (Fig. 10.5)
	Acute disseminated encephalomyelitis (ADEM)
	Neuromyelitis optica
	Wallerian degeneration
	Marchiafava-Bignami disease
Idiopathic	
	Cytotoxic lesions of the CC (CLOCCs) (former transient splenial lesion)
	Post-shunt decompression
	Posterior reversible encephalopathy syndrome (PRES)
	CC impingement syndrome
Congenital or genetic	CC agenesis/dysgenesis
	Lipoma
	Leukodystrophies (Krabbe disease, adrenoleukodystrophy, metachromatic leukodystrophy, mucopolysaccharidosis
Autoimmune	Susac syndrome
Traumatic or toxic	Diffuse axonal injury
	Traumatic contusion
	Toxic leukoencephalopathy
Emboli	Fat emboli

11 Prenatal Diagnosis of Anomalies of the Corpus Callosum with Three-Dimensional Ultrasound, Transvaginal Sonography, and Fetal MRI

Fedi Ercan, Ali Çağlar Turgut, and Funda Köylüoğlu

11.1 Introduction

Morphological evaluation of the cerebral anatomy provides detection of midline defects such as neural tube defects and to suspect agenesis of the corpus callosum (CC). The axial transthalamic view is the pathognomonic image in which the fetal head circumference (HC) and biparietal diameter (BPD) are measured. On this view, evaluation of the cavum septum pellucidum (CSP), the thalami, and the cerebral hemispheres is also possible without any difficulty (Fig. 11.1). Some midline malformations should be considered in the absence of CSP. The CC is a midline structure, developmental disorders of which can be evaluated prenatally.

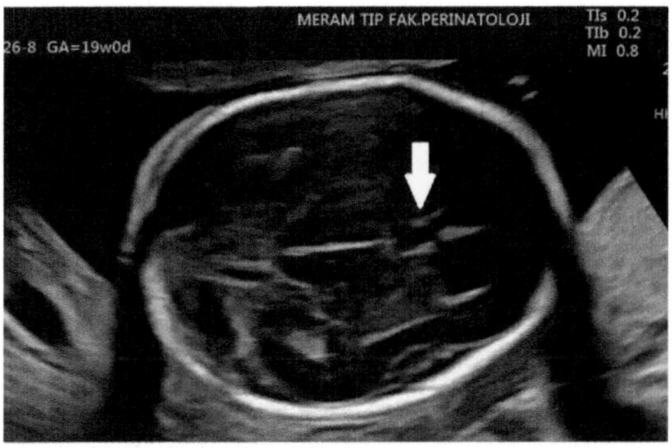

Fig. 11.1 Image of the fetal head in the transthalamic section during the intrauterine period. The arrow points to the cavum septum pellucidum

The axial transcerebellar view is obtained lower than the transventricular section and with a slight posterior slope. This section allows to evaluate the cerebellum (vermis) and cisterna magna in the posterior fossa (Fig. 11.2). The landmark points for measuring the cisterna magna are the distance between the posterior edge of the cerebellar vermis and the inner edge of the occipital bone. As a general rule, the width of this forehead should be between 2 and 10 mm.

CC is best viewed in the midsagittal plane and therefore CC pathologies are best understood in this plane. However, it is not always easy to obtain the midsagittal plane during fetal examination. On the other hand, axial sections provide important information.

F. Ercan (✉)
Department of Obstetrics and Gynecology, Division of Perinatology, Aydin Adnan Menderes University School of Medicine, Efeler, Aydın, Turkey

A. C. Turgut
Department of Radiology, Ege University School of Medicine, İzmir, Turkey

Department of Histology and Embryology, Aydın Adnan Menderes University Health Sciences Institute, Efeler, Aydın, Turkey

F. Köylüoğlu
Department of Obstetrics and Gynecology, Division of Gynecologic Oncology, Aydin Maternity Hospital, Efeler, Aydın, Turkey

M. Turgut et al. (eds.), *The Corpus Callosum*, https://doi.org/10.1007/978-3-031-38114-0_11

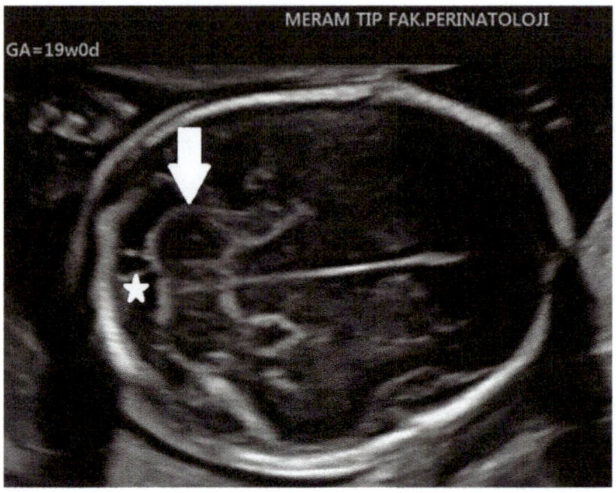

Fig. 11.2 Transcerebellar view of the fetal head during the intrauterine period. The arrow points to the cerebellum and the star to the cisterna magna

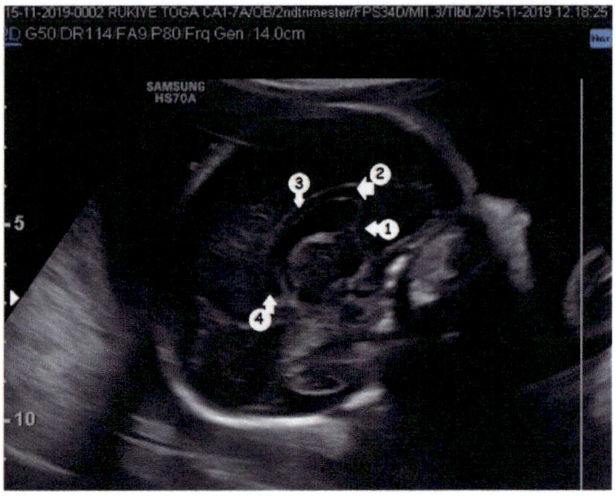

Fig. 11.3 Cross-section of the fetal head in the midsagittal plane during the intrauterine period. The arrow points to the CC and the star to the cerebellar vermis

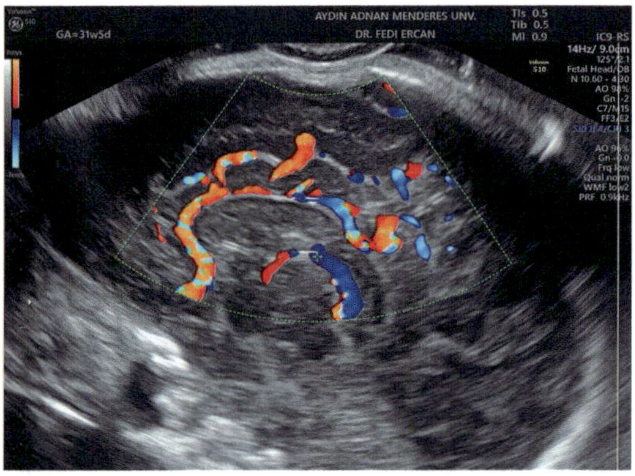

Fig. 11.4 Cross-section of fetal head in midsagittal plane obtained transvaginally during the intrauterine period. The arrow points to the pericallosal artery extending over the CC

In this chapter, prenatal diagnosis of CC pathologies is explained in detail.

11.2 Normal Sonographic Anatomy of the Corpus Callosum

The CC is the major telencephalic commissure [1]. Evaluation of fetal intracranial anatomy with ultrasonography (US) can be done using two (2D) or three (3D) dimension US [2]. However, direct viewing of CC using 2D US requires hard-to-obtain image planes. Due to its shape, it is not possible to view the whole in the axial and coronal planes. Only certain parts can be viewed. However, the whole can be visualized in the midsagittal plane, and its parts and thickness can be evaluated separately [3].

Anatomically, the CC is a hypoechoic structure located between the cavum septi pellucidi and superior part of the cingulate gyrus. It consists of four parts: rostrum, genu, body, and splenium [4] (Fig. 11.3). When color Doppler is used, the pericallosal artery is visualized above the CC (Fig. 11.4). A normal CC can be seen at 18–20 weeks of gestation [5]. Its final shape is completed around 20–22 weeks of gestation [4]. If CC agenesis is suspected, US examination with color Doppler is performed. This can be used from the 18th week of pregnancy. Seeing the pericallosal artery is an indirect sign that a normal CC will develop [6].

11.3 Ultrasonographic Findings of Corpus Callosum Agenesis

CC agenesis gives clues during routine screening at 20–22 weeks of gestation. The most important clues for CC agenesis in US findings are as follows:

1. Absence of cavum septum pellucidum
2. Enlargement of the lateral ventricles (ventriculomegaly >10 mm)

When CC agenesis is suspected, the CC is not directly visualized. However, the diagnosis can also be made indirectly from US findings.

Direct US finding of CC agenesis [7, 8]; complete absence of CC and cavum septum pellucidum in the midsagittal plane.

Indirect US findings of CC agenesis [9, 10]:

- Absence of normal pericallosal artery on color Doppler
- Teardrop-shape of the lateral ventricles (colpocephaly) (Fig. 11.5)
- Dilated third ventricle (Fig. 11.6)

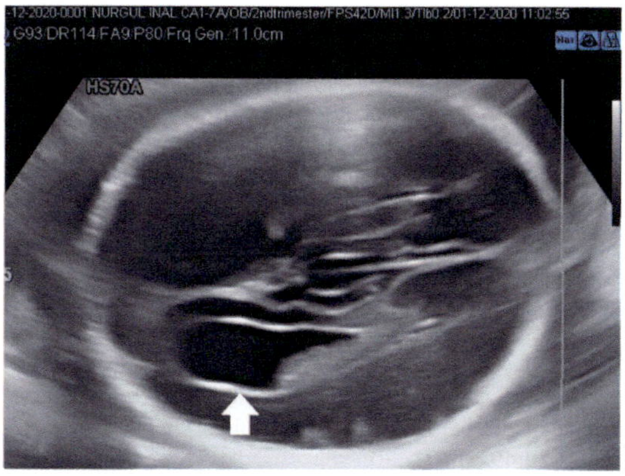

Fig. 11.5 Teardrop-shaped lateral ventricle in fetus with CC agenesis (arrowed)

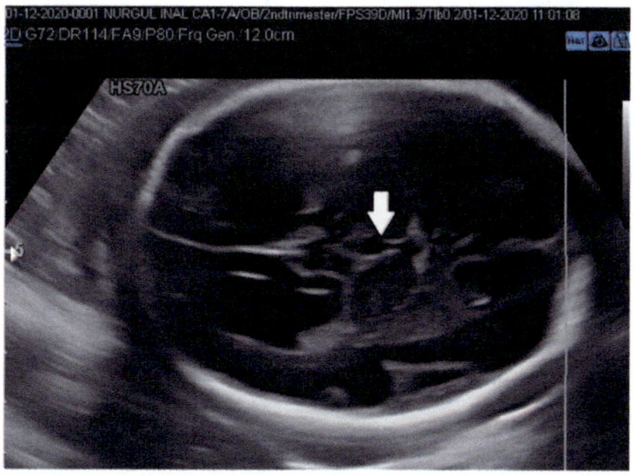

Fig. 11.6 Dilated third ventricle in fetus with CC agenesis (arrowed)

- In the coronal plane, falx cerebri and a large interhemispheric fissure
- The resemblance of the anterior horns of the lateral ventricle to the "Viking's helmet"

11.4 The Role of 3D Ultrasound

Clear US imaging of the CC requires scan planes that are difficult to obtain with 2D US. The coronal plane allows only a small portion of the CC to be visualized [11]. However, owing to the arch structure of the CC, the midsagittal plane can be visualized fully. Three-dimensional US affords many advantages for imaging fetal brain anatomy. It is not only used to obtain a superficial view of the fetal face, but it also makes it possible to evaluate the intracranial anatomy quickly and easily [12, 13].

11.5 The Role of MRI

US is the primary option for imaging fetal intracranial anatomy. Magnetic resonance imaging (MRI), on the other hand, is used as a complementary imaging method to obtain additional information and obtain difficult image plans [14]. Evaluation with US may be affected by maternal obesity and fetal position. MRI does not carry these disadvantages. It also provides better parenchymal imaging.

Fetal brain MRI is performed with T2-weighted sequences. These images are made by combining 2D images (in slices a few millimeters thick) obtained in three planes. The CC displayed in these sections is then compared to the reference values [15]. However, a major disadvantage of fetal MRI is maternal and fetal movement. Planes can be inclined. This can make it difficult to define anatomical landmarks. Despite these disadvantages, midsagittal planes that cannot be obtained with US are much easier to obtain with MRI. Therefore, it provides useful information for a small number of patients, though not all patients. Today, the tools used in new studies indicate that prognosis at the neurodevelopmental level shows great heterogeneity even in isolated CC agenesis [16].

11.6 Conclusion

It is concluded that prenatal imaging of fetuses with CC anomalies is very important. CC pathologies are best viewed in the midsagittal plane. However, at the neurodevelopmental level, prognosis is very heterogenous even in cases of isolated CC agenesis. Nevertheless, a prenatal diagnosis allows accompanying syndromes to be analyzed early with further genetic tests, including microarray analysis.

References

1. Malinger G, Paladini D, Haratz KK, Monteagudo A, Pilu GL, Timor-Tritsch IE. ISUOG Practice Guidelines (updated): sonographic examination of the fetal central nervous system. Part 1: performance of screening examination and indications for targeted neurosonography. Ultrasound Obstet Gynecol. 2020;56(3):476–84.
2. Rossi AC, Prefumo F. Additional value of fetal magnetic resonance imaging in the prenatal diagnosis of central nervous system anomalies: a systematic review of the literature. Ultrasound Obstet Gynecol. 2014;44(4):388–93.
3. Palmer EE, Mowat D. Agenesis of the corpus callosum: a clinical approach to diagnosis. Am J Med Genet C Semin Med Genet. 2014;166C:184.
4. Edwards TJ, Sherr EH, Barkovich AJ, Richards LJ. Clinical, genetic and imaging findings identify new causes for corpus callosum development syndromes. Brain. 2014;137(Pt 6):1579–613.
5. Lockwood CJ, Ghidini A, Aggarwal R, Hobbins JC. Antenatal diagnosis of partial agenesis of the corpus callosum: a benign cause of ventriculomegaly. Am J Obstet Gynecol. 1988;159:184.
6. Díaz-Guerrero L, Giugni-Chalbaud G, Sosa-Olavarría A. Assessment of pericallosal arteries by color Doppler ultrasonography at 11-14 weeks: an early marker of fetal corpus callosum development in normal fetuses and agenesis in cases with chromosomal anomalies. Fetal Diagn Ther. 2013;34(2):85–9.
7. Conturso R, Contro E, Bellussi F, Youssef A, Pacella G, Martelli F, Rizzo N, Pilu G, Ghi T. Demonstration of the pericallosal artery at 11-13 weeks of gestation using 3D ultrasound. Fetal Diagn Ther. 2015;37(4):305–9.
8. Li Y, Estroff JA, Khwaja O, Mehta TS, Poussaint TY, Robson CD, Feldman HA, Ware J, Levine D. Callosal dysgenesis in fetuses with ventriculomegaly: levels of agreement between imaging modalities and postnatal outcome. Ultrasound Obstet Gynecol. 2012;40(5):522–9.
9. Atlas SW, Shkolnik A, Naidich TP. Sonographic recognition of agenesis of the corpus callosum. AJR Am J Roentgenol. 1985;145(1):167–73.
10. Pilu G, Sandri F, Perolo A, Pittalis MC, Grisolia G, Cocchi G, Foschini MP, Salvioli GP, Bovicelli L. Sonography of fetal agenesis of the corpus callosum: a survey of 35 cases. Ultrasound Obstet Gynecol. 1993;3(5):318–29.
11. Malinger G, Lerman-Sagie T, Viñals F. Three-dimensional sagittal reconstruction of the corpus callosum: fact or artifact? Ultrasound Obstet Gynecol. 2006;28(5):742–3.
12. Merz E. Targeted depiction of the fetal corpus callosum with 3D-ultrasound. Ultraschall Med. 2010;31(5):441.
13. Bornstein E, Monteagudo A, Santos R, Keeler SM, Timor-Tritsch IE. A systematic technique using 3-dimensional ultrasound provides a simple and reproducible mode to evaluate the corpus callosum. Am J Obstet Gynecol. 2010;202(2):201.e1–5.
14. Prayer D, Malinger G, Brugger PC, Cassady C, De Catte L, De Keersmaecker B, et al. ISUOG Practice Guidelines: performance of fetal magnetic resonance imaging. Ultrasound Obstet Gynecol. 2017;49(5):671–80.
15. Tilea B, Alberti C, Adamsbaum C, Armoogum P, Oury JF, Cabrol D, Sebag G, Kalifa G, Garel C. Cerebral biometry in fetal magnetic resonance imaging: new reference data. Ultrasound Obstet Gynecol. 2009;33(2):173–81.
16. D'Antonio F, Pagani G, Familiari A, Khalil A, Sagies TL, Malinger G, Leibovitz Z, et al. Outcomes associated with isolated agenesis of the corpus callosum: a meta-analysis. Pediatrics. 2016;138(3):e20160445.

12 Volume Measurements of the Corpus Callosum Volume Using MRI

Niyazi Acer, Ali Çağlar Turgut, and Adem Tokpınar

12.1 Introduction

The corpus callosum (CC) is the greatest commissural pathway in the human brain, containing more than 300 million fibers [1]. There are studies showing that learning a second language affects the anatomical structure of the brain. Broca emphasized that the left hemisphere of the human brain is the dominant hemisphere in language processing [2]. Also, on this side, there are lateral and medial longitudinal striae consisting of longitudinal white matter fibers. On the lower face, the septum pellucidum is attached anteriorly and the fornix body posteriorly. The outer part of the lower surface of the CC forms the roof of the lateral ventricles. The fibers pass through the genu forward on the sides and extend into the frontal lobe [3].

The CC, the interconnection of white matter fibers in our brain, not only connects the cerebral hemispheres but also provides motor and sensory integration. Anatomically, it has four main parts: the splenium, the trunk, the rostrum, and the genu [4]. These compartments are connected to the prefrontal, premotor, and additional motor cortical areas, respectively [5]. Fibers arising from the sensorimotor cortex are accepted to cross the CC through the posterior and middle body, while compartments of the posterior CC, including the splenium and the isthmus, connect to temporal, occipital, and parietal cortical regions. No macroscopic anatomical structure specifically delimits these callosal regions. Imaging studies are performed to obtain more precise local measurements to divide the CC into anatomical and functional areas [5].

Today, it is widely accepted that the CC changes during aging. Studies have shown that CC development increases rapidly during infancy, while the increase during youth is more gradual. Although development of CC is complete by 4 years of age, growth continues, much more slowly, until the thirties [4]. Studies have revealed a negative relationship between age and CC volume. Examining the morphometric development of the CC shows that it is related to the maturation of various brain functions [6]. The development of the CC changes during growth. Normally, it is completed by age four. Some studies show a negative relationship between CC development and age. While the increase during infancy is rapid, it became gradual during youth. In addition to age, gender and race appear to affect the CC [4].

N. Acer (✉)
Department of Anatomy, Arel University School of Medicine, İstanbul, Turkey

A. Ç. Turgut
Department of Radiology, Ege University School of Medicine, Izmir, Turkey

Department of Histology and Embryology, Aydın Adnan Menderes University Health Sciences Institute, Aydin, Turkey

A. Tokpınar
Department of Anatomy, Ordu University School of Medicine, Ordu, Turkey

M. Turgut et al. (eds.), *The Corpus Callosum*, https://doi.org/10.1007/978-3-031-38114-0_12

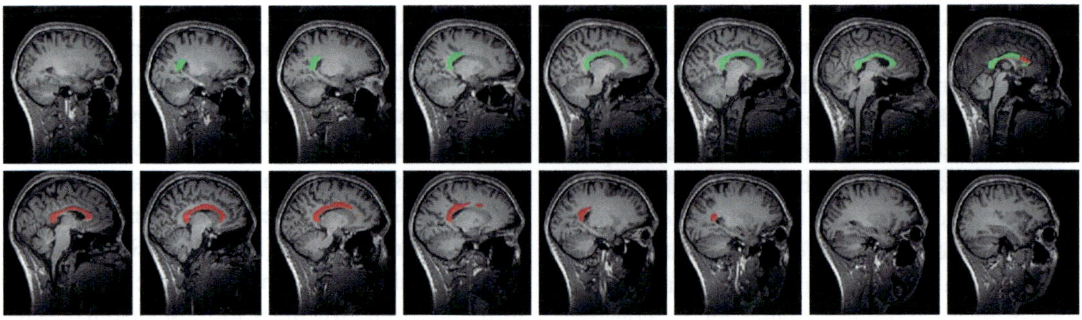

Fig. 12.3 CC displayed on sagittal section showing the insula on T1-weighted images

us to examine the subregions of CC by providing us with the volumes of the genu, body, and splenium [4].

12.5 Discussion

MRI measures brain regions that are crucial in studies of the pathophysiology and pathogenesis of neurological disorders [17]. Many volumetric brain studies reveal neuroanatomical abnormalities in MRI. However, some structures are small, and their intensity is nonuniform and noncontrasted [18].

Differences in anatomical sections can affect structural MRI studies investigating CC morphology in mental diseases and lead to conflicting results. Despite much research on CC morphology since the second half of the twentieth century, there is still no general consensus on its involvement in the pathophysiology of psychiatric phenomena. Mental disorders are often characterized to various degrees by overlapping clinical manifestations [19]. In another study, Guenette et al. calculated the CC volume of National Football League players manually as 121.96 cm^3 [20].

Studies emphasize differences in CC microstructure, volume, and shape in various psychiatric disorders such as schizophrenia, obsessive-compulsive disorder, and depressive disorders. Evidence of CC involvement in the pathophysiology of schizophrenia comes from early ultrastructural investigations and new studies using advanced MRI methods [20]. An updated review suggests that first-episode drug-naive patients diagnosed with schizophrenia have a reduced area of the CC. The CC is important in the pathophysiology of mood disorders and reveals the disconnection between cortical regions related with disturbances in emotional processing [21].

The CC has a vital role in the integration of cognitive, sensory, and motor performances. It also has important roles in verbal working memory, eye movement and vision, maintaining balance, attention, and control [22]. In chromosome 22q11.2 deletion syndrome, also called "velocardiofacial syndrome," poor social skills and cognitive deficits are seen. Risks of schizophrenia spectrum disorders and bipolar disorder increase with advancing age. In addition, there are volumetric changes in cortical and subcortical regions in children with 22q11.2 deletion syndrome [7, 23]. These are associated with impairments in thought (e.g., reasoning and abstract thinking), complex behaviors (e.g., social interactions and relationships), high cognitive processes (e.g., attention and memory), mood regulation (e.g., mania or depression), perception (e.g., illusions and hallucinations), and motor functions (e.g., psychomotor agitation/retardation, catatonia, compulsions) [24, 25].

The vast majority of CC volumetric studies have been focused on differences among special groups, gender differences, and right-left side differences [26]. Acer et al. found that most of the CC volume was smaller in musicians than in a control group, except for the left genu and right splenium, but the difference was not statistically significant ($p > 0.05$) [27, 28]. More recently, Soysal et al. stated that all volumes of the CC were significantly greater in males than females and the genu and splenium volumes were significantly greater in the left hemisphere in both sexes [4]. Arnone et al. found reduced CC volumes using structural MRI [29]. Vannucci et al. reported that the cerebral hemisphere length/height ratio is also greater in women than men, which means that during infancy the female brain (and its component, the splenium of the CC) is relatively longer than the male brain [30]. Fan et al. investigated the relationship between axonal metrics, CC area, and regional gray matter volume [31]. They found a significant increase in the axon diameter index with advancing age over the entire CC [31]. Investigations of the lower regions of the CC showed that age-related changes in axon density and axon diameter index were most prominent in the genus and were relatively small in the splenium [29]. Most recently, Işıklar et al. studied age- and sex-related changes in volumetric development and asymmetry of CC from birth to 18 years old (696 subjects) [16]. They used 3D T1-weighted sequence for their retrospective study and they calculated the genu, body, splenium, and total volume of CC using MRICloud [16]. They found that the total CC volume has six developmental periods 0 years, 1, 2–4, 5–9, 10–16, and 17–18 years; also, they found that all volume measurements of total CC were highly positively correlated with age according to the correlation coefficient analysis [16].

The CC is important in many diseases such as schizophrenia, Alzheimer's disease, bipolar disorder, epilepsy, autism, and unipolar depression [24, 32]. Hence, it is thought that correct measurements of CC volume will be different in the disease states [30]. Structural MRI studies revealed reductions in CC volume, signal intensity, and microstructural integrity in patients with bipolar disorder, suggesting that the interhemispheric connectivity of bilateral homologous cortical areas is altered [20].

Studies have focused on the relationship between area measurements of midsagittal CC and cortical lateralization or asymmetries of various brain structures. Luders et al. performed thickness and area measurements for CC asymmetry analysis on 1 mm thick parasagittal sections 6 mm away from the midsagittal CC [33]. According to the asymmetry coefficient of the area measurements of the CC, they found right lateralization in the region up to the isthmus and left lateralization in the isthmus and splenium [34]. However, Oh et al. suggested that the cross-sectional evaluation of the parasagittal structure may lead to potential inconsistencies and inaccuracies, however, three-dimensional evaluation of the CC [35]. Ardekani et al. reported significant right-left-large asymmetry in the genu, splenium, and body using DTI and voxel-based analysis methods [29].

Fryer et al. typically associated language skills, psychomotor, and visuospatial structuring skills with the integrity of the CC [36]. Yokota et al. stuided whole-brain multiple regression analysis to investigate the association between fractional anisotropy (FA). They found that FA was positively correlated with full-scale IQ in bilateral genu and the splenium of CC and were also positively correlated with performance IQ [37]. Surprisingly, negative correlations with thickness of the CC and area intelligence in children and adolescents have also been reported [38]. Yu et al. reported that mental retardation patients showed lower CC integrity than healthy controls [39]. It is stated that atypical CC development may be associated with developmental disorders such as dyslexia, attention deficit/hyperactive disorder, and autism [16].

All of the brain regions, including from the anterior prefrontal cortex of the brain to the posterior occipital cortex area of the brain, have various roles in these processes for proper functioning. Since the CC provides the physiological substrate for such homotopic interhemispheric cortical synchronization, it was proposed that CC injury is differently involved in the pathophysiological mechanisms underlying different mental disorders [31].

In major depression patients, there are CC abnormalities from the early stages of the disease to chronic stages. CC abnormalities have received increasing attention recently in obsessive-compulsive disorder, with evidence of brain morphological abnormalities [13].

12.6 Conclusion

Comparison of the anatomical features of the CC in obsessive-compulsive disorder, depressive disorder, and schizophrenia, using a hypersensitive and highly localized morphometric technique, has provided us to characterize the morphometric features of the CC finely in these four major mental disorders. Compared to the differences among these patients and from healthy people and other major disorders, a thinned CC is a potential biomarker for obsessive-compulsive disorder. These results give new insights into the involvement of the major brain commissural fiber tract in the pathophysiology of each of these mental disorders. Accurate measurement of the volume of the CC in healthy humans can provide standard values for diagnosing these diseases. We believe that quantitative structural MRI data about the CC are vital for understanding human brain diseases.

References

1. Jarbo K, Verstynen T, Schneider W. In vivo quantification of global connectivity in the human corpus callosum. NeuroImage. 2012;59(3):1988–96.
2. Negin E, Farahani S, Jalaie S, Barootian SS, Pourjavid A, Eatemadi M, et al. Effect of bilingualism on volume of corpus callosum. PNAS. 2016;112(5):1334–7.
3. Di Ieva A, Fathalla H, Cusimano MD, Tschabitscher M. The indusium griseum and the longitudinal striae of the corpus callosum. Cortex. 2015;62:34–40.
4. Soysal H, Acer N, Özdemir M, Eraslan Ö. A volumetric study of the corpus callosum in the Turkish population. J Neurol Surg B Skull Base. 2022;83(4):443–50.
5. Peters M, Oeltze S, Seminowicz D, Steinmetz H, Koeneke S, Jäncke L. Division of the corpus callosum into subregions. Brain Cogn. 2002;50(1):62–72.
6. Tanaka-Arakawa MM, Matsui M, Tanaka C, Uematsu A, Uda S, Miura K, et al. Developmental changes in the corpus callosum from infancy to early adulthood: a structural magnetic resonance imaging study. PLoS One. 2015;10(3):e0118760.
7. Sotgiu MA, Piga G, Mazzarello V, Zarbo IR, Carta A, Saderi L, et al. Corpus callosum volumetrics and clinical progression in early multiple sclerosis. Eur Rev Med Pharmacol Sci. 2022;26(1):225–31.
8. Mori S, Wu D, Ceritoglu C, Li Y, Kolasny A, Vaillant MA, Faria AW, Oishi K, Miller MI. MRICloud: delivering high-throughput MRI neuroinformatics as cloud-based software as a service. Comput Sci Eng. 2016;18(5):21–35.
9. Keller SS, Gerdes JS, Mohammadi S, Kellinghaus C, Kugel H, Deppe K, et al. Volume estimation of the thalamus using freesurfer and stereology: consistency between methods. Neuroinformatics. 2012;10(4):341–50.
10. Poretti A, Mall V, Smitka M, Grunt S, Risen S, Toelle SP, et al. Macrocerebellum: significance and pathogenic considerations. Cerebellum. 2012;11(4):1026–36.
11. Ashburner J, Friston KJ. Voxel-based morphometry—the methods. NeuroImage. 2000;11(6 Pt 1):805–21.
12. Rolls ET, Joliot M, Tzourio-Mazoyer N. Implementation of a new parcellation of the orbitofrontal cortex in the automated anatomical labeling atlas. NeuroImage. 2015;122:1–5.
13. Piras F, Vecchio D, Kurth F, Piras F, Banaj N, Ciullo V, et al. Corpus callosum morphology in major mental disorders: a magnetic resonance imaging study. Brain Commun. 2021;3(2):fcab100.
14. Djamanakova A, Tang X, Li X, Faria AV, Ceritoglu C, Oishi K, et al. Tools for multiple granularity analysis of brain MRI data for individualized image analysis. NeuroImage. 2014;101:168–76.
15. Bruner E, la Cuétara JM, De CR, Martin-Loeches M. Gen- der-based differences in the shape of the human corpus callosum are associated with allometric variations. J Anat. 2012;220:417–21.
16. Isıklar S, Ozdemir ST, Ozkaya G, Ozpar R. Three dimensional development and symmetry of the corpus callosum in the 0-18 age group: aretrospective magnetic resonance imaging study. Clin Anat. 2022;36(4):581–98.
17. Yilmaz S, Tokpinar A, Acer N, Degirmencioglu L, Ates S, Gray SBJEMJ. Evaluation of cerebellar volume in adult. Turkish male individuals: comparison of three methods in magnetic resonance imaging. Erciyes Med J. 2020;42(4):405–11.
18. Tokpınar A, Ülger H, Yılmaz S, Acer N, Ertekin T, Görkem SB, Güler H. Examination of inclinations of the spine at childhood and adolescence. Folia Morphol (Warsz). 2019;78(1):47–53.
19. Unalmis D, Acer N, Yilmaz S, Tokpinar A, Dogan S, Demir HJEMJ. The calculation of the femoral condyle cartilage volume and surface area in patients with osteoarthritis. Erciyes Med J. 2020;42(2):178–85.
20. Guenette JP, Stern RA, Tripodis Y, Chua AS, Schultz V, Sydnor VJ, et al. Automated versus manual segmentation of brain region volumes in former football players. Neuroimage Clin. 2018;18:888–96.
21. Anticevic A, Hu S, Zhang S, Savic A, Billingslea E, Wasylink S, Pittenger C, et al. Global resting-state functional magnetic resonance imaging analysis identifies frontal cortex, striatal, and cerebellar dys-

connectivity in obsessive-compulsive disorder. Biol Psychiatry. 2014;75(8):595–605.
22. Gonçalves ÓF., Sousa, S., Maia, L., Carvalho, S., Leite, J., Ganho, A., .et al (2015). Inferior frontal gyrus white matter abnormalities in obsessive-compulsive disorder. Neuroreport 26(9):495–500.
23. Sarrazin S, Poupon C, Linke J, Wessa M, Phillips M, Delavest M, et al. A multicenter tractography study of deep white matter tracts in bipolar I disorder: psychotic features and interhemispheric disconnectivity. JAMA Psychiatry. 2014;71(4):388–96.
24. Igual L, Soliva JC, Hernández-Vela A, Escalera S, Jiménez X, Vilarroya O, Radeva P. A fully-automatic caudate nucleus segmentation of brain MRI: application in volumetric analysis of pediatric attention-deficit/hyperactivity disorder. Biomed Eng Online. 2011;10:105.
25. Xu K, Jiang W, Ren L, Ouyang X, Jiang Y, Wu F, et al. Impaired interhemispheric connectivity in medication-naive patients with major depressive disorder. J Psychiatry Neurosci. 2013;38(1):43–8.
26. Sexton CE, Mackay CE, Ebmeier KP. A systematic review of diffusion tensor imaging studies in affective disorders. Biol Psychiatry. 2009;66(9):814–23.
27. Acer N, Bastepe-Gray S, Sagiroglu A, Gumus KZ, Degirmencioglu L, Zararsiz G, Ozic MU. Diffusion tensor and volumetric magnetic resonance imaging findings in the brains of professional musicians. J Chem Neuroanat. 2018;88:33–40.
28. Sakai T, Mikami A, Suzuki J, Miyabe-Nishiwaki T, Matsui M, Tomonaga M, et al. Developmental trajectory of the corpus callosum from infancy to the juvenile stage: comparative MRI between chimpanzees and humans. PLoS One. 2017;12(6):e0179624.
29. Arnone D, McIntosh AM, Chandra P, Ebmeier KP. Meta-analysis of magnetic resonance imaging studies of the corpus callosum in bipolar disorder. Acta Psychiatr Scand. 2008;118(5):357–62.
30. Vannucci RC, Barron TF, Vannucci SJ. Development of the corpus callosum: an MRI study. Dev Neurosci. 2017;39(1–4):97–106.
31. Fan Q, Tian Q, Ohringer NA, Nummenmaa A, Witzel T, Tobyne SM, et al. Age-related alterations in axonal microstructure in the corpus callosum measured by high-gradient diffusion MRI. NeuroImage. 2019;191:325–36.
32. Shashi V, Francis A, Hooper SR, Kranz PG, Zapadka M, Schoch K, et al. Increased corpus callosum volume in children with chromosome 22q11.2 deletion syndrome is associated with neurocognitive deficits and genetic polymorphisms. Eur J Hum Genet. 2012;20(10):1051–7.
33. Luders E, Rex DE, Narr KL, Woods RP, Jancke L, Thompson PM, Mazziotta JC, Toga AW. Relationships between Sulcal asymmetries and corpus callosum size: gender and handedness effects. Cereb Cortex. 2003;13:1084–93.
34. Luders E, Narr KL, Zaidel E, Thompson PM, Jancke L, Toga AW. Parasagittal asymmetries of the corpus callosum. Cereb Cortex. 2006;16:346–54.
35. Oh JS, Song IC, Lee JS, Kang H, Park KS, Kang E, Lee DS. Tractography-guided statistics (TGIS) in diffusion tensor imag- ing for the detection of gender difference of fiber integrity in the mid- sagittal and parasagittal corpora callosa. NeuroImage. 2007;36:606–16.
36. Fryer SL, Frank LR, Spadoni AD, Theilmann RJ, Nagel BJ, Schweinsburg AD, Tapert SF. Microstructural integrity of the corpus callosum linked with neuropsychological performance in adolescents. Brain Cogn. 2008;67:225–33.
37. Yokota S, Takeuchi H, Asano K, Asano M, Sassa Y, Taki Y, Kawashima R. Sex interaction of white matter microstructure and verbal IQ in corpus callosum in typically developing children and adolescents. Brain Dev. 2022;44:531–9.
38. Ganjavi H, Lewis JD, Bellec P, MacDonald PA, Waber DP, Evans AC, Karama S. Negative associations between cor- pus callosum midsagittal area and IQ in a representative sample of healthy children and adolescents. PLoS One. 2011;6:e19698.
39. Yu C, Li J, Liu Y, Qin W, Li Y, Shu N, Jiang T, Li K. White matter tract integrity and intelligence in patients with mental retarda- tion and healthy adults. NeuroImage. 2008;40:1533–41.

Part II

Neurophysiology of the Human Corpus Callosum

Editor's Summary

Apart from the anatomy of the corpus callosum (CC), clinicians must have a good working knowledge of the function of this part of the human brain. Such information is important when diagnosing patients who are found to have malfunction of this part of the brain. Chapters within this part of the textbook cover callosal disconnections, in general, the split brain, handedness, and the CC and what role does the CC have in decision making. Taken together, these offerings will improve the reader's understanding of what is currently known of the function of the CC.

segments of the CC include the rostrum, body, genu, isthmus, and splenium and are arched over the ventricular system in a longitudinal fashion, terminating just over the midbrain [3]. The splenium is the thickest portion with the entirety of the CC varying from 1½ cm to just greater than 1 cm in thickness and 6.5 cm in length [3]. The fibers of the CC are topographically organized with anterior to posterior location containing modality-specific pathways, for example, the rostrum and genu connect the frontal hemispheres including the motor, premotor, anterior cingulate, and insular regions [3]. Given this topographic organization, the anterior portion of the CC is thought to be essential for generalization of tonic and tonic-clonic convulsions as well as atonic seizures and thus is the primary target of a corpus callosotomy [3, 9]. Persistent epileptiform activity following corpus callosotomy suggests there are additional networks, such as the dorsal hippocampal commissure, which may also be involved in contralateral spread [10].

13.3 Surgical Indications

The original indications for corpus callosotomy included medically refractory generalized epilepsy that by definition did not localize to a specific focus, multifocal epilepsy, and epilepsy associated with drop attacks [11]. In general, callosotomy is indicated as a palliative surgical procedure in patients that have medically refractory generalized seizures or partial seizures with rapid secondary generalization that have an unidentifiable or multifactorial seizure focus precluding a focal resection [3]. This most notably includes drop attacks as well as medically refractory mixed seizures types and generalized seizures in adults or children [12]. In addition to drop attacks, callosal sectioning is indicated for West syndrome and Lennox-Gastaut syndrome with tonic, atonic, or tonic-clonic seizures [1, 3, 13]. Over time, the indications have broadened to include other seizure types including recurrent status epilepticus, generalized tonic-clonic seizures, absence seizures, and complex partial seizures with suboptimal medication control where focal loci resection is not an option [3, 14].

13.4 Preoperative Evaluation

As with any neurosurgical preoperative evaluation, the patient's comorbidities must be taken into account. A full medical history including all medications and anticoagulation agents, prior surgical procedures, and substance abuse history should be noted. Most importantly, a detailed history of all prior antiepileptic medications and treatments must be well documented. A full seizure history including seizure type, frequency, and their impact on daily function should be obtained from the patient or parent/guardian, in conjunction with evaluation from an epileptologist. Specifically, the impact of the patient's seizures on their quality of life, mood and behavior, as well as family interaction should be discussed. If available, neuropsychiatric testing should be obtained. A consultation with the patient or family should include the anticipation of some recurrent seizures following the procedure in addition to possible transient or rarely permanent changes to memory, speech, or personality. Following a full medical history, a detailed physical exam should take place followed by a contrasted magnetic resonance image (MRI) that includes stealth sequences for intraoperative navigation of the brain and video electroencephalogram (EEG) [3, 13].

13.5 Open Corpus Callosotomy

13.5.1 Surgical Positioning [3, 15]

- Supine, head secured with three-point pin fixation.
- Lounge chair position with 10–20° of torso elevation with neck moderately flexed.
- Secure body with several straps to allow intraoperative manipulation of bed positioning.
- Frameless stereotactic neuronavigation with preoperative MRI stealth imaging.
- Neuromonitoring: EEG, SSEP, motor evoked potentials.

13.5.2 Key Surgical Steps [3, 4, 15]

- Antibiotics, dexamethasone, and mannitol administration prior to skin incision.
- Trapdoor or bicoronal incision at the coronal suture, biased to the nondominant side with ellipsoid or zigzag incision for optimal cosmesis.
- Anterior-posterior measurement of the bone flap ranging from 8 cm.
- Lateral dimension 4–6 cm biased to nondominant hemisphere.
- Multiple (4+) burr holes are made on either side of the superior sagittal sinus to allow for a safe craniotomy; the most dangerous cut with the craniotomy across the superior sagittal sinus is made last with adequate stripping of the dura prior to bone flap removal.
- Hemostatic agents (Gelfoam, Surgicel) and cotton patties prepared on standby prior to bone flap removal with preparedness for sinus injury and repair.
- Dural tack-up sutures are placed around the periphery of the bone flap to ensure epidural hemostasis.
- U-shaped dural opening made and reflected toward the superior sagittal sinus with the goal for 1.5–2 cm window among bridging veins.
- Interhemispheric fissure visualized with the goal to minimize the use of self-retaining retractors.
- Telfa or cotton strips placed on the mesial frontal lobe, interhemispheric approach with arachnoid dissection with exposure from the genu to the posterior extent of the resection.
- Pericallosal arteries are identified with ideal trajectory between the arteries in the midline avascular plane.
- Stimulation of interhemispheric cortex to identify lower limb motor area.
- Disconnection using bipolar and suction from the posterior to anterior until ependyma is visible, stopping at the anterior commissure:

  The anterior commissure is used as a marker of the anterior extent given proximity to the anterior cerebral arteries and neighboring perforators.

  Arachnoid over the vein of Galen is the posterior limit, approximately 5 cm from the anterior genu.
- Ventricle perforation must be sealed to prevent intraventricular hemorrhage and cerebrospinal fluid (CSF) leak.
- Meticulous hemostasis, irrigation, and standard craniotomy closure should follow the above steps.
- Intensive care unit observation is recommended for one night following the procedure.

13.6 Radiosurgery

13.6.1 Surgical Positioning [16, 17]

- Leksell frame fitted to the patient's head and fixed to the skull with local anesthetic.
- Thin sliced stereotactic MRI imaging was obtained with axial and coronal 1 mm T1-weighted sequences for volumetric reconstruction.
- Three to five isocenters planned over the rostrum, genu, and body of the callosum (or posterior target if applicable) are planned using a software.
- Maximal dose typical ranges from 110 Gy to 170 Gy and marginal dose (50% isodose curve) at 55–85 Gy.
- Consideration for general anesthesia if patient compliance with immobility during treatment is a potential issue.

13.6.2 Key Surgical Steps

- Registration in the Gamma Knife operating room and placement of the first isocenter in the correct coordinates.
- First exposure started, patient closely monitored with audio and video equipment.

Procedure is repeated stepwise for all planned isocenters.
- Monitoring of patient for 24 h for posttreatment edema, headache, and reversible neurologic deficit.

13.7 Laser Ablation

Laser interstitial thermal therapy (LITT) has more recently been used for minimally invasive corpus callosotomy with the aid of real-time MRI thermography. In a retrospective case series of ten patients, Roland et al. describe a robotic-assisted approach to corpus callosotomy in young children with medically refractory epilepsy. Not only did these patients have a short hospital stay of average 2 days but LITT also minimized the perceived morbidity of an open cranial procedure that often prevents families from seeking care, a barrier identified by the authors [4].

The hallmarks of LITT therapy with MRI thermography include a gradient-recalled echo (GRE) sequence which measures the proton resonance frequency shift, resulting in real-time thermography reflecting tissue destruction. These calculations are based off of the Arrhenius equation for chemical reactions and temperature dependence [4].

While this procedure was initially approved for tumors and more recently used for mesial temporal lobe epilepsy, the ability to selectively destroy tissue with thermal energy is promising for the future of the corpus callosotomy [4]. Barriers to performing this procedure include the availability of MRI technicians, a dedicated and experienced anesthesia team able to stay for several hours in MRI, high operative cost, stereotactic equipment for placement of the laser, surgeon experience, and availability of industry support.

13.7.1 Surgical Positioning and Considerations [4, 18]

- Preoperative MRIs (T1 axial postcontrast and T2 axial/coronal/sagittal sequences) were analyzed and two to four trajectories used for surgical planning using a software.
- Trajectories were planned to avoid crossing sulci or large vessels, with particular care to avoid the plane of the pericallosal arteries and lateral ventricle margin.
- Positioned supine with the head mildly flexed or in a neutral position anchored in a stereotactic frame.

13.7.2 Key Surgical Steps

- Registration MRI performed with thin 1 mm slices and confirmed with preoperative MRI.
- Small stab incisions and burr holes placed at planned entry sites.
- Dura coagulated and guiding bolts are inserted over the guiding probe; probes are removed, and laser fibers are inserted and secured.
- Patient is transferred to the MRI suite and localizing image obtained to confirm trajectory.
- Laser thermal lesioning is performed under MR-guided thermography to monitor changes in temperature with post-lesioning MRI imaging performed to gauge initial response.
- Laser fibers and bolts subsequently removed and incisions closed with absorbable suture.
- Patient transferred to the intensive care unit for 24 h post-procedure.

13.8 MIS/Endoscopic

Endoscopic corpus callosotomy first described by Bahuleyan et al. in cadaveric specimens aims to limit retraction injury as well as skin incision and craniotomy size [19]. Further potential advantages are improved visualization around critical neurovascular structures near the rostrum and genu of the CC. The one-handed technique has a potential drawback of difficulty in controlling possible hemorrhage; bimanual dissection and consideration of a three-handed technique have been adapted where the endoscope is held by an assistant or holder. A clamp may be used in adaptation to allow suction in conjunction with the hand controlling the endoscope movement [15]. Due to limited reported results and lack of a randomized controlled trial, no direct comparison of results is available; however, improved visibility and reduced occurrence of infection are potential advantages to this procedure.

13.8.1 Surgical Positioning and Considerations [15, 19]

- Similar to open procedure with positioning and skull fixation.
- Frameless stereotactic neuronavigation with preoperative MRI stealth imaging.
- EEG neuromonitoring.

13.8.2 Key Surgical Steps

- Pre-coronal craniotomy planned to optimize trajectory.
- Antibiotics, dexamethasone, and mannitol administration prior to skin incision.
- Incision 1.5 cm lateral to midline, 1.5 in. in length along the sagittal plane.
- Baby mastoid retractor to retract the scalp, subgaleal dissection.
- Burr hole in the midline over the sagittal sinus. D-shaped craniotomy with the straight portion over the midline for a 2–3 cm micro-craniotomy.
- Dura opened with flap based on the sagittal sinus.
- The endoscope is introduced along the falx cerebri; in the interhemispheric fissure, the cingulate gyrus is dissected and CC is exposed. Dissection is carried out anterior to posterior between the pericallosal arteries.
- The ultrasonic aspirator is used to remove the CC from anterior to posterior, avoiding entry to the ventricle and bleeding controlled with suction and bipolar cautery.
- Intensive care unit observation is recommended for one night following the procedure.

13.9 Outcomes

13.9.1 Pediatric

According to a retrospective literature review on outcomes of corpus callosotomy in children, favorable prognostic factors included a younger age of seizure onset and of surgery and seizure semiology, most importantly the presence of drop attacks [11, 13]. Diagnostic results that proved favorable results after surgery included slow spike wave patterns on ictal and interictal electroencephalogram as well as a normal MRI of the brain and IQ >50 [11]. Perhaps most importantly, significant benefit was found, as expected, in patients whose postoperative epileptogenic discharges had reduced synchronicity [11]. In a case series of 76 children undergoing this procedure, complete arrest of drop attacks occurred in 90% who had complete section of the CC and 67% with partial anterior section [20]. Of these patients, one died of pneumonia, and four had surgical complications: one hydrocephalus, one subdural CSF collection, and two epidural bleeds. Of note, quality of life was subjectively improved in 77% of patients by family assessment with satisfaction of surgical results in 97% of families [20]. In another large case series of 74 patients with age range of 1–20 years (median 8.1 years), 80% with Lennox-Gastaut syndrome followed for 2 years after surgery, corpus callosotomy had a > 50% reduction in seizure frequency in 66.2% of patients with 18.9% seizure freedom rate with medications [9]. In another pediatric series, drop attacks were the most responsive to treatment with >90% seizure reduction in 85% of these patients with total callosotomy more effective than anterior two-third callosotomy [13]. In a systematic review, worthwhile reduction of seizures using the Engel classification was increased in total callosotomy vs. anterior callosotomy from 88.2 to 58.6% [11]. An improvement was also noted in drop attacks for overall seizure outcome with complete callosotomy. Other studies recommended initial resection of the anterior two-third of the CC with consideration of a second staged procedure if there is no significant improvement [21].

The advantage of improved seizure control must be weighed against the possible increased operative risk of single-stage complete callosotomy [12]. Extended single operation complete callosotomy may be considered in appropriately selected patients with severe symptoms at the surgeon's discretion [22, 23].

Thus far, the only surgical alternative to corpus callosotomy besides focal resection is vagal nerve stimulation which has not been proven to have similar benefit compared to complete callosotomy but is comparable for anterior callosotomy [11]. A further consideration is increased current cost of VNS compared to corpus callosotomy limiting its use in developing nations. One systematic review directly comparing atonic seizures and drop attacks treated with callosotomy versus VNS found a 58.0% vs. 21.1% chance of seizure freedom and 88.6% vs. 52.6% chance of adequate seizure control, demonstrating the superiority of callosotomy in these seizure types [24]. Certain lateralizing pathologies with generalized seizures including Sturge-Weber syndrome, hemimegalencephaly, tuberous sclerosis, and focal cortical dysplasia may be considered for functional hemispherectomy in appropriately selected patients [20].

13.9.2 Adult

Corpus callosotomy is an appropriate and therapeutic palliative option for adults and children alike with medically refractory epilepsy. One series of 95 patients with multiple seizure types and preop EEG confirming multifocal and generalized interictal epileptiform discharges, with ages ranging from 3 to 60 (median age 24.0) with follow-up >5 years, underwent corpus callosotomy. The seizure types best treated often that had complete cessation after surgery were drop attacks and generalized tonic-clonic seizures followed by simple partial seizures with the poorest results in complex partial seizures. 11.5% of patients had a complication including one epidural hematoma, one intracranial abscess, one subgaleal hematoma, one superficial suture dehiscence, four transient neurologic deficits with contralateral leg weakness or decreased verbal output that resolved by 2 weeks, and three non-neurosurgical complications including a deep vein thrombosis, pneumonia, and fever [21]. This operation is tolerated well with low complication rates among pediatric and adult patients. Presently, insignificant studies have compared minimally invasive endoscopic surgery as well as LITT and radiosurgery with traditional open callosotomy to determine differences in surgical outcomes.

13.9.3 Seizure Freedom

Very few patients, under 10%, of those undergoing corpus callosotomy become completely seizure-free following surgery; however as discussed earlier, the most benefit is seen in patients with tonic and atonic seizures as well as children under the age of sixteen [3]. As discussed earlier, while performed less commonly, the total corpus callosotomy had a 10% higher response rate for all seizures compared to anterior corpus callosotomy only and may be considered as an initial or secondary procedure following anterior two-third callosotomy [3, 4, 13, 25]. Specifically, drop attacks were very effectively treated with corpus callosotomy with satisfactory reduction occurring in 94% of patients with total callosotomy versus 65% with partial callosotomy in one center [13]. Another series comparing anterior two-third and complete callosotomy found a worthwhile seizure reduction or better (Engel class I–III) in 75% of single-stage anterior two-third callosotomy and 90% of complete callosotomy patients with a 10% total morbidity rate [23]. In another systematic review examining outcomes after corpus callosotomy for drop attacks and generalized seizures, a worthwhile seizure reduction (Engel Class 3, or better) was reported in 67.4% of drop attack patients and 39.5% of generalized seizure predominant patients [11].

13.10 Complications

Common surgical complications include CSF leakage, epidural hematoma, subdural hygroma, hydrocephalus, infection, and venous embolism or infarction [3, 4, 11]. The most concerning and common complication, as discussed in the literature however, being disconnection syndrome, which will be discussed below. Prophylactic measures to prevent these complications include perioperative antibiotics, mannitol, dexamethasone, and reduction in end tidal pCO_2 during dissection [3].

Dejerine was first to describe disconnection syndrome in 1892 through his description of a patient with a posterior cerebral artery infarct [1]. Disconnection syndrome was described at this time as alexia without agraphia given the patient had infarcted their ipsilateral visual cortex and posterior commissural fibers [1]. Lower rates of disconnection syndrome occur with partial callosotomy although with potentially inferior results in seizure control [3, 4]. The current definition has broadened to describe objects presented to the non-language dominant hemisphere not being verbally recognized by the patient and the nondominant hand no longer responding to verbal commands [3]. This syndrome may be transient or persist for months or even years in select cases. In a systematic review looking at seizure outcome after callosotomy, disconnection syndrome occurred in 20 patients (12.5%) of the total callosotomy, and none of the anterior callosotomy patients was always transient and was most prominent in children with LGS or intellectual disability [11]. In contrast, a less common phenomena, alien hand syndrome, can occur after surgery in which the nondominant hand, and sometimes ipsilateral leg, acts without the patient's guidance or direction [3].

In addition to disconnection syndrome, severe language and memory impairments can be seen following a corpus callosotomy [3]. Memory impairment is most commonly seen if the hippocampal commissure is disrupted or sectioned or if the resection extends into the posterior CC; however, more extensive sectioning also has the benefit of potentially reducing antiepileptic medication requirements, improving cognition [3, 13].

13.11 Conclusion

Surgical connection of the CC has an extensive history within neurosurgery and has evolved significantly since its inception. While an open cranial procedure is still the most common method for corpus callosotomy, newer technologies have allowed for minimally invasive approaches requiring shorter hospital stays, smaller incisions, and less morbidity for the patient. Although the debate regarding the extent of resection continues, it is clear that this procedure has benefit for young and old patients with medically intractable epilepsy and that this debate should continue to improve neurosurgical patient care.

References

1. Vaddiparti A, et al. The evolution of Corpus callosotomy for epilepsy management. World Neurosurg. 2021;145:455–61.
2. Wagenen V, William P, Yorke Herren R. Surgical division of commissural pathways in the corpus callosum: relation to spread of an epileptic attack. Arch Neurol Psychiatr. 1940;44(4):740–59.
3. Asadi-Pooya AA, et al. Corpus callosotomy. Epilepsy Behav. 2008;13(2):271–8.
4. Roland JL, et al. Corpus callosotomy performed with laser interstitial thermal therapy. J Neurosurg. 2019;134:1–9.
5. Bogen JE. "The neurosurgeon's interest in the corpus callosum." A history of neurosurgery. Park Ridge Am Assoc Neurol Surg. 1997;1:489–98.
6. Luessenhop AJ. Interhemispheric commissurotomy: (the split brain operation) as an alternate to hemispherectomy for control of intractable seizures. Am Surg. 1970;36(5):265–8.
7. Tripathi M, et al. Radiosurgical Corpus callosotomy: a review of literature. World Neurosurg. 2021;145:323–33.
8. Kaas JH. The organization of callosal connections in primates. In: Epilepsy and the Corpus callosum 2. Boston, MA: Springer; 1995. p. 15–27.
9. Wong TT, et al. Corpus callosotomy in children. Childs Nerv Syst. 2006;22(8):999–1011.

This chapter describes the functional outcomes of the split brain and will cover studies of patients with congenital agenesis of the CC and patients post-corpus callosotomy. The focus will be on the corpus callosotomy, as patients' pre- and posttreatment function can be compared, allowing direct investigation of the consequences of destruction of the CC.

14.2 Function in Patients Post-Corpus Callosotomy

Function in patients post-callosotomy is quite variable and is affected by age as well as anatomic and neuropsychosocial development preoperation. Postoperation, some patients have no neurological deficits and demonstrate improved quality of life, while others have varying problems with cognition, sensory processing, language, memory, and other cerebral functions. Disconnection syndrome is a common complication of callosotomy. Patients with congenital agenesis of the CC also present with a wide variety of functional deficits; however, analysis of these patients is less informative of the role of the CC because these patients have no baseline to compare their function without a CC. Additionally, analysis of these patients is often confounded by the presence of other congenital brain pathologies.

Callosotomy is often an extremely beneficial surgery in patients with refractory seizures. Additionally, in some cases, there exist minimal functional impairment and even improvement in other aspects of mental function. In ten patients with seizures and bihemispheric malformations of cortical development, callosotomy resulted in disappearance of seizures in eight patients and reduction to 10% of baseline in one patient. In all ten patients, there were no signs of significant and persistent neurological deficits [3]. Callosotomy can result in improvements in quality of life, behavior, intelligence/development quotient [4], alertness, and responsiveness [5]. Improvements in social adjustment have been observed, particularly in young patients [6]. Additionally, personality, intelligence, and emotion are widely unaffected [7].

Disconnection syndrome is a common complication of callosotomy and occurs due to the lack of communication between the two cerebral hemispheres. This phenomenon was discovered by Sperry in the 1960s during experiments on four of the first patients to undergo callosotomies. These experiments form the basis for modern understanding of the split brain and lateralization of cerebral function. In the vision test, patients were shown a horizontal row of lights, and the experimenters would randomly turn on lights. Patients were only able to say that the right side lit up but could point to lights all across the row that lit up. The authors concluded both hemispheres can process visual input but only the left the can verbalize what is seen. In the tactile test, patients were given an object in the left or right hand. Patients could only identify the object verbally when in the right hand. They could pick out object from a group when it was in the left hand but could not verbally describe or identify the object. In the combination test, if the right hemisphere was shown an image, patients could not verbally name but could identify the object and related objects through left hand touch [7]. This study was instrumental in demonstrating lateralization of cerebral function.

Other studies have since further characterized this phenomenon, with one author observing disconnection syndrome in 12.5% of postop patients [4] and another observing in 4/20 patients [8]. One study only observed disconnection syndrome in postpuberty patients, hypothesizing increased plasticity in young patients accounts for lack of impairment.

Callosotomy can result in a wide variety of other deficits. Common complications include deficits in bimanual coordination and apraxia of the nondominant hand to verbal commands [9]. Language deficits are observed mainly in patients with crossed dominance [9]. Mutism is a common complication [10]. Patients often struggle with memory post-callosotomy, particularly between knowledge they already know and information that they have only inferred [11].

Additionally, patients suffer self-recognition and facial processing deficits [12].

Callosotomy can have varying effects on cognition. Studies have observed variable change of development quotient in young patients [13]. Other studies have shown negative effect on patients' performance intelligence quotient (IQ) but only in patients with at least average performance levels prior to surgery [14].

14.3 Function in Patients with Congenital Agenesis of the Corpus Callosum

Patients with congenital agenesis of the CC present with widely varying function and development. The status of these patients is often confounded by additional congenital brain malformations, and analysis is less powerful due to lack of ability to compare the status of patients with and without the CC, as is possible in callosotomies.

Kim Peek is a well-known person born without a CC. He was a savant featured in the film Rain Man. He was notable for his ability to memorize almost anything as well as his ability to read extremely fast. However, his IQ was 87, he had autism, and he struggled to button his own shirt [15].

One study found children with isolated callosal agenesis were asymptomatic or suffered only mild hypotonia [16]. One study observed normal neurologic examination in 54% of patients and only clumsiness in 39%. A study observed that ~50% of children experienced general intellectual, academic, executive, social, and/or behavioral difficulties and ~ 20% were normal [17]. Studies have observed autism in 5% of children, 35% of adolescents, and 18% of adults with congenital agenesis of the CC [18]. Other studies have found deficits in complex emotional processing and social interaction [19], visual and spatial reasoning (43% of patients), language (54–100% of patients), information processing speed [20] (50% of patients), attention (100% of patients), executive function (40–50% of patients), memory (50–60% of patients), and academic skills (10–86% of patients) [21, 22]. Another study found deficits in proverb comprehension [23]. Patients often suffer deficits in behavioral regulation and metacognition [24]. Commonly, patients have normal motor function [25], with one author finding normal motor development in 84% of patients [21]. There are conflicting reports of IQ in patients born without the CC. One author reports below average IQs [22], while another reports normal intellectual ability (IQ > 85) in approximately two thirds and borderline ability in just over a quarter of patients [26]. Another author reports 10/12 patients had normal or borderline IQ [21].

14.4 Conclusion

The CC is a large nerve tract connecting the left and right cerebral hemispheres. Split brain occurs when the hemispheres do not communicate as a result of corpus callosotomy surgery or congenital agenesis of the CC.

Studies of split-brain patients reveal the functional significance of the CC. The CC assists cerebral function and may be necessary for full neurological function in many patients, especially adults. Patients with a split brain who lack the CC can suffer from a wide variety of deficits in cerebral function, ranging from asymptomatic to severe disability.

A striking feature of split-brain patients is the wide variety of deficits incurred. Any cerebral function may be affected, most commonly somatosensory, language, and memory. Some patients are completely asymptomatic. The majority of patients retain motor function, personality, intelligence, and emotion. Older patients tend to suffer more functional consequences compared to prepubertal patients post-callosotomy. These findings demonstrate the complexity, uniqueness, and plasticity of the human brain.

References

1. Ballardini E, Marino P, Maietti E, Astolfi G, Neville AJ. Prevalence and associated factors for agenesis of corpus callosum in Emilia Romagna (1981-2015). Eur J Med Genet. 2018;61(9):524–30.
2. Markosian C, Patel S, Kosach S, Goodman RR, Tomycz LD. Corpus callosotomy in the modern era: origins, efficacy, technical variations, complications, and indications. World Neurosurg. 2022;159:146–55.
3. Kawai K, Shimizu H, Yagishita A, Maehara T, Tamagawa K. Clinical outcomes after corpus callosotomy in patients with bihemispheric malformations of cortical development. J Neurosurg. 2004;101(1 Suppl):7–15.
4. Graham D, Tisdall MM, Gill D. Corpus callosotomy outcomes in pediatric patients: a systematic review. Epilepsia. 2016;57(7):1053–68.
5. Gilliam F, Wyllie E, Kotagal P, Geckler C, Rusyniak G. Parental assessment of functional outcome after corpus callosotomy. Epilepsia. 1996;37(8):753–7.
6. Lassonde M, Sauerwein C. Neuropsychological outcome of corpus callosotomy in children and adolescents. J Neurosurg Sci. 1997;41(1):67–73.
7. Gazzaniga MS. The Split brain in man. Sci Am. 1967;217(2):24–9.
8. Andersen B, Rogvi-Hansen B, Kruse-Larsen C, Dam M. Corpus callosotomy: seizure and psychosocial outcome. A 39-month follow-up of 20 patients. Epilepsy Res. 1996;23(1):77–85.
9. Sauerwein HC, Lassonde M. Neuropsychological alterations after split-brain surgery. J Neurosurg Sci. 1997;41(1):59–66.
10. Sussman NM, Gur RC, Gur RE, O'Connor MJ. Mutism as a consequence of callosotomy. J Neurosurg. 1983;59(3):514–9.
11. Metcalfe J, Funnell M, Gazzaniga MS. Right-hemisphere memory superiority: studies of a Split-brain patient. Psychol Sci. 1995;6(3):157–64.
12. Turk DJ, Heatherton TF, Kelley WM, Funnell MG, Gazzaniga MS, Macrae CN. Mike or me? Self-recognition in a split-brain patient. Nat Neurosci. 2002;5(9):841–2.
13. Honda R, Baba H, Adachi K, Koshimoto R, Ono T, Toda K, et al. Developmental outcome after corpus callosotomy for infants and young children with drug-resistant epilepsy. Epilepsy Behav. 2021;117:107799.
14. Westerhausen R, Karud CMR. Callosotomy affects performance IQ: a meta-analysis of individual participant data. Neurosci Lett. 2018;665:43–7.
15. Treffert DA, Christensen DD. Inside the mind of a savant. Sci Am. 2005;293(6):108–13.
16. Francesco P, Maria-Edgarda B, Giovanni P, Dandolo G, Giulio B. Prenatal diagnosis of agenesis of corpus callosum: what is the neurodevelopmental outcome? Pediatr Int. 2006;48(3):298–304.
17. Siffredi V, Anderson V, McIlroy A, Wood AG, Leventer RJ, Spencer-Smith MM. A neuropsychological profile for agenesis of the corpus callosum? Cognitive, academic, executive, social, and behavioral functioning in school-age children. J Int Neuropsychol Soc. 2018;24(5):445–55.
18. Lau YC, Hinkley LB, Bukshpun P, Strominger ZA, Wakahiro ML, Baron-Cohen S, et al. Autism traits in individuals with agenesis of the corpus callosum. J Autism Dev Disord. 2013;43(5):1106–18.
19. Anderson LB, Paul LK, Brown WS. Emotional intelligence in agenesis of the corpus callosum. Arch Clin Neuropsychol. 2017;32(3):267–79.
20. Marco EJ, Harrell KM, Brown WS, Hill SS, Jeremy RJ, Kramer JH, et al. Processing speed delays contribute to executive function deficits in individuals with agenesis of the corpus callosum. J Int Neuropsychol Soc. 2012;18(3):521–9.
21. Romaniello R, Marelli S, Giorda R, Bedeschi MF, Bonaglia MC, Arrigoni F, et al. Clinical characterization, genetics, and long-term follow-up of a large cohort of patients with agenesis of the corpus callosum. J Child Neurol. 2016;32(1):60–71.
22. Siffredi V, Anderson V, Leventer RJ, Spencer-Smith MM. Neuropsychological profile of agenesis of the corpus callosum: a systematic review. Dev Neuropsychol. 2013;38(1):36–57.
23. Rehmel JL, Brown WS, Paul LK. Proverb comprehension in individuals with agenesis of the corpus callosum. Brain Lang. 2016;160:21–9.
24. Mangum RW, Miller JS, Brown WS, Nolty AAT, Paul LK. Everyday executive function and self-awareness in agenesis of the corpus callosum. J Int Neuropsychol Soc. 2021;27(10):1037–47.
25. Folliot-Le Doussal L, Chadie A, Brasseur-Daudruy M, Verspyck E, Saugier-Veber P, Marret S. Neurodevelopmental outcome in prenatally diagnosed isolated agenesis of the corpus callosum. Early Hum Dev. 2018;116:9–16.
26. des Portes V, Rolland A, Velazquez-Dominguez J, Peyric E, Cordier MP, Gaucherand P, et al. Outcome of isolated agenesis of the corpus callosum: a population-based prospective study. Eur J Paediatr Neurol. 2018;22(1):82–92.

Handedness and the Corpus Callosum

15

Viktoriya Grayson and R. Shane Tubbs

15.1 Introduction

The corpus callosum (CC) is the major connection composed of myelinated fibers between the two cerebral hemispheres and thus functions as a highway system for communication between the right and left hemispheres. The size of the CC reflects connectivity between the right and left hemispheres and has also been proposed to be related to lateralization of different functions, such as language, handedness, etc. Many have tried to measure behavioral lateralization using handedness, analyzing whether a person or animal is more likely to use right or left hand to complete a task. This has been assessed using questionnaires or physical tasks to identify handedness. In this chapter, the literature regarding the relationship of callosal anatomy and handedness is reviewed and discussed in relation to biological sex.

V. Grayson
Department of Neurosurgery, Tulane University School of Medicine, New Orleans, LA, USA
e-mail: vgrayson@tulane.edu

R. S. Tubbs (✉)
Department of Neurosurgery, Tulane School of Medicine, New Orleans, LA, USA
e-mail: rtubbs@tulane.edu

15.2 Handedness and Corpus Callosum Morphology

It has been a long-standing controversy whether differences in size of CC and/or its different subdivisions exist in relation to handedness, lateralization, and sex. Witelson [1] studied postmortem sections of CC from 42 subjects whose age ranged from 25 to 65 at the time the brain samples were obtained. She was the first to report a difference and concluded that people who were mixed-handers (MH) or left-handers (LH) had a 11% increase in size of midsagittal area of CC compared to right-handers (RH). The difference in size of CC was observed in the anterior and posterior halves of the CC; however, the splenium specifically was not significantly different between the groups. Witelson [1] hypothesized that these variations in CC between LH and RH would allow for greater interhemispheric sharing of function and less functional asymmetry in LH. In other words, it was proposed that LH would have greater myelination and higher fiber density with more fiber crossing and communicating to the other hemisphere. The findings of a study by Wang et al. [2] were consistent with this hypothesis. This study included 50 college students of which 13 were identified as LH, 24 as RH, and 13 as MH. Handedness was determined with a questionnaire, and the subjects underwent resting state functional magnetic resonance imaging (fMRI), and voxel-mirrored homotropic con-

M. Turgut et al. (eds.), *The Corpus Callosum*, https://doi.org/10.1007/978-3-031-38114-0_15

nectivity was measured. Wang et al. [2] found that subjects who scored higher on the continuum toward left-handedness showed stronger functional interconnectivity between the right and left hemisphere compared to those who were identified as RH based on their score in the handedness questionnaire. Using diffusion-tensor MRI, Westerhausen et al. [3] determined that LH subjects had a higher relative anisotropy (RA) in the CC, reflecting the degree of integrity of the CC. Additionally, LH subjects had lower mean diffusion (MD) parameters, which signifies decreased diffusion barriers such as cell membranes or myelin sheaths, further strengthening the reliability of the findings. These findings were supported in vivo using MRI by Habib et al. [4] who examined 53 normal subjects and found the anterior body to be larger in left-handed subjects.

Josse et al. [5] discovered that RH have a smaller midsagittal area in CC compared to subjects who are not RH. These findings are also consistent with CC size and handedness relationship in capuchin monkeys [6]. Midsagittal MRI studies from 14 adult capuchin monkeys were taken, and handedness was tested using a bimanual task. Phillips et al. [6] found that left-handed capuchin monkeys had a larger genu, while right-handed monkeys had a larger splenium. These findings suggest that the CC and handedness relationship is not limited to Homo sapiens. In a different study using 67 chimpanzees, Hopkins et al. [7] determined that LH chimpanzees had an overall smaller CC size as well as subdivisions of CC compared to RH chimpanzees. This finding is contradictory to the findings of smaller CC in RH compared to LH [1, 4–6].

Others have not found an effect of CC on functional lateralization, such as handedness [8–10]. Preuss et al. [8] conducted a study with 46 men who identified as RH of which 32 tested as RH and 14 were labeled "non-consistent right-handers" after undergoing the handedness dominance test. Preuss et al. [8] used MRI and did not find any differences between handedness and CC size or its subregions. These findings were also supported by a study that included both males and females with RH, LH, and MH subgroups and also used MRI to examine the CC and concluded that there was no correlation between CC size and handedness [9]. Westerhausen and Papadatou-Pastou [10] conducted a meta-analysis that included studies which analyzed CC morphology in relation to handedness in healthy subjects and determined that there was no significant difference in hand preference in relation to CC size. By nature of analysis, meta-analytical studies provide more robust results; thus, the lack of findings suggests that the initial research from Witelson [1] suggesting that CC size is increased in left-handed individuals and thus is associated with increased interhemispheric connectivity is not supported by the meta-analytic findings. Luders et al. [11] examined the relationship between CC size and sulcal asymmetries and also did not find any handedness effects or biological sex effects on CC size; however, there were significant morphological differences between males and females.

15.3 Role of Biological Sex on Corpus Callosum Morphology and Handedness

The role of biological sex in CC morphology in relation to handedness has shown notable differences. De Lacoste-Utamsing and Holloway [12] pioneered the research of sex differences in CC size, noting that females have a larger midsagittal area and splenium compared to males. Habib et al. [4] used a handedness questionnaire and MRIs and determined that the CC was larger in non-consistent RH, especially the anterior half in males. Furthermore, it was found that the posterior midbody of the CC was larger in females. Tuncer et al. [13] analyzed MRI from 40 females and 40 males, with 20 RH and 20 LH in each group, using Witelson's method to divide the subregions. The rostrum and isthmus of the CC were larger in RH males than RH females [13]. The areas of the posterior body of the CC were also larger in males than females. These findings demonstrate that even between MRI findings, there is not a consensus on the role sex and handedness play in CC morphology.

The CC size was also significantly different between capuchin monkeys [6]. Females had a larger CC including all subdivisions (rostral, isthmus, posterior, splenium) than male capuchin monkeys. Similarly to humans, right-handed male monkeys had a smaller size CC in the anterior midbody compared to right-handed females [6]. Other studies have not found a significant handedness effect in relation to CC morphology and biological sex; however, morphological differences were still noted between sexes [9, 11]. Luders et al. [11] found that the anterior and posterior regions of CC differed between males and females, in addition to males having more surface variability. Steinmetz et al. [9] also found a sex difference, in a study with 52 healthy subjects of which half were female, noting that females having a larger isthmus subdivision of CC compared to males. The isthmus is a subdivision of CC that functions to join regions involved in functional asymmetry; thus, a larger isthmus is suggestive of decreased lateralization and increased bihemispheric distribution of function. Contrary, Welcome et al. [14] did not find any evidence of a relationship between sex and handedness; rather, a significant effect of sex and consistency of handedness was found. Specifically, females with consistent handedness had a larger rostrum compared to females who showed mixed-handedness. In the posterior body of the CC, males with mixed-handedness had larger areas than males with consistent handedness, while females did not show a difference. This finding is consistent with a previous finding by Tuncer et al. [13] who also found a male sex difference in the posterior body, with LH males having a larger area than RH males. Together, these findings suggest that interhemispheric connectivity and functional behavioral lateralization are affected by sex.

15.4 Conclusion

Overall, the discrepancy in results regarding handedness and the size of CC can be due to several factors, contributing to conflicting results including but not limited to methodological differences such as using postmortem [1] vs. fMRI in vivo to analyze CC size [2, 8], lack of standardization in the type of MRI data used for analysis, methods used for subdivision of CC, subject population, age of subjects, and methods for assessing handedness (functional tests vs. questionnaires). In a study with 117 patients who were between 15 and 75 years old, it was discovered that the size of CC decreased with age, more specifically the body of the CC [15]. These findings suggest that differences in age of subjects can contribute to the size of CC and can also contribute to the conflicting results in relation to handedness and CC size. Furthermore, individual differences in callosal morphology can also affect the results. The lack of agreement in findings suggest that there is not one single relationship that describes the CC anatomy and behavioral lateralization. Luders et al. [16] proposed that the size of CC should be examined in relation to the degree of handedness lateralization rather than left- or right-handedness. In a study with 361 subjects, they found that larger callosal size in anterior and posterior midbody was negatively associated with a weaker handedness lateralization. In other words, the size of CC was increased in individuals with less lateralization regardless of whether they were right- or left-handers. When they specifically examined callosal size in relation to right- and left-handedness, there were no significant findings. These results support the notion that increased functional lateralization examined by handedness is associated with decreased CC size or thickness, indicating less interhemispheric communication. However, it also highlights the importance of examining the degree of handedness lateralization as a whole rather than the direction of handedness in subjects.

In summary, there are many factors that should be taken into account and methods of standard-

order to establish effective communication to achieve brain function [2]. This proposed action of the CC serves as the basis for its role in decision-making.

The absence of the connections of the CC has been studied in people in whom the structure has been severed, collectively called split-brain patients [4]. From observations of individuals with this condition, two hypotheses of "will" have been distinguished, one being the existence of two free wills and the other the existence of two counteracting wills in the brain [4]; this means that in some cases the two hemispheres need to act independently of each other or in opposition to each other. Using computational models and oculomotor paradigms, a behavioral disorder was observed in split-brain patients in which the left arm acted in a manner to counteract the actions of the right arm [4]. This observation supported the counteracting wills hypothesis, the role of the CC being to activate and inhibit regions simultaneously [4]. This behavioral disorder was termed diagnostic dyspraxia. Other cases show that this conflict is not limited to intermanual behavior; it also applies to cognitive processes [4].

Various methods have been used explore the role of the CC in decision-making. These methods have included diffusion tensor imaging, Iowa Gambling Task, functional magnetic resonance imaging (fMRI), and the expectancy-valence model [1, 3, 5–8].

Diffusion tensor imaging (DTI) is critical for studying the axonal integrity of the CC's white matter; its values reveal a predictive and informative pattern of decision-making activities in the structure [1, 5, 6, 8]. DTI is a imaging tool that uses the three-dimensional properties of water in axons to explore the CC, mapping the white matter pathways using directionality and the rate of water diffusion [6, 8]. Four critical parameters are used in DTI to measure the integrity of white matter bundles: fractional anisotropy (FA), mean diffusivity (MD), axial diffusivity (AD), and radial diffusivity (RD) [1, 6]. FA measures the coherence of white matter integrity directly: the higher the FA, the more coherently organized the white matter [1, 6]. Mean diffusivity measures the rate of diffusion: the higher its values, the lower the white matter integrity [1, 6]. Axial and radial diffusivity measure longitudinal and perpendicular water movements, elucidating the microstructure of axons and myelin [6]. Greater axial diffusivity and lower radial diffusivity are correlated with greater white matter integrity [6].

Iowa Gambling Task is also employed in the study of the CC and its role in decision-making [1, 9]. The task was initially designed to measure decision-making in patients with lesions in the ventromedial prefrontal cortex but has proved very valuable in measuring decision-making in studies of the CC [1, 9]. Its aim is to create real-life decision-making scenarios in a game providing uncertainty in premises, outcomes, monetary reward, and punishment [1, 9]. The Iowa Gambling Task can be analyzed using a computational model, the expectancy-valence (EV) model, which highlights the cognitive and motivation processes behind Iowa Gambling Task behavior [9]. This model isolates four parameters: learning or recency, to measure the focus on recent outcomes; motivation, highlighting gains and losses; attention, choice consistency, to elucidate the consistency of decision-making strategies; and finally, goodness of fit [9].

16.3 Outcomes of Studies on the Corpus Callosum and Decision-Making

To understand decision-making processes further or outline the huge contribution of the CC therein, certain conditions affecting individual patients have been proven useful. Of note, individuals with substance use disorders have patterns of poor decision-making, emphasizing impulsivity, high risk, high reward, short-term gains, undersensitivity to loss, and limited ability to integrate information given numerous trials [3]. Individuals with alcohol dependence disorder, MDMA use, and cocaine use are collectively placed in the substance use disorder category [3, 5]. In some studies, DTI shows reduced FA in the rostral body of the CC in cocaine users, while alcohol dependents have reduced FA in the splenium of the CC [3]. MDMA users performing the Iowa Gambling

Task showed impaired decision-making by preferring a disadvantageous deck and lacking the ability and flexibility for advantageous decks; also, DTI in MDMA users showed lower unfavorable scores in the rostral or anterior CC when correlated with the Iowa Gambling Task [5].

Individuals with agenesis of the CC show a different pattern of decision-making from individuals with substance abuse disorders or with lesions [2, 9]. In a study combining the Iowa Gambling Task with the expectancy-valence model, individuals with agenesis of the CC showed normal range intelligence similar to controls but were slower in understanding novel problems and contingencies of tasks. This implied a rigid framework for novel problem-solving and the need for information to be presented slowly and with ample time to integrate it in order to solve problems fully [2, 9]. These individuals showed heightened cautiousness and lower consistency, reliance on recent outcomes to form strategy, and aversion to negative outcomes. This result can be interpreted to further the notion that the CC is critical in large integrative networks that form the basis for a decision-making strategy [2, 9].

References

1. Zorlu N, et al. Abnormal white matter integrity and decision-making deficits in alcohol dependence. Psychiatry Res. 2013;214(3):382–8. https://doi.org/10.1016/j.pscychresns.2013.06.014.
2. Ferreira Furtado LM, Bernardes HM, de Souza Félix Nunes FA, Gonçalves CA, Filho JADCV, de Miranda AS. The role of neuroplasticity in improving the decision-making quality of individuals with agenesis of the corpus callosum: a systematic review. Cureus. 2022;14(6):e26082. https://doi.org/10.7759/cureus.26082.
3. Lane SD, et al. Diffusion tensor imaging and decision making in cocaine dependence. PLoS One. 2010;5(7):e11591. https://doi.org/10.1371/journal.pone.0011591.
4. Pouget P, Pradat-Diehl P, Rivaud-Péchoux S, Wattiez N, Gaymard B. An oculomotor and computational study of a patient with diagonistic dyspraxia. Cortex. 2011;47(4):473–83. https://doi.org/10.1016/j.cortex.2010.04.001.
5. Moeller FG, et al. Diffusion tensor imaging in MDMA users and controls: association with decision making. Am J Drug Alcohol Abuse. 2007;33(6):777–89. https://doi.org/10.1080/00952990701651564.
6. Estella NM, et al. Brain white matter microstructure in obese women with binge eating disorder. Eur Eat Disord Rev. 2020;28(5):525–35. https://doi.org/10.1002/erv.2758.
7. Yip SW, et al. Reduced genual corpus callosal white matter integrity in pathological gambling and its relationship to alcohol abuse or dependence. World J Biol Psychiatry. 2013;14(2):129–38. https://doi.org/10.3109/15622975.2011.568068.
8. Garibotto V, et al. Disorganization of anatomical connectivity in obsessive compulsive disorder: a multi-parameter diffusion tensor imaging study in a subpopulation of patients. Neurobiol Dis. 2010;37(2):468–76. https://doi.org/10.1016/j.nbd.2009.11.003.
9. Brown WS, Anderson LB, Symington MF, Paul LK. Decision-making in individuals with agenesis of the corpus callosum: expectancy-valence in the Iowa gambling task. Arch Clin Neuropsychol. 2012;27(5):532–44. https://doi.org/10.1093/arclin/acs052.

Part III

Congenital and Acquired Neuropathology of the Corpus Callosum

Editor's Summary

Physicians who diagnose and treat patients with brain disorders must be familiar with the various pathologies that can occur due to direct injury or pathology of the corpus callosum (CC) or from lesions around its periphery. Common neuropathological entities that might be encountered include patients with missing or thinned areas of the CC, vascular lesions within the CC, metabolic diseases that affect the CC such as hypoglycemia, surrounding diseases that affect the CC such as hydrocephalus, and inflammatory and infectious diseases that affect this structure such as multiple sclerosis and Epstein-Barr virus, respectively. Finally, this part offers insight into what is found of the CC in patients with dementia, schizophrenia, and autism spectrum disorder.

Agenesis or Hypoplasia of the Corpus Callosum

17

Seçil Oktay and Huriye Berra Ertuğrul

17.1 Introduction

The corpus callosum (CC) is a white matter structure that connects the right and left hemispheres of the brain and contains about 200 million axons. These axonal connections integrate sensory, motor, and cognitive functions (language, integration of complex sensory information). Agenesis of the CC (AgCC) is a congenital malformation of the CNS presenting as partial or complete absence of the CC [1, 2]. It can occur without somatic CNS findings or accompany other CNS malformations. There are different degrees of callosal defects depending on the time during embryogenesis when they arise: hypoplasia (thin, underdeveloped), hypogenesis (partial agenesis), and dysgenesis (malformation) [3]. CC formation begins during the ninth prenatal week. During the 11th–12th weeks of gestation, the fibers cross the midline. CC development is completed during gestational weeks 18–20 [4–6]. The CC is assembled from front to back: It starts with the genu (front) and ends towards the splenium (back). Myelination and orientation of neurons, which reach a certain number at birth, continue postnatally. The completion of CC development takes up to the third decade. There are sex differences in the thickness of the CC and the sizes of axon fibers and bundles during childhood and adolescence [6, 7]. During the completion of CC development, the most important areas are the formation of the sulcus medianus telencephali medii (the median groove in the dorsal lamina reuniens) and the massa commissuralis. Events during this process can lead to AgCC [8].

17.2 Epidemiology

AgCC is among the most widespread congenital malformations of the CNS. Although it varies among populations and with diagnostic criteria, its prevalence is 0.3–0.7% [6]. The prevalence in children with neurodevelopmental delay is 2.3%, and its incidence at autopsy is 1/19,000 [6, 9]. Neonatal and prenatal scanning studies show that AgCC happens in 1:4000 live births, and it is detected in 3–5% of people evaluated for neurodevelopmental retardation [10]. AgCC can occur in unaccompanied or to be together with other developmental abnormalities of the CNS [6, 11].

S. Oktay (✉)
Department of Pediatric Neurology, Aydın Adnan Menderes University Faculty of Medicine, Aydın, Turkey

H. B. Ertuğrul
Department of Pediatrics, Aydın Adnan Menderes University Faculty of Medicine, Aydın, Turkey

M. Turgut et al. (eds.), *The Corpus Callosum*, https://doi.org/10.1007/978-3-031-38114-0_17

17.3 Etiology

Abnormal development of the CC occurs prenatally at 5–16 weeks; the cause can be genetic, infectious, vascular, or toxic. Although the etiology is uncertain, the occurrence of AgCC and other CC disorders is mediated by heterogeneous factors [2, 6]. Genetic factors (chromosomal errors, genetic mutations, etc.), prenatal infections and injuries, prenatal toxin exposure (maternal alcohol use during pregnancy), structural disorders of the CNS (a structural cyst that prevents CC formation), and metabolic disorders are among the possible causes [3, 12, 13].

17.3.1 Genetic Factors

Genetic factors often accompany with various syndromes (Table 17.1). There are chromosomal abnormalities in approximately 20% of AgCC cases, particularly mosaic trisomies 18, 13, and 8 (Chromosomal anomalies, e.g., trisomies, del(4)(p16), del(6)(q23), dup(8)(p21p23), dup(11)(q23qter)) [9, 12, 13]. Mutations of the disrupted-in-schizophrenia 1 (*DISC1*) have recently been implicated in agenesis of the CC. DISC1 gene is located at 1q42.1. AgCC is a prosencephalic midline developmental abnormality, and its pathophysiology has not yet been clarified. Autosomal dominant mutations, for example, Apert syndrome and basal cell nevus syndrome, and autosomal recessive mutations such as Joubert syndrome and X-linked mutations such as Aicardi syndrome accompany more than 200 congenital syndromes (Table 17.2). Although these syndromes have different phenotypes, their common feature is CC abnormality [6, 14].

Table 17.1 Aneuploidy and non-aneuploidy syndrome

Aneuploidy syndrome	Non-aneuploidy syndrome
Trisomy 18	Aicardi syndrome
Trisomy 13	Apert syndrome
Trisomy 8	Bickers Adams Edwards syndrome
	Coffin-Siris syndrome
	Fetal alcohol syndrome
	Fryns syndrome
	Gorlin syndrome
	Hydrolethalus syndrome
	Lowe syndrome
	Zellweger syndrome

Table 17.2 Syndromes associated with agenesis of the corpus callosum

Syndromes associated with AgCC
Acrocallosal syndrome
Aicardi syndrome
Andermann syndrome
Donnai-Barrow syndrome
Dwarfism
FG syndrome
L1CAM syndrome
Microcephalic osteodysplastic primordial dwarfism type II
Mowat-Wilson syndrome
Oculocerebrocutaneous syndrome
Saal Bulas syndrome
Septo-optic dysplasia (optic nerve hypoplasia)
Shapiro syndrome
Vici syndrome

17.3.2 Metabolic Disorders

AgCC can accompany metabolic diseases such as congenital lactic acidosis due to a mitochondrial respiratory chain defect, pyruvate metabolism disorders, nonketotic hyperglycemia, mucolipidoses, and mucopolysaccharidoses. When the etiology of AgCC is investigated, metabolic tests should be performed.

17.3.3 Other CNS Associations

AgCC is accompanied by hydrocephalus in 30% of cases. The trigones and posterior horns of the lateral ventricles take on the appearance of colpocephaly. Dandy-Walker spectrum accompanies 11% of cases and Chiari II malformation 7%. Developmental anomalies of the CNS (gray matter heterotopia, polymicrogyria, porencephaly, schizencephaly), interhemispheric cysts, and intracranial lipoma are other potentially coexisting conditions [6, 12, 14].

17.4 Histopathology

Depending on the affected part of the CC, AgCC is classed as "complete agenesis" or "partial agenesis." Partial agenesis is hypogenesis or dysgenesis without a splenium and/or rostrum [15]. In primary agenesis, the CC is completely absent. In "secondary dysgenesis," the CC begins normally but is destroyed as a consequence of encephalomalacia secondary to traumatic, toxic, or ischemic events. This is an intrauterine developmental anomaly. One exception is holoprosencephaly. In this situation, callosal dysgenesis is atypical; anterior parts of the CC are absent [15–17].

17.5 Radiology

17.5.1 Morphology

There are two morphological types of AgCC. In type 1, axons do not cross over the midline but form wide abnormal "Probst bundles" through the medial hemispheric walls (Fig. 17.1). Because commissural axons cannot form in type 2, Probst bundle fibers are not found. Axons extends parallel to the medial walls of the ventricles and, at the level of the frontal horns, dilate and deform the medial borders of the lateral ventricles. Thus, AgCC-specific images emerge. The wide spindle-shaped neurons located in the anterior cingulate cortex and frontoinsular cortex in AgCC are "von Economo neurons." A decrease in their number has been described in recent years. In partial AgCC, there are "sigmoid bundles" that connect the frontal lobe asymmetrically to the contralateral occipitoparietal cortex [15–18] (Fig. 17.2). The anterior commissure is usually enlarged. The structure of the hippocampal formations is usually hypoplastic, so the temporal horns of the lateral ventricles are dilated. CC development starts anteriorly and progresses posteriorly. MR tractography and recent studies show that the ventral trunk occurs first and continues bidirectionally and the ventral parts (genu) developing earlier/more prominently than the posterior parts (splenium). The process is not fully understood and studies continue [19, 20]. The obvious exception to the rule is "holoprosencephaly," in which the ventral parts of the CC are absent (Fig. 17.3). This has been termed "atypical callosal dysgenesis" [21]. Holoprosencephaly is a congenital brain malformation in AgCC. The CC is myelinated in the opposite direction, from the splenium forward [19–21].

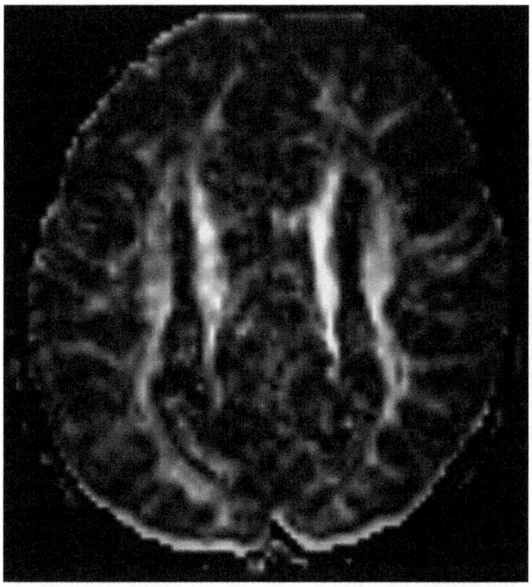

Fig. 17.1 DTI tractography of the Probst bundle fibers

17.5.2 Ultrasonography

Ultrasonography (US) can be used for prenatal diagnosis CC abnormalities by 18–20 weeks of gestation. CC formation is completed during the 20th prenatal week. Detection of the cavum septi pellucidi and ventriculomegaly (lateral ventricles > 10 mm) in routine US fetal anomaly screening at 20–22 weeks of gestation necessitates further examination to detect CC anomalies [22].

17.5.3 Magnetic Resonance Imaging

MRI helps after the prenatal 20th week, detailing developmental and morphological abnormalities of the CNS accompanying AgCC. In MRI find-

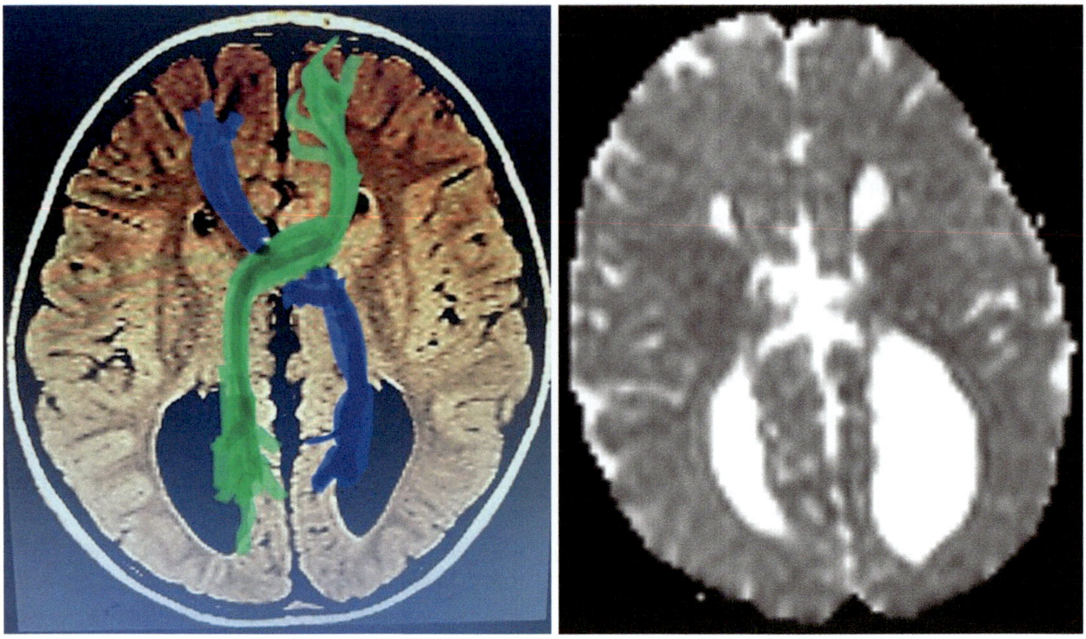

Fig. 17.2 DTI tractography of the sigmoid bundles (in green and blue)

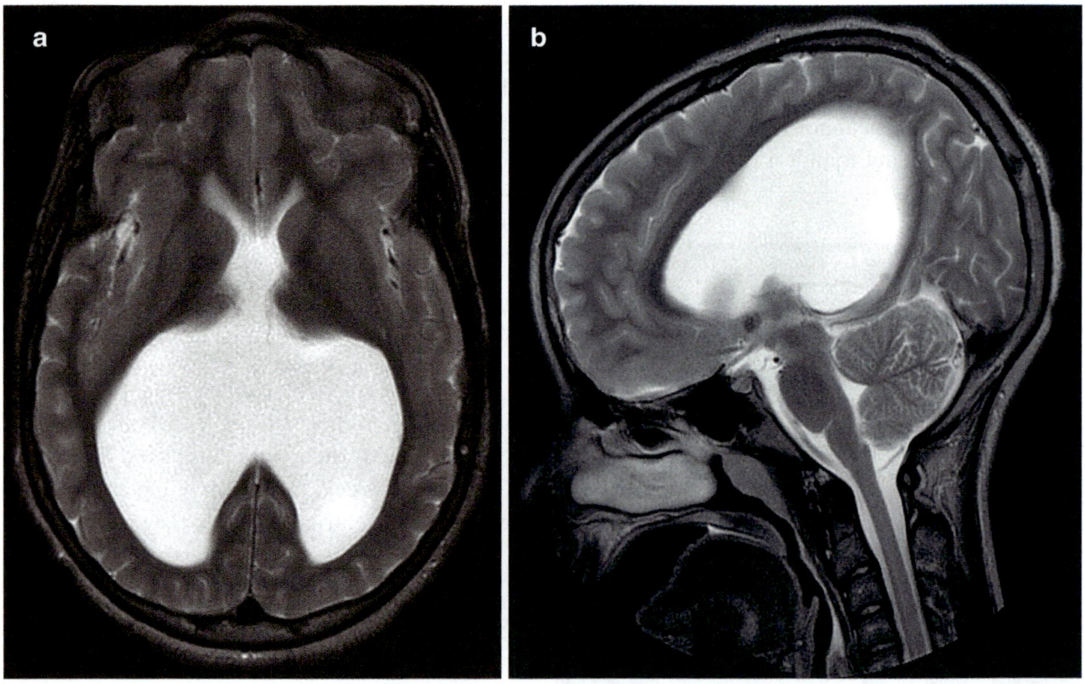

Fig. 17.3 (**a**) T2 axial, (**b**) T2 sagittal holoprosencephaly

ings of AgCC, the anterior horns appear narrow and laterally displaced in axial section and the occipital horns slightly dilated to "teardrop" shapes (colpocephaly). Colpocephaly is a widespread finding in imaging but is not the most characteristic one; a "racing car sign," formed by ventriculomegaly and the absence of CC with intervening Probst bundles, is a typical neuroim-

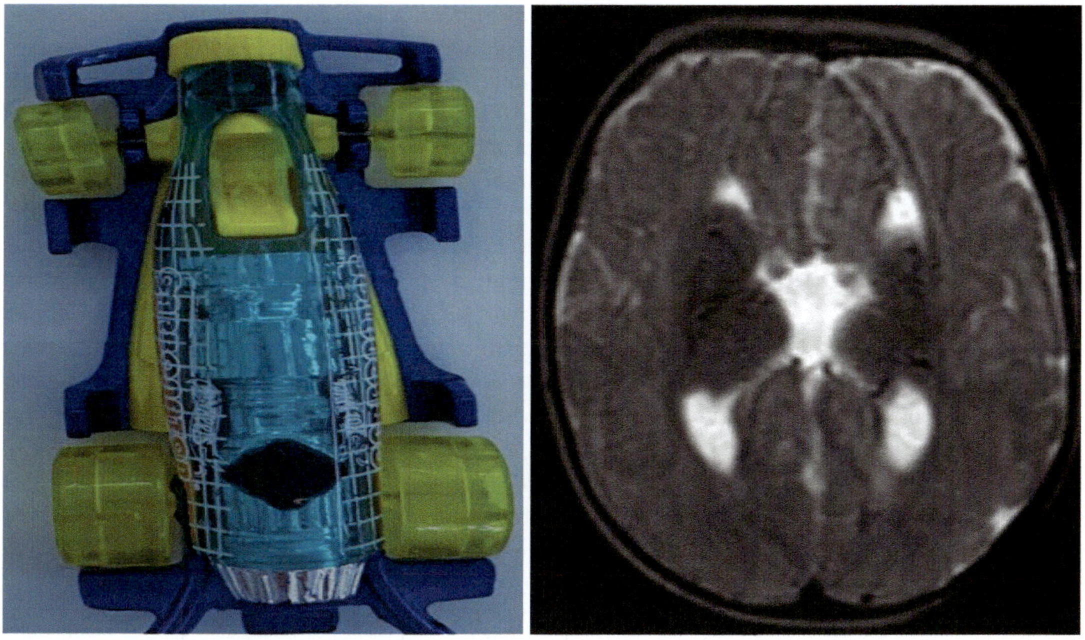

Fig. 17.4 Racing car sign axial T2 image photography

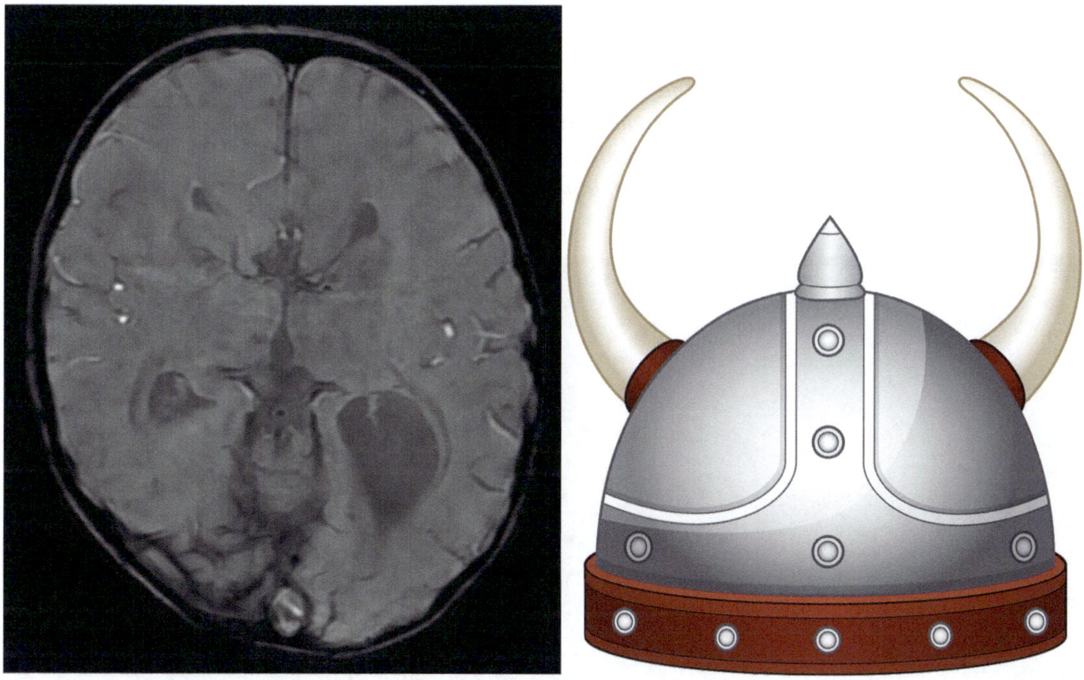

Fig. 17.5 Viking's helmet axial SWI image

aging presentation of this developmental defect [23] (Fig. 17.4). On coronal section, the falx cerebri can be seen in a broad interhemispheric fissure that meets the third ventricle; the lateral ventricles are widely separated in a "Viking's helmet" shape (Fig. 17.5). The thalami are widely separated because the third ventricle is dilated [24]. Tractography, a special MRI technique, is a very supportive imaging method for CC anomalies. The trajectory of axons revealed by tractography, diffusion-weighted MRI, represents the white matter corticospinal tracts in the brain. Advances in tractography will be very helpful in AgCC [25].

17.6 Clinical Findings

Clinically, AgCC is often asymptomatic. Asymptomatic cases, although IQ is normal, show decreased sensory and motor interhemispheric transmission, decreased cognitive processing speed, and possible deficiencies in complex reasoning and new problem-solving. This group of several cognitive deficits is called "core syndrome" [2, 26]. Symptomatic cases usually present during the first 2 years of life as mental retardation, vision problems, speech delay, seizures, feeding problems, attention deficit hyperactivity disorder, and autism [13, 27].

Case 17.1 *A 4-year-old female patient had a history of arrest during delivery. Postnatal transfontanelle US revealed total AgCC, and she was admitted to our pediatric neurology outpatient clinic for follow-up of her neuromotor development. Her neurological examination was normal. Social cognitive, language, and gross-fine motor development was appropriate for her age. Although total AgCC was detected in her cranial MRI, it was remarkable that the clinical presentation was completely normal* (Fig. 17.6).

Case 17.2 *An 11-year-old girl was followed up after attention deficit hyperactivity disorder and type 1 diabetes mellitus were diagnosed. Cranial MRI performed to investigate the etiology of the specific learning difficulties revealed partial AgCC, the posterior part and the cavum septum pellucidum vergae being absent* (Fig. 17.7). *Neurological examination was normal. There was no epilepsy. The patient was illiterate and underwent special education.*

Case 17.1 had total AgCC. Although case 17.2 had partial AgCC, it was accompanied by comorbidities such as attention deficit hyperactivity disorder. Neuropsychological evaluation of all patients with AgCC is mandatory. Treatment is mainly symptomatic and supportive. The prognosis is determined primarily by associated malformations. Children with isolated AgCC and no significant neurological sequelae have the best prognosis. AgCC associated with neuronal migration disorder has the worst prognosis. If

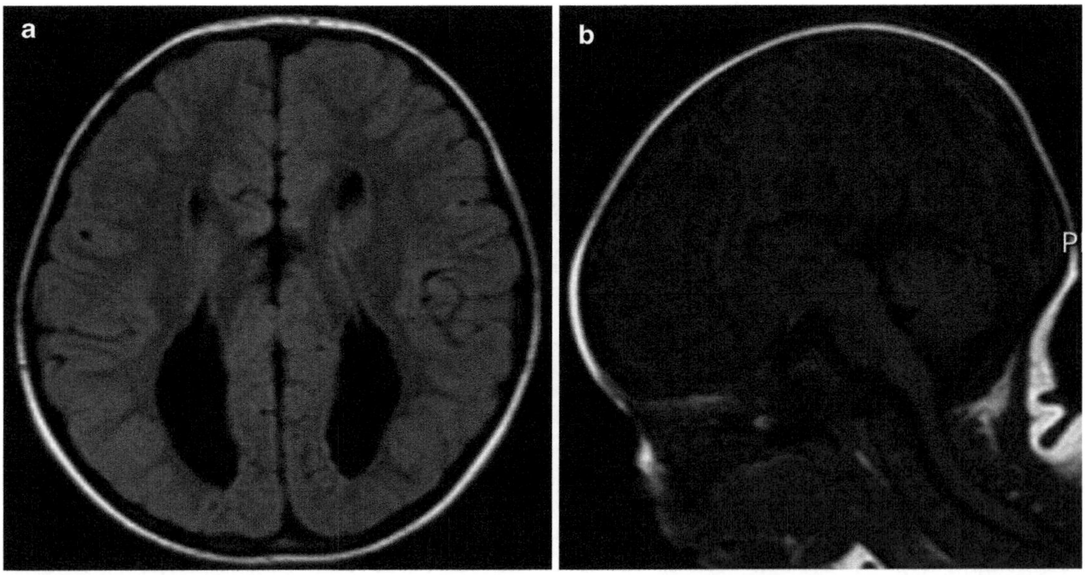

Fig. 17.6 Case 17.1 (**a**) axial T1 image (**b**) sagittal T1 image

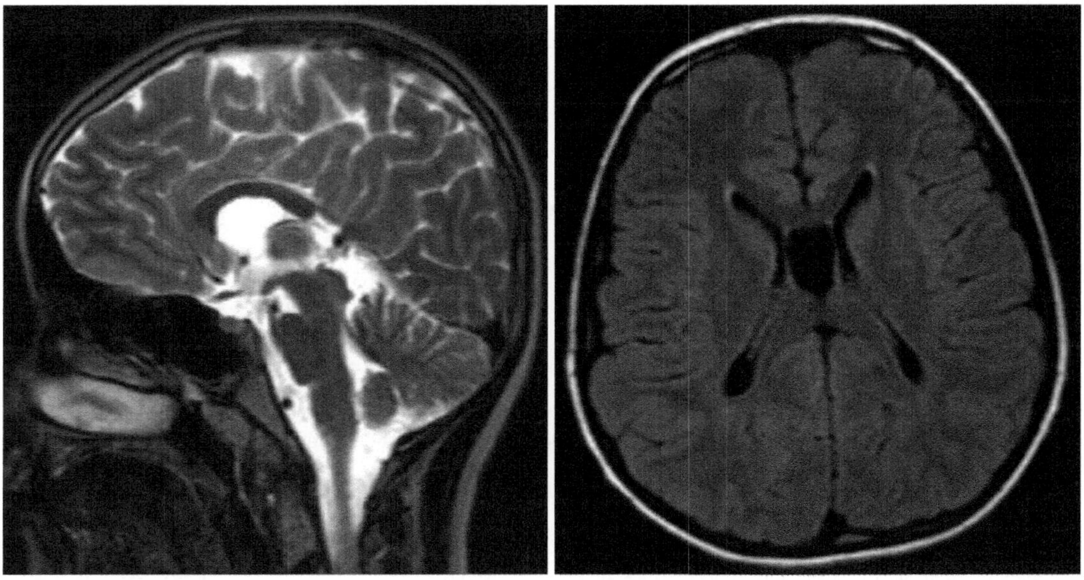

Fig. 17.7 Case 17.2 (**a**) sagittal T2 image (**b**) axial flair image of cavum septum pellucidum vergae

complications such as aspiration pneumonia, seizures, schizophrenia, or cerebral palsy accompany AgCC, the prognosis is poor.

17.7 Key Diagnostic Definitions in AgCC

For young neuroscientists, several diagnostic definitions regarding AgCC are the following:

Complete AgCC: CC is completely absent congenitally.

CC hypogenesis: Congenital partial absence of the CC. Absence should be detectable prenatally and should not be degenerative.

CC hypoplasia: CC is completely formed but thinner than expected for age and gender.

Isolated AgCC: This description indicates complete absence of the CC. In isolated AgCC, colpocephaly and Probst bundles are often found.

Primary AgCC: Describes intact intellectual function, IQ $\geq$ 80.

Anterior commissure: Bundles of white fibers that connect the right and left hemispheres of the brain. The anterior commissure connects the temporal lobes and is located at the base of the fornix.

Probst bundle: This runs parallel to the interhemispheric fissure and defines misdirected callosal axons. It can be seen in cases of full and partial AgCC.

Sigmoid bundle: The heterotopic commissural pathway that connects the anterior lobe to the contralateral occipitoparietal cortex.

Colpocephaly: Posterior longitudinal dilatation of the lateral ventricles. It usually includes the temporal horns and is a reduction of the ipsilateral cortical junctional pathways.

17.8 Conclusion

The CC is a wide white matter tract that connects the right and left hemispheres of the brain. It is a very important structural and functional part of the brain. It allows us to perceive depth and enables the two sides of our brain to communicate. Most patients with AgCC live normal lives and their intelligence is average. However, minor differences in higher cognitive functions from peers of the same age group and education are detected by detailed mental tests. AgCC is common among developmental disorders affecting the brain. It can be severe or mild, clinically noticeable from the neonatal period. Sometimes, it can be asymptomatic enough to be detected only incidentally. It requires further research with a multidisciplinary approach and further examination by pediatric neurologists, neuroradiologists, embryologists, pediatric neurosurgeons, and genetic counselors.

References

1. Hofman J, Hutny M, Sztuba K, Paprocka J. Corpus callosum agenesis: an insight into the etiology and spectrum of symptoms. Brain Sci. 2020;10(9):625.
2. Tang P, Bartha A, Norton M, Barkovich A, Sherr E, Glenn O. Agenesis of the corpus callosum: an MR imaging analysis of associated abnormalities in the fetus. Am J Neuroradiol. 2009;30(2):257–63.
3. Wang L, Huang C-C, Yeh T. Major brain lesions detected on sonographic screening of apparently normal term neonates. Neuroradiology. 2004;46(5):368–73.
4. Edwards TJ, Sherr EH, Barkovich AJ, Richards LJ. Clinical, genetic and imaging findings identify new causes for corpus callosum development syndromes. Brain. 2014;137(6):1579–613.
5. Hernanz-Schulman M, Dohan F, Jones T, Cayea P, Wallman J, Teele RL. Sonographic appearance of callosal agenesis: correlation with radiologic and pathologic findings. Am J Neuroradiol. 1985;6(3):361–8.
6. Knezović V, Kasprian G, Štajduhar A, Schwartz E, Weber M, Gruber G, et al. Underdevelopment of the human hippocampus in callosal agenesis: an in vivo fetal MRI study. Am J Neuroradiol. 2019;40(3):576–81.
7. Tovar-Moll F, Moll J, de Oliveira-Souza R, Bramati I, Andreiuolo PA, Lent R. Neuroplasticity in human callosal dysgenesis: a diffusion tensor imaging study. Cereb Cortex. 2007;17(3):531–41.
8. Kier EL, Truwit CL. The normal and abnormal genu of the corpus callosum: an evolutionary, embryologic, anatomic, and MR analysis. Am J Neuroradiol. 1996;17(9):1631–41.
9. Osbun N, Li J, O'Driscoll MC, Strominger Z, Wakahiro M, Rider E, et al. Genetic and functional analyses identify DISC1 as a novel callosal agenesis candidate gene. Am J Med Genet A. 2011;155(8):1865–76.
10. Ketonen LM, Hiwatashi A, Sidhu R, Westesson P-L. Pediatric brain and spine: an atlas of MRI and spectroscopy, vol. 33. Berlin: Springer; 2005. p. 138.
11. Das MJ, Geetha R. Corpus callosum agenesis. In: StatPearls. Treasure Island, FL: StatPearls Publishing; 2022.
12. Lábadi B, Beke AM. Behavioral and cognitive profile of corpus callosum agenesia-review. Ideggyogy Sz. 2016;69(11–12):373–9.
13. Paul LK, Brown WS, Adolphs R, Tyszka JM, Richards LJ, Mukherjee P, et al. Agenesis of the corpus callosum: genetic, developmental and functional aspects of connectivity. Nat Rev Neurosci. 2007;8(4):287–99.
14. Dimond D, Rohr CS, Smith RE, Dhollander T, Cho I, Lebel C, et al. Early childhood development of white matter fiber density and morphology. NeuroImage. 2020;210:116552.
15. Kornienko VN, Pronin IN. Diagnostic neuroradiology. Berlin: Springer Science & Business Media; 2008.
16. Al-Hashim AH, Blaser S, Raybaud C, MacGregor D. Corpus callosum abnormalities: neuroradiological and clinical correlations. Dev Med Child Neurol. 2016;58(5):475–84.
17. Brown WS, Paul LK. The neuropsychological syndrome of agenesis of the corpus callosum. J Int Neuropsychol Soc. 2019;25(3):324–30.
18. Song JW, Gruber GM, Patsch JM, Seidl R, Prayer D, Kasprian G. How accurate are prenatal tractography results? A postnatal in vivo follow-up study using diffusion tensor imaging. Pediatr Radiol. 2018;48(4):486–98.
19. Kaplan P. X linked recessive inheritance of agenesis of the corpus callosum. J Med Genet. 1983;20(2):122–4.
20. Oba H, Barkovich AJ. Holoprosencephaly: an analysis of callosal formation and its relation to development of the interhemispheric fissure. Am J Neuroradiol. 1995;16(3):453–60.
21. Barkovich AJ, Norman D. Anomalies of the corpus callosum: correlation with further anomalies of the brain. Am J Neuroradiol. 1988;9(3):493–501.
22. Volpe P, Paladini D, Resta M, Stanziano A, Salvatore M, Quarantelli M, et al. Characteristics, associations and outcome of partial agenesis of the corpus callosum in the fetus. Ultrasound Obstet Gynecol. 2006;27(5):509–16.
23. Kumar N. Sevoflurane in child with carpus callosum agenesis syndrome, a series of four cases. Pediatr Anesth Crit Care J. 2014;2:48–51.
24. Li Y, Estroff J, Khwaja O, Mehta T, Poussaint T, Robson C, et al. Callosal dysgenesis in fetuses with ventriculomegaly: levels of agreement between imaging modalities and postnatal outcome. Ultrasound Obstet Gynecol. 2012;40(5):522–9.
25. Huang H, Zhang J, Wakana S, Zhang W, Ren T, Richards LJ, et al. White and gray matter development in human fetal, newborn and pediatric brains. NeuroImage. 2006;33(1):27–38.
26. Jiang Y, Qian Y-Q, Yang M-M, Zhan Q-T, Chen Y, Xi F-F, et al. Whole-exome sequencing revealed mutations of MED12 and EFNB1 in fetal agenesis of the corpus callosum. Front Genet. 2019;10:1201.
27. Paul LK, Corsello C, Kennedy DP, Adolphs R. Agenesis of the corpus callosum and autism: a comprehensive comparison. Brain. 2014;137(6):1813–29.

18 Thick Fetal Corpus Callosum

Ayhan Kanat

18.1 Introduction

The corpus callosum (CC) is a group of white matter tracts that span the left and right cerebral hemispheres [1]. It is situated in the center of the human brain [2] and forms a bridge between the cerebral hemispheres in which axonal fibers from both hemispheres cross [3]. It contains about 200 million axons connecting the left and the right hemispheres and has a fundamental role in integrating sensory, motor, visuomotor, and cognitive processes [4].

The CC is named for its compactness [5]. The word commissure derives from the Latin "commissure," which means joining or connecting [6]. In the brain, a "commissure" is defined as a junction at which two anatomical parts meet [6]. The CC is the largest commissure connecting the cerebral hemispheres [7]. There are five major commissures in the brain: the anterior commissure, the posterior commissure, the hippocampal commissure, the habenular commissure, and the CC [6].

The CC has four segments: rostrum, genu, body, and splenium. The narrowing between the body and splenium is called the isthmus [4]. The rostrum/lamina rostralis (beak) extends anteriorly from the anterior commissure to the posterior inferior aspect of the genu [5]. The genu (knee) is a thickened part of the CC, so named because of the abrupt change in orientation it marks between the lamina rostralis and the callosal body [5]. The callosal body (corpus) is the horizontal portion that extends from the genu to the point where the fornix abuts the undersurface of the CC [5]. The isthmus usually appears as a slight focal narrowing where the fornix joins the CC [5]. The splenium (spleen) is the thickest portion of the CC [5]. Figure 18.1 shows the thickness and the anatomical parts of the CC.

The CC connects the frontal, parietal, temporal, occipital, insular, and limbic lobes and the basal ganglia of the hemispheres to each other [6]. Its cross-sectional area is twice the sum of all the other commissural structures in the adult brain [2]. In the medical literature, several fetal callosal anomalies have been reported [8]. Agenesis (total absence), partial agenesis (because of missing segments), or hypoplasia (short and thin but completely formed) have frequently been documented [8]. However, the thick CC anomaly has seldom been investigated [9]. An abnormally thick CC can be seen in children with mental retardation or autism: megalencephaly [8].

The appearance of the CC is one of the most distinctive modifications in placental mammals [6]. Its embryology is not yet well known [2]. During embryogenesis, the hippocampi and hippocampal commissure develop before the CC [3]. Numerous CC pathologies are revealed during computed tomography (CT) and magnetic

A. Kanat (✉)
Department of Neurosurgery, Medical Faculty, Recep Tayyip Erdogan University, Rize, Turkey

M. Turgut et al. (eds.), *The Corpus Callosum*, https://doi.org/10.1007/978-3-031-38114-0_18

References

1. Vaddiparti A, Huang R, Blihar D, et al. The evolution of corpus callosotomy for epilepsy management. World Neurosurg. 2021;145:455–61.
2. Fitsiori A, Nguyen D, Karentzos A, et al. The corpus callosum: white matter or terra incognita. Br J Radiol. 2011;84:5–18.
3. Blaauw J, Meiners LC. The splenium of the corpus callosum: embryology, anatomy, function and imaging with pathophysiological hypothesis. Neuroradiology. 2020;62:563–85.
4. Edwards TJ, Sherr EH, Barkovich AJ, Richards LJ. Clinical, genetic and imaging findings identify new causes for corpus callosum development syndromes. Brain. 2014;137:1579–613.
5. Raybaud C. The corpus callosum, the other great forebrain commissures, and the septum pellucidum: anatomy, development, and malformation. Neuroradiology. 2010;52:447–77.
6. Shah A, Jhawar S, Goel A, Goel A. Corpus callosum and its connections: a fiber dissection study. World Neurosurg. 2021;151:e1024–35.
7. Bardin R, Leibovitz Z, Mashiach R, et al. Short and thick corpus callosum - the thin border between a minor anatomical variant to very poor outcome. J Matern Fetal Neonatal Med. 2020;35:3305–8.
8. Izzo G, Toto V, Doneda C, et al. Fetal thick corpus callosum: new insights from neuroimaging and neuropathology in two cases and literature review. Neuroradiology. 2021;63:2139–48.
9. Kanat A, Tsianaka E, Gasenzer E, Drosos E. Some interesting points of competition of X-ray using during the Greco-ottoman war in 1897 and development of neurosurgical radiology: a reminiscence. Turk Neurosurg. 2022;32:877. https://doi.org/10.5137/1019-5149.JTN.33484-20.3.
10. Hetts SW, Sherr EH, Chao S, et al. Anomalies of the corpus callosum: an MR analysis of the phenotypic spectrum of associated malformations. AJR Am J Roentgenol. 2006;187:1343–8.
11. Göçmen R, Oğuz KK. Mega corpus callosum and caudate nuclei with bilateral hippocampal malformation. Diagn Interv Radiol. 2008;14:69–71.
12. D'Antonio F, Pagani G, Familiari A, et al. Outcomes associated with isolated agenesis of the corpus callosum: a meta-analysis. Pediatrics. 2016;138:e20160445. https://doi.org/10.1542/peds.2016-0445.
13. Velut S, Destrieux C, Kakou M. Morphologic anatomy of the corpus callosum. Neurochirurgie. 1998;44:17–30.
14. Glenn OA, Goldstein RB, Li KC, et al. Fetal magnetic resonance imaging in the evaluation of fetuses referred for sonographically suspected abnormalities of the corpus callosum. J Ultrasound Med. 2005;24:791–804.
15. Witelson SF, Kigar DL, Harvey T. The exceptional brain of Albert Einstein. Lancet. 1999;353:2149–53.
16. Men W, Falk D, Sun T, et al. The corpus callosum of Albert Einstein's brain: another clue to his high intelligence? Brain. 2014;137:e268.
17. Kanat A. Wrong-site craniotomy. J Neurosurg. 2013;119(4):1079–80; United States.
18. Gasenzer ER, Kanat A, Ozdemir V, Neugebauer E. Analyzing of dark past and bright present of neurosurgical history with a picture of musicians. Br J Neurosurg. 2018;32:303. https://doi.org/10.1080/02688697.2018.1467000.
19. Kanat A, Yazar U, Kazdal H, et al. Neurosurgery is a profession. Neurol Neurochir Pol. 2009;43:286.
20. Polat HB, Kanat A, Celiker FB, et al. Rationalization of using the MR diffusion imaging in B12 deficiency. Ann Indian Acad Neurol. 2020;23:72–7.
21. Gasenzer ER, Kanat A, Neugebauer E. Neurosurgery and music; effect of Wolfgang Amadeus Mozart. World Neurosurg. 2017;102:313–9.
22. Andronikou S, Pillay T, Gabuza L, et al. Corpus callosum thickness in children: an MR pattern-recognition approach on the midsagittal image. Pediatr Radiol. 2015;45:258–72.
23. Gasenzer ER, Kanat A, Nakamura M. The influence of music on neurosurgical cases: a neglected knowledge. J Neurol Surg A Cent Eur Neurosurg. 2021;82:544. https://doi.org/10.1055/s-0040-1721017.
24. Vaquero L, Hartmann K, Ripolles P, et al. Structural neuroplasticity in expert pianists depends on the age of musical training onset. NeuroImage. 2016;126:106–19.
25. Schlaug G, Jancke L, Huang Y, et al. Increased corpus callosum size in musicians. Neuropsychologia. 1995;33:1047–55.

19 Vascular Lesions of the Corpus Callosum

Grace Posey and R. Shane Tubbs

19.1 Introduction

Bilaterally, the pericallosal artery, a branch of the anterior cerebral artery, perfuses the anterior portion of the corpus callosum (CC), while the posterior pericallosal artery, a branch of the posterior cerebral artery, perfuses the splenium [1]. Although well preserved, pathology from vascular lesions (i.e., infarction in the CC) does occur and often presents as diffusion restriction on magnetic resonance imaging (MRI). However, there are several other potential suspects that should be kept in mind upon presentation of restricted diffusion findings [2]. Some have found that nonvascular cases of diffusion-restricted lesions in the CC are seen more in younger patients and are more likely to occur in conjunction with findings on imaging such as enhancement and edema [2]. In the study of Wilson et al., several risk factors such as hypertension, diabetes, and atrial fibrillation were associated with identification of vascular vs. nonvascular causes of diffusion restriction on MRI [2]. Vasculitis, cocaine vasculopathy, vasospasm in the setting of subarachnoid hemorrhage, and hypercoagulable states associated with systemic malignancy or autoimmune diseases were listed among the atypical mechanisms of such infarctions.

There are, of course, manifestations of vascular lesions of the CC that occur as consequences of interruption of the interhemispheric connections of its white matter tracts. In a study examining the cognitive implications of vascular lesions of the CC, it was noted that the presence of such lesions was associated with reduction in the integrity of the microstructure of white matter not only within the CC but also the white matter of the whole brain [3]. Furthermore, study participants with such damage to the CC were found to have deficits in all cognitive domains, with the language saved, compared to controls. Specifically, the executive functioning, processing speed and attention, memory, language, and visuospatial processing capabilities of the participants were evaluated [3]. Similarly, the CC has been shown to become atrophic in the setting of cerebrovascular diseases—specifically, large vessel occlusive disease (LVOD)—as marked by correlation with reduction in cortical oxygen metabolism and benzodiazepine receptor binding. This is not the case, however, in Binswanger's disease, a form of small vessel dementia characterized by fibrohyalinosis and extensive white matter lesions, as the CC is generally spared aside from a reported decrease in nerve fiber density [4].

G. Posey (✉)
Tulane University School of Medicine, New Orleans, LA, USA
e-mail: gposey@tulane.edu

R. S. Tubbs
Department of Neurosurgery, Tulane School of Medicine, New Orleans, LA, USA
e-mail: rtubbs@tulane.edu

M. Turgut et al. (eds.), *The Corpus Callosum*, https://doi.org/10.1007/978-3-031-38114-0_19

19.1.1 Ischemic Lesions

Studies have indicated that infarction (ischemia) involving the CC makes up a mere 3–8% of all neurological occurrences, with some asserting that the genu and body of the CC are the most common sites at which this occurs while others say the splenium [2]. It is generally considered difficult to definitively state the true rate of infarction limited to the CC itself because these are easy to overlook as a result of relatively subtle manifestations in clinical symptoms [5]. Despite their likely rarer occurrence, instances of ischemic lesions in the CC are much more common compared to hemorrhagic lesions and may be seen in diseases of the small vessels such as cerebral autosomal dominant arteriopathy with subcortical infarcts and leukoencephalopathy (CADASIL), small vessel vasculitis, Susac syndrome, and autosomal dominant hereditary cerebrovascular disease due to *COL4A1* point mutations. Typically, they are central and unilateral, without crossing the midline in the CC because of the aforementioned bilateral blood supply. However, watershed infarction—occlusion of the anterior cerebral artery, both posterior cerebral arteries, or third small arteries that originate from the anterior communicating artery—can lead to midline ischemic CC lesions [1]. Ischemic injury of the CC has also been reported in instances of emboli, stenosis of the anterior or posterior cerebral arteries and their branches secondary to atheromatosis or vasospasm, and cerebral venous thrombosis. In the case of ischemia of the splenium specifically, it has been asserted that the ipsilateral thalamus and mesial-posterior region of the temporal lobe would also likely have suffered ischemic injury due to the shared blood source of the ipsilateral posterior cerebral artery. By the same reasoning, ischemia of the area perfused by the anterior cerebral artery would be suspected upon discovery of ischemic lesions in the more anterior portion of the CC [6].

Some have documented cases of CC infarction (ischemia) in patients suffering from acute noncommunicating hydrocephalus (ANCH) at a rate which suggests that CC infarction may occur more frequently than the current consensus allows. In their 2019 study aimed at gaining a better understanding of the underlying pathophysiology and mechanism of CC infarction in patients with this condition, Hirono et al. completed an institutional review of patients with ANCH, ultimately finding that 26% of these patients had suffered CC splenium infarctions. The authors postulated that the observed enlargement in the bilateral lateral ventricles created an environment in which the superior branch of the posterior callosal artery (PCaA) could be compressed [5]. Of note, COVID-19 has also been documented as a cause of widespread microvascular injury and, consequently, ischemic lesions of the CC [7]. Additionally, there has even been at least one reported case of spontaneous CC hematoma as a complication of severe COVID-19 pneumonia and ARDS in a 46-year-old male. It is thought that endothelial dysfunction occurred in this patient due to the highly inflammatory state induced by cytokine storm [8].

19.1.2 Hemorrhagic Lesions

Hemorrhagic lesions in the CC are uncommon but have been reported in association with a number of pathologies. These include but are not limited to large anterior communicating artery or distal anterior cerebral artery aneurysms, arteriovenous malformations, and cavernomas, all of which often present with intraventricular hemorrhage [1]. As far as the category of ruptured aneurysms, the anterior cerebral artery and the pericallosal artery seem to be the culprit most often [6]. Interestingly, iatrogenic hemorrhagic injury of the CC can occur in the setting of ECMO therapy. In such instances, multiple small hemorrhages have been identified [4]. Ventriculostomies in patients with abnormal coagulation and even nonlethal high-altitude cerebral edema may cause microhemorrhages within the CC [6]. Lastly, at least one case of CC hemorrhage in a patient with sickle cell disease (SCD) has been recorded in the literature. While spontaneous intracranial hemorrhage is unfortunately a well-known, serious complication of this hemoglobinopathy, it typically results in the development of an epidural hematoma [9]. In one case, a

42-year-old female suffered a ruptured left pericallosal artery aneurysm and CC hemorrhage and presented with alien hand syndrome (AHS). This syndrome is consistently associated with the disruption of callosal connections as seen in such vascular derangements. In this case, utilization of neuroimaging techniques—diffusion tensor tractography (DTT) and diffusion tensor imaging (DTI)—allowed illustration of what the authors believed were atypical neural connections that function to compensate as much as possible for CC fiber interruption. These tract changes were observed between CC fibers in the temporal lobes and inferior fronto-occipital fasciculus (IFOF) of both hemispheres made via the anterior commissure. Many other studies have also documented disruption of CC fibers following removal of hemorrhagic arteriovenous malformations, etc., but the findings in this study may be promising in that they gave reason to postulate that neural recovery—at least to a certain extent—is possible in patients who undergo such injuries [10].

It is also worth mentioning that a study aimed at elucidating the effects of traumatic brain injury (TBI) in the pediatric population demonstrated that simulated closed-head injuries in mouse models induce numerous changes in the quality of the vasculature of the ipsilateral CC, including vessel density, length, and number of junctions. Specifically, increased vascular density—observed via changes in measurement of the apparent diffusion coefficient—likely negatively impacts the functionality of the microvasculature of the white matter, thus inducing persistent changes in the CC. In other words, the CC is susceptible to microstructural changes and lesions because of traumatic injury and the ensuing long-term changes in its associated microvasculature [11].

19.2 Conclusions

The CC has been implicated in a wide array of conditions ranging from ubiquitous to exceedingly rare. Because of its lofty anatomical significance, it is imperative that we understand the mechanisms of potential injury to the CC as well as the clinical manifestations that are typically seen when damage does transpire. This way, preservation of this structure in the context of various vascular pathologies may be possible, thus enabling better outcomes for patients.

References

1. Renard D, Castelnovo G, Campello C, et al. An MRI review of acquired corpus callosum lesions. J Neurol Neurosurg Psychiatry. 2014;85(9):1041–8. https://doi.org/10.1136/jnnp-2013-307072.
2. Wilson CA, Mullen MT, Jackson BP, Ishida K, Messé SR. Etiology of Corpus callosum lesions with restricted diffusion. Clin Neuroradiol. 2017;27(1):31–7. https://doi.org/10.1007/s00062-015-0409-8.
3. Freeze WM, Zanon Zotin MC, Scherlek AA, et al. Corpus callosum lesions are associated with worse cognitive performance in cerebral amyloid angiopathy. Brain Commun. 2022;4(3):fcac105. https://doi.org/10.1093/braincomms/fcac105; Published 2022 Apr 26.
4. Tomimoto H, Lin JX, Matsuo A, Ihara M, Ohtani R, Shibata M, Miki Y, Shibasaki H. Different mechanisms of corpus callosum atrophy in Alzheimer's disease and vascular dementia. J Neurol. 2004;251(4):398–406. https://doi.org/10.1007/s00415-004-0330-6.
5. Hirono S, Kawauchi D, Kobayashi M, et al. Mechanism of corpus callosum infarction associated with acute hydrocephalus: clinical, surgical, and radiological evaluations for pathophysiology. World Neurosurg. 2019;127:e873–80. https://doi.org/10.1016/j.wneu.2019.03.288.
6. Fitsiori A, Nguyen D, Karentzos A, Delavelle J, Vargas MI. The corpus callosum: white matter or terra incognita. Br J Radiol. 2011;84(997):5–18. https://doi.org/10.1259/bjr/21946513.
7. Conklin J, Frosch MP, Mukerji S, et al. Cerebral microvascular injury in severe COVID-19. medRxiv. 2020:2020.07.21.20159376. https://doi.org/10.1101/2020.07.21.20159376; Published 2020 Jul 24. Preprint.
8. Hamad MK, et al. Corpus callosum hematoma, as a rare complication of COVID-19. Clin Case Rep. 2021;9(12):e05178.
9. Kotey SN, Dike NO, Nani E, Nyame K. Spontaneous epidural and corpus callosum hemorrhage in sickle cell disease—an unusual presentation in a Ghanaian patient. Cureus. 2020;12(12):e12292. https://doi.org/10.7759/cureus.12292; PMID: 33520496; PMCID: PMC7834524.
10. Jang SH, Yeo SS, Chang MC. Unusual compensatory neural connections following disruption of Corpus callosum fibers in a patient with Corpus callosum hemorrhage. Int J Neurosci. 2013;123(12):892–5.
11. Wendel KM, et al. Corpus callosum vasculature predicts white matter microstructure abnormalities after pediatric mild traumatic brain injury. J Neurotrauma. 2018;36(1):152–64.

20 Toxic Lesions of the Corpus Callosum

Fayize Maden Bedel and Nagehan Bilgeç

20.1 Introduction

The corpus callosum (CC) is a white matter structure containing more than 250 million axons that connects cerebral cortical areas. It consists of three parts: genu, splenium, and body. Various diseases can affect the CC, including congenital anomalies, traumatic lesions, ischemic diseases, tumors, and congenital acquired lesions including metabolic, toxic, degenerative, and demyelinating diseases.

Conditions that cause damage to the CC are grouped under the heading of cytotoxic lesions of the CC. In the pathophysiology of cytotoxic damage, macrophages become active after exposure to any toxic agent and release the inflammatory cytokines interleukin 1 (IL-1) and IL-6, initiating the cascade that leads to cytokinopathy. By releasing IL-1 and IL-6 from monocytes, T cells are activated. T cells stimulate endothelial cells to release TNF-alpha. With this mechanism, the blood-brain barrier is damaged. Astrocytes, in turn, are stimulated by IL-1 to release glutamate and block glutamate reuptake, thereby increasing extracellular glutamate. Microglia, which are macrophages of the central nervous system (CNS), are activated, and cytokine release is increased, which can cause demyelination. The result of this cytokinopathy has greatly increased the amounts of glutamate in the extracellular space at levels 100 times the normal level or more. The CC, which contains many glutamate receptors containing oligodendrocytes, is also likely to be affected, especially in the splenium.

In diffusion-weighted magnetic resonance imaging (MRI), these cytotoxic CC areas are manifested as low diffusion areas. Involvement of the CC typically shows one of the three patterns: (a) a small round or oval lesion located in the center of the splenium, (b) a lesion centered on the splenium and spreading to the white matter, and (c) a lesion located in the posterior part of the CC but spreading forward as well [1].

These cytotoxic lesions occur due to secondary causes such as drug therapy, malignancy, infections, trauma, alcohol, and carbon monoxide poisoning. In this section, we will talk about toxic CC lesions due to Marchiafava-Bignami disease, addictive substance use such as cocaine and heroin, and carbon monoxide poisoning.

20.2 Marchiafava-Bignami Disease

Marchiafava-Bignami disease (MBD) is a quite uncommon disorder characterized by demyelination and necrosis of the CC. Italian pathologists Ettore Marchiafava and Amico Bignami first described this disease in 1903 by showing CC necrosis in autopsies of a case with red wine

F. M. Bedel (✉) · N. Bilgeç
Department of Pediatric Genetics, Meram Medical Faculty, Necmettin Erbakan University, Konya, Turkey

M. Turgut et al. (eds.), *The Corpus Callosum*, https://doi.org/10.1007/978-3-031-38114-0_20

92.8% alcoholism and 39.8% malnutrition, MBD mimic cases did not have alcoholism or malnutrition. Altered mental state which included confusion, delirium, unconsciousness, impaired memory, or disorientation on admission; impaired walking; dysarthria; mutism; signs of disconnection or split brain syndrome; pyramidal signs; primitive reflexes; rigidity; incontinence; sensory symptoms; and gaze palsy or diplopia, although were seen in both groups, were also seen significantly more frequently in MBD cases. In MBD mimic cases, these clinical findings are milder, and the findings regressed within a few weeks. Although seizures were in both groups, they were detected more frequently in MBD mimic cases. Hemiparesis or tetraparesis was seen with the same frequency in MBD and MBD mimic cases. According to the onset of symptoms, the cases were categorized as acute phase for less than 2 weeks, subacute phase for more than 2 weeks, and chronic phase 3 months after hospitalization, and the location and characteristics of the CC lesions observed on diffusion-weighted imaging (DWI) were examined. In MBD cases, 1/3 of the splenium, genu, and body of the CC; 1/3 single splenium; and 1/3 splenium and genu/body involvement were present, whereas in MBD mimic cases, over 90% only was the splenium part involved. After admission, lesions in the CCof MBD mimic cases completely disappeared within 1 week, while only 4.3% had residual MRI lesions. In MBD cases, the CC lesions disappeared in 27.9% completely and 37.7% partially in MRI after the acute phase. DWI reveals the earliest signs of lesions and predicts the development of callosal necrosis, giving information about prognosis. While callosal atrophy or necrosis is frequently reported in MBD cases, no such finding was detected in MBD mimic cases. Deep white matter lesions were more common in MBD cases as 49.6% and cortical lesions 30.8% and deep white matter lesions in MBD mimic cases as 21.4% and cortical lesions 3.7%. In this study, they also observed that the prognosis of patients with MBD was related to the location of the lesion and the time of initiation of thiamine therapy [17].

Disease management is similar to that of Wernicke-Korsakoff syndrome or alcohol abuse disorder. Most MBD case reports have demonstrated a favorable response to intravenous administration of thiamine, folate, and vitamin B complexes as well as high-dose corticosteroids [2, 18]. Although clinical features may be variable and nonspecific, Marchiafava-Bignami disease should be considered, and early treatment should be initiated in patients with chronic alcohol abuse or malnutrition presenting with neurological symptoms.

20.3 The Abuse of Drugs Such as Cocaine and Heroin

Cocaine is an illegal substance obtained from the leaves of the coca plant, which grows in Central and South America, and can become addictive in a very short time. The coca plant goes through various processes, takes the form of powder, and is most used by sniffing. Apart from this, it can be smoked like a cigarette or taken into the body by intravenous injection. It causes excessive and continuous stimulation of neurons by affecting the reuptake of norepinephrine and serotonin, especially dopamine in synapses. In addition, it is known that it increases dopamine transmission in the mesocorticolimbic dopamine pathway, and its reinforcing effects appear in this way. Although there are many studies on the toxic effects of chronic use on the brain, the mechanism still has not been clearly elucidated [19].

Many studies have been conducted to date to define cocaine-induced white matter anomalies. It was observed that white matter changes were more common in cocaine users compared to nonusers. At the same time, these changes in white matter are also associated with duration of cocaine use [20].

CC rostral body changes observed in cocaine addicts may cause impairment in executive functions since this region contains crossed fibers from the prefrontal cortex and frontal cortex [21]. The isthmus contains transcallosal fibers from the parietal and temporal regions, so changes in the isthmus may affect the normal function of these regions. Cocaine-related functional differences have been reported in the parietal and temporal cortices (Witelson 1989) [22]. In addition, the passage of sensorimotor fibers from the isthmus suggests that changes in the isthmus may cause a decrease in the performance of the sensorimotor system. In conclusion, recent studies in animal models and human subjects suggest that the sensorimotor system may be affected by cocaine abuse [23].

Diffusion tensor imaging (DTI) is a method that shows microstructural changes and measures the limited diffusion of water in cell membranes and organelles, which disrupts the structural integrity of the brain. Fractional anisotropy (FA) is also one of the most widely used measurements of DTI [24]. In preclinical studies, lower FA was measured in rats exposed to chronic cocaine use. Although these studies support the hypothesis that FA is reduced in white matter in humans as well, not all cocaine users experience white matter changes.

On the other hand, there are studies on some genetic variants that may be associated with white matter changes in cocaine use. Twenty-one variants were detected in 17 candidate genes, with the most significant DTI findings associated with GAD1a and GAD1b polymorphisms. GAD1a polymorphisms were more consistent with typical forebrain white matter changes including the CC, corona radiata, internal capsule, and thalamus, with cocaine use disorder. These changes were also affected by variables such as duration of cocaine use and gender. Male sex was found to be more protective. Long-term use was also more likely to cause white matter changes [20].

In the study to detect microstructural changes in the CC due to cocaine use, thin-section and more sensitive DTI was used, low FA values were determined, and evidence supporting white matter pathology was obtained [8].

Heroin, one of the leading semisynthetic narcotic analgesics, is a highly addictive opium derivative with 4–10 times more analgesic effects than morphine [25]. Heroin is the strongest of drugs and therefore the most harmful. Heroin addiction is a major public health problem worldwide and is one of the most abused drugs. Heroin addiction is a complex brain disease, causing euphoric feelings and pain relief. Withdrawal symptoms include severe muscle and bone pain, runny nose, diarrhea, abdominal cramps, agitation, and anxiety.

Classical and molecular genetic studies have provided important information on the probability of individuals becoming addicted. The polygenic risk score based on the results of the genome-wide association study (GWAS) may be promising to evaluate the relationship between phenotype and genetic variants [26]. Chromosomes 2 and 17 and part of chromosome 14q have been found to be related to opioid addiction [27, 28]. However, it has been determined that several SNPs located on chromosome 2 may be associated with addiction [29].

With repeated opioid use, the opioid receptor develops tolerance, and the cAMP/protein kinase A cycle is upregulated at the cellular level. The change in glutaminergic projection in the prefrontal cortex and basal ganglia also decreases self-control and increases the desire for drugs [30].

White matter changes in heroin addicts were investigated using diffusion tensor imaging (DTI). DTI is a noninvasive MRI technique sensitive to diffusion of water molecules. Microstructural changes in white matter can be evaluated using DTI parameters. In a study with heroin addicts in the early and late stages of abstinence, DTI was used, and it was shown that the integrity of the CC was impaired in the early period of abstinence. It has been observed that the integrity of the corrupted CC is recovered after a long period of heroin abstinence. This study also suggested that the use of heroin may cause microstructural deterioration in white matter fibers [31].

Another study found that characteristic imaging and brain biopsy findings occur, especially with inhalation of heroin vapor. Symmetrical white matter hyperintense signal changes were detected in the posterior cerebrum, posterior arm of the internal capsule, splenium of the CC, medial lemniscus, and lateral brain stem with

MRI. After inhaling roinpyrolisate, spongiform leukoencephalopathy shows specific radiological findings with high-signal-intensity symmetrical lesions on T2-WI in the cerebellar and occipital white matter, posterior arm of the internal capsule, splenium of the CC, and medial lemniscus [32].

The CC volumes were compared in the study conducted with 45 heroin addicts and 35 nonaddicted individuals. Volume reduction was detected in the CC splenium and genu region in heroin-dependent individuals. It was determined that the duration of heroin use and impulsivity and CC genu and splenium volume were negatively correlated [9].

As a result, diffusion restriction in the CC in white matter areas with addictive substances was determined as the main pathology. In addition to this, a decrease in the volume of the CC has also been determined by some studies. The dose, duration, and mode of using the addictive substance also affect the white matter changes, especially in the CC.

20.4 Acute Carbon Monoxide Intoxication and the Corpus Callosum

Carbon monoxide (CO) poisoning is one of the most frequently reported toxicological causes of death. Poisoning occurs when CO is released as a result of incomplete combustion of carbon-bearing compounds.

CO is a colorless, tasteless, and odorless gas. CO is of endogenous and exogenous origin. Normally, serum carboxyhemoglobin (COHb) level is 0.4–0.7%. Since the ability of CO to bind to hemoglobin is 200–300 times greater than that of oxygen, it causes tissue hypoxia. CO shifts the oxyhemoglobin dissociation curve to the left and changes it from sigmoidal to hyperbolic in shape. As a result of all this, anaerobic glucose increases, and lactic acidosis occurs.

The clinical findings of CO poisoning are very variable and do not have specific findings. Moderate to mild CO poisoning can be confused with acute viral syndromes. Organs with high oxygen demand, such as the heart and brain, can easily become dysfunctional with CO poisoning. Symptoms and signs that develop due to CO poisoning become more evident with the increase in neurological and myocardial oxygen requirements. These patients state that after the onset of the first symptoms at rest, the symptoms increase with movement. Neurologic findings ranging from headache, dizziness, weakness, lethargy, confusion, agitation, depression, syncope, seizure, visual impairment, memory and gait disturbances, and coma may be detected. The most common symptom is headache. Findings vary according to serum carboxyhemoglobin level (Table 20.1).

The role of physical examination in diagnosis is limited. When poisoning is suspected, blood carboxyhemoglobin levels should be checked. The relationship between the severity of poisoning and the blood CO level is not strong, but it can be a good indicator for monitoring the response to treatment. However, blood gas evaluation, biochemical parameters (BUN, creatinine, electrolytes, creatinine phosphokinase), electrocardiogram, lung X-ray, computed tomography (CT), and cranial magnetic resonance imaging (MRI) should be requested.

CT can be used to detect accompanying findings such as intracranial hemorrhage in patients

Table 20.1 Symptoms and signs associated with HgbCO levels [33]

HgbCO levels	Symptoms and signs
10%–20%	Nausea, fatigue, emotional imbalance, confusion
21%–30%	Headache, exertional dyspnea, angina, visual impairment, mild inability to adapt to the environment, weakness in reacting to danger, mild weakness, weakening of senses
31%–40%	Dizziness, drowsiness, vomiting, visual disturbances, inability to make decisions
41%–50%	Syncope, changes in consciousness, forgetfulness, cardiac dysrhythmias
51%–60%	Seizures, coma, pulmonary edema, severe metabolic acidosis
>60%	Death

Table 20.2 Apparent diffusion coefficient (ADC) values in acute carbon monoxide (CO) intoxication patients and the control group [35]

Region	Control	Acute
CC—genu	768.9 ± 70.4	759.9 ± 122.7
CC—body	653.8 ± 49.0	768.1 ± 68.9
CC—splenium	743.6 ± 70.3	737.1 ± 68.7

with neurological deficits [1]. Cranial MRI is more effective in detecting lesions in the globus pallidus and white matter. Even in recent studies, diffusion-weighted MRI has been found to be superior to conventional MRI [34]. CO poisoning usually causes ischemic lesions in the bilateral globus pallidus of the basal ganglia as well as various regions including the CC, thalamus, hippocampus, periventricular white matter, and cerebral cortex. Therefore, brain MRI with diffusion-weighted imaging plays an important role in detecting areas of interest due to acute CO poisoning.

In the study in which the clinical, laboratory, and imaging findings of seven patients were evaluated, the carboxyhemoglobin level of a 31-year-old male patient with clouding of consciousness was found to be 8.3. In this patient, diffusion restriction was detected in bilateral globus pallidus, hippocampus, midbrain, cerebellum, splenium, right temporal cortex, and left temporal white matter on diffusion-weighted MRI [34].

In another study in which diffusion-based MRI and diffusion kurtosis imaging (DKI) findings of CO poisoning were defined, diffusion kurtosis MRI was used, and the mean kurtosis (MK) values were found to decrease in the CC. This decrease in MK value may be associated with degenerative changes and neuronal shrinkage; therefore, they suggested that the MK value is more sensitive in detecting encephalopathy that may occur in the early phase of CO poisoning. It has been stated that the damage to periventricular white matter structures such as the CC and cingulum is related to cognitive function deficits and memory disorders [34].

From another point of view, the response of the tissues to CO poisoning in the brain may not be the same. The changes in the diffusion coefficient (ADC) in the globus pallidus and CC were different from the other tissues and the control group (Table 20.2). As a result, the CC and globus pallidus were evaluated as the most sensitive area among all gray and white matter areas. In CO poisoning, the damage to the CC myelin is constant and varies from discrete to diffuse myelinopathy. In contrast to the white matter areas that remained unchanged in volume, general atrophy of the CC was observed in as little as 6 months. These ADC values in the CC were found to be correlated with cognitive test scores, suggesting that the damage in the CC is not clinically silent [35].

As a result, diffusion restriction is observed in the CC, which is more sensitive in acute CO poisoning. This finding is thought to be correlated with the decrease in cognitive functions.

20.5 Conclusion

The CC is one of the most important structures of the brain. Unfortunately, one of the areas most affected by the addictions of toxic substances such as alcohol, opioids, and carbon monoxide is the CC. Although diffusion restriction has generally been detected in studies conducted so far, more studies are needed to understand the microstructural changes related to the CC in drug and alcohol abusers.

References

1. Starkey J, Kobayashi N, Numaguchi Y, Moritani T. Cytotoxic lesions of the corpus callosum that show restricted diffusion: mechanisms, causes, and manifestations. Radiographics. 2017;37(2):562–76.
2. Tian TY, PescadorRuschel MA, Park S, Liang JW. Marchiafava-Bignami disease. In: StatPearls; 2022.

3. Bachar M, Fatakhov E, Banerjee C, Todnem N. Rapidly resolving nonalcoholic marchiafava-bignami disease in the setting of malnourishment after gastric bypass surgery. J Investig Med High Impact Case Rep. 2018;6:1–4.
4. Boutboul D, Lidove O, Aguilar C, Klein I, Papo T. Marchiafava-Bignami disease complicating SC hemoglobin disease and plasmodium falciparum infection. Presse Med. 2010;39(9):990–9.
5. Cui Y, Zheng L, Wang X, Zhang W, Yuan D, Wei Y. Marchiafava-Bignami disease with rare etiology: a case report. Exp Ther Med. 2015;9(4):1515–7.
6. Heinrich A, Runge U, Khaw AV. Clinicoradiologic subtypes, of Marchiafava-Bignami disease. J Neurol. 2004;251:1050–9.
7. Dong X, Bai C, Nao J. Clinical and radiological features of Marchiafava-Bignami disease. Medicine (Baltimore). 2018;97(5):e9626.
8. Ma L, Hasan KM, Steinberg JL, Narayana PA, Lane SD, Zuniga EA, Kramer LA, Moeller FG. Diffusion tensor imaging in cocaine dependence: regional effects of cocaine on corpus callosum and effect of cocaine administration route. Drug Alcohol Depend. 2009;104(3):262–7.
9. Qiu YW, Jiang GH, Ma XF, Su HH, Lv XF, Zhuo FZ. Aberrant interhemispheric functional and structural connectivity in heroin-dependent individuals. Addict Biol. 2017;22(4):1057–67.
10. Tung CS, Wu SL, Tsou JC, Hsu SP, Kuo HC, Tsui HW. Marchiafava-Bignami disease with widespread lesions and complete recovery. Am J Neuroradiol. 2010;31(8):1506–7.
11. Baker KG, Harding AJ, Halliday GM, Kril JJ, Harper CG. Neuronal loss in functional zones of the cerebellum of chronic alcoholics with and without Wernicke's encephalopathy. Neuroscience. 1999;91(2):429–38.
12. Fernandes LMP, Bezerra FR, Monteiro MC, Silva ML, de Oliveira FR, Lima RR, Fontes-Júnior EA, Maia CSF. Thiamine deficiency, oxidative metabolic pathways and ethanol-induced neurotoxicity: how poor nutrition contributes to the alcoholic syndrome, as Marchiafava-Bignami disease. Eur J Clin Nutr. 2017;71:580.
13. He X, Sullivan EV, Stankovic RK, Harper CG, Pfefferbaum A. Interaction of thiamine deficiency and voluntary alcohol consumption disrupts rat corpus callosum ultrastructure. Neuropsychopharmacology. 2007;32(10):2207–16.
14. Kumar A, LaVoie HA, DiPette DJ, Singh US. Ethanol neurotoxicity in the developing cerebellum: underlying mechanisms and implications. Brain Sci. 2013;3(2):941–63.
15. Zieve L. Influence of magnesium deficiency on the utilization of thiamine. Ann N Y Acad Sci. 1969;162(2):732–43.
16. Costin BN, Miles MF. Molecular and neurologic responses to chronic alcohol use. Handb Clin Neurol. 2014;125:157–71.
17. Hillbom M, Saloheimo P, Fujioka S, Wszolek ZK, Juvela S, Leone MA. Diagnosis and management of Marchiafava-Bignami disease: a review of CT/MRI confirmed cases. J Neurol Neurosurg Psychiatry. 2014;85(2):168–73.
18. Namekawa M, Nakamura Y, Nakano I. Cortical involvement in marchiafava-bignami disease can be a predictor of a poor prognosis: a case report and review of the literature. Intern Med. 2013;52(7):811–3.
19. Frazer KM, Richards Q, Keith DR. The long-term effects of cocaine use on cognitive functioning: a systematic critical review. Behav Brain Res. 2018;348(1):241–62.
20. Alballa T, Boone EL, Ma L, Snyder A, Moeller FG. Exploring the relationship between white matter integrity, cocaine use and GAD polymorphisms using Bayesian model averaging. PLoS One. 2021;16(7):e0254776.
21. Lee SH, Kim SS, Kim SH, Lee SY. Acute marchiafava-bignami disease with selective involvement of the precentral cortex and splenium: a serial magnetic resonance imaging study. Neurologist. 2011;17(4):213–7.
22. Witelson SF. Handandsexdifferences in theisthmusandgenu of thehumancorpuscallosum. A postmortem morphological study. Brain. 1989;112(Pt 3): 799–835.
23. Tomasi D, Goldstein RZ, Telang F, Maloney T, Alia-Klein N, Caparelli EC, Volkow ND. Thalamo-cortical dysfunction in cocaine abusers: implications in attention and perception. Psychiatry Res. 2007;155(3):189–201.
24. Taylor DG, Bushell MC. The spatial mapping of translational diffusion coefficients by the NMR imaging technique. Phys Med Biol. 1985;30(4):345–9.
25. Malow RM, West JA, Williams JL, Sutker PB. Personality disorders classification and symptoms in cocaine and opioid addicts. J Consult Clin Psychol. 1989;57(6):765–7.
26. Wang SC, Chen YC, Lee CH, Cheng CM. Opioid addiction, genetic susceptibility, and medical treatments: a review. Int J Mol Sci. 2019;20(17):4294.
27. Wetherill L, et al. Association of substance dependence phenotypes in the COGA sample. Addict Biol. 2015;20(3):617–27.
28. Volkow ND, Morales M. The brain on drugs: from reward to addiction. Cell. 2015;162(4):712–25.
29. Shen Y, Wang E, Wang X, Lou M. Disrupted integrity of white matter in heroin-addicted subjects at different abstinent time. J Addict Med. 2012;6(2):172–6.
30. Blasel S, Hattingen E, Adelmann M, Nichtweiß M, Zanella F, Weidauer S. Toxic leukoencephalopathy after heroin abuse without heroin vapor inhalation: MR imaging and clinical features in three patients. Clin Neuroradiol. 2010;20(1):48–53.
31. Chenoweth JA, Albertson TE, Greer MR. Carbon monoxide poisoning. Crit Care Clin. 2021;37(3):657–72.
32. Kim DM, Lee IH, Park JY, Hwang SB, Yoo DS, Song CJ. Acute carbon monoxide poisoning: MR imag-

ing findings with clinical correlation. Diagn Interv Imaging. 2017;98(4):299–306.

33. Chen NC, Huang CW, Lui CC, Lee CC, Chang WN, Huang SH, Chen C, Chang CC. Diffusion-weighted imaging improves prediction in cognitive outcome and clinical phases in patients with carbon monoxide intoxication. Neuroradiology. 2013;2013: 107–15.
34. Lane SD, Moeller FG, Steinberg JL, Buzby M, Kosten TR. Performance of cocaine dependent individuals and controls on a response inhibition task with varying levels of difficulty. Am J Drug Alcohol Abuse. 2007;33(5):717–26.
35. Silver DAT, Cross M, Fox B, Paxton RM. Computed tomography of the brain in acute carbon monoxide poisoning. Clin Radiol. 1996;51(7):480–3.

Table 21.1 The infections of the corpus callosum

Infection	Sign and symptoms	MRI findings of the CC
HHV-6	Acute encephalopathy accompanied by biphasic seizures and late reduced diffusion (AESD)	Hyperintensities in the CC
	Acute confusion and reduced LOC	
EBV	Headache	Reversible signal intensities in the splenium and anterior parts of the CC
	Disorientation and confusion	
	Ataxia	
	Seizures	
Influenza	Delirium and reduced LOC	MERS
	Hallucination	Additional callosal lesions that can be extending to the white matter
Rotavirus	Seizure	MERS
	Confusion	Focal or diffuse splenial lesions
	Agitation	
	Apnea	
	Mutism	
HIV	Dementia	Reduction in the CC thickness
	Motor speed loss	Reported diffusion abnormalities in the splenium
SARS-CoV-2	Deteriorated mental state	Small, oval-shaped, hyperintense splenial lesion in the CC
	Lethargy	Severe abnormal T2-weighted hyperintense signal
	Dysarthria	Restricted diffusion in the entire CC
	Ataxia	
	Auditory hallucinations	
	Restlessness	
	Suicidal ideations	
Measles virus	Tremor	Uniform hyperintensity of the splenium of CC on DWI
	Rigidity	
	Bradykinesia	
Legionella pneumophila serogroup 1	Dysarthria	Hypointensity and slight edema in the CC on T1-weighted images without enhancement
	Aphasia	Hyperintense on T2 weighted
	Limb apraxias	
	Tremor	
	Dysmetria on finger–nose testing	
Puumala virus	Headache	Hyperintense splenial lesion in the CC
Acute hepatitis	Confusion	MERS
	Dysarthria	
Staphylococcus aureus	Consciousness disturbance	MERS
E. coli	Consciousness disturbance	MERS
Salmonella enteritidis	Consciousness disturbance	MERS
	Hallucination	
Streptococcus	Severe headache	Hypointense lesions in the CC
	Nuchal rigidity	
Mycoplasma pneumoniae	Consciousness disturbance	MERS
Enterococcus faecalis	Consciousness disturbance	MERS
	Seizure	

Accordingly, infections in CNS can cause neural damage through the mentioned events [4–7].

In this chapter, we review the literature on the infectious pathogens that can affect the CC and discuss their pathogenesis (Table 21.1).

21.2 Viral Infections

21.2.1 Human Herpesviruses-6 (HHV-6)

The human herpesvirus (HHV) or herpes simplex virus (HSV) has the tendency to invade the nervous system and can cause human neurological pathologies. HHV-6 was first found in 1986 and was considered a T-lymphotropic virus. It is a member of the Herpesviridae family and Betaherpesvirinae subfamily. Two types were identified for this virus: HHV-6 A and B. Both of these were first found in the peripheral blood mononuclear cells of patients with lymphoproliferative diseases [8, 9]. These two variants have different genetic content and antigenic characteristics [10]. HHV-6 tends to involve the lymphoid tissue, epithelial and endothelial cells, and the CNS. It affects tissues via binding to its receptor which is the CD46, a widely expressed molecule on the nucleated cells and enters the cells. Like all of the Herpesviridae family, HHV-6 causes primary infection and also is capable of establishing a lifelong latent infection. Accordingly, it can cause intermittent reactivation, which may lead to different clinical manifestations [11–13]. Interestingly, it was reported that HHV-6 has been associated with the development of different neurologic pathologies, such as ataxia, hypersomnia, mild dementia, seizures, encephalitis, mesial temporal lobe epilepsy (MTLE), and multiple sclerosis (MS) [12, 14–17]. It was demonstrated that HHV-6 can cause neurological disorders through the state of latency after primary exposure and was concluded that, later, reactivation may lead to neurological symptoms. In addition to different factors associated with the host, differences in gene variations and antigenic specificity in the A and B variants contribute to manifesting various neurological symptoms [12]. Primary infection with HHV-6 is the second most common cause of acute encephalopathy after influenza [18].

The HHV-6 can involve the CC. It is commonly related to acute encephalopathy accompanied by biphasic seizures and late reduced diffusion (AESD). The concurrent mild AESD and mild reversible lesions in the splenium, which is the thickest and most posterior portion of the CC, were found [19, 20]. Transient lesions of the splenium can be detected on MRI with two patterns: well-circumscribed, small, oval lesions in the midline and more extensive and irregular lesions that are extended from the splenium into the hemispheres [21]. In a case report by Kato et al., reversible lesions on MRI of the cerebellum and the splenium of the CC in a 2-year-old girl with acute cerebellitis were reported. Titers of IgG and IgM as anti-HHV-6 were increased in her serum which was a sign of primary infection with HHV-6. She was treated with mannitol, dexamethasone, and acyclovir to target acute encephalitis. The lesion of the splenium was cured and disappeared 72 hours after the first MRI [22]. Barigou et al. reported a 41-year-old male with a previous HIV-1 infection who was admitted to their medical center due to acute confusion. On physical examination, confusion, periods of loss of contact, and a bilateral Babinski sign in the absence of any focal neurological deficit were observed. Hyperintensities in the CC were detected on the brain MRI. The PCR of the CSF was reported positive for HHV-6 and HHV-6 encephalitis was confirmed as the diagnosis. After receiving ganciclovir, in addition to oral levetiracetam and lacosamide, the patient's health condition improved, and HHV-6 viral load in both CSF and blood was decreased below the detection threshold. On brain MRI, the hyperintensity of the CC disappeared as well [23]. Kubo et al. also reported a 15-month-old girl with acute necrotizing encephalopathy (ANE). She presented with high fever and loss of appetite and had microcephaly, pachygyria, and absence of the CC. On brain MRI, symmetrically distributed lesions in the thalamus, cerebral white matter, basal nuclei, and brain stem were detected, which are characteristics of ANE. PCR of the serum was positive for HHV-6 and serum IgM for HHV-6 was also positive. Accordingly, she was diagnosed with HHV-6-associated ANE [24].

Fang et al. assessed the etiology of mild ence phalitis/encephalopathy that was accompanied by a reversible splenial lesion of the CC (MERS) in 29 children who had symptoms of acute encephalopathy. Eighteen out of 29 patients presented with seizures and loss of consciousness. Several pathogens were identified in the sera of patients which mostly were viruses. HSV was reported in 12.5% of cases. Patients received ganciclovir for 7 days, and after being discharged, they fully recovered without any sequelae, and neurological symptoms were absent [25].

Based on other investigations, HHV can be responsible for splenial lesions within the CC and related neurological symptoms.

21.2.2 Epstein–Barr Virus (EBV)

The Epstein–Barr virus (EBV) is a herpesvirus. The genome which is a linear DNA is surrounded by a nucleocapsid, and a viral lipoprotein envelope is present around this whole structure [26]. EBV was discovered 58 years ago by examining electron micrographs of cells cultured from Burkitt lymphoma tissue by Epstein, Achong, and Barr and was considered the first human tumor virus due to its ability to stimulate proliferation and cause cancer such as lymphoma [27, 28]. It can infect over 90% of humans and can be persistent for the rest of the patient's life due to the presence of memory B cells which are considered to be the site of persistence of EBV within the body [29]. B lymphocytes are major targets of EBV, and its pathogenesis occurs through the attachment of its glycoprotein gp350 to the CD21 receptor on the surface of B cells. Also, the binding of glycoprotein gp42, to human leukocyte antigen (HLA) CLASS II, which is considered as a co-receptor, assists this process [30, 31]. It can be transmitted by contact with oral secretions [32, 33]. Infection via EBV can cause a spectrum of diseases and clinical manifestations, which are associated with the host immune response. Mostly, infections with EBV are symptomless, particularly in children. Usually, in adults, it causes infectious mononucleosis, which presents with low-grade fever, generalized lymphadenopathy, sore throat, malaise, anorexia, tiredness, splenomegaly, and abnormal liver function tests [34]. The nervous system can be involved in EBV in 0.5 ± 7.5% of patients, especially in immunocompromised cases, and can result in CSF abnormalities. The neurological manifestations due to EBV infection include meningitis, encephalitis, cranial nerve palsies, myelitis, polyradiculitis, and neuropathies [34]. There are several studies that reported the pathogenesis of EBV within the CC. Hagemman et al. reported a 21-year-old male presented with a history of febrile infection, neuropsychologic symptoms, disorientation, ataxia, and generalized tonic–clonic seizures. The brain MRI showed reversible signal intensities in the splenium of the CC and posterior cerebral hemispheres. The serologic response was positive for EBV. PCR detected EBV DNA in peripheral mononuclear cells, as well, and EBV DNA PCR was also positive in the CSF. The EBV encephalitis was established as the diagnosis of this case. They concluded that EBV infection can be an important differential diagnosis of reversible splenial lesions in the CC [35]. In another study, a case of acute encephalitis/encep halopathy related to EBV was reported, which is a serious neurological complication of EBV. The symptoms included headache, recurrent tonic–clonic seizures, confusion, vomiting, and fever. The results of antibody testing and the quantitative RT-PCR assay of CSF were positive for EBV. The brain MRI revealed widespread hyperintensities in the splenium of the CC. After treating this patient with methylprednisolone pulse and ganciclovir, she completely recovered. Brain lesions and neurological sequelae were not detected [36]. A 13-year-old boy with EBV infection who presented with deteriorated mental status, hepatosplenomegaly, and hypercytokinemia was reported. He was treated with cyclosporine. MRI showed a hyperintense splenial lesion and mild involvement of the anterior CC. Accordingly, he was diagnosed with infection-associated cytotoxic lesions of the CC (CLOCCs). On histopathology assessment, infiltration of hemophagocytic histiocytes and atypical lymphocytes were detected which are consistent with EBV-associated hemophagocytic lymphohistiocytosis [37]. CLOCCs are secondary lesions related to various entities including drug therapy, malignancy, infection, subarachnoid hemorrhage, metabolic disorders, trauma, and other entities. It occurs following cell–cytokine interactions and causes significantly high levels of cytokines and extracellular glutamate. This can lead to injury and dysfunction of the callosal neurons and microglia [6]. Ishikura et al. also reported another patient with infection-associated CLOCC who was a 6-year-old girl that had a fever, hepatic dysfunction, and histiocytosis with hemophagocytosis. EBV-associated hemophagocytic lymphohistiocytosis was diagnosed in this patient. Axial diffusion-weighted MRI demonstrated an abnormality of the whole CC which was spread into the hemispheric white matter. This case report highlights the role of EBV in lesion formation within the CC and its outcomes [38]. Accordingly, EBV can be considered a pathogen responsible for lesions within the CC.

21.2.3 Influenza

Influenza viruses are important human pathogens that belong to the Orthomyxoviridae family and have a segmented RNA genome. It has three types, A, B, and C [39]. Influenza causes annual epidemics, worldwide outbreaks, and pandemics at irregular intervals. The frequent antigenic change in influenza results in new seasonal influenza which can be controlled by antigenically matched vaccines and medications against this virus. The attachment of a hemagglutinin surface glycoprotein of the virus to sialic acid on cells is the beginning of the pathogenesis of influenza [40, 41]. Sialic acid receptors are spread on cells of the respiratory tract, and the upper respiratory tract is the first site that influenza infects [42]. It mostly spreads via airborne droplets [43]. Influenza can be an asymptomatic infection or cause mild to severe infection in the respiratory tract or even systemic symptoms. The incubation period is 1–5 days. Nonspecific symptoms including fever, vomiting, diarrhea, cough, and rhinorrhea can be seen in children, and it can also cause febrile convulsions in the pediatric population [44, 45]. Influenza can have CNS involvement, particularly in children, which could be manifested by a variety of syndromes. The main CNS impairment is encephalitis or encephalopathy [46]. Myelitis and Guillain–Barre syndrome also are rare presentations caused by the influenza virus [47, 48]. Influenza is the most important pathogen of acute encephalopathy that has the highest incidence among the viruses that cause encephalopathy. It can cause different types of encephalopathy; especially, is is responsible for clinical manifestations in 40–50% of patients with Reye-like syndrome and ANE that have a high rate of mortality and neurological sequelae. Influenza encephalopathy was also responsible for 30% death in patients before the year 2000, but declined to about 15% after 2000. However, currently, influenza encephalopathy is a major cause of mortality and morbidity among children [18, 49]. There are several reports on the attribution of influenza viruses in the formation of lesions within the CC that are responsible for encephalitis/encephalopathy. Influenza A and B were reported to be the main cause of induction of MERS and delirious behavior [50, 51]. Hoshino et al., in a large case series, included 153 patients who were detected with transient splenial lesions in the CC; it was reported that the most common associated pathogen was influenza (34.4%) [20]. Takanashi et al. assessed 11 patients with influenza that were diagnosed with delirium. Accordingly, five patients had a reversible splenial lesion in the CC that had reduced diffusion on brain MRI [51]. Both focal and diffuse splenial lesions with additional callosal lesions that can extend into the white matter have been described in influenza A (H1N1) infection [52, 53]. Fluss et al., in a case study, reported a patient who had symptoms and signs of encephalopathy and had a positive nasopharyngeal swab for influenza A. The brain MRI suggested a focal lesion in the splenium and involvement of the right dentate nucleus as well [54]. Takanashi et al. reported a 7-year-old girl with mild encephalitis and transient splenial lesions of the CC that was suggested to be due to influenza. She presented with consciousness disturbance and hallucination on day 2 and recovered completely on day 4. The splenial lesions were completely improved on day 10 [55]. Takatsu et al. reported a 31-year-old woman who presented with decreased consciousness and olfactory disturbance following influenza A infection. On the brain MRI, presence of lesion in the splenium of CC was detected. She was diagnosed with MERS. The neurological impairment and the splenial lesion disappeared within 14 days without therapy. It was proposed that the neurological symptoms after the influenza infection could be due to different diagnoses, for example, viral encephalitis, medication-related encephalopathy, or MRES [56].

Accordingly, influenza viruses can be considered a pathogen that can involve the CC and cause neurological symptoms, particularly encephalitis (Fig. 21.1).

21.2.4 Rotavirus

Rotaviruses are viruses with non-enveloped double-stranded RNA (dsRNA). Rota in Latin means wheel; hence, it was assigned to this virus because of its morphology. Rotaviruses are considered an important pathogen that is responsible for life-threatening diarrhea in children under the age of 5 worldwide. The transmission of this virus is majorly through the fecal–oral route and by close contact [58]. *Rotavirus* binds to host cells by its outer capsid protein VP4 and binding partners on the host cell surface, such as sialoglycans and histo-blood group antigens (HBGAs) [59, 60]. This virus majorly causes localized infection in the intestine and gastrointestinal symptoms. However, it can infiltrate the blood, cause systemic infection, and be found in sites other than the intestines including blood, CSF, CNS, heart, endothelial cells, liver, and kidney [61]. It was reported that *Rotavirus* infections can cause CNS complications. Salmi et al. reported a 3-year-old girl with symptoms of acute gastroenteritis. Two days after the beginning of the symptoms, she had clonic symmetrical convulsions in the upper extremities, confusion, tetanic cramps, and aphasia. In the electron microscopy of the fecal specimen, rotaviruses were detected. After 3 months, the movements gradually improved. This result confirms the involvement of CNS in patients with *Rotavirus* [62]. A few studies assessed the white matter injury of neonates that had seizures or apnea related to the *Rotavirus* infection via MRI. Verboon et al. in their study assessed 8 infants with seizures during *Rotavirus* infection within 6 weeks after birth. Six of them had late-onset cystic periventricular leukomalacia, which is the most severe intracranial lesion that occurs in the periventricular white matter of preterm infants and causes adverse neurologic sequelae. They concluded that *Rotavirus* can result in neurological symptoms by causing necrosis within the white matter that is indistinguishable from cystic periventricular leukomalacia [63]. In another study by Nishimura et al., the detection of *Rotavirus* RNA in the blood and CSF of eight children with concurrent symptoms of gastroenteritis and convulsions in the acute stage was reported, which demonstrates that *Rotavirus* can penetrate the CNS through the blood vessels [64]. There is evidence that *Rotavirus* can involve the CC. The pattern of this involvement can be different and the splenial lesions can be focal or diffuse and cause cerebellitis [37, 65–67]. Kobata et al. reported a 2-year-old Japanese girl with diarrhea, vomiting, and fever who manifested generalized tonic–clonic convulsions, confusion, and agitation. The brain MRI showed a focal lesion in the central part of the splenium of the CC. Only *Rotavirus* particles were detected by electron microscopy of the stool specimen, and RT-PCR for group A rotavirus serotype G1 genomic RNA was positive. Recent infections by HSV and adenoviruses were ruled out via viral titer tests. Acute encephalopathy related to rotavirus gastroenteritis was the final diagnosis. On day 6, all abnormal findings on MRI disappeared and symptoms of CNS impairment were improved [68]. Yeom et al. in a case series of 18 neonates with seizures or apnea who showed periventricular white matter and CC inju-

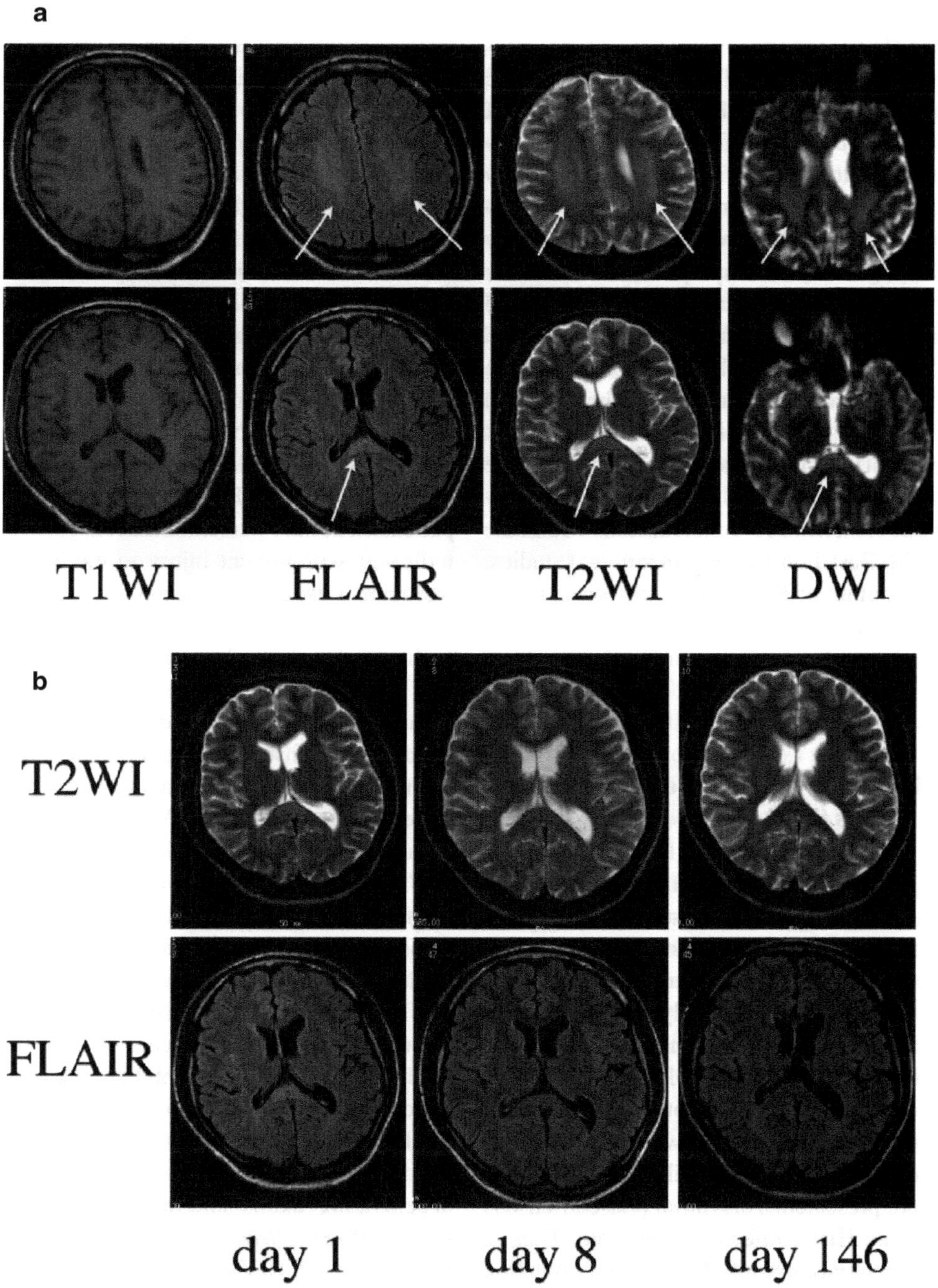

Fig. 21.1 Brain MRI of a 35-year-old patient with influenza type A virus infection who presented with acute progressive tetraplegia, transcortical motor aphasia, and a mild loss of consciousness. T2-weighted and diffusion-weighted images demonstrate lesions in the central splenium of the CC and symmetric bilateral white matter (**a**), and the lesion was disappeared in follow up MRI images (**b**) [57]. (Reprinted from Kimura E, Okamoto S, Uchida Y, Hirahara T, Ikeda T, Hirano T, Uchino M. A reversible lesion of the corpus callosum splenium with adult influenza-associated encephalitis/encephalopathy: a case report. J Med Case Rep. 2008 Jun 28;2:220. doi: 10.1186/1752-1947-2-220. PMID: 18588700; PMCID: PMC2474850. (Published under license to BioMed Central Ltd. This is an Open Access article distributed under the terms of the Creative Commons Attribution License (http://creativecommons.org/licenses/by/2.0)))

Barré syndrome, seizure, vascular events, and neurodegenerative disorders [105, 106].

There is growing evidence for the involvement of CC as one of the neurological complications of COVID-19. Gaur et al. reported two cases with COVID-19 related cytotoxic lesions of the CC. The first case was a 12-year-old boy with positive COVID-19 serology, who presented with fever, lethargy, severe headache, and gastrointestinal symptoms. Other infectious pathogens were ruled out. Consequently, a diagnosis of COVID-19 was established. The brain MRI demonstrated a small, oval-shaped, hyperintense splenial lesion in the CC that showed restricted diffusion. The second case was a 9-year-old boy with a positive SARS-CoV-2 RT-PCR, who presented with fever, deteriorated mental state, and lethargy. Dysarthria and ataxia were detected in the physical exam. On his brain MRI, a more severe abnormal T2-weighted hyperintense signal and restricted diffusion in the entire CC was detected. He rapidly recovered and the follow-up MRI showed a virtual complete fade of the splenial lesion [107]. Rasmussen et al. reported a 66-year-old COVID-19 patient with symptoms of fevers, chills, and severe headache with associated blurry vision, severe shortness of breath, and chest pain. On day 27 after admission, she was lethargic and aphasic with no neurologic recovery. In electroencephalography, a triphasic morphology, most consistent with toxic encephalopathy, was detected. On brain MRI, several sites with reduced diffusion within the CC, mostly in the splenium, with associated T2-FLAIR hyperintensities and areas of micro-hemorrhage were detected. They proposed that the cytokine storm associated with COVID-19 could be the main cause of involvement of CC and the subsequent hemorrhage in this area was a sign of stroke [108]. In another case series by Sparr et al., the ischemic infarction of the CC was reported in four confirmed COVID-19 patients who had clinical presentations of encephalopathy. They stated that the CC involvement in COVID-19 provided a window into the complexity of neurological impairments in this disease [109]. Edjlali et al. also reported two COVID-19 cases with acute encephalopathy who had lesions with T2-FLAIR hyperintensity and restricted diffusion in the splenium of the CC which was considered a cytotoxic lesion of the CC. They explained the detected lesions of CC by vulnerability of the splenium of CC to cytokinopathy followed by coronavirus infection [110]. Elkhaled et al. reported a confirmed COVID-19 case with no previous history of medical and psychiatric disorders presented with auditory hallucinations, restlessness, and suicidal ideations. On brain MRI, an isolated oval-shaped lesion in the splenium of CC was detected. They have mentioned the possibility of occurring atypical manifestations in SARS-CoV-2 infection and considering CLOCC as a differential diagnosis in patients with SARS-CoV-2 infection who have neurological and neuropsychiatric symptoms (Fig. 21.2) [111]. Charra et al. reported a COVID-19 patient with spontaneous hematoma in the CC. He had psychomotor agitation in the absence of any focal neurological deficit. They proposed that the history of recurrent thrombophlebitis, hematologic dysfunction, and hemodialysis and also the virus itself may be the cause of his intracranial hemorrhage [112]. These studies highlighted the influence of coronavirus on the CC.

21.2.7 Other Viruses

There are several studies that reported uncommon viral infections that were responsible for the CC pathology.

Malhotra et al., in a case series, reported a 10-year-old boy with a history of high-grade fever, rash, headache, and vomiting who was admitted with tremor, rigidity, and bradykinesia predominantly involving the upper limb. The history of seizures or altered sensorium was absent. MRI of the brain demonstrated T2WI hyperintense signals of bilateral substantia nigra and the uniform hyperintensity of the splenium of CC on DWI. After workups of the patient, raised IgM antibody titers for measles virus in the blood were reported. After receiving pramipexole, the extrapyramidal symptoms subsided, and the complete disappearance of altered signals in splenium was revealed on the MRI performed 10 days after admission [21].

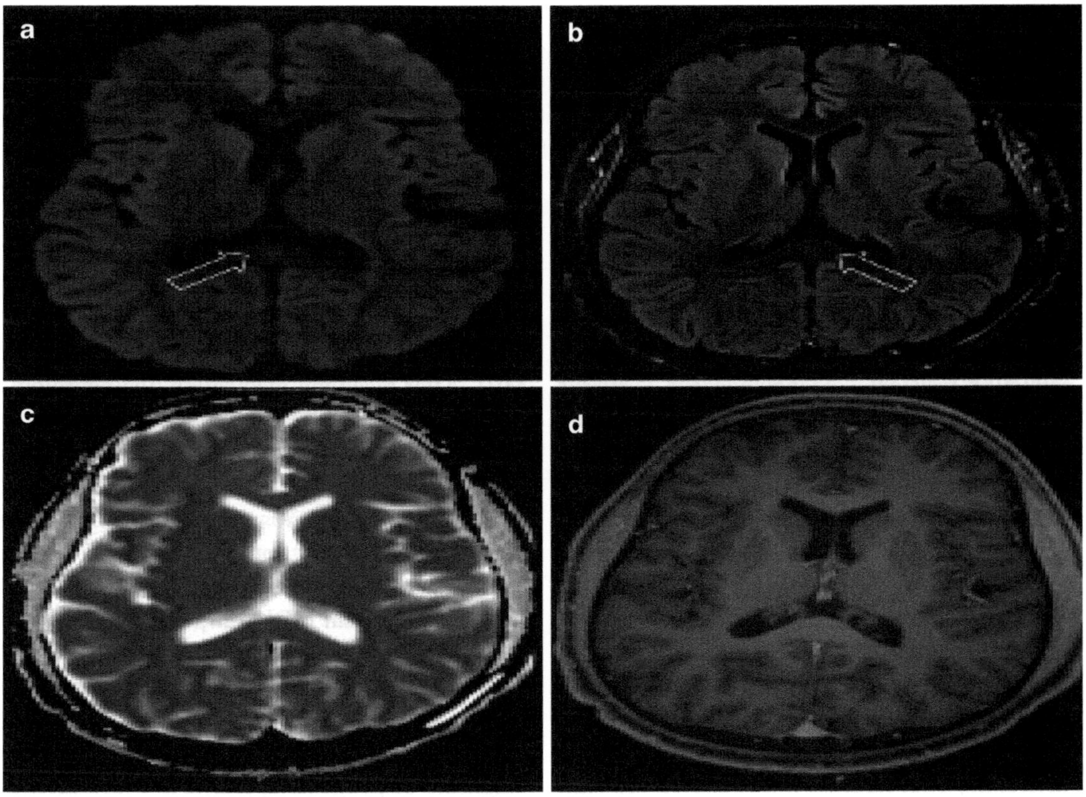

Fig. 21.2 Brain MRI of a 23-year-old case infected with SARS-CoV-2 presented with auditory hallucinations, restlessness, suicidal ideations, fever, malaise, headache, dizziness, and vomiting. A hyperintense signal in the splenium of the CC in diffusion-weighted (**a**) and fluid-attenuated inversion recovery (**b**) imaging, loss of signal on apparent diffusion coefficient maps (**c**) corresponding to restricted diffusion, and an isointense signal without contrast enhancement on T1-weighted images with contrast (**d**) are the sign of the presence of a cytotoxic lesion within the splenium of CC [111]. (Reprinted from Elkhaled W, Ben Abid F, Akhtar N, Abukamar MR, Ibrahim WH. A 23-Year-Old Man with SARS-CoV-2 Infection Who Presented with Auditory Hallucinations and Imaging Findings of Cytotoxic Lesions of the Corpus Callosum (CLOCC). Am J Case Rep. 2020 Dec 14;21:e928798. doi: 10.12659/AJCR.928798. PMID: 33315854; PMCID: PMC7749447. (Published under Creative Common Attribution-NonCommercial-NoDerivatives 4.0 International (CC BY-NC-ND 4.0)))

Steiner et al. reported a 38-year-old female presented with acute onset of headache, fever, and generalized fatigue. A hyperintense splenial lesion in the CC was detected on her cranial MRI. Puumala virus RNA was positive in the patient's blood sample. The Puumala hantavirus infection in humans predominantly follows the inhalation of contaminated aerosolized droplets contaminated with infected bank voles' excreta. After one week, the symptoms disappeared, and no neurological deficit was present. The 2 weeks' follow-up brain MRI suggested a regression of the splenial lesion [113].

Also, it has been reported MERS in a patient with acute hepatitis A presented with confusion, dysarthria, and fever. Intensified signal in the CC was seen on brain MRI. The mental status improved after 5 days, and the CC lesion disappeared after 15 days [114].

Tada et al., in a case series of patients with signs of encephalitis/encephalopathy and lesions within the CC, investigated the pathogens that were responsible for them. They reported that detected pathogens included influenza A, mumps virus, varicella-zoster virus, and adenovirus [115].

immediate medical care. Further, potentially fatal complications might arise in case of systemic infection. This bacteria can also cause extraintestinal manifestations [137]. One of the sites that can be affected by this pathogen is the CNS. Meningitis, brain abscess, and infected hematomas were reported as CNS involvements of *Salmonella* [138–141]. There is evidence that *Salmonella* can involve the CC. Previous studies reported CC changes as a result of typhoid fever [142]. Ahmed et al. reported a case who presented with high-grade fever, altered mental status, and seizures. Blood culture was positive for *Salmonella*. Brain MRI showed bilateral diffuse symmetrical hyperintense signal in the centrum semiovale, periventricular and deep white matter, and splenium of the CC which was a sign of demyelination [143]. However, there are few reports of encephalopathies and changes in the CC following *Salmonella enteritidis*. Martin et al. [144] reported a 15-year-old girl presented with fever, diarrhea and vomiting, confusion, and incontinency. Her CSF was normal but stool analysis showed *S. enteritidis* phage type 4. They believed the confusion could be a result of endotoxemia. In 2003, there was a report of reversible changes in the splenium of the CC and bilateral parietal subcortical white matter in a disorientated patient with irritation and hallucination following *S. enteritidis* gastroenteritis [145]. All MRI changes disappeared in 10 days. The authors proposed cytotoxic edema as the underlying pathophysiology.

21.3.4 *Streptococcus*

Streptococci are gram-positive catalase-negative cocci that are anaerobes and are members of the normal flora. Various classifications have been considered for streptococci based on the form of the colony, hemolysis, biochemical reactions, and serologic specificity. According to the type of hemolysis on blood agar, three types have been proposed: β-hemolytic, α hemolytic, and γ hemolytic. Based on the variations in cell wall carbohydrates, cell wall pili-associated structures, and the polysaccharide capsule, they were divided into groups A to V streptococci [146].

Allard et al. exposed pregnant rats to group B streptococcus (GBS) and demonstrated the offspring developed IUGR which persisted beyond adolescence and a demyelinated white matter in the CC adjacent to thinner primary motor cortices was detected [147]. Morii et al. reported a 47-year-old patient who presented with severe headache and fever. In neurological exams, nuchal rigidity was the only finding. CSF culture was positive for *Streptococcus pneumoniae*. Streptococcal meningitis was diagnosed in this patient. On brain MRI, an ovoid bright signal in the splenium of the CC was revealed. On the apparent diffusion coefficient map, the lesion was distinguished as hypointensity. After treatment, full recovery without any neurological sequelae was obtained for this case [148]. Choi et al. assessed the ischemic injury due to neonatal GBS meningitis in brain images of eight infants. On brain MRI, they found high-signal-intensity lesions in the genu and splenium of the CC of three infants with confirmed GBS [149]. Okumura et al. reported two infants with confirmed GBS meningitis who presented with fever, and one of them had seizures. Diffusion-weighted images of both cases demonstrated restricted water diffusion in the CC [150]. Fallata et al. reported a 17-day-old case with confirmed late-onset neonatal group B streptococcal (LOGBS) disease who presented with fever, lethargy, and convulsions. Brain MRI suggested moderate-to-diffuse bilateral acute subarachnoid hemorrhage and acute ischemic alterations with bilateral diffuse topography within the cortical gray matter, the CC, and internal capsules [151]. Based on the aforementioned studies, *Streptococcus* is considered another pathogen that can involve the CC and lead to neurological manifestations (Fig. 21.3).

21.3.5 *Mycoplasma pneumoniae* (MP)

Mycoplasmas are the smallest self-replicating prokaryotes, and because of the absence of the cell wall, they are pleomorphic and cannot be

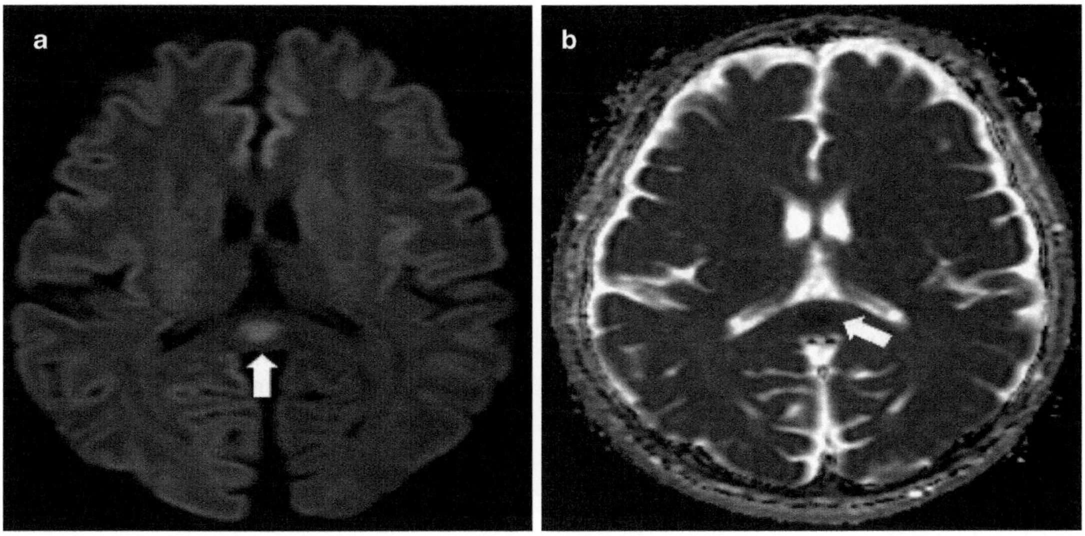

Fig. 21.3 MERS in the brain MRI of a 47-year-old man who was diagnosed with *Streptococcus pneumoniae*. (**a**) Axial diffusion-weighted image. (**b**) Axial apparent diffusion coefficient map [148]. (Reprinted from Morii K, Kogita Y, Takata M, Yamamoto T, Kishida H, Okushin H, Uesaka K. Reversible splenial lesion of the corpus callosum associated with bacterial meningitis. Int J Infect Dis. 2014 Feb;19:107-8. doi: 10.1016/j.ijid.2013.09.017. Epub 2013 Nov 1. PMID: 24188992. (open access article distributed under the terms of the Creative Commons CC-BY license))

detected on a gram stain [152]. However, they are considered to be related to the gram-positive bacterial group [153]. More than 200 types of mycoplasma species have been discovered so far. However, MP is the most prevalent in humans. MP is the leading cause of community-acquired pneumonia among children and young adults [154].

This infection results in different clinical manifestations and can be asymptomatic and cause fatal pneumonia or extrapulmonary symptoms [154]. The variety in symptoms is associated with the interaction between this pathogen and the host tissue. This interaction between MP and host cells takes place via adhesion molecules [155]. There is evidence that MP can cause a variety of extrapulmonary involvement, particularly in immunocompromised patients [156, 157]. Invasive CNS infection with MP in some cases has been reported. The involvement of the CC associated with MP has been reported as well. Akbar et al. described the first reported patient in North America who was diagnosed with MERS due to *Mycoplasma* infection [158]. Ueda et al. reported two cases with MP-associated MERS and also reviewed 13 MP patients with MERS who were reported in the literature and were diagnosed via serology tests. The most prevalent neurological manifestations were drowsiness, ataxia, seizure, delirium, confusion, tremor, hallucination, irritability, muscle weakness, and facial nerve paralysis. They reported lesions in brain MRI resolved after 18 days in most of the patients [159]. Yuan et al. described two children with MP-induced MERS who were diagnosed with MP infection via PCR and serum antibodies. Both of them manifested mild encephalopathy and had lesions in the central part of the splenium of the CC, which completely disappeared on the eighth day [160]. Dong et al. reported four cases of acute MP-associated encephalitis with consistent lesions in the splenium of the CC with a hyperintensity on DWI and hypointensities on T1WI, which were disappeared after treating with azithromycin or combined treatment with immunomodulators. They concluded that neuroimaging should be considered in patients with MP-associated encephalitis [161]. Shibuya et al.

in a case report also presented a 30-year-old patient who was infected with MP and experienced fever and loss of consciousness. On diffusion-weighted and T2-weighted brain MRI, high-density lesions within the splenium of the CC were detected which disappeared after 1 week [162].

21.3.6 *Enterococcus faecalis*

Enterococci are gram-positive coccus and normal flora of the gastrointestinal tract, the oral cavity, and the vagina. They can cause different diseases including infecting the urinary tract, bloodstream, endocardium, abdomen, burn wounds, CNS, lung, soft tissues, periodontal tissue, paranasal sinuses, eye, and ear. The pathogen of about 90% of human enterococcal infections is considered to be *Enterococcus faecalis* [163]. There are studies demonstrating infection of the CC via this pathogen. Kometani et al. reported two cases who were diagnosed with acute focal bacterial nephritis (AFBN) and presented fever, delirium, hyponatremia, and a significant increase in interleukin-6 in the CSF. The brain imaging suggested MERS that was proposed to be associated with AFBN. The result of urine culture showed *Enterococcus faecalis*. After antibiotic therapy, their mental status improved without any neurological sequelae. The intensified signals on brain MRI in the splenium of the CC were solved as well. They concluded the increased level of interleukin-6 within the CSF is a sign of remote activation of intracerebral immune response which might have a key role in the pathophysiology of MERS [164]. Cappellari et al., in a recent case report, described a 6-year-old boy with the chief complaint of prolonged seizure with fever, followed by loss of consciousness. Brain MRI findings revealed MERS within the splenium of the CC. Urinary tract infection by *Enterococcus faecalis* was detected. After treatment with antibiotics, neurological symptoms and brain MRI became normal. Inflammation of the CC and intramyelinic edema were proposed to be occurred due to infection with *Enterococcus faecalis* [165]. Further investigation is needed to confirm the association of this pathogen and the CC involvement.

21.4 Conclusion

The CC is the largest white matter structure within the brain, is located in the center of the brain, and has an essential role in the integration of information between the left and right brain hemispheres [1]. Different disorders and pathologies can target the CC such as congenital abnormalities, tumors, neurodegenerative disorders, inflammation, infections, metabolic disorders, trauma, vascular and hemorrhagic pathologies, and toxic agents [2]. MRI is the modality of choice for studying the CC and its pathological entities [2]. The lesions of the CC have been shown to be associated with various infectious diseases. In this chapter, we presented and summarized the major differential considerations for infections involving the CC. Two types of lesions can be detected in the CC due to the infection: (1) CLOCCs are secondary lesions related to various entities such as infection. It occurs following cell–cytokine interactions and causes significantly high levels of cytokines and extracellular glutamate. This can lead to injury and dysfunction of the callosal neurons and microglia [6]. (2) MERS is defined as mild encep halitis/encephalopathy that was accompanied by a reversible splenial lesion of the CC. MERS is predominantly caused by viral infections. The important pathogens that were reported to cause lesions within the CC were discussed in this chapter; HHV-6, influenza, *Rotavirus*, HIV, EBV, SARS-CoV-2, *Staphylococcus aureus*, *Escherichia coli*, *Salmonella enteritidis*, *Streptococcus*, *Mycoplasma pneumoniae*, and *Enterococcus faecalis*.

References

1. Goncalves-Ferreira A, Herculano-Carvalho M, Melancia J, Farias J, Gomes L. Corpus callosum: microsurgical anatomy and MRI. Surg Radiol Anat. 2002;23:409–14.

2. Fitsiori A, Nguyen D, Karentzos A, Delavelle J, Vargas MI. The corpus callosum: white matter or terra incognita. Br J Radiol. 2011;84:5–18.
3. Mirsattari SM, Lee DH, Jones MW, Blume WT. Transient lesion in the splenium of the corpus callosum in an epileptic patient. Neurology. 2003;60:1838–41.
4. Kim YS, Honkaniemi J, Sharp FR, Täuber MG. Expression of proinflammatory cytokines tumor necrosis factor-A and interleukin-1β in the brain during experimental group B Streptococcal meningitis. Mol Brain Res. 2004;128:95–102.
5. Phelps CP, Korneva EA. Cytokines and the brain. Amsterdam: Elsevier; 2008.
6. Starkey J, Kobayashi N, Numaguchi Y, Moritani T. Cytotoxic lesions of the corpus callosum that show restricted diffusion: mechanisms, causes, and manifestations. Radiographics. 2017;37:562–76.
7. Tisoncik JR, Korth MJ, Simmons CP, Farrar J, Martin TR, Katze MG. Into the eye of the cytokine storm. Microbiol Mol Biol Rev. 2012;76:16–32.
8. Lopez C, Pellett P, Stewart J, Goldsmith C, Sanderlin K, Black J, Warfield D, Feorino P. Characteristics of human herpesvirus-6. J Infect Dis. 1988;157:1271–3.
9. Salahuddin SZ, Ablashi DV, Markham PD, Josephs SF, Sturzenegger S, Kaplan M, Halligan G, Biberfeld P, Wong-Staal F, Kramarsky B. Isolation of a new virus, HBLV, in patients with lymphoproliferative disorders. Science. 1986;234:596–601.
10. Ablashi D, Agut H, Berneman Z, Campadelli Fiume G, Cardigan D, Ceccherini-Nelli L, Chandran B, Chou S, Collandre H, Cone R. Human herpesvirus-6 strain group: a nomenclature. Arch Virol. 1993;129:363.
11. De Bolle L, Naesens L, De Clercq E. Update on human herpesvirus 6 biology, clinical features, and therapy. Clin Microbiol Rev. 2005;18:217–45.
12. Yao K, Crawford JR, Komaroff AL, Ablashi DV, Jacobson S. Review part 2: human herpesvirus-6 in central nervous system diseases. J Med Virol. 2010;82:1669.
13. Zerr DM. Human herpesvirus 6: a clinical update. Herpes. 2006;13:20–4.
14. Birnbaum T, Padovan CS, Sporer B, Rupprecht TA, Ausserer H, Jaeger G, Pfister HW. Severe meningoencephalitis caused by human herpesvirus 6 type B in an immunocompetent woman treated with ganciclovir. Clin Infect Dis. 2005;40:887–9.
15. Cameron B, Flamand L, Juwana H, Middeldorp J, Naing Z, Rawlinson W, Ablashi D, Lloyd A. Serological and virological investigation of the role of the herpesviruses EBV, CMV and HHV-6 in post-infective fatigue syndrome. J Med Virol. 2010;82:1684–8.
16. Dewhurst S. Human herpesvirus type 6 and human herpesvirus type 7 infections of the central nervous system. Herpes. 2004;11:105a–11a.
17. Fotheringham J, Jacobson S. Human herpesvirus 6 and multiple sclerosis: potential mechanisms for virus-induced disease. Brain. 2005;78:36.
18. Mizuguchi M, Yamanouchi H, Ichiyama T, Shiomi M. Acute encephalopathy associated with influenza and other viral infections. Acta Neurol Scand. 2007;115:45–56.
19. Hatanaka M, Kashiwagi M, Tanabe T, Nakahara H, Ohta K, Tamai H. Overlapping MERS and mild AESD caused by HHV-6 infection. Brain Dev. 2015;37:334–8.
20. Hoshino A, Saitoh M, Oka A, Okumura A, Kubota M, Saito Y, Takanashi J-I, Hirose S, Yamagata T, Yamanouchi H. Epidemiology of acute encephalopathy in Japan, with emphasis on the association of viruses and syndromes. Brain Dev. 2012;34:337–43.
21. Malhotra HS, Garg RK, Vidhate MR, Sharma PK. Boomerang sign: clinical significance of transient lesion in splenium of corpus callosum. Ann Indian Acad Neurol. 2012;15:151.
22. Kato Z, Kozawa R, Kondo N, Hashimoto K. Transient lesion in the splenium of the corpus callosum in acute cerebellitis. J Child Neurol. 2003;18:291–2.
23. Barigou M, Garnier C, Debard A, Mengelle C, Dumas H, Porte L, Delobel P, Bonneville F, Marchou B, Martin-Blondel G. Favorable outcome of severe human herpes virus-6 encephalitis in an HIV-infected patient. AIDS. 2016;30:532–4.
24. Kubo T, Sato K, Kobayashi D, Motegi A, Kobayashi O, Takeshita S, Nonoyama S. A case of HHV-6 associated acute necrotizing encephalopathy with increase of Cd56bright NKcells. Scand J Infect Dis. 2006;38:1122–5.
25. Fang Q, Chen L, Chen Q, Lin Z, Yang F. Clinically mild encephalitis/encephalopathy with a reversible splenial lesion of corpus callosum in Chinese children. Brain Dev. 2017;39:321–6.
26. Kieff E. Epstein-Barr virus and its replication. In: Fields virology. Philadelphia: Lippincott Williams & Wilkins; 2007. p. 2603–54.
27. De-Thé G, Geser A, Day N, Tukei P, Williams E, Beri D, Smith P, Dean A, Bornkamm G, Feorino P. Epidemiological evidence for causal relationship between Epstein-Barr virus and Burkitt's lymphoma from Ugandan Prospective Study. Nature. 1978;274:756–61.
28. Epstein MA, Achong BG, Barr YM. Virus particles in cultured lymphoblasts from Burkitt's lymphoma. Lancet. 1964;283:702–3.
29. Babcock GJ, Decker LL, Volk M, Thorley-Lawson DA. EBV persistence in memory B cells in vivo. Immunity. 1998;9:395–404.
30. Borza CM, Hutt-Fletcher LM. Alternate replication in B cells and epithelial cells switches tropism of Epstein–Barr virus. Nat Med. 2002;8:594–9.
31. Nemerow G, Mold C, Schwend VK, Tollefson V, Cooper N. Identification of gp350 as the viral glycoprotein mediating attachment of Epstein-Barr virus (EBV) to the EBV/C3d receptor of B cells: sequence homology of gp350 and C3 complement fragment C3d. J Virol. 1987;61:1416–20.

32. Anagnostopoulos I, Hummel M, Kreschel C, Stein H. Morphology, immunophenotype, and distribution of latently and/or productively Epstein-Barr virus-infected cells in acute infectious mononucleosis: implications for the interindividual infection route of Epstein-Barr Virus. Blood. 1995;85:744.
33. Yao Q, Rickinson A, Epstein M. A re-examination of the Epstein-Barr virus carrier state in healthy seropositive individuals. Int J Cancer. 1985;35:35–42.
34. Portegies P, Corssmit N. Epstein-Barr virus and the nervous system. Curr Opin Neurol. 2000;13:301–4.
35. Hagemann G, Mentzel HJ, Weisser H, Kunze A, Terborg C. Multiple reversible MR signal changes caused by Epstein-Barr virus encephalitis. AJNR Am J Neuroradiol. 2006;27:1447–9.
36. Zhang S, Feng J, Shi Y. Transient widespread cortical and splenial lesions in acute encephalitis/encephalopathy associated with primary Epstein–Barr virus infection. Int J Infect Dis. 2016;42:7–10.
37. Moritani T, Capizzano A, Kirby P, Policeni B. Viral infections and white matter lesions. Radiol Clin. 2014;52:355–82.
38. Ishikura R, Ando K, Hirota S, Okamoto N, Fatterpekar G, Sacher M. Callosal and diffuse white matter lesions with restricted water diffusion in hemophagocytic syndrome. Magn Reson Med Sci. 2010;9:91–4.
39. Kendal A, Harmon MW. Orthomyxoviridae: the influenza viruses. Laboratory diagnosis of infectious diseases principles and practice. Berlin: Springer; 1988.
40. Matrosovich M, Zhou N, Kawaoka Y, Webster R. The surface glycoproteins of H5 influenza viruses isolated from humans, chickens, and wild aquatic birds have distinguishable properties. J Virol. 1999;73:1146–55.
41. Yamada S, Suzuki Y, Suzuki T, Le MQ, Nidom CA, Sakai-Tagawa Y, Muramoto Y, Ito M, Kiso M, Horimoto T. Haemagglutinin mutations responsible for the binding of H5N1 influenza A viruses to human-type receptors. Nature. 2006;444:378–82.
42. Matrosovich MN, Matrosovich TY, Gray T, Roberts NA, Klenk H-D. Human and avian influenza viruses target different cell types in cultures of human airway epithelium. Proc Natl Acad Sci. 2004;101:4620–4.
43. Tellier R. Review of aerosol transmission of influenza A virus. Emerg Infect Dis. 2006;12:1657–62.
44. Brocklebank J, Court S, Mcquillin J, Gardner P. Influenza-A infection in children. Lancet. 1972;300:497–500.
45. Price D, Postlethwaite R, Longson M. Influenzavirus A2 infections presenting with febrile convulsions and gastrointestinal symptoms in young children. Clin Pediatr. 1976;15:361–7.
46. Koskiniemi M, Rantalaiho T, Piiparinen H, Von Bonsdorff C-H, Färkkilä M, Järvinen A, Kinnunen E, Koskiniemi S, Mannonen L, Muttilainen M. Infections of the central nervous system of suspected viral origin: a collaborative study from Finland. J Neurovirol. 2001;7:400–8.
47. Jacobs B, Rothbarth P, Van Der Meché F, Herbrink P, Schmitz P, De Klerk M, Van Doorn P. The spectrum of antecedent infections in Guillain-Barré syndrome: a case-control study. Neurology. 1998;51:1110–5.
48. Salonen O, Koskiniemi M, Saari A, Myllyla V, Pyhälä R, Airaksinen L, Vaheri A. Myelitis associated with influenza A virus infection. J Neurovirol. 1997;3:83–5.
49. Mizuguchi M. Acute necrotizing encephalopathy of childhood: a novel form of acute encephalopathy prevalent in Japan and Taiwan. Brain Dev. 1997;19:81–92.
50. Takanashi J-I. Two newly proposed infectious encephalitis/encephalopathy syndromes. Brain Dev. 2009a;31:521–8.
51. Takanashi J-I, Tada H, Kuroki H, Barkovich AJ. Delirious behavior in influenza is associated with a reversible splenial lesion. Brain Dev. 2009;31:423–6.
52. Iwata A, Matsubara K, Nigami H, Kamimura K, Fukaya T. Reversible splenial lesion associated with novel influenza A (H1N1) viral infection. Pediatr Neurol. 2010;42:447–50.
53. Iype M, Ahamed S, Thomas B, Kailas L. Acute encephalopathy with a lesion of the splenium of the corpus callosum—a report of two cases. Brain Dev. 2012;34:322–4.
54. Fluss J, Ferey S, Menache-Starobinski C, Delavelle J, Van Bogaert P, Vargas MI. Mild influenza-associated encephalopathy/encephalitis with a reversible splenial lesion in a Caucasian child with additional cerebellar features. Eur J Paediatr Neurol. 2010;14:97–100.
55. Takanashi J-I, Barkovich AJ, Yamaguchi K-I, Kohno Y. Influenza-associated encephalitis/encephalopathy with a reversible lesion in the splenium of the corpus callosum: a case report and literature review. Am J Neuroradiol. 2004;25:798–802.
56. Takatsu H, Ishimaru N, Ito M, Kinami S. Mild encephalitis/encephalopathy with a reversible splenial lesion in an adult patient with influenza. Intern Med. 2017:8997–17.
57. Kimura E, Okamoto S, Uchida Y, Hirahara T, Ikeda T, Hirano T, Uchino M. A reversible lesion of the corpus callosum splenium with adult influenza-associated encephalitis/encephalopathy: a case report. J Med Case Rep. 2008;2:220.
58. Knipe D, Howley P, Griffin D, Lamb R, Martin M, Roizman B, Straus S. Fields virology, vol. 1 and 2. Philadelphia: Lippincott Williams & Wilkins; 2013.
59. Arias CF, Silva-Ayala D, López S. Rotavirus entry: a deep journey into the cell with several exits. J Virol. 2015;89:890–3.
60. López S, Arias CF. Multistep entry of rotavirus into cells: a versaillesque dance. Trends Microbiol. 2004;12:271–8.
61. Blutt SE, Matson DO, Crawford SE, Staat MA, Azimi P, Bennett BL, Piedra PA, Conner ME. Rotavirus antigenemia in children is associated with viremia. PLoS Med. 2007;4:E121.

62. Salmi TT, Arstila P, Koivikko A. Central nervous system involvement in patients with rotavirus gastroenteritis. Scand J Infect Dis. 1978;10:29–31.
63. Verboon-Maciolek MA, Truttmann AC, Groenendaal F, Skranes J, Døllner H, Hunt RW, Hayman M, Dippersloot RJ, Van Loon AM, De Vries LS. Development of cystic periventricular leukomalacia in newborn infants after rotavirus infection. J Pediatr. 2012;160:165–168.e1.
64. Nishimura S, Ushijima H, Nishimura S, Shiraishi H, Kanazawa C, Abe T, Kaneko K, Fukuyama Y. Detection of rotavirus in cerebrospinal fluid and blood of patients with convulsions and gastroenteritis by means of the reverse transcription polymerase chain reaction. Brain Dev. 1993;15:457–9.
65. Jang YY, Lee KH. Transient splenial lesion of the corpus callosum in a case of benign convulsion associated with rotaviral gastroenteritis. Korean J Pediatr. 2010;53:859.
66. Kato Z, Orii KE, Morimoto M, Sasai H, Funato M, Sawatari S, Kuwabara H, Kondo N. A transient lesion in the corpus callosum during rotavirus infection. Pediatr Neurol. 2009;41:467–9.
67. Mazur-Melewska K, Jonczyk-Potoczna K, Szpura K, Biegański G, Mania A, Kemnitz P, Służewski W, Figlerowicz M. Transient lesion in the splenium of the corpus callosum due to rotavirus infection. Childs Nerv Syst. 2015;31:997–1000.
68. Kobata R, Tsukahara H, Nakai A, Tanizawa A, Ishimori Y, Kawamura Y, Ushijima H, Mayumi M. Transient MR signal changes in the splenium of the corpus callosum in rotavirus encephalopathy: value of diffusion-weighted imaging. J Comput Assist Tomogr. 2002;26:825–8.
69. Yeom JS, Kim Y-S, Seo J-H, Park JS, Park ES, Lim J-Y, Woo H-O, Youn H-S, Choi DS, Chung J-Y, Han T-H, Park C-H. Distinctive pattern of white matter injury in neonates with rotavirus infection. Neurology. 2015;84:21–7.
70. Shiihara T, Watanabe M, Honma A, Kato M, Morita Y, Ichiyama T, Maruyama K. Rotavirus associated acute encephalitis/encephalopathy and concurrent cerebellitis: report of two cases. Brain Dev. 2007;29:670–3.
71. Chiu I-M, Yaniv A, Dahlberg JE, Gazit A, Skuntz SF, Tronick SR, Aaronson SA. Nucleotide sequence evidence for relationship of AIDS retrovirus to lentiviruses. Nature. 1985;317:366–8.
72. Levy JA. HIV pathogenesis: knowledge gained after two decades of research. Adv Dent Res. 2006;19:10–6.
73. Nilsson J, Kinloch-De-Loes S, Granath A, Sönnerborg A, Goh L-E, Andersson J. Early immune activation in gut-associated and peripheral lymphoid tissue during acute HIV infection. AIDS. 2007;21:565–74.
74. Fang RHT, Colantonio AD, Uittenbogaart CH. The role of the thymus in HIV infection: a 10 year perspective. AIDS. 2008;22:171–84.
75. Berger EA, Murphy PM, Farber JM. Chemokine receptors as HIV-1 coreceptors: roles in viral entry, tropism, and disease. Annu Rev Immunol. 1999;17:657–700.
76. Heaton RK, Franklin DR, Ellis RJ, Mccutchan JA, Letendre SL, Leblanc S, Corkran SH, Duarte NA, Clifford DB, Woods SP. HIV-associated neurocognitive disorders before and during the era of combination antiretroviral therapy: differences in rates, nature, and predictors. J Neurovirol. 2011;17:3–16.
77. Lescure F-X, Omland LH, Engsig FN, Roed C, Gerstoft J, Pialoux G, Kronborg G, Larsen CS, Obel N. Incidence and impact on mortality of severe neurocognitive disorders in persons with and without HIV infection: a Danish Nationwide Cohort Study. Clin Infect Dis. 2011;52:235–43.
78. Levy RM, Bredesen DE, Rosenblum ML. Neurological manifestations of the acquired immunodeficiency syndrome (AIDS): experience at UCSF and review of the literature. J Neurosurg. 1985;62:475–95.
79. Sailasuta N, Ross W, Ananworanich J, Chalermchai T, Degruttola V, Lerdlum S, Pothisri M, Busovaca E, Ratto-Kim S, Jagodzinski L. Change in brain magnetic resonance spectroscopy after treatment during acute HIV infection. PLoS One. 2012;7:E49272.
80. Valcour V, Chalermchai T, Sailasuta N, Marovich M, Lerdlum S, Suttichom D, Suwanwela NC, Jagodzinski L, Michael N, Spudich S. Central nervous system viral invasion and inflammation during acute HIV infection. J Infect Dis. 2012;206:275–82.
81. Bell J. An update on the neuropathology of HIV in the HAART era. Histopathology. 2004;45:549–59.
82. Navia BA, Cho ES, Petito CK, Price RW. The AIDS dementia complex: II. Neuropathology. Ann Neurol. 1986;19:525–35.
83. Filippi CG, Uluğ AM, Ryan E, Ferrando SJ, Van Gorp W. Diffusion tensor imaging of patients with HIV and normal-appearing white matter on MR images of the brain. Am J Neuroradiol. 2001;22:277–83.
84. Pomara N, Crandall DT, Choi SJ, Johnson G, Lim KO. White matter abnormalities in HIV-1 infection: a diffusion tensor imaging study. Psychiatry Res Neuroimaging. 2001;106:15–24.
85. Ances BM, Ortega M, Vaida F, Heaps J, Paul R. Independent effects of HIV, aging, and HAART on brain volumetric measures. J Acquir Immune Defic Syndr. 2012;59:469.
86. Northam GB, Liégeois F, Chong WK, Wyatt JS, Baldeweg T. Total brain white matter is a major determinant of IQ in adolescents born preterm. Ann Neurol. 2011;69:702–11.
87. Thompson PM, Dutton RA, Hayashi KM, Lu A, Lee SE, Lee JY, Lopez OL, Aizenstein HJ, Toga AW, Becker JT. 3D mapping of ventricular and corpus callosum abnormalities in HIV/AIDS. NeuroImage. 2006;31:12–23.

88. Waller AV. Experiments on the section of the glossopharyngeal and hypoglossal nerves of the frog, and observations of the alterations produced thereby in the structure of their primitive fibres. Abstracts of the papers communicated to The Royal Society Of London, 1851. The Royal Society London. p. 924–5.
89. Wu Y, Storey P, Cohen BA, Epstein LG, Edelman RR, Ragin AB. Diffusion alterations in corpus callosum of patients with HIV. AJNR Am J Neuroradiol. 2006;27:656–60.
90. Thompson P, Moussai J, Zohoori S, Goldkorn A, Khan A, Mega M, Small G, Cummings J, Toga A. Cortical variability and asymmetry in normal aging and Alzheimer's disease. Cereb Cortex. 1998;8:492–509.
91. Thompson PM, Hayashi KM, De Zubicaray G, Janke AL, Rose SE, Semple J, Herman D, Hong MS, Dittmer SS, Doddrell DM. Dynamics of gray matter loss in Alzheimer's disease. J Neurosci. 2003;23:994–1005.
92. Leite SCB, Corrêa DG, Doring TM, Kubo TTA, Netto TM, Ferracini R, Ventura N, Bahia PRV, Gasparetto EL. Diffusion tensor MRI evaluation of the corona radiata, cingulate gyri, and corpus callosum in HIV patients. J Magn Reson Imaging. 2013;38:1488–93.
93. Wohlschlaeger J, Wenger E, Mehraein P, Weis S. White matter changes in HIV-1 infected brains: a combined gross anatomical and ultrastructural morphometric investigation of the corpus callosum. Clin Neurol Neurosurg. 2009;111:422–9.
94. Garvey LJ, Pavese N, Politis M, Ramlackhansingh A, Brooks DJ, Taylor-Robinson SD, Winston A. Increased microglia activation in neurologically asymptomatic HIV-infected patients receiving effective art. AIDS. 2014;28:67–72.
95. Hoare J, Fouche J-P, Phillips N, Joska JA, Paul R, Donald KA, Thomas KGF, Stein DJ. White matter micro-structural changes in art-naive and art-treated children and adolescents infected with HIV in South Africa. AIDS. 2015;29:1793–801.
96. Gosztonyi G, Artigas J, Lamperth L, Webster HD. Human immunodeficiency virus (HIV) distribution in HIV encephalitis: study of 19 cases with combined use of in situ hybridization and immunocytochemistry. J Neuropathol Exp Neurol. 1994;53:521–34.
97. Ksiazek TG, Erdman D, Goldsmith CS, Zaki SR, Peret T, Emery S, Tong S, Urbani C, Comer JA, Lim W. A novel coronavirus associated with severe acute respiratory syndrome. N Engl J Med. 2003;348:1953–66.
98. Li Q, Guan X, Wu P, Wang X, Zhou L, Tong Y, Ren R, Leung KS, Lau EH, Wong JY. Early transmission dynamics in Wuhan, China, of novel coronavirus–infected pneumonia. N Engl J Med. 2020;382:1199.
99. Ciotti M, Ciccozzi M, Terrinoni A, Jiang W-C, Wang C-B, Bernardini S. The Covid-19 pandemic. Crit Rev Clin Lab Sci. 2020;57:365–88.
100. Huang C, Wang Y, Li X, Ren L, Zhao J, Hu Y, Zhang L, Fan G, Xu J, Gu X. Clinical features of patients infected with 2019 novel coronavirus in Wuhan, China. Lancet. 2020;395:497–506.
101. Hamming I, Timens W, Bulthuis M, Lely A, Navis GV, Van Goor H. Tissue distribution of Ace2 protein, the functional receptor for SARS coronavirus. A first step in understanding SARS pathogenesis. J Pathol. 2004;203:631–7.
102. Renu K, Prasanna PL, Gopalakrishnan AV. Coronaviruses pathogenesis, comorbidities and multi-organ damage—a review. Life Sci. 2020;255:117839.
103. Wu Y, Xu X, Chen Z, Duan J, Hashimoto K, Yang L, Liu C, Yang C. Nervous system involvement after infection with Covid-19 and other coronaviruses. Brain Behav Immun. 2020;87:18–22.
104. Xiang P, Xu X, Gao L, Wang H, Xiong H, Li R. First case of 2019 novel coronavirus disease with encephalitis, China, XIV, 2020, p. 202003**,** 00015.
105. Nagu P, Parashar A, Behl T, Mehta V. CNS implications of Covid-19: a comprehensive review. Rev Neurosci. 2021;32:219–34.
106. Sadeghmousavi S, Rezaei N. Covid-19 and multiple sclerosis: predisposition and precautions in treatment. SN Comprehens Clin Med. 2020;2:1802–7.
107. Gaur P, Dixon L, Jones B, Lyall H, Jan W. Covid-19-associated cytotoxic lesions of the corpus callosum. Am J Neuroradiol. 2020;41:1905–7.
108. Rasmussen C, Niculescu I, Patel S, Krishnan A. Covid-19 and involvement of the corpus callosum: potential effect of the cytokine storm? Am J Neuroradiol. 2020;41:1625–8.
109. Sparr SA, Bieri PL. Infarction of the splenium of the corpus callosum in the age of Covid-19. Stroke. 2020;51:E223–6.
110. Edjlali M, Le Gal A, Louvet M, Matt M, Leveque C, Diffre C, Orlikowski D, Annane D, Carlier R-Y, Garches COVID-19 Collaborative Group. Teaching neuroimages: cytotoxic lesions of the corpus callosum in encephalopathic patients with Covid-19. Neurology. 2020;95:1021–2.
111. Elkhaled W, Ben Abid F, Akhtar N, Abukamar MR, Ibrahim WH. A 23-year-old man with Sars-Cov-2 infection who presented with auditory hallucinations and imaging findings of cytotoxic lesions of the corpus callosum (CLOCC). Am J Case Rep. 2020;21:E928798.
112. Charra B, Ellouadghiri A, Kebbou T, Ettouki O, El Benna N, Afif MH, Gharbi MB. Acute spontaneous hematoma of the corpus callosum in a Covid-19 patient: a case report. Pan Afr Med J. 2021;38:263.
113. Steiner T, Ettinger J, Peng Z, Hofmann J, Hartmann M, Burkhardt U, Schnitzler P. Hyperintense lesion in the corpus callosum associated with Puumala hantavirus infection. J Neurol. 2012;259:1742–5.
114. Ko SY, Kim BK, Kim DW, Kim JH, Choe WH, Seo HY, Kwon SY. Reversible splenial lesion on the corpus callosum in nonfulminant hepatitis a

presenting as encephalopathy. Clin Mol Hepatol. 2014;20:398–401.

115. Tada H, Takanashi J, Barkovich AJ, Oba H, Maeda M, Tsukahara H, Suzuki M, Yamamoto T, Shimono T, Ichiyama T, Taoka T, Sohma O, Yoshikawa H, Kohno Y. Clinically mild encephalitis/encephalopathy with a reversible splenial lesion. Neurology. 2004;63:1854–8.
116. Lowy FD. Staphylococcus aureus infections. N Engl J Med. 1998;339:520–32.
117. Sheagren JN. Staphylococcus aureus: the persistent pathogen. N Engl J Med. 1984;310:1437–42.
118. Deleo FR, Diep BA, Otto M. Host defense and pathogenesis in Staphylococcus aureus infections. Infect Dis Clin N Am. 2009;23:17–34.
119. Tong SY, Davis JS, Eichenberger E, Holland TL, Fowler VG Jr. Staphylococcus aureus infections: epidemiology, pathophysiology, clinical manifestations, and management. Clin Microbiol Rev. 2015;28:603–61.
120. Jensen AG, Espersen F, Skinhøj P, Rosdahl VT, Frimodt-Møller N. Staphylococcus aureus meningitis: a review of 104 nationwide, consecutive cases. Arch Intern Med. 1993;153:1902–8.
121. Pedersen M, Benfield TL, Skinhoej P, Jensen AG. Haematogenous Staphylococcus aureus meningitis. A 10-year nationwide study of 96 consecutive cases. BMC Infect Dis. 2006;6:1–7.
122. Aguilar J, Urday-Cornejo V, Donabedian S, Perri M, Tibbetts R, Zervos M. Staphylococcus aureus meningitis: case series and literature review. Medicine. 2010;89:117–25.
123. Kosami K, Kenzaka T, Sagara Y, Minami K, Matsumura M. Clinically mild encephalitis/encephalopathy with a reversible splenial lesion caused by methicillin-sensitive Staphylococcus aureus bacteremia with toxic shock syndrome: a case report. BMC Infect Dis. 2016;16:160.
124. Yang S-L, Zhu M, Wang F, Liu W-J, Wang Y, Wu Y-C. An unusual presentation of infective endocarditis: transient encephalopathy with reversible lesions in corpus callosum and white matter. Neurol Asia. 2017;22:373–6.
125. Fukagawa K, Izumi M, Higuchi K, Adachi Y. Reversible splenial lesion associated with Staphylococcus aureus endocarditis. Intern Med. 2013;52:1147–8.
126. Shimozono K, Korenaga H, Mawatari R, Tsukishima N. A case of Staphylococcus aureus meningitis associated with cryoglobulin-related renal failure and clinically mild encephalitis/encephalopathy with a reversible splenial lesion. Rinsho Shinkeigaku. 2016;56:318–22.
127. Cassels F, Wolf M. Colonization factors of diarrheagenic E. coli and their intestinal receptors. J Ind Microbiol Biotechnol. 1995;15:214–26.
128. Kaper JB, Nataro JP, Mobley HLT. Pathogenic Escherichia coli. Nat Rev Microbiol. 2004;2:123–40.
129. Keller R, Ordoñez JG, De Oliveira RR, Trabulsi LR, Baldwin TJ, Knutton S. Afa, a diffuse adherence fibrillar adhesin associated with enteropathogenic Escherichia coli. Infect Immun. 2002;70:2681–9.
130. Johnson JR, Russo TA. Extraintestinal pathogenic Escherichia coli: "the other bad E coli". J Lab Clin Med. 2002;139:155–62.
131. Ephros M, Cohen D, Yavzori M, Rotman N, Novic B, Ashkenazi S. Encephalopathy associated with enteroinvasive Escherichia coli 0144:NM infection. J Clin Microbiol. 1996;34:2432–4.
132. Löbel U, Eckert B, Simova O, Meier-Cillien M, Kluge S, Gerloff C, Röther J, Magnus T, Fiehler J. Cerebral magnetic resonance imaging findings in adults with haemolytic uraemic syndrome following an infection with Escherichia coli, subtype O104:H4. Clin Neuroradiol. 2014;24:111–9.
133. Ogura H, Takaoka M, Kishi M, Kimoto M, Shimazu T, Yoshioka T, Sugimoto H. Reversible MR findings of hemolytic uremic syndrome with mild encephalopathy. AJNR Am J Neuroradiol. 1998;19:1144–5.
134. Yeom JS, Koo CM, Park JS, Seo JH, Park ES, Lim JY, Woo HO, Youn HS. Mild encephalopathy with a reversible splenial lesion in a girl with acute pyelonephritis. Korean J Pediatr. 2018;61:64–7.
135. Yıldız AE, Maraş Genç H, Gürkaş E, Akmaz Ünlü H, Öncel İH, Güven A. Mild encephalitis/encephalopathy with a reversible splenial lesion in children. Diagn Interv Radiol. 2018;24:108–12.
136. Ochman H, Groisman E. The origin and evolution of species differences in Escherichia coli and Salmonella typhimurium. In: Molecular ecology and evolution: approaches and applications. Berlin: Springer; 1994.
137. Fierer J, Guiney DG. Diverse virulence traits underlying different clinical outcomes of Salmonella infection. J Clin Invest. 2001;107:775–80.
138. Ahmed T, Ahmed T. Salmonella meningitis complicated by ventriculitis and brain abscesses in an HIV positive adult patient. Cureus. 2020;12:e11223.
139. Al-Yaqoobi M, Al-Khalili S, Mishra GP. Salmonella brain abscess in an infant. Neurosci J. 2018;23:250–3.
140. Cohen JI, Bartlett JA, Corey GR. Extra-intestinal manifestations of Salmonella infections. Medicine. 1987;66:349–88.
141. Zhao J, Hua C, Zhou M, Wang H, Xie Y, Wang G. A case report of Salmonella meningitis in three Chinese children and A literature review; 2020.
142. Chopra P, Bhatia RS, Chopra R. Mild encephalopathy/encephalitis with reversible splenial lesion in a patient with Salmonella typhi infection: an unusual presentation with excellent prognosis. Indian J Crit Care Med. 2019;23:584–6.
143. Ahmed M, Sureka J, Mathew V, Jakkani RK, Abhilash K. Magnetic resonance imaging findings in a fatal case of Salmonella typhi-associated encephalopathy: a case report and literature review. Neurol India. 2011;59:270.
144. Martin K, Sharland M, Davies EG. Encephalopathy associated with Salmonella enteritidis infection. Scand J Infect Dis. 1994;26:486–8.

145. Kobuchi N, Tsukahara H, Kawamura Y, Ishimori Y, Ohshima Y, Hiraoka M, Hiraizumi Y, Ueno M, Mayumi M. Reversible diffusion-weighted MR findings of Salmonella enteritidis-associated encephalopathy. Eur Neurol. 2003;49:182–4.
146. Patterson MJ. Streptococcus. Medical microbiology. 4th ed. Galveston: University of Texas Medical Branch at Galveston; 1996.
147. Allard MJ, Brochu ME, Bergeron JD, Segura M, Sébire G. Causal role of group B Streptococcus-induced acute chorioamnionitis in intrauterine growth retardation and cerebral palsy-like impairments. J Dev Orig Health Dis. 2019;10:595–602.
148. Morii K, Kogita Y, Takata M, Yamamoto T, Kishida H, Okushin H, Uesaka K. Reversible splenial lesion of the corpus callosum associated with bacterial meningitis. Int J Infect Dis. 2014;19:107–8.
149. Choi SY, Kim JW, Ko JW, Lee YS, Chang YP. Patterns of ischemic injury on brain images in neonatal group B Streptococcal meningitis. Korean J Pediatr. 2018;61:245–52.
150. Okumura A, Kitamura Y, Abe S, Ikeno M, Shimizu T. Transiently restricted diffusion in the corpus callosum in bacterial meningitis. J Pediatr Infect Dis. 2011;6:51–3.
151. Fallata EM, Bokhary NA, Bugshan AS, Hakami MH. Intracranial haemorrhage in late-onset neonatal group B streptococcal disease: a case report. J Taibah Univ Med Sci. 2021;16:771–5.
152. Krafft C, Christy C. Mycoplasma pneumonia in children and adolescents. Pediatr Rev. 2020;41:12.
153. Maniloff J. Mycoplasma. In: Encyclopedia of molecular biology. New York: Wiley; 2002.
154. Atkinson TP, Balish MF, Waites KB. Epidemiology, clinical manifestations, pathogenesis and laboratory detection of mycoplasma pneumoniae infections. FEMS Microbiol Rev. 2008;32:956–73.
155. Waldo Iii RH, Krause DC. Synthesis, stability, and function of cytadhesin P1 and accessory protein B/C complex of mycoplasma pneumoniae. J Bacteriol. 2006;188:569–75.
156. Gelfand EW. Unique susceptibility of patients with antibody deficiency to mycoplasma infection. Clin Infect Dis. 1993;17(Suppl 1):S250–3.
157. O'sullivan MV, Isbel NM, Johnson DW, Rencken K, Byrne S, Graham MM, Playford EG. Disseminated pyogenic mycoplasma pneumoniae infection in a renal transplant recipient, detected by broad-range polymerase chain reaction. Clin Infect Dis. 2004;39:E98–9.
158. Akbar A, Ahmad S. Atypical case of mild encephalopathy/encephalitis with reversible splenial lesion of the corpus callosum (MERS) associated with Mycoplasma pneumoniae infection in a paediatric patient. BMJ Case Rep. 2021;14:E242791.
159. Ueda N, Minami S, Akimoto M. Mycoplasma pneumoniae-associated mild encephalitis/encephalopathy with a reversible splenial lesion: report of two pediatric cases and a comprehensive literature review. BMC Infect Dis. 2016;16:671.
160. Yuan Z-F, Shen J, Mao S-S, Yu Y-L, Xu L, Jiang P-F, Gao F, Xia Z-Z. Clinically mild encephalitis/encephalopathy with a reversible splenial lesion associated with mycoplasma pneumoniae infection. BMC Infect Dis. 2016;16:230.
161. Dong X, Cong S. Reversible splenial lesion syndrome associated with acute Mycoplasma pneumoniae-associated encephalitis: a report of four cases and literature review. Exp Ther Med. 2018;16:2152–9.
162. Shibuya H, Osamura K, Hara K, Hisada T. Clinically mild encephalitis/encephalopathy with a reversible splenial lesion due to Mycoplasma pneumoniae infection. Intern Med. 2012;51:1647–8.
163. Jett BD, Huycke MM, Gilmore MS. Virulence of enterococci. Clin Microbiol Rev. 1994;7:462–78.
164. Kometani H, Kawatani M, Ohta G, Okazaki S, Ogura K, Yasutomi M, Tanizawa A, Ohshima Y. Marked elevation of interleukin-6 in mild encephalopathy with a reversible splenial lesion (MERS) associated with acute focal bacterial nephritis caused by Enterococcus faecalis. Brain Dev. 2014;36:551–3.
165. Cappellari AM, Cozzi L, Gandini C, Bruschi G, Costa A. Mild encephalitis/encephalopathy with reversible splenial lesion due to enterococcus Faecalis in children. Turk Arch Pediatr. 2021;56:527–9.

Inflammatory and Demyelinating Diseases of the Corpus Callosum

22

Keaton Ott and R. Shane Tubbs

22.1 Introduction

The corpus callosum (CC) is the largest of the major commissures and the largest white matter structure in the brain. Composed of hundreds of millions of densely packed nerve fibers connecting the two cerebral hemispheres and richly supplied by both anterior and posterior cerebral circulations, it is understandably a common site of both demyelinating and inflammatory processes. In this chapter, we review the various acquired demyelinating and non-demyelinating inflammatory diseases that affect the CC.

22.2 Demyelinating Diseases

22.2.1 Multiple Sclerosis

Multiple sclerosis (MS) is the most common acquired autoimmune demyelinating disease of the central nervous system (CNS) and a principal pathology affecting the CC. Its etiology is multifactorial, involving polygenic inheritance with the HLA-DRB1*15:01 allele conferring the greatest risk, as well as several environmental risk factors, namely infection with Epstein-Barr virus, vitamin D insufficiency, smoking, and childhood obesity [1]. It mainly affects young women, with a female/male ratio of 3:1 [2]. Although not entirely understood, MS has historically been classified as a T-cell-mediated autoimmune disease targeting oligodendrocytes; however, the important role of B-cells in its pathogenesis has been elucidated in recent years with the effective advent of B-cell depleting therapy [3]. While three clinical phenotypes (relapsing-remitting, secondary progressive, and primary progressive) encompass the variability of disease progression, the pathological changes in MS form a continuum. Its chronic course involves focal inflammation and demyelination primarily characterizing the early relapsing stage of disease, while neuronal/axonal degeneration and atrophy primarily characterize the late progressive stage of disease [4–6]. It is a notoriously polysymptomatic disease on account of the varied locations throughout the CNS where MS plaques may arise. More typical locations for lesions include the optic nerves, spinal cord, brainstem, cerebellum, and juxtacortical and periventricular white matter [7].

The CC and pericallosal region are frequently affected in MS, appearing abnormal on MRI in anywhere between 55% and 95% of MS patients [8]. MS lesions most characteristically involve the inferior surface of the CC, also referred to as the ventral surface, subcallosal surface, ventricu-

K. Ott (✉) · R. S. Tubbs
Department of Neurosurgery, Tulane School of Medicine, New Orleans, LA, USA
e-mail: kott1@tulane.edu; rtubbs@tulane.edu

M. Turgut et al. (eds.), *The Corpus Callosum*, https://doi.org/10.1007/978-3-031-38114-0_22

lar surface, or callosal-septal interface, among other names throughout the literature. Lesions tend to be well-defined with clearly demarcated borders on MRI, with T2-weighted and fluid-attenuated inversion recovery (FLAIR) imaging especially sensitive for detecting these focal white matter lesions. Radiating perpendicularly from the ependymal surface of the lateral ventricles into the CC and pericallosal/periventricular white matter are two well-studied and described T2-hyperintense lesions: thin linear lesions called subependymal or subcallosal striations and ovoid lesions. Considering (1) perivenular inflammation is known to be critical in the pathogenesis and development of demyelinating MS lesions, and (2) periventricular cerebral veins follow the same perpendicularly radiating course as these two lesion types, subcallosal striations and ovoid lesions are thus believed to represent perivenular inflammation and demyelination [2, 9–11]. The smaller, thinner subcallosal striations are understood to represent early inflammatory changes, while the larger ovoid lesions are the radiological correlate of (and often referred to interchangeably with) Dawson's fingers, the histopathological eponym for the fingerlike periventricular perivenular inflammatory lesions first described by histologist James W. Dawson [9–12]. These perivenular inflammatory lesions are highly indicative of MS, prompting visualization of a central vein within a suspected MS lesion (the so-called central vein sign) as a proposed novel MRI biomarker for distinguishing MS from its mimics [8, 10, 11, 13, 14]. The CC may also possess T1-hypointense "black hole" lesions, which represent the chronic phase of previously active gadolinium-enhancing lesions. These persistent lesions have been reported to be more specific markers for profound demyelination, axonal loss, and inflammatory tissue damage and have been shown to be better correlated with clinical status and disability [15–18]. Progressive atrophy of otherwise normal-appearing white matter has also been implicated in disease progression, with special attention paid to the CC since neuroimaging measures of callosal atrophy have long been used to estimate overall brain atrophy and as a marker of lesion load, cognitive function, and disability [19–24]. One study has even shown CC atrophy to be disproportionally greater compared to other global atrophy measures in patients with MS [25].

22.2.2 Neuromyelitis Optica Spectrum Disorder

Neuromyelitis optica spectrum disorder (NMOSD), previously referred to as Devic's disease, is a spectrum of uncommon antibody-mediated autoimmune CNS disease which was previously believed to be a more severe form of MS seen typically in Asian countries called opticospinal MS. The revelation that most NMOSD patients possessed serum antibodies against aquaporin-4 (AQP4), a water channel primarily located on the foot processes of astrocytes in regions contacting cerebrospinal fluid, first allowed for reliable distinction between MS and NMOSD. Presence of aquaporin-4 immunoglobulin G antibodies (AQP4-IgG) is highly specific for NMOSD; however, seronegativity is possible with rates varying among reported studies [26, 27]. For such patients without detected AQP4-IgG or with unknown AQP4-IgG status, diagnostic criteria are more stringent [28]. In contrast to MS, a gradually progressive course of disease is very uncommon in NMOSD, with the accrual of disability instead due to typically severe relapses [26]. Classic clinical presentations include longitudinally extensive transverse myelitis, optic neuritis, area postrema syndrome, acute brainstem syndromes, and symptomatic narcolepsy or acute diencephalic syndrome, with these presentations correlating with regions of the CNS most abundantly expressing AQP4: spinal cord, optic nerves, dorsal medulla, and thalamus/hypothalamus [26, 28]. Of these classic clinical presentations, the most suggestive of NMOSD are longitudinally extensive complete transverse myelitis with paroxysmal tonic spasms, simultaneous bilateral longitudinally extensive optic neuritis involving the optic chiasm with severe vision loss, and area postrema syndrome involving intractable nausea, vomiting, and/or hiccups [26, 28].

While early reports suggested little to no brain involvement on MRI, AQP4-IgG assays have allowed for brain MRI abnormalities in NMOSD to be better appreciated. The most common brain lesions on MRI are clinically silent, nonspecific punctate hyperintensities in subcortical and deep white matter on T2-weighted and FLAIR imaging [29, 30]. More characteristic brain lesions are typically periependymal, coinciding with the high density of AQP4 expression in the ependymal lining of the ventricular system, and commonly include the ependymal surface of the diencephalon, brainstem (such as the area postrema in the dorsal medulla), and the CC [30, 31]. With the CC being a common lesion site in both NMOSD and MS, callosal lesions present an additional method of distinguishing the two diseases. Compared to the well-defined, typically ovoid lesions perpendicularly oriented to the lateral ventricles in MS, callosal lesions in NMOSD typically extensively follow the ependymal lining, affecting most of its length, and emerge diffusely through the CC toward its upper surface as an edematous and heterogenous "marbled pattern" with blurred margins [29, 31–33]. Lesions of the splenium of the CC are also more common in NMOSD than MS, sometimes involving the full thickness of the splenium, producing an "arch-bridge pattern" [29, 33]. Other brain MRI abnormalities which may be seen in NMOSD include lesions of the corticospinal tracts, rather unusual considering AQP4 is not highly expressed along this area, as well as extensive tumefactive lesions (>2 cm in longest diameter) in subcortical white matter [29]. With gadolinium enhancement on T1-weighted images, lesions can exhibit cloud-like enhancement described as patchy with poorly-defined margins, and the ependymal surface of the lateral ventricles can show thin linear enhancement termed pencil-thin enhancement [29, 31].

22.2.3 Acute Disseminated Encephalomyelitis

Acute disseminated encephalomyelitis (ADEM) is a classically monophasic autoimmune demyelinating disease of the CNS, typically preceded by a febrile illness or vaccination and most commonly affecting young children and adolescents. As much as 50–75% of cases have been shown to follow viral or bacterial infections, most typically nonspecific upper respiratory tract infections, and vaccinations most frequently associated with ADEM include measles, mumps, rubella, and rabies [34, 35]. Despite this postinfectious/postimmunization association, the immunopathogenesis of ADEM remains poorly understood, with a primary proposed mechanism being molecular mimicry [34, 36]. As its name suggests, ADEM involves disseminated multifocal CNS demyelination, characteristically producing acute or subacute encephalopathy accompanied by varied symptoms based on lesioned areas. Apart from altered mental status, common symptomatology includes ataxia, pyramidal signs, optic neuritis or other cranial nerve involvement, impaired speech, seizures, extended fever and headaches more often in childhood cases, and motor/sensory deficits more often in adult cases [34, 35, 37]. Its typically monophasic, rapid disease course provides an important distinction from the relapsing or progressive course of MS. However, possible also is multiphasic disseminated encephalomyelitis (MDEM), defined as two episodes consistent with ADEM separated by at least three months; also, cases of three or more ADEM-like episodes are no longer considered MDEM but instead a chronic relapsing demyelinating disorder, often prompting a diagnosis of ADEM followed by some other demyelinating disorder such as optic neuritis, NMOSD, or MS [35]. Recurrent episodes have been demonstrated in 25–33% of ADEM patients [34]. How these recurrent ADEM cases differ from other relapsing demyelinating disorders remains a point of debate. This uncertainty, alongside variable clinical manifestations and the absence of well-defined specific biomarkers, maintains ADEM as a diagnosis of exclusion. A hyperacute form of ADEM called acute hemorrhagic leukoencephalitis, also known as Weston-Hurst syndrome or Hurst disease, is characterized by fulminant inflammation of white matter, perivascular hemorrhage, and microvascular necrosis and is often fatal [35, 36, 38]. Otherwise, the prognosis for patients with ADEM is favorable, with most patients recovering fully or with minor residual disability within weeks of onset [34, 35].

Perivenular inflammation is a pathological hallmark of ADEM, similar to MS. However, the pattern of demyelination this perivenular inflammation produces differs between the two demyelinating disorders. While MS is characterized by sharp-edged plaques of confluent demyelination, ADEM is characterized by sleeves of demyelination restricted to the perivascular area called perivenular demyelination [34, 39]. Unique to ADEM as well are multifocal cortical microglial aggregates in non-demyelinated areas, postulated to represent the pathological correlate of the altered level of consciousness typically present in patients [39]. On MRI, the disseminated CNS demyelination in ADEM produces widespread, bilateral, multifocal, poorly marginated T2/FLAIR hyperintense lesions typically in the subcortical white matter, central white matter, cerebellum, brainstem, and spinal cord [35]. Lesions in the thalami and basal ganglia may also be seen, typically at gray-white junctions [34]. Compared to MS and NMOSD, lesions in periventricular white matter and the CC are generally less common in ADEM, as well as usually not seen at the callosal-septal interface [8, 35, 40]. However, when present, callosal lesions have been observed primarily in the splenium, as usually larger than typical MS lesions, and often as extensions of large lesions from adjacent white matter [41, 42]. With its traditionally monophasic course, most patients have complete or partial resolution of MRI abnormalities on serial MRIs three months after disease onset [35].

22.2.4 Marchiafava-Bignami Disease

Marchiafava-Bignami disease (MBD) is a rare demyelinating disease which primarily affects the CC. Originally described in wine drinkers, it has characteristically been associated with chronic alcoholism; however, nonalcoholic cases have also been reported. These nonalcoholic patients typically suffer from severe malnutrition or prolonged vomiting and generally have a more favorable prognosis compared to alcoholic cases [43]. Believed to be of toxic or nutritional etiology, the exact pathogenic mechanisms and why the CC is primarily targeted are unknown. A deficiency of B vitamins, particularly vitamin B_1 (thiamine), is thought to be a primary factor involved considering that those treated with thiamine have been shown to have overall better outcomes, especially if treated in the acute phase [43, 44]. Clinical presentation is variable but frequently involves altered mental status, loss of consciousness, impaired gait, dysarthria, impaired memory, signs of disconnection or split-brain syndrome, and pyramidal signs [37, 43, 44]. Less common is a chronic form of MBD characterized by progressive mild dementia [44]. Differing statements have been made regarding which areas of the CC are most commonly involved in MBD; however, one extensive review of 122 case reports noted that lesions were most often observed across the entire CC (splenium, body, and genu) [43]. MBD preferentially affects the center of the CC while generally sparing the periphery, producing a "sandwich-like" appearance on MRI [14, 40]. In the acute phase, lesions are T2/FLAIR hyperintense and T1 hypointense and possess variable contrast enhancement [40]. In the chronic phase, the central CC becomes necrotic and atrophic, producing cystic/cavitary lesions [14, 43, 44]. Less frequently affected CNS structures include other white matter tracts, cerebral cortex, and middle cerebellar peduncles [31, 40, 43, 44].

22.2.5 Osmotic Demyelination Syndrome

Osmotic demyelination syndrome (ODS), the consolidated name for two previous entities known as central pontine myelinolysis (CPM) and extrapontine myelinolysis (EPM), is a demyelinating condition of the CNS classically associated with rapid iatrogenic correction of hyponatremia. ODS is characterized by noninflammatory loss of myelin with preservation of neuronal cell bodies and axons (myelinolysis), which is elicited by rapid changes in serum osmolality leading to alterations in cellular volume control in the brain [45]. With changes to chronic hypoosmolar states, rapidly increasing serum osmolality leads to water shifts out of glial cells responsible for maintaining normal water regulation in the brain, eliciting cellular damage, apoptosis, and disruption of tight junctions [46]. These processes are believed to cause disruption of the blood-brain barrier, with proposed resultant vasogenic edema, fiber tract compression, intramyelinic edema, and oligodendrocyte degeneration, leading to myelinolysis [45, 46]. While correction of chronic hyponatremia is the classic and most common cause of ODS, any electrolyte imbalance or other cause of osmotic gradient changes may elicit the same effects. Common conditions associated with development of ODS besides rapid correction of hyponatremia include alcoholism, malnutrition, and liver transplantation, while less common conditions include other causes of hypoosmolar/hyperosmolar states, such as cirrhosis, severe burns, renal failure, dialysis disequilibrium syndrome, hyperemesis, hypernatremia, or hyperglycemia [45–48]. The classic presentation of ODS involving the pons (CPM) consists of altered mental status and rapidly progressive quadriparesis with associated dysphagia, dysarthria, and other pseudobulbar symptoms; however, clinical presentation can vary when other CNS structures are concomitantly or instead affected (EPM), with symptomatology commonly involving movement disorders, psychiatric disorders, or cognitive deficits [46, 49]. The pons is most susceptible to the demyelinating lesions of ODS, while extrapontine lesions can occur in the cerebellum, lateral geniculate body, external and extreme capsule, basal ganglia, gray-white junction of the cerebral cortex, hippocampus, and CC, among other sites [45]. When the CC is involved, lesions are typically located in the splenium [47, 48, 50]. On MRI, lesions are hypointense on T1-weighted imaging, hyperintense on T2-weighted, FLAIR, and diffusion-weighted imaging, and do not enhance with contrast [46, 49].

22.3 Non-demyelinating Inflammatory Diseases

22.3.1 Susac Syndrome

Susac syndrome (SS) is a rare, presumed autoimmune-mediated endotheliopathy of microvasculature in the brain, retina, and inner ear. In line with this putative etiology, SS primarily affects women, with a female/male ratio of 3.5:1 [51]. Considered pathognomonic of SS is the clinical triad of CNS dysfunction/encephalopathy, branch retinal artery occlusions (BRAOs), and hearing loss; however, rarely do patients initially present with the full triad, presumably a major cause of misdiagnosis. In a comprehensive review of 304 published cases of SS, only 13% of patients presented with the full clinical triad at disease onset, and the average delay between first symptom and complete presentation of the triad was around five months [51]. CNS symptoms are varied and include migrainous-like headaches, cognitive impairment (loss of memory, concentration, and executive functions), confusion/disorientation, psychiatric disturbances, cranial nerve disorders, and dementia [51, 52]. Ophthalmological findings resulting from BRAOs include blurred vision, altitudinal defects, central or paracentral scotomas, and photopsia, or patients may be asymptomatic if such occlusions occur in the peripheral retina [52]. Hearing loss is sensorineural, typically bilateral, frequently accompanied by tinnitus and vertigo, and usually permanent [51–53]. In most cases, the disease course of SS is monocyclic, defined as fluctuating disease that self-limits within two years, while less frequently it presents with a polycyclic or chronic continuous course [51].

Primary diagnostic procedures crucial to diagnosing SS are MRI, retinal fluorescein angiography, and audiometry. The most frequently lesioned CNS structure is the CC, the characteristic callosal lesion being "snowball" lesions on T2/FLAIR imaging described as multifocal small and round lesions in the central CC which represent microinfarcts due to occlusion of callosal microvasculature [8, 51, 54]. Chronically, these lesions evolve into central callosal "holes" on T1-weighted imaging, also known as "punched-out" lesions [54]. When the periphery of the CC is involved, the roof is more frequently affected by microinfarcts that produce linear "spoke" lesions and wedge-shaped "icicle" lesions which extend into the CC [8, 54]. Other areas less frequently affected include the periventricular white matter, deep gray matter structures, cerebellum, and leptomeninges [51, 52]. Fluorescein angiography of the retina is also vital in diagnosing SS as it detects BRAOs, seen in 99% of SS patients tested, and arterial wall hyperfluorescence, which indicates disease activity and can be considered pathognomonic for retinal involvement in SS [51, 52]. Findings of patchy damage of inner retinal layers on optical coherence tomography have also been demonstrated and used for diagnosis [52].

22.3.2 Transient Splenial Lesions

Transient isolated lesions of the splenium of the CC exist as a nonspecific radiological finding described in a wide spectrum of diseases and conditions. On MRI, they appear as typically round/ovoid lesions of the central splenium that are T2/FLAIR hyperintense and T1 hypointense, lack contrast enhancement, show homogenously reduced diffusion on diffusion-weighted imaging, and either disappear or significantly improve on follow-up [55, 56]. Greater reporting of such lesions in recent years, owing to improvements in MRI technology, has led to designation of a clinico-radiological syndrome referred to as reversible splenial lesion syndrome (RESLES), whose criteria include the presence of neurological deficits alongside lesions of the splenium on MRI which resolve or significantly improve over time [55, 57]. Etiologies found to be associated with transient splenial lesions and thus RESLES include antiepileptic drug withdrawal and use, seizures, infections leading to encephalitis/encep halopathy, hypoglycemia, hyponatremia, hypernatremia, high-altitude cerebral edema (HACE), systemic lupus erythematosus (SLE), and other miscellaneous conditions, with the most commonly reported causes being related to seizures/antiepileptic drugs and infections [55, 58]. In accordance with the varying etiologies associated with transient splenial lesions, neurological deficits in patients with RESLES vary widely.

When RESLES is determined to be secondary to infection, this entity is generally instead referred to as mild encephalitis/encephalopathy with reversible splenial lesion (MERS). MERS is a clinico-radiological syndrome characterized by acute neurological symptoms (typically headaches, altered consciousness, or seizures) accompanied by an isolated lesion of the splenium of the CC observed on MRI, which both resolve within approximately one month [59, 60]. It is more common in children/adolescents and most cases have been reported in Asia, especially in Japan [60, 61]. Viral pathogens, particularly influenza viruses, are the predominant etiologic agent; however, other identified pathogens include rotavirus, measles virus, mumps virus, Epstein-Barr virus, varicella-zoster virus, adenovirus, *Escherichia coli*, Salmonella species, *Legionella pneumophila*, and *Mycoplasma pneumoniae*, among others [60, 61]. Splenial lesions in MERS are identical to those in other cases of RESLES, although lesions of both the splenium and extracallosal structures, as well as of the entire CC, have been reported in some MERS cases [61].

The pathogenesis of these callosal lesions, their transient nature, and their selectivity for the splenium are poorly understood. Several hypotheses for their pathophysiology exist, including intramyelinic edema, inflammatory infiltrates and oxidative stress, hyponatremia and fluid imbalance, and axonal damage [56, 60, 61]. However, the diversity of etiologies associated with these transient splenial lesions makes a single specific mechanism unlikely.

22.3.3 Other Inflammatory Diseases

Several other inflammatory diseases of the CNS have demonstrated CC involvement, although generally to a less frequent and less significant degree. Some of these conditions include systemic lupus erythematosus (SLE), Sjögren syndrome, primary angiitis of the central nervous system (PACNS), neuro-Behçet disease, and chronic lymphocytic inflammation with pontine perivascular enhancement responsive to steroids (CLIPPERS).

In some cases of SLE and Sjögren syndrome, both well-recognized autoimmune diseases, lesions of the CC as well as callosal atrophy have been reported [62–67]. The splenium of the CC seems to be particularly affected in both of these diseases. PACNS, also known as primary CNS vasculitis, is an inflammatory vasculopathy involving transmural inflammation of and subsequent injury to small, medium, or large vessels exclusively in the brain and spinal cord. This inflammation can be granulomatous, lymphocytic, or necrotizing [68]. Callosal lesions have been reported in both primary (PACNS) and secondary angiitis of the CNS [66, 69]. Neuro-Behçet disease is the neurologic manifestation of Behçet disease, a systemic vasculitis classically characterized by recurrent oral ulcerations, genital ulcerations, and uveitis. Clinical manifestations can vary depending upon where pathology occurs in the CNS, yet patients most often present with migraine- or tension-type headaches [70]. Although rare, CC involvement in neuro-Behçet disease has been reported [70, 71]. CLIPPERS is a relatively recently described inflammatory disorder of the CNS affecting primarily the pons, characterized by brainstem symptoms, perivascular T-cell infiltrates on biopsy, and punctate and curvilinear gadolinium-enhancing lesions "peppering" the pons on MRI [72]. Variable extension of these lesions to supratentorial regions, basal ganglia, thalamus, cerebral white matter, medulla, spinal cord, or the CC has been reported, although lesions are typically smaller and less numerous with increasing distance from the pons [72, 73].

22.4 Conclusion

Although affected by pathology of numerous types, the CC is notably susceptible to inflammatory processes. Demyelinating and non-demyelinating inflammatory diseases alike have been implicated in lesions of the CC. Callosal involvement is varied among these inflammatory diseases: some diseases affect the periphery of the CC, some affect the central CC, some preferentially affect specific parts of the CC such as the splenium, some affect the entire CC, some cause callosal atrophy, some produce necrotic or cystic lesions, some produce focal well-defined lesions, some produce heterogenous diffuse lesions, some almost exclusively affect the CC, some have callosal and extracallosal involvement, some cause transient pathology, some cause permanent residual damage, some elicit mild symptoms, and some lead to severe neurological deficits. With greater innovations in imaging modalities, greater understanding of the pathogenesis of CNS inflammation, and greater advancements in medicine as a whole, more is to be elucidated about the diseases discussed in this chapter, and more neuroinflammatory diseases are likely to be linked to callosal pathology.

References

1. Hollenbach JA, Oksenberg JR. The immunogenetics of multiple sclerosis: a comprehensive review. J Autoimmun. 2015;64:13–25. https://doi.org/10.1016/j.jaut.2015.06.010.
2. Dobson R, Giovannoni G. Multiple sclerosis - a review. Eur J Neurol. 2019;26(1):27–40. https://doi.org/10.1111/ene.13819.
3. Greenfield AL, Hauser SL. B-cell therapy for multiple sclerosis: entering an era. Ann Neurol. 2018;83(1):13–26. https://doi.org/10.1002/ana.25119.
4. Coles AJ, Cox A, Le Page E, Jones J, Trip SA, Deans J, et al. The window of therapeutic opportunity in multiple sclerosis: evidence from monoclonal antibody therapy. J Neurol. 2006;253(1):98–108. https://doi.org/10.1007/s00415-005-0934-5.

5. Faissner S, Plemel JR, Gold R, Yong VW. Progressive multiple sclerosis: from pathophysiology to therapeutic strategies. Nat Rev Drug Discov. 2019;18(12):905–22. https://doi.org/10.1038/s41573-019-0035-2.
6. Leray E, Yaouanq J, Le Page E, Coustans M, Laplaud D, Oger J, et al. Evidence for a two-stage disability progression in multiple sclerosis. Brain. 2010;133(Pt 7):1900–13. https://doi.org/10.1093/brain/awq076.
7. Popescu BF, Pirko I, Lucchinetti CF. Pathology of multiple sclerosis: where do we stand? Continuum (Minneap Minn). 2013;19(4):901–21. https://doi.org/10.1212/01.CON.0000433291.23091.65.
8. Garg N, Reddel SW, Miller DH, Chataway J, Riminton DS, Barnett Y, et al. The corpus callosum in the diagnosis of multiple sclerosis and other CNS demyelinating and inflammatory diseases. J Neurol Neurosurg Psychiatry. 2015;86(12):1374–82. https://doi.org/10.1136/jnnp-2014-309649.
9. Ge Y, Law M, Herbert J, Grossman RI. Prominent perivenular spaces in multiple sclerosis as a sign of perivascular inflammation in primary demyelination. AJNR Am J Neuroradiol. 2005;26(9):2316–9.
10. Horowitz AL, Kaplan RD, Grewe G, White RT, Salberg LM. The ovoid lesion: a new MR observation in patients with multiple sclerosis. AJNR Am J Neuroradiol. 1989;10(2):303–5.
11. Palmer S, Bradley WG, Chen DY, Patel S. Subcallosal striations: early findings of multiple sclerosis on sagittal, thin-section, fast FLAIR MR images. Radiology. 1999;210(1):149–53. https://doi.org/10.1148/radiology.210.1.r99ja38149.
12. Hashemi RH, Bradley WG Jr, Chen DY, Jordan JE, Queralt JA, Cheng AE, et al. Suspected multiple sclerosis: MR imaging with a thin-section fast FLAIR pulse sequence. Radiology. 1995;196(2):505–10. https://doi.org/10.1148/radiology.196.2.7617868.
13. Filippi M, Rocca MA, Barkhof F, Brück W, Chen JT, Comi G, et al. Association between pathological and MRI findings in multiple sclerosis. Lancet Neurol. 2012;11(4):349–60. https://doi.org/10.1016/S1474-4422(12)70003-0.
14. Uchino A, Takase Y, Nomiyama K, Egashira R, Kudo S. Acquired lesions of the corpus callosum: MR imaging. Eur Radiol. 2006;16(4):905–14. https://doi.org/10.1007/s00330-005-0037-9.
15. Sahraian MA, Radue EW, Haller S, Kappos L. Black holes in multiple sclerosis: definition, evolution, and clinical correlations. Acta Neurol Scand. 2010;122(1):1–8. https://doi.org/10.1111/j.1600-0404.2009.01221.x.
16. Thaler C, Faizy T, Sedlacik J, Holst B, Stellmann JP, Young KL, et al. T1-thresholds in black holes increase clinical-radiological correlation in multiple sclerosis patients. PLoS One. 2015;10(12):e0144693. https://doi.org/10.1371/journal.pone.0144693.
17. Thaler C, Faizy TD, Sedlacik J, Holst B, Stürner K, Heesen C, et al. T1 recovery is predominantly found in black holes and is associated with clinical improvement in patients with multiple sclerosis. AJNR Am J Neuroradiol. 2017;38(2):264–9. https://doi.org/10.3174/ajnr.A5004.
18. van Walderveen MA, Kamphorst W, Scheltens P, van Waesberghe JH, Ravid R, Valk J, et al. Histopathologic correlate of hypointense lesions on T1-weighted spin-echo MRI in multiple sclerosis. Neurology. 1998;50(5):1282–8. https://doi.org/10.1212/wnl.50.5.1282.
19. De Stefano N, Matthews PM, Filippi M, Agosta F, De Luca M, Bartolozzi ML, et al. Evidence of early cortical atrophy in MS: relevance to white matter changes and disability. Neurology. 2003;60(7):1157–62. https://doi.org/10.1212/01.wnl.0000055926.69643.03.
20. Evangelou N, Esiri MM, Smith S, Palace J, Matthews PM. Quantitative pathological evidence for axonal loss in normal appearing white matter in multiple sclerosis. Ann Neurol. 2000;47(3):391–5.
21. Granberg T, Bergendal G, Shams S, Aspelin P, Kristoffersen-Wiberg M, Fredrikson S, et al. MRI-defined corpus callosal atrophy in multiple sclerosis: a comparison of volumetric measurements, corpus callosum area and index. J Neuroimaging. 2015;25(6):996–1001. https://doi.org/10.1111/jon.12237.
22. Granberg T, Martola J, Bergendal G, Shams S, Damangir S, Aspelin P, et al. Corpus callosum atrophy is strongly associated with cognitive impairment in multiple sclerosis: results of a 17-year longitudinal study. Mult Scler. 2015;21(9):1151–8. https://doi.org/10.1177/1352458514560928.
23. Yaldizli Ö, Atefy R, Gass A, Sturm D, Glassl S, Tettenborn B, et al. Corpus callosum index and long-term disability in multiple sclerosis patients. J Neurol. 2010;257(8):1256–64. https://doi.org/10.1007/s00415-010-5503-x.
24. Yaldizli Ö, Penner IK, Frontzek K, Naegelin Y, Amann M, Papadopoulou A, et al. The relationship between total and regional corpus callosum atrophy, cognitive impairment and fatigue in multiple sclerosis patients. Mult Scler. 2014;20(3):356–64. https://doi.org/10.1177/1352458513496880.
25. Klawiter EC, Ceccarelli A, Arora A, Jackson J, Bakshi S, Kim G, et al. Corpus callosum atrophy correlates with gray matter atrophy in patients with multiple sclerosis. J Neuroimaging. 2015;25(1):62–7. https://doi.org/10.1111/jon.12124.
26. Huda S, Whittam D, Bhojak M, Chamberlain J, Noonan C, Jacob A. Neuromyelitis optica spectrum disorders. Clin Med (Lond). 2019;19(2):169–76. https://doi.org/10.7861/clinmedicine.19-2-169.
27. Ömerhoca S, Akkaş SY, İçen NK. Multiple sclerosis: diagnosis and differential diagnosis. Noro Psikiyatr Ars. 2018;55(Suppl 1):S1–9. https://doi.org/10.29399/npa.23418.
28. Wingerchuk DM, Banwell B, Bennett JL, Cabre P, Carroll W, Chitnis T, et al. International consensus diagnostic criteria for neuromyelitis optica spectrum disorders. Neurology. 2015;85(2):177–89. https://doi.org/10.1212/WNL.0000000000001729.

29. Kim HJ, Paul F, Lana-Peixoto MA, Tenembaum S, Asgari N, Palace J, et al. MRI characteristics of neuromyelitis optica spectrum disorder: an international update. Neurology. 2015;84(11):1165–73. https://doi.org/10.1212/WNL.0000000000001367.
30. Solomon JM, Paul F, Chien C, Oh J, Rotstein DL. A window into the future? MRI for evaluation of neuromyelitis optica spectrum disorder throughout the disease course. Ther Adv Neurol Disord. 2021;14:1–18. https://doi.org/10.1177/17562864211014389.
31. Dutra BG, da Rocha AJ, Nunes RH, Júnior ACMM. Neuromyelitis optica spectrum disorders: spectrum of MR imaging findings and their differential diagnosis. Radiographics. 2018;38:169–93. https://doi.org/10.1148/rg.2018170141.
32. Cai MT, Zhang YX, Zheng Y, Fang W, Ding MP. Callosal lesions on magnetic resonance imaging with multiple sclerosis, neuromyelitis optica spectrum disorder and acute disseminated encephalomyelitis. Mult Scler Relat Disord. 2019;32:41–5. https://doi.org/10.1016/j.msard.2019.04.019.
33. Makino T, Ito S, Mori M, Yonezu T, Ogawa Y, Kuwabara S. Diffuse and heterogeneous T2-hyperintense lesions in the splenium are characteristic of neuromyelitis optica. Mult Scler. 2013;19(3):308–15. https://doi.org/10.1177/1352458512454772.
34. Menge T, Hemmer B, Nessler S, Wiendl H, Neuhaus O, Hartung HP, et al. Acute disseminated encephalomyelitis: an update. Arch Neurol. 2005;62(11):1673–80. https://doi.org/10.1001/archneur.62.11.1673.
35. Pohl D, Alper G, Van Haren K, Kornberg AJ, Lucchinetti CF, Tenembaum S, et al. Acute disseminated encephalomyelitis: updates on an inflammatory CNS syndrome. Neurology. 2016;87(9 Suppl 2):S38–45. https://doi.org/10.1212/WNL.0000000000002825.
36. Wender M. Acute disseminated encephalomyelitis (ADEM). J Neuroimmunol. 2011;231(1–2):92–9. https://doi.org/10.1016/j.jneuroim.2010.09.019.
37. Fitsiori A, Nguyen D, Karentzos A, Delavelle J, Vargas MI. The corpus callosum: white matter or terra incognita. Br J Radiol. 2011;84(997):5–18. https://doi.org/10.1259/bjr/21946513.
38. Wingerchuk DM. The clinical course of acute disseminated encephalomyelitis. Neurol Res. 2006;28(3):341–7. https://doi.org/10.1179/016164106X98251.
39. Young NP, Weinshenker BG, Parisi JE, Scheithauer B, Giannini C, Roemer SF, et al. Perivenous demyelination: association with clinically defined acute disseminated encephalomyelitis and comparison with pathologically confirmed multiple sclerosis. Brain. 2010;133(2):333–48. https://doi.org/10.1093/brain/awp321.
40. Renard D, Castelnovo G, Campello C, Bouly S, Le Floch A, Thouvenot E, et al. An MRI review of acquired corpus callosum lesions. J Neurol Neurosurg Psychiatry. 2014;85(9):1041–8. https://doi.org/10.1136/jnnp-2013-307072.
41. Liu JG, Qiao WY, Dong QW, Zhang HL, Zheng KH, Qian HR, et al. Clinical features and neuroimaging findings of 12 patients with acute disseminated encephalomyelitis involved in corpus callosum. Zhonghua Yi Xue Za Zhi. 2012;92(43):3036–41.
42. Singhi PD, Ray M, Singhi S, Kumar Khandelwal N. Acute disseminated encephalomyelitis in north Indian children: clinical profile and follow-up. J Child Neurol. 2006;21(10):851–7. https://doi.org/10.1177/08830738060210100201.
43. Hillbom M, Saloheimo P, Fujioka S, Wszolek ZK, Juvela S, Leone MA. Diagnosis and management of Marchiafava-Bignami disease: a review of CT/MRI confirmed cases. J Neurol Neurosurg Psychiatry. 2014;85(2):168–73. https://doi.org/10.1136/jnnp-2013-305979.
44. Gambini A, Falini A, Moiola L, Comi G, Scotti G. Marchiafava-Bignami disease: longitudinal MR imaging and MR spectroscopy study. AJNR Am J Neuroradiol. 2003;24(2):249–53.
45. Alleman AM. Osmotic demyelination syndrome: central pontine myelinolysis and extrapontine myelinolysis. Semin Ultrasound CT MR. 2014;35(2):153–9. https://doi.org/10.1053/j.sult.2013.09.009.
46. King JD, Rosner MH. Osmotic demyelination syndrome. Am J Med Sci. 2010;339(6):561–7. https://doi.org/10.1097/MAJ.0b013e3181d3cd78.
47. Guerrero WR, Dababneh H, Nadeau SE. Hemiparesis, encephalopathy, and extrapontine osmotic myelinolysis in the setting of hyperosmolar hyperglycemia. J Clin Neurosci. 2013;20(6):894–6. https://doi.org/10.1016/j.jocn.2012.05.045.
48. Saroja AO, Naik KR, Mali RV, Kunam SR. 'Wine Glass' sign in recurrent postpartum hypernatremic osmotic cerebral demyelination. Ann Indian Acad Neurol. 2013;16(1):106–10. https://doi.org/10.4103/0972-2327.107719.
49. Lambeck J, Hieber M, Dreßing A, Niesen WD. Central pontine myelinolysis and osmotic demyelination syndrome. Dtsch Arztebl Int. 2019;116(35–36):600–6. https://doi.org/10.3238/arztebl.2019.0600.
50. Madjedi K, Wiegers S, Chan RK, Andreychuk T, Elliott M, Camara-Lemarroy CR. Osmotic demyelination syndrome: a new mime in the circus of neurology. Can J Neurol Sci. 2022;49(2):270–1. https://doi.org/10.1017/cjn.2021.66.
51. Dörr J, Krautwald S, Wildemann B, Jarius S, Ringelstein M, Duning T, et al. Characteristics of Susac syndrome: a review of all reported cases. Nat Rev Neurol. 2013;9(6):307–16. https://doi.org/10.1038/nrneurol.2013.82.
52. Pereira S, Vieira B, Maio T, Moreira J, Sampaio F. Susac's syndrome: an updated review. Neuroophthalmology. 2020;44(6):355–60. https://doi.org/10.1080/01658107.2020.1748062.
53. Aubart-Cohen F, Klein I, Alexandra JF, Bodaghi B, Doan S, Fardeau C, et al. Long-term outcome in Susac syndrome. Medicine (Baltimore). 2007;86(2):93–102. https://doi.org/10.1097/MD.0b013e3180404c99.

54. Rennebohm R, Susac JO, Egan RA, Daroff RB. Susac's syndrome--update. J Neurol Sci. 2010;299(1–2):86–91. https://doi.org/10.1016/j.jns.2010.08.032.
55. Garcia-Monco JC, Cortina IE, Ferreira E, Martínez A, Ruiz L, Cabrera A, et al. Reversible splenial lesion syndrome (RESLES): what's in a name? J Neuroimaging. 2011;21(2):e1–14. https://doi.org/10.1111/j.1552-6569.2008.00279.x.
56. Tada H, Takanashi J, Barkovich AJ, Oba H, Maeda M, Tsukahara H, et al. Clinically mild encephalitis/encephalopathy with a reversible splenial lesion. Neurology. 2004;63(10):1854–8. https://doi.org/10.1212/01.wnl.0000144274.12174.cb.
57. Zhang S, Ma Y, Feng J. Clinicoradiological spectrum of reversible splenial lesion syndrome (RESLES) in adults: a retrospective study of a rare entity. Medicine (Baltimore). 2015;94(6):e512. https://doi.org/10.1097/MD.0000000000000512.
58. Takanashi J, Tada H, Maeda M, Suzuki M, Terada H, Barkovich AJ. Encephalopathy with a reversible splenial lesion is associated with hyponatremia. Brain Dev. 2009;31(3):217–20. https://doi.org/10.1016/j.braindev.2008.04.002.
59. Pan JJ, Zhao YY, Lu C, Hu YH, Yang Y. Mild encephalitis/encephalopathy with a reversible splenial lesion: five cases and a literature review. Neurol Sci. 2015;36(11):2043–51. https://doi.org/10.1007/s10072-015-2302-2.
60. Shi BC, Li J, Jiang JW, Li MX, Zhang J, Shang XL. Mild encephalitis/encephalopathy with a reversible splenial lesion secondary to encephalitis complicated by hyponatremia: a case report and literature review. Medicine (Baltimore). 2019;98(47):e17982. https://doi.org/10.1097/MD.0000000000017982.
61. Yuan J, Yang S, Wang S, Qin W, Yang L, Hu W. Mild encephalitis/encephalopathy with reversible splenial lesion (MERS) in adults—a case report and literature review. BMC Neurol. 2017;17(1):103. https://doi.org/10.1186/s12883-017-0875-5.
62. Akasbi M, Berenguer J, Saiz A, Brito-Zerón P, Pérez-De-Lis M, Bové A, et al. White matter abnormalities in primary Sjögren syndrome. QJM. 2012;105(5):433–43. https://doi.org/10.1093/qjmed/hcr218.
63. Appenzeller S, Faria A, Marini R, Costallat LT, Cendes F. Focal transient lesions of the corpus callosum in systemic lupus erythematosus. Clin Rheumatol. 2006;25(4):568–71. https://doi.org/10.1007/s10067-005-0174-1.
64. Appenzeller S, Rondina JM, Li LM, Costallat LT, Cendes F. Cerebral and corpus callosum atrophy in systemic lupus erythematosus. Arthritis Rheum. 2005;52(9):2783–9. https://doi.org/10.1002/art.21271.
65. Fogel B, Cardenas D, Ovbiagele B. Magnetic resonance imaging abnormalities in the corpus callosum of a patient with neuropsychiatric lupus. Neurologist. 2006;12(5):271–3. https://doi.org/10.1097/01.nrl.0000237119.00864.d7.
66. Kim SS, Richman DP, Johnson WO, Hald JK, Agius MA. Limited utility of current MRI criteria for distinguishing multiple sclerosis from common mimickers: primary and secondary CNS vasculitis, lupus and Sjogren's syndrome. Mult Scler. 2014;20(1):57–63. https://doi.org/10.1177/1352458513491329.
67. Tzarouchi LC, Tsifetaki N, Konitsiotis S, Zikou A, Astrakas L, Drosos A, et al. CNS involvement in primary Sjogren syndrome: assessment of gray and white matter changes with MRI and voxel-based morphometry. AJR Am J Roentgenol. 2011;197(5):1207–12. https://doi.org/10.2214/AJR.10.5984.
68. Beuker C, Schmidt A, Strunk D, Sporns PB, Wiendl H, Meuth SG, et al. Primary angiitis of the central nervous system: diagnosis and treatment. Ther Adv Neurol Disord. 2018;11:1–16. https://doi.org/10.1177/1756286418785071.
69. Cho TA, Jones A. CNS vasculopathies: challenging mimickers of primary angiitis of the central nervous system. Best Pract Res Clin Rheumatol. 2020;34(4):1–16. https://doi.org/10.1016/j.berh.2020.101569.
70. Borhani-Haghighi A, Kardeh B, Banerjee S, Yadollahikhales G, Safari A, Sahraian MA, et al. Neuro-Behcet's disease: an update on diagnosis, differential diagnoses, and treatment. Mult Scler Relat Disord. 2019;39:1–15. https://doi.org/10.1016/j.msard.2019.101906.
71. Albayram S, Saip S, Hasiloglu ZI, Teke M, Ceyhan E, Tutuncu M, et al. Evaluation of parenchymal neuro-Behçet disease by using susceptibility-weighted imaging. AJNR Am J Neuroradiol. 2011;32(6):1050–5. https://doi.org/10.3174/ajnr.A2477.
72. Pittock SJ, Debruyne J, Krecke KN, Giannini C, van den Ameele J, De Herdt V, et al. Chronic lymphocytic inflammation with pontine perivascular enhancement responsive to steroids (CLIPPERS). Brain. 2010;133(9):2626–34. https://doi.org/10.1093/brain/awq164.
73. Nemani T, Udwadia-Hegde A, KeniKarnavat P, Kashikar R, Epari S. CLIPPERS spectrum disorder: a rare pediatric neuroinflammatory condition. Child Neurol Open. 2019;6:1–9. https://doi.org/10.1177/2329048X19831096.

Metabolic Pathologies of the Corpus Callosum

23

Hayriye Nermin Keçeci, Abdullah Canbal, and Burcu Çalışkan

23.1 Introduction

The corpus callosum (CC), consisting of white matter tracts responsible for interhemispheric communication and coordination, is the primary telencephalic commissure. It comprises the rostrum, the genu, the body, and the splenium [1].

Metabolic conditions such as acute renal failure, alcoholism, extrapontine myelinolysis, central pontine myelinolysis, hepatic encephalopathy, hyperammonemia, hypernatremia, hypoglycemia, hyponatremia, malnutrition, Marchiafava-Bignami disease, Wernicke encephalopathy, and Wilson's disease may cause changes in the CC.

Hypoglycemia can have a huge impact on people's lives as it has the potential to impair cerebral function. Profound hypoglycemia can cause structural and functional disturbances in both the central and the peripheral nervous systems. The use of glucose is responsible for almost all of the oxygen consumptions in the brain. The survival of the brain depends on continuous glucose support as glucose is not synthesized in the brain and there is only enough glycogen storage for a few minutes [2]. Recurrent hypoglycemia in infancy, where rapid brain development and differentiation are experienced, causes neurological sequelae, mental retardation, and convulsions in the long term. Neurological manifestations caused by hypoglycemia range from reversible focal deficits and transient encephalopathy to irreversible coma or death. Immediate recognition and effective treatment are essential to prevent the devastating effects of hypoglycemia [3].

Osmotic demyelination syndrome (ODS), which is one of the most important causes of central pontine myelinolysis (CPM), is higher than the rates reported with cases diagnosed by autopsy in neurology and intensive care clinics [4]. It may occur as a result of rapid deterioration and rapid correction in serum sodium values [5]. In ODS, the pons as well as the CC can be affected. CPM should be suspected, and radiological imaging should be performed when an unexpected change in the patient's clinic and findings cannot be explained. In this section, hypoglycemia and CPM, which are metabolic pathologies affecting the CC, will be discussed in detail.

23.2 Hypoglycemia

The body needs energy to function normally, and energy is provided only from appropriate fuels. Glucose is the main source of energy storage consisting of glycogen, fat, and protein and plays

H. N. Keçeci (✉)
Necmettin Erbakan University Faculty of Medicine, Department of Pediatric Genetics, Necmettin Erbakan University, Konya, Turkey

A. Canbal · B. Çalışkan
Necmettin Erbakan University Faculty of Medicine, Department of Pediatric Neurology, Necmettin Erbakan University, Konya, Turkey
e-mail: abdullahcanbal@yaani.com

M. Turgut et al. (eds.), *The Corpus Callosum*, https://doi.org/10.1007/978-3-031-38114-0_23

a major role in fuel use [2]. There are two sources of glucose to keep the blood glucose level within the normal range. These are glucose ingested with food and glucose produced by glycogenolysis and gluconeogenesis in the liver. Glycogen is the main storage form of carbohydrates. It is found in the liver and muscle. Liver glycogen acts as a depot for the body during fasting and is almost depleted after 12–18 h of fasting. Muscle glycogen is depleted only after prolonged and intense exercise [6].

Hypoglycemia is most common in the neonatal period. Discontinuation of glucose support provided through the placenta in intrauterine life at birth creates a risky environment for hypoglycemia. Prematurity, intrauterine growth retardation, sepsis, asphyxia, and hypothermia are factors that facilitate hypoglycemia. In addition, genetic defects of enzymes or hormones are also risk factors. In childhood, hypoglycemia is less common than hyperglycemia [2, 7].

Clinically, hypoglycemia is defined as a plasma glucose concentration that is low enough to cause symptoms and signs of brain dysfunction [2]. There is no single threshold for hypoglycemia-specific brain responses. Alternative substrates such as ketone and lactate can change the threshold. Normal plasma glucose values in the postabsorptive stage in infants, children, and adults are between 70 and 100 mg/dL [8].

The Whipple triad is used to define symptomatic hypoglycemia in adults. In Whipple's triad, there are typical symptoms of hypoglycemia, the blood glucose measured during the symptom is below 50 mg/dL, and the symptoms are relieved by raising the blood glucose level [9]. This definition is also useful for children who can describe the symptoms of hypoglycemia. But it is not useful for babies and young children [2].

According to the protocol of the American Academy of Pediatrics, the threshold value in the first 24 h is 40 mg/dL if the baby is symptomatic, 25–40 mg/dL in the first 4 h when the baby is asymptomatic, and 35–45 mg/dL in the 4–24-h range [10].

The causes of hypoglycemia can be divided into transient and resistant, and these causes are mentioned in Table 23.1 [2, 6, 9, 11, 12].

Table 23.1 Etiology of hypoglycemia

Transient hypoglycemia	Intrauterine growth retardation
	Prematurity
	Diabetic mother baby
	Rh incompatibility
	Exchange blood
	Babies of pregnant women given IV glucose at birth
	Birth asphyxia
	Hypothermia
	Sepsis
	Polycythemia
	Heart failure and congenital heart disease
Resistant hypoglycemia	**Hyperinsulinism**
	Congenital hyperinsulinism (*ABCC8, KCNJ11, GLUD1, GCK, HADH, SLC16A1, UCP2, HNF1A, HNF4A, HK1, PGM1, PMM2, FOXA2*, CACNA1D, and *EIF2S3*mutations)
	Insulinoma
	Functional beta cell disorders (nesidioblastosis)
	Noninsulinoma pancreatogenous hypoglycemia
	Post gastric bypass hypoglycemia
	Insulin autoimmune hypoglycemia
	Antibody to insulin
	Antibody to insulin receptor
	Insulin secretagogue

Table 23.1 (continued)

	Nonislet cell tumor
	Counter-regulatory hormone deficiencies
	Cortisol
	Glucagon and epinephrine
	Metabolic diseases
	Glycogen storage diseases, types 0, 1, 3, 6, 9, and 11
	Gluconeogenesis disorders
	Galactosemia
	Hereditary fructose intolerance
	Fatty acid oxidation defect
	Organic acidemias
	Ketotic hypoglycemia
	Drugs
	Insulin, alcohol, salicylate, quinine, and indomethacin
	Critical illnesses
	Hepatic, renal, or cardiac failure
	Sepsis (including malaria)
	Factitious hypoglycemia
	Munchausen syndrome

The clinical features of hypoglycemia are divided into two groups. The first is the symptoms related to the release of epinephrine, which occurs with the activation of the autonomic nervous system. It usually occurs when blood glucose drops rapidly. The second group of symptoms is related to neuroglycopenia and accompanies slow-developing, long-lasting, and more severe hypoglycemia [8]. Adrenergic and neuroglycopenic symptoms of hypoglycemia are indicated in Table 23.2 [6, 9, 11]. Older children may have these classic symptoms, whereas newborns and infants may have vague symptoms [13]. There is not always a correlation between blood glucose level and clinical findings in newborns. Low blood glucose and impaired brain glucose consumption may occur in the absence of symptoms. Hypoglycemia should always be kept in mind as the underlying cause of cyanosis, pallor, apnea, hypothermia, bradycardia, feeding difficulties, hypotonia, spasms, and convulsions [2, 13].

The main energy source of the human brain is glucose. Glucose entry into the brain is mediated by diffusion facilitated by glucose transporters 1 and 3 (GLUT 1 and GLUT3), which act as insulin-independent carrier proteins. A decrease in blood glucose or a defect in the transport mechanism causes brain damage by lowering the glucose concentration in the brain and in the cerebrospinal fluid [14].

Table 23.2 Symptoms and signs of hypoglycemia

Autonomic nervous system effects due to epinephrine release	Neuroglycopenic effects	Clinical features in newborns and infants
Sweating	Headache	Startle
Tachycardia	Confusion	Tremor
Pallor	Restlessness	Apathy
Anxiety	Irritability	Cyanosis
Tremor	Personality change	Convulsion
Weakness	Altered state of consciousness	Apnea-tachypnea attacks
Hunger	Speech disorder	Brachycardia
Nausea	Mood disorder	Weak or high-pitched crying
Vomiting	Dizziness	Hypotonia
	Memory loss	Sudden pallor
	Ataxia	Hypothermia

Glucose is important in brain metabolism as well as a source of lipids in the cell membrane. It contributes to the production of structural proteins and myelin formation in the brain. In severe and continuous hypoglycemia conditions, structures containing lipids and proteins are broken down into products such as lactate, pyruvate, keto acid, and amino acid, which can be used as energy [2]. The brain is vulnerable to hypoglycemia. Mitochondrial damage is one of the earliest consequences of hypoglycemia and leads to decreased ATP levels. An imbalance in electrolyte homeostasis occurs by causing an increase in Ca^{2+} in the neuron. The permeability of mitochondrial inner membranes changes and mitochondria swell. Cytochrome *c*, which plays a role in the initiation of apoptosis by activating caspases, increases and neuron damage occurs [15]. The most important long-term sequelae of severe and prolonged hypoglycemia are mental retardation and epilepsy [3, 16]. Hypoglycemia can cause mood changes and impair cognitive functions such as verbal skills, attention, reaction time, learning, and memory [17]. Symmetrical distal sensorimotor neuropathy including numbness in the extremities, weakness, muscle atrophy in the hands and feet, difficulty walking, and painful distal paresthesia may develop in patients with recurrent hypoglycemia [18].

Permanent neurological sequelae occur in 25–50% of patients younger than 6 months with recurrent severe hypoglycemia [19]. The cerebral cortex, and especially the white matter in the parietal and occipital lobes, is the most vulnerable to hypoglycemia in newborns [20]. Severe hypoglycemia in newborns and T2-weighted (T2W) imaging cause hyperintensity in the occipital and parietal lobes, and diffusion restriction is observed in the splenium of corpus callosum (SCC). If left untreated, localized thinning of SCC may occur [21]. Figure 23.1 shows the magnetic resonance images (MRI) of the patient with seizures secondary to hypoglycemia in the neonatal period.

High-signal-intensity lesions have been reported on diffusion-weighted imaging (DWI) brain MRI SCC in adults experiencing hypoglycemia. This condition, called reversible splenial lesion syndrome (RESLES), has been documented in adults and children. RESLES can occur due to various reasons such as mild enceph alitis/encephalopathy (Kawasaki disease, rotavirus, Japanese encephalitis virus, influenza, *Mycoplasma pneumoniae*, bacterial urinary tract

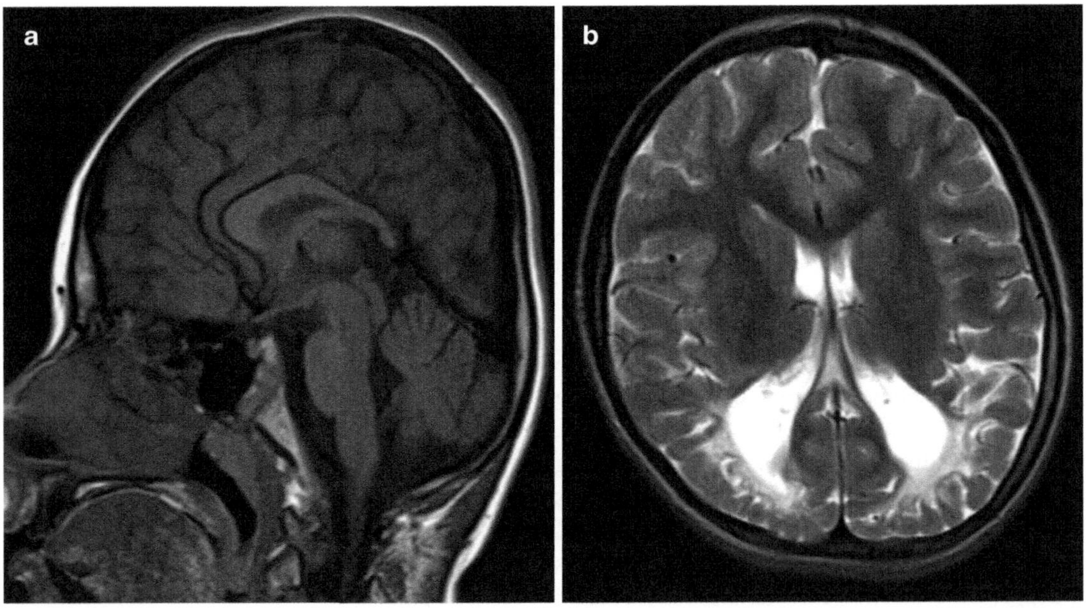

Fig. 23.1 Magnetic resonance imaging in a 7-year-old child with cerebral palsy who had a history of hypoglycemia during the neonatal period. (**a**) T1-sagittal imaging shows thinning of the isthmus and splenium of the CC. (**b**) T2-axial imaging shows bilateral symmetrical hyperintensities, gliotic changes, and atrophy in the occipital lobes. The occipital horns of both lateral ventricles are also enlarged secondary to atrophy

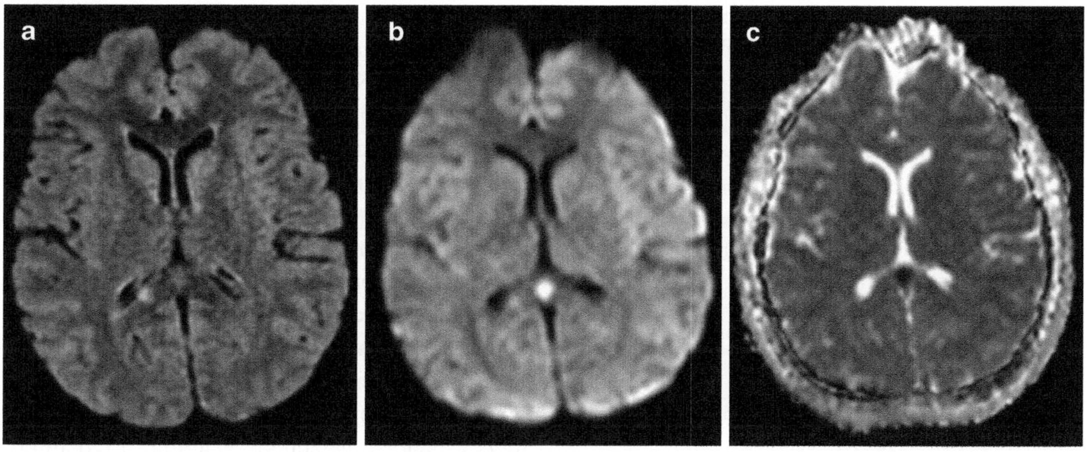

Fig. 23.2 Magnetic resonance imaging in a 29-year-old female patient with type 1 diabetes mellitus who was admitted with sleepiness. She was diagnosed with a reversible splenial lesion syndrome secondary to hypoglycemia. (Courtesy of Prof. Salih Hattapoğlu, Dicle University, Diyarbakır, Turkey). (**a**) Fluid-attenuated inversion recovery imaging shows hyperintensity in the splenium of CC. (**b**) Diffusion-weighted magnetic resonance imaging showing corresponding hyperintensity in the CC. (**c**) Apparent diffusion coefficient (ADC) map at the same levels as (**a**, **b**) shows decreased ADC in the lesion compared with normal white matter

infection), metabolic disorders (hypoglycemia or hypernatremia), epilepsy patients taking antiepileptic drugs (especially phenytoin and carbamazepine), and autoimmune disease [22, 23]. SCC has been hypothesized to have a certain susceptibility to glutamate-related excitotoxic damage in metabolic diseases [24, 25]. Although it has been suggested that hypoglycemia results in cytotoxic edema by disrupting cellular fluid mechanics in SCC, the mechanism of the pathogenesis of RESLES remains unclear [26]. Brain MRI, particularly DWI, is useful in diagnosing hypoglycemic encephalopathy and assessing neurological prognosis. Images suggestive of RESLES in a patient with hypoglycemic encephalopathy are shared in Fig. 23.2. Restricted water diffusion areas elicited high signal intensity in the DWI with significant diffusion coefficient reductions shown for cytotoxic edema. It has been noted that this may be due to energy depletion caused by hypoglycemia followed by membrane ionic pump failure. In many case series, lesions regressed with correction of hypoglycemia. The timing of neurological recovery is related to the duration and severity of the hypoglycemic injury. Therefore, patients may have permanent neurological deficits if not promptly treated with glucose uptake [27–29].

23.3 Central Pontine Myelinolysis

CPM is a neurological disease caused by loss of myelin in the pons. It is usually located symmetrically at the base of the pons and has similar histological structures. It is called extrapontine myelinolysis (EPM) when the basal ganglia, lateral geniculate body, external and internal capsule, and CC are affected, especially in regions such as the splenium and cerebellum [30, 31]. CPM was first described as a disorder in 1959. In this article, the authors describe four cases who died and their autopsy findings. These patients were alcohol dependent and had malnutrition findings. Adams et al. used the term "myelinolysis" to describe the location of the lesion in the pons and to emphasize that myelin was affected. The authors deliberately avoided using the term "demyelination" to describe the condition to distinguish it from multiple sclerosis, acute disseminated encephalomyelitis, and other neuroinflammatory disorders [32, 33].

Table 23.3 Conditions that increase the risk for central pontine myelinolysis and extrapontine myelinolysis

Alcoholism (chronic)
Amyotrophic lateral sclerosis
Burns
Coagulopathies
Diabetic hyperosmolarity
Diabetic ketoacidosis
Hypernatremia (acute/chronic)
Hyponatremia (acute/chronic)
Hypothalamic tumors
Liver failure (acute/chronic)
Liver transplantation
Ornithine carbamoyl transferase deficiency
Pancreatitis (acute hemorrhagic)
Pineal region tumors
Polydipsia
Renal failure (acute/chronic)
Sepsis (bacterial)
Sickle cell crisis
Viral infections (mumps, hepatitis)
Vomiting, persistent
Wilson's disease

The etiology of central pontine myelinolysis and extrapontine myelinolysis is not fully known. Increased risk factors causing CPM and EPM are given in Table 23.3 [30]. Although many studies have shown that osmotic stress due to rapid correction of hyponatremia plays a role in the etiology, it can be seen in acute hypernatremia [33, 34].

ODS is used for both CPM and EPM [31]. ODS is an acute demyelination process that usually occurs within a few days following the rapid increase in serum osmolality. ODS is a rare condition, reported in 0.06% of hospital admissions. Disruption of the blood-brain barrier (BBB) plays an important role in the pathogenesis of ODS. Rapid correction of low serum sodium or acute elevation of sodium causes water to pass into the extracellular space, shrinkage of brain vascular endothelial cells and disruption of glial cells' tight junctions in the BBB, resulting in lymphocytes, cytokines, and vasoactive amines entering the central nervous system, causing inflammatory demyelination and axonal damage [35]. It is accepted that the pathophysiology of ODS is multifactorial and depends on the underlying cause. Rapid correction of hyperammonemia causes aquaporin-4 dysregulation, osmotic stress, and ODS [5].

In ODS, clinical signs appear a few days after rapid serum sodium correction. Clinically, it can cause various findings such as progressive spastic quadriparesis, pseudobulbar palsy, ataxia, nystagmus, dysarthria, and dysphagia ophthalmoplegia. In advanced cases, locked-in syndrome and death may occur. In EPM, symptoms occur according to the affected area. Variable mental status, seizures, dystonia, myoclonus, stool incontinence, and deafness may be seen. Variable mental status, seizures, dystonia, myoclonus, stool incontinence, and deafness may be seen. Clinical findings of CPM and EPM are given in Table 23.4 [30].

Table 23.4 Clinical findings of central pontine myelinolysis and extrapontine myelinolysis

Central pontine myelinolysis	Ataxia
	Coma
	Depressed/absent reflexes
	Dysarthria
	Dysphasia
	Lethargy
	Ophthalmoplegia
	Quadriparesis
Extrapontine myelinolysis	Akinesis
	Ataxia
	Catatonia
	Choreoathetosis
	Cogwheel rigidity
	Disorientation
	Dysarthria
	Dystonia
	Emotional lability
	Extrapyramidal symptoms
	Gait disturbance
	Movement disorders
	Mutism
	Myoclonus
	Myokymia
	Parkinsonism
	Rigidity
	Tremor

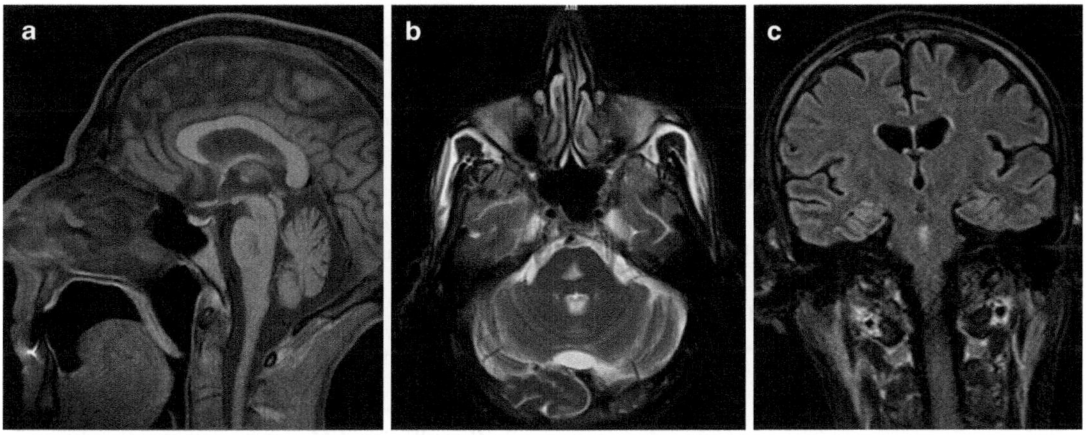

Fig. 23.3 Magnetic resonance imaging of a 48-year-old male patient who was followed for the diagnosis of gastric cancer and was admitted with confusion. The patient had signs of dehydration and his serum sodium was measured at 169 mEq/L. He was diagnosed with CPM developing after acute hypernatremia. (**a**) Sagittal T1W imaging shows hypointensity in the central pons region. (**b**) Axial T2W imaging shows hyperintensity in the central pons region. (**c**) Coronal fluid-attenuated inversion recovery imaging shows hyperintensity in the central pons region

The best radiological diagnosis method of CPM is cranial MRI. Weekly or monthly consecutive MRI may be performed depending on clinical symptoms. On MRI T2W images, the central pons, tegmentum, corticospinal tract, and ventrolateral pons are hyperintense. They appear as hypointense areas on T1-weighted (T1W) images that do not have a mass effect (Fig. 23.3). Contrast enhancement is usually not seen on contrast imaging. Pons lesions are usually symmetrical. Extrapontine involvement is seen as hyperintensity in the SCC section, cerebellar peduncle, putamen, globus pallidus, internal and external capsule, and thalamus supratentorial area on MRI T2W images [4, 35]. Lesions that are pontine and extrapontine hyperintense on DWI images may have low or normal ADC values. A decrease in ADC early in the disease suggests the presence of a cytotoxic edema [36, 37]. Low colin (cho), *N*-acetyl aspartate levels, and increased cho-creatine ratios in magnetic resonance spectroscopy imaging in CPM help diagnosis [38].

Current treatment of ODS is a supportive treatment and reduction of possible risks. Hyponatremia should be corrected at a rate not exceeding 10 mmol/L/24 h, 0.5 mEq/L h, or 18 mEq/L/48 h to minimize the risk from rapid correction of the most common cause, hyponatremia [39]. There is no specific treatment after the onset of osmotic demyelination. The potential benefits of steroids and intravenous immunoglobulins have been identified in pediatric and adult patients but are not routinely used in clinics [35, 40]. Among other potential treatments, thyroid-releasing hormone, minocycline, lovastatin, and myoinositol are under investigation [41, 42].

23.4 Other Metabolic Conditions

The CC may also be affected in some inherited metabolic diseases. X-linked adrenoleukodystrophy, in its classical childhood form, is a metabolic disease caused by a peroxisomal enzyme defect in beta oxidation leading to demyelination of the central nervous system and shows a typical pattern of involvement that originates and progresses from the splenium of the CC [43].

Wilson's disease is an autosomal recessive copper metabolism disorder that can affect the anterior and posterior portions of the CC. More extensive brain lesions, more severe neurological dysfunctions, and psychiatric symptoms were observed in patients with CC involvement [44].

The mucopolysaccharidoses lead to the accumulation of glycosaminoglycans in many tissues, such as brain parenchyma. The perivascular spaces can enlarge, and this is usually followed by white matter changes and atrophy [45].

Urea cycle disorders cause focal thinning in the CC. Cobalamin C deficiency is a treatable disorder that causes marked volume loss of the CC body and hypomyelination [21].

Diffusion restriction has been demonstrated in the SCC in metabolic diseases such as argininosuccinic aciduria and saposin B deficiency [46].

23.5 Conclusion

Lesions can be seen in CC secondary to drug therapy, infections, trauma, malignancy, and metabolic disorders. Hypoglycemia, hyponatremia, and hypernatremia are common due to metabolic causes, and CC lesions may be reversible with timely and appropriate treatment.

References

1. Filippi CA, Cauley KA. Lesions of the corpus callosum and other commissural fibers: diffusion tensor studies. Semin Ultrasound CT MR. 2014;35:445–58.
2. Thornton PS, Stanley CA, De Leon DD, Harris D, Haymond MW, Hussain K, Levitsky LL, Murad MH, Rozance PJ, Simmons RA, Sperling MA, Weinstein DA, White NH, Wolfsdorf JI. Pediatric Endocrine Society. Recommendations from the Pediatric Endocrine Society for Evaluation and Management of Persistent Hypoglycemia in neonates, infants, and children. J Pediatr. 2015;167:238–45.
3. Menni F, de Lonlay P, Sevin C, Touati G, Peigné C, Barbier V, Nihoul-Fékété C, Saudubray JM, Robert JJ. Neurologic outcomes of 90 neonates and infants with persistent hyperinsulinemic hypoglycemia. Pediatrics. 2001;107:476–9.
4. Bansal LR, Zinkus T. Osmotic demyelination syndrome in children. Pediatr Neurol. 2019;97:12–7.
5. Pan CF, Zhu SM, Zheng YY. Ammonia induces upregulation of aquaporin-4 in neocortical astrocytes of rats through the p38 mitogen-activated protein kinase pathway. Chin Med J. 2010;123:1888–92.
6. Mayes PA, Bender DA. Glycogen metabolism. In: Murray RK, Granner DK, Mayes P, Rodwell VW, editors. Harper's illustrated biochemistry. 26th ed. New York: McGraw-Hill; 2003. p. 145–52.
7. Langdon DR, Stanley CS, Sperling MA. Hypoglycemia in the toddler and child. In: Sperling MA, editor. Pediatric endocrinology. 4th ed. Philadelphia: Elsevier Saunders; 2014. p. 920–55.
8. De Leon DD, Thoronton PS, Stanley CS, Sperling MA. Hypoglycemia in neonates and infants. In: Sperling MA, editor. Pediatric endocrinology. 4th ed. Philadelphia: Elsevier Saunders; 2014. p. 157–85.
9. Whipple AO, Frantz VK. Adenoma of islet cells with hyperinsulinism: a review. Ann Surg. 1935;101:1299–335.
10. Committee on Fetus and Newborn, Adamkin DH. Postnatal glucose homeostasis in late-preterm and term infants. Pediatrics. 2011;127:575–9.
11. Cryer PE, Axelrod L, Grossman AB, Heller SR, Montori VM, Seaquist ER, F John Service, Endocrine Society. Evaluation and management of adult hypoglycemic disorders: an Endocrine Society Clinical Practice Guideline. J Clin Endocrinol Metab. 2009;94:709–28.
12. Zhang W, Sang YM. Genetic pathogenesis, diagnosis, and treatment of short-chain 3-hydroxyacyl-coenzyme a dehydrogenase hyperinsulinism. Orphanet J Rare Dis. 2021;16:467.
13. Tas E, Garibaldi L, Muzumdar R. Glucose homeostasis in newborns: an endocrinology perspective. NeoReviews. 2020;2:e14–29.
14. Nehlig A. Cerebral energy metabolism, glucose transport and blood flow: changes with maturation and adaptation to hypoglycaemia. Diabetes Metab. 1997;23:18–29.
15. Mohseni S. Neurologic damage in hypoglycemia. Handb Clin Neurol. 2014;126:513–32.
16. Wickström R, Skiöld B, Petersson G, Stephansson O, Altman M. Moderate neonatal hypoglycemia and adverse neurological development at 2-6 years of age. Eur J Epidemiol. 2018;33:1011–20.
17. Hershey T, Lillie R, Sadler M, White NH. Severe hypoglycemia and long-term spatial memory in children with type 1 diabetes mellitus: a retrospective study. J Int Neuropsychol Soc. 2003;9:740–50.
18. Puri V, Garg N, Kumar N, Tatke M. Hypoglycaemic neuropathy: a case report. Neurol India. 2000;48:263–5.
19. Meissner T, Wendel U, Burgard B, Schaetzle S, Mayatepek E. Long-term follow-up of 114 patients with congenital hyperinsulinism. Eur J Endocrinol. 2003;149:43–51.
20. Murakami Y, Yamashita Y, Matsuishi T, Utsunomiya H, Okudera T, Hashimoto T. Cranial MRI of neurologically impaired children suffering from neonatal hypoglycaemia. Pediatr Radiol. 1999;29:23–7.
21. Andronikou S, Pillay T, Gabuza L, Mahomed N, Naidoo J, Hlabangana LT, De Plessis V, Prabhu SP. Corpus callosum thickness in children: an MR pattern-recognition approach on the midsagittal image. Pediatr Radiol. 2015;45:258–72.
22. Chan R, Erbay S, Oljeski S, Thaler D, Bhadelia R. Case report: hypoglycemia and diffusion-weighted imaging. J Comput Assist Tomogr. 2003;27:420–3.
23. Doherty MJ, Jayadev S, Watson NF, Konchada RS, Hallam DK. Clinical implications of splenium magnetic resonance imaging signal changes. Arch Neurol. 2005;62:433–7.

24. Kang EG, Jeon SJ, Choi SS, Song CJ, Yu IK. Diffusion MR imaging of hypoglycemic encephalopathy. AJNR Am J Neuroradiol. 2010;31:559–64.
25. Zhang S, Ma Y, Feng J. Clinicoradiological spectrum of reversible splenial lesion syndrome (RESLES) in adults: a retrospective study of a rare entity. Medicine (Baltimore). 2015;94:e512.
26. Lee SH, Kang CD, Kim SS, Tae WS, Lee SY, Kim SH, Koh SH. Lateralization of hypoglycemic encephalopathy: evidence of a mechanism of selective vulnerability. J Clin Neurol. 2010;6:104–8.
27. Hasegawa Y, Formato JE, Latour LL, Gutierrez JA, Liu KF, Garcia JH, Sotak CH, Fisher M. Severe transient hypoglycemia causes reversible change in the apparent diffusion coefficient of water. Stroke. 1996;27:1648–56.
28. Kashiwagi M, Tanabe T, Shimakawa S, Nakamura M, Murata S, Shabana K, Shinohara J, Odanaka Y, Matsumura H, Maki K, Okumura K, Okasora K, Tamai H. Clinico-radiological spectrum of reversible splenial lesions in children. Brain Dev. 2014;36:330–6.
29. Malik AM. The reversible corpus callosum splenium lesion associated with hypoglycemic encephalopathy. Neurohospitalist. 2013;3:169.
30. Brown WD. Osmotic demyelination disorders: central pontine and extrapontine myelinolysis. Curr Opin Neurol. 2000;13:691–7.
31. Lampl C, Yazdi K. Central pontine myelinolysis. Eur Neurol. 2002;47:3–10.
32. Adams RD, Victor M, Mancall EL. Central pontine myelinolysis. AMA Arch Neurol Psychiatry. 1959;81:154–72.
33. Yoon B, Shim YS, Chung SW. Central pontine and extrapontine myelinolysis after alcohol withdrawal. Alcohol Alcohol. 2008;43:647–9.
34. Han MJ, Kim DH, Kim YH, Yang MI, Park HJ, Hong MK. A case of osmotic demyelination presenting with severe hypernatremia. Electrolyte Blood Press. 2015;13:30–6.
35. Bansal L. Therapeutic effect of steroids in osmotic demyelination of infancy. Child Neurol Open. 2018;5:2329048X18770576.
36. Dervisoglu E, Yegenaga I, Anik Y, Sengul E, Turgut T. Diffusion magnetic resonance imaging may provide prognostic information in osmotic demyelination syndrome: report of a case. Acta Radiol. 2006;47:208–12.
37. Shin HW, Song D, Sohn YH. Normal diffusion-weighted MR imaging predicts a good prognosis in extrapontine myelinolysis-induced parkinsonism. Mov Disord. 2009;24:1701–3.
38. Aslan H, Donmez FY, Hekimoglu OK, Tore HG. The magnetic resonance spectroscopy findings of extrapontine myelinolysis in a patient with acute lymphoblastic leukemia. Neurol Sci. 2012;33:391–4.
39. Hoorn EJ, Zietse R. Diagnosis and treatment of hyponatremia: compilation of the guidelines. J Am Soc Nephrol. 2017;28:1340–9.
40. Murthy SB, Izadyar S, Dhamne M, Kass JS, Goldsmith CE. Osmotic demyelination syndrome: variable clinical and radiologic response to intravenous immunoglobulin therapy. Neurol Sci. 2013;34:581–4.
41. Suzuki H, Sugimura Y, Iwama S, Suzuki H, Nobuaki O, Nagasaki H, Arima H, Sawada M, Oiso Y. Minocycline prevents osmotic demyelination syndrome by inhibiting the activation of microglia. J Am Soc Nephrol. 2010;21:2090–8.
42. Zein EF, Karaa SE, Rollot F, Blanche P, Chemaly RE. Treatment of central pontine myelinolysis with thyrotropin-releasing hormone. Presse Med. 2006;35:618–20.
43. Patel PJ, Kolawole TM, Malabarey TM, Al-Herbish AS, Al-Jurrayan NA, Saleh M. Adrenoleukodystrophy: CT and MRI findings. Pediatr Radiol. 1995;25:256–8.
44. Zhou ZH, Wu YF, Cao J, Hu JY, Han YZ, Hong MF, Wang GQ, Liu SH, Wang XM. Characteristics of neurological Wilson's disease with corpus callosum abnormalities. BMC Neurol. 2019;19:85.
45. Kara S, Sherr EH, Barkovich AJ. Dilated perivascular spaces: an informative radiologic finding in Sanfilippo syndrome type A. Pediatr Neurol. 2008;38:363–6.
46. Aksu Uzunhan T, Maras Genc H, Kutlubay B, Kalın S, Bektas G, Yapici O, Ciraci S, Sozen HG, Sevketoglu E, Palabiyik F, Gor Z, Cakar NE, Kara B. Cytotoxic lesions of the corpus callosum in children: etiology, clinical and radiological features, and prognosis. Brain Dev. 2021;43:919–30.

Traumatic Axonal Lesions of the Corpus Callosum

24

Robert Sumkovski and Ivica Kocevski

24.1 Introduction

Traumatic axonal injury (TAI) was first described in the mid-twentieth century, as diffuse microscopic pathological changes to the brain tissue [1], and was initially named diffuse axonal injury (DAI). Nowadays, what was formerly known as DAI is now referred to as TAI, which is considered more accurate [2, 3].

The term itself implicates that there is diffuse topographic distribution of traumatic findings [4]. However, studies showed that the distribution of traumatic lesions in DAI has a predisposition for white matter tracts in the midline of the brain, including the corpus callosum (CC), internal capsule, cerebral peduncles, brainstem, and the gray-white junction of the cerebral cortex [2, 5].

Worldwide, trauma causes 10% of all deaths, and trauma is the leading cause of permanent disability. Most traumatic events occur in individuals between 5 and 44 years old [6]. Traumatic brain injury (TBI) is a major economic and social problem worldwide as it leads to significant cognitive, physical, and emotional disabilities [7–9]. Traumatic axonal injury (TAI) is commonly found in traumatic brain injury (TBI), and it has been related to prognosis and the patient's quality of life. TAI lesions can occur in the cerebral hemispheres, CC, basal ganglia, thalamus, brainstem, and cerebellum [2, 10].

DAI describes a process of widespread axonal damage in the aftermath of acute or repetitive TBI, leading to deficits in cerebral connectivity that may or may not recover over time. It is a component of injury in 40–50% of hospital admissions for TBI and one of the most common pathologies in all closed-head trauma [11].

The spectrum of TBI encompasses three distinct grades: mild, moderate, and severe TBI. Often there is confusion in the nosology of TBI especially in relation to mild TBI (mTBI), a term that implicitly refers to the TBI event consistent with acute concussion. TBI is categorized according to the clinical pillars of post-traumatic amnesia (PTA) and/or a disturbance of consciousness—either alteration of consciousness (AOC) or loss of consciousness (LOC). These clinical features, although correlated, allow for independent diagnosis of TBI severity [12]. See Table 24.1.

A study by Cicuendez et al. [13] found a mean age of 33 years, ranging from 23 to 40. Patient population was predominantly male (75% males and 25% females). The CC lesions were divided subjectively into mild (38%), moderate (46%), and severe (15%). Nonhemorrhagic lesions were more frequent (65%) than hemorrhagic lesions (35%). The authors classified CC lesions according to their location inside the CC. The most frequent location was the splenium (56%), followed by 23% of lesions affecting both the splenium

R. Sumkovski (✉) · I. Kocevski
Clinic for Neurosurgery, University Clinical Centre Mother Teresa, Skopje, North Macedonia

M. Turgut et al. (eds.), *The Corpus Callosum*, https://doi.org/10.1007/978-3-031-38114-0_24

Table 24.1 The spectrum of traumatic brain injury

Glasgow coma scale (GCS)	Loss of consciousness (LOC)	Post-traumatic amnesia (PTA)	TBI
13–15	<1 h	<24 h	Mild or mTBI
9–12	>1 and <24 h	>24 h and <7 days	Moderate TBI
3–8	>24 h	>7 days	Severe TBI

and body. In 21% of patients, the lesion was just in the body of the CC, and only 1 patient presented a lesion at the genu of the CC. According to the same study, the maximum number of lesions in the CC counted in the same patient was 3.

In a study by Vieira et al., 89.7% of patients with diffuse axonal lesion/injury DAI were male and 43.6% between 18 and 28 years of age, with a mean age of 32 years. The main cause of DAI in this study was traffic accidents, with motorcyclists accounting for 43.6% of these cases. Individuals in coma (GCS ≤8) constituted 75.7% of the patients. In this study, no patients were classified as having mild TBI, while 19.2% had moderate and 80.8% severe TBI. Out of the total, 53.8% underwent surgery, and 19.2% required repeat surgeries. Mortality after 6 months was 30.8%. Among survivors, 88.2% achieved recovery consistent with independent life 6 months post DAI [14]. In the study by Adams et al., the incidence of traffic accidents in patients with DAI was much higher than that in patients without DAI, and the incidence of DAI was much lower in patients whose TBI resulted from a fall [15].

The incidence of DAI across all TBI patients is estimated to be roughly 33%. According to Adams et al., the incidence of DAI in closed TBI is 29%, and that is likely to be an underestimate, since DAI may be camouflaged by other types of brain injury, like severe contusions, brain hematomas, brain injury secondary to intracranial hypertension, brain herniation leading to hemorrhage and infarction of the brainstem, and severe hypoxic injury [2]. In moderate to severe TBI, the prevalence of DAI is approximately 72% [16].

According to the National Institutes of Health Common Data Elements, traumatic axonal injury (TAI) is a condition characterized as multiple, scattered, small hemorrhagic, and/or nonhemorrhagic lesions, alongside brain swelling, in a more confined white matter distribution on imaging studies, together with impaired axoplasmic transport, axonal swelling, and disconnection with more than three such foci present on imaging studies [17]. Numerous studies showed that the distribution of traumatic lesions in DAI has a predisposition for white matter tracts in the midline of the brain, including the CC, internal capsule, cerebral peduncles, brainstem, and the gray-white junction of the cerebral cortex [2, 5]. TAI is thought to be caused by a variety of traumatic mechanisms most of which involve fast/rapid acceleration and/or deceleration [3], or axonal injury is caused by mechanical strain related to rotational acceleration/deceleration forces [18].

Today CC lesions were classified as (1) hemorrhagic or nonhemorrhagic; (2) mild, moderate, or severe lesions (according to subjective measurement of CC lesion volume); (3) genu, splenium, or body involvement; (4) total number of lesions; and (5) total volume of the lesions [13].

The CC is one of the common sites of TAI, and its involvement is an indicator of worse prognosis. It has been reported that reducing the integrity of the CC after TBI may disrupt the connectivity between bilateral frontoparietal neural networks underlying cognitive functions, such as working memory, bimanual coordination, and executive function processes [19, 20].

The CC is thought to serve several key roles in motor control, namely, transfer of lateralized information (i.e., verbal input from the left hemisphere) to the opposite hemisphere to guide unilateral movement, transfer of information to coordinate bilateral movements, and transfer of information to inhibit contralateral movement during a unilateral motor activity [21]. CC lesions are thought to be a marker of the severity of diffuse TBI [22, 23].

The CC is the main tract connecting the hemispheres and forms the largest commissural white matter bundle in the brain, with some 200 million axons. CC lesions are of serious concern in TBI owing to the structural role of the CC in the interhemispheric transfer of auditory, visual, sensory, and motor information that is relevant to multiple cognitive processes [24]. In addition, the CC is particularly vulnerable to TBI due to its unique location and composition [24].

24.2 Biomechanics of Craniocerebral Trauma (Biomechanical Features)

There are two major mechanisms involved in head trauma: direct impact and accelerative and decelerative (a/d) forces [25]. Sudden head movement produces a force vector inside the intracranial cavity, resulting in shearing and strain injury. Shear and tear of axonal fibers can cause axonal damage, resulting in TAI [3]. The duration of the a/d forces is decisive for the type of injury. Slower a/d forces with a relatively long duration (20–25 ms) will mainly cause TAI, whereas shorter duration of a/d forces will likely cause acute subdural hematoma (ASDH) through shearing of bridging veins [3, 26]. a/d forces in the coronal plane are primarily associated with the occurrence of TAI [3, 27, 28]. The combination of translational and angular acceleration [29] and also rotational acceleration [30] has also been suggested as the most prominent cause of TAI. The fact that TAI can be caused by traumas with relatively low rates of acceleration is of importance in forensic pathology. Deadly TAI can occur even if the initial impact force is not strong enough to cause fracturing of the skull or evident macroscopic pathology to the brain [27, 31].

24.3 Pathophysiology of TBI

24.3.1 Pathogenesis of Diffuse Axonal Injury

Most of our present knowledge of the pathogenesis of axonal injury has been derived mainly from experimental studies in non-primate species. According to Gennarelli et al., evidence has accumulated from different groups that makes clear that whatever the precise mechanism involved in axonal injury, the severity of DAI is variable, forming a continuum that can be divided into four well-defined stages of increasing severity: (1) membrane injury, (2) reversible cytoskeletal damage, (3) secondary axotomy, and (4) primary axotomy [32, 33]. In brief, the stages are as follows:

Stage I: Axonal membrane injury and alterations in ionic fluxes due to the special structure of the CNS axons. Their weakest part is the node of Ranvier and the paranodal regions [33]. Experiments in which single axons were stretched revealed a relationship between the amount of axonal stretch and the severity of axonal dysfunction [33]. At the lowest level of strain (5% or less), ionic influx changes and causes a temporary failure in the generation and propagation of action potentials. The most significant alteration at this stage is an increase in cytosolic free Ca++. However, ionic alterations are restored in a matter of minutes in such minor injuries [33]. Recently, the concept of mechanoporation has been introduced to describe a phenomenon that is characterized by a transient defect in the cell membrane as a direct result of mechanical deformation [33]. Mechanoporation may be the mechanism by which defects or pores in the axoplasmic membrane are produced, thus increasing its permeability to ionic fluxes.

Stage II: Reversible cytoskeletal damage when the magnitude of stretch is higher (5–10% range). Ionic disturbances are greater and frequently associated with local swelling of the axon as well as abnormalities in axoplasmic transport. These abnormalities result in axonal varicosities but not, at this stage, in axonal disconnection [33]. In this form of moderate DAI, Gennarelli et al. hypothesized that restitution of axonal structure and function could generally be expected [33].

Stage III: Secondary axotomy in experimental models. This level of injury has been associated with a 15–20% strain on the axons [33]. Severe alterations in ionic fluxes are produced first; these ionic disturbances evolve to structural abnormalities that end in axonotmesis. An influx of calcium appears to be the most important single event that activates neutral proteases (calpains) and phospholipases. Proteolytic enzymes induce a cascade of events that injures the axon. At the beginning of this process, calpains alter all the cytoskeletal proteins, thus provoking proteolysis of the micro tubules and neurofilaments, structures that support the axon. Phospholipases attack membrane fatty acids activating free-radical species and various inflammatory mediators [33]. The structural weakness of the axoplasmic membrane and the ionic disturbances usually evolve to axotomy within 24–72 h after injury [33]. As stated by Gennarelli, "if the process leading to secondary or delayed axotomy can be interrupted, then perhaps reparative mechanisms can rescue the axon from permanent disruptive damage" [33] (see Fig. 24.1).

Stage IV: Primary axotomy is the most severe form of DAI and consists of a tearing of the axon provoked by severe strains immediately on impact. This type of injury corresponds to the traditional view of DAI as initially described by Adams and co-workers [34–37]. Axonal tearing is experimentally related to axonal strains above 20%. Gennarelli et al. have suggested that primary axotomy occurs less frequently in human head injury than stages I, II, and III of axonal damage [33]. However, this belief has not been confirmed in human studies, in which the magnitude of acceleration can be very different to that in animal models. A potential source of bias is that mechanical loading may be higher in patients who die almost immediately after impact and who are never seen at hospitals than in those who survive long enough to be transferred to hospitals [38].

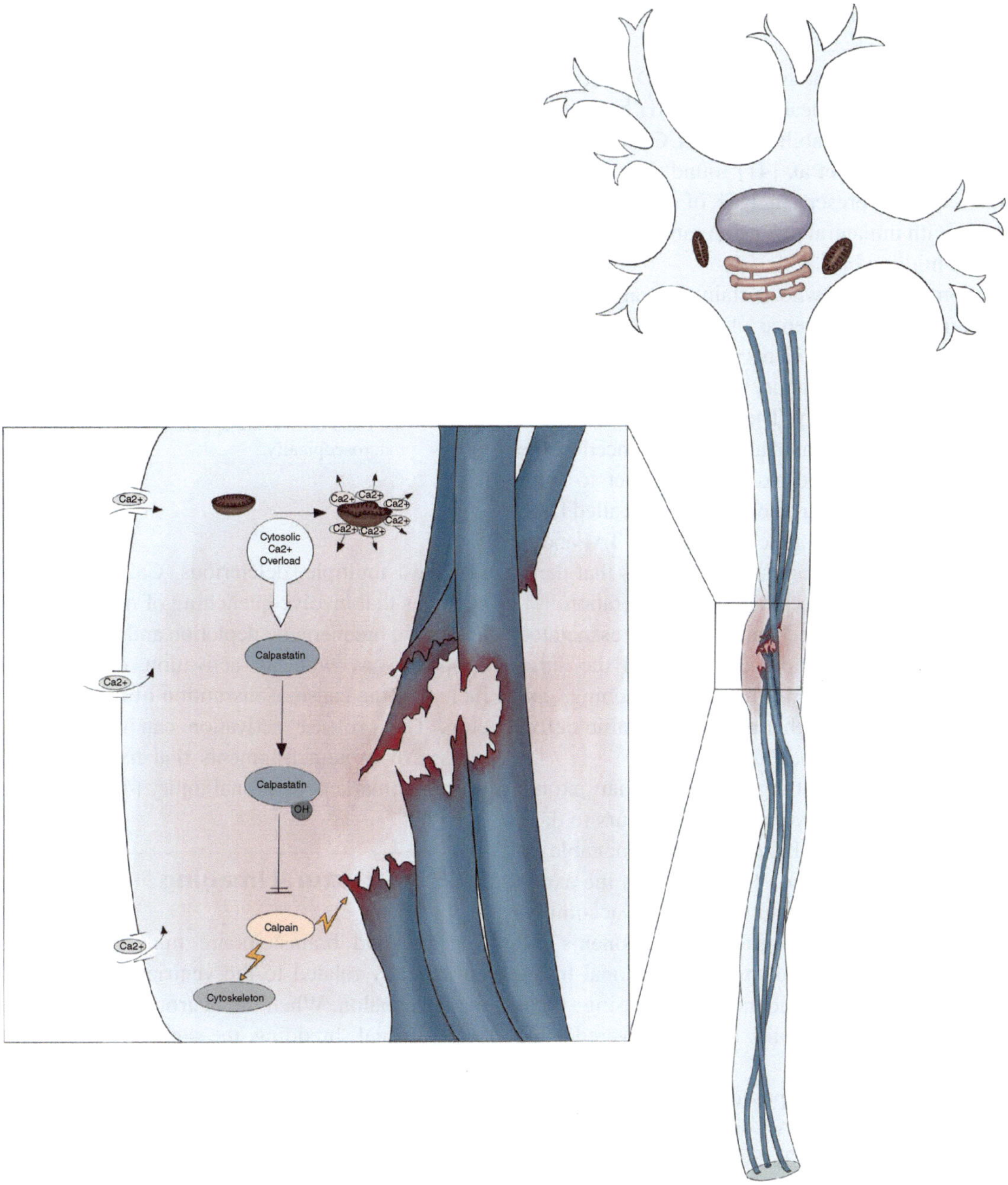

Fig. 24.1 Primary and secondary axotomy pathophysiological processes (left-hand inlay). (Courtesy of M. W. T. van Bilsen, MD). The figure illustrates an axon upon which shear and rotational forces act. The microtubules (blue) become progressively stiffer and eventually break, leading to a disruption of the axonal transport of molecules. Calcium accumulates in the cell, both through the mechanical opening of calcium channels and through the disruption of mitochondria. Through hydrolyzation of calpastatin (which normally inhibits calpain), calpain is activated and it in turn hydrolyzes the cytoskeleton and microtubules. This cascade leads to apoptosis and axon disconnection. (From Bruggeman GF, Haitsma IK, Dirven CMF, Volovici V. Traumatic axonal injury (TAI): definitions, pathophysiology and imaging-a narrative review. Acta Neurochir (Wien). 2021 Jan;163(1):31–44)

24.4 Neuropathology

Several neuropathologic reports of DAI in adults sustaining closed head injury (CHI) have documented the vulnerability of the CC [35, 36, 39–42]. Lindenberg et al. [41] found that lesions of the CC were present in 16% of fatal CHI, typically with unilateral extension rather than limited to the midline [41].

Some patients who sustain TBI are severely impaired in the absence of gross lacerations or hematomas. These patients are generally comatose from the instant of injury and subsequently have only limited recovery. They have sustained widespread microscopic axonal injury evidenced by the presence of ruptured axons that retract to form spheroids. This syndrome was initially called DAI, but it is more appropriately designated TAI and is believed to result from shearing forces that damage axons during acceleration or deceleration.

The principal mechanical loading associated with TAI is rotational acceleration of the unrestricted head, resulting in shear, stretching, and compressive strains that produce dynamic deformation of brain tissue [32, 43].

There is compelling evidence that axonal alterations mature over a period of hours to days and may therefore be potentially amenable to therapeutic intervention [44]. Initially, the axons appear normal, but soon there is focal accumulation of axoplasmic cytoskeletal components and organelles indicating disruption of axonal transport. After several hours of increasing focal swelling, the axon splits and the severed ends retract.

Adams et al. proposed a histopathological grading system for TAI, dividing TAI into three different subgroups [2]. See Table 24.2.

Table 24.2 Grading of traumatic axonal injury, according to the histopathologic study of Adams et al. [2]

Grade 1	Evidence of axonal damage in the white matter of the cerebral hemispheres including the CC, in the brainstem, and occasionally in the cerebellum: this damage can only be identified microscopically.
Grade 2	Focal lesion in the CC in addition to diffuse axonal damage: the focal lesion was identifiable only microscopically. It can be said to be severe if the focal lesions are apparent macroscopically.
Grade 3	Focal lesions both in the CC and the dorsolateral quadrant of the rostral brainstem: the focal lesion was identifiable only microscopically. It can be said to be severe if the focal lesions are apparent macroscopically.

Axonal deformation at the moment of injury results in a focal impairment of axoplasmic transport and subsequent focal swelling of the axon because of abnormal accumulation of neurofilaments and membranous organelles, and over the next 6–12 h, there is disconnection of the proximal axonal segment from the distal segment. The separated distal segment undergoes Wallerian degeneration in the weeks to months following the injury. The transient focal disruption of the axonal membrane allows an influx of $Ca2+$, which activates multiple deleterious $Ca2+$-dependent cascades that involve quenching of mitochondria, leading to bioenergetic depletion and activation of apoptosis, as well as activation of proteases including the cause of disruption of the cytoskeleton. The protease activation can result in the release of protein fragments that have potential use as biomarkers of axonal injury [45–53].

24.5 Structural Imaging

The CC and brainstem are midline structures intrinsically related to the ventricular system of the encephalon. When the neurons of those areas suffer axonal shredding, the same shearing strain causes rupture of the midline superficial vessels. This mechanism produces intraventricular hemorrhage (IVH) and a midline subarachnoid hemorrhage (interhemispheric and peri-mesencephalic). Furthermore, the severity of the IVH is related to the severity of the DAI, because the greater the forces of the accident, the greater the shearing forces, both for axons and small vessels [54].

According to the International Mission for Prognosis and Clinical Trial study, accurate neuroradiologic diagnostic evaluation represents a potential prognostic factor in TBI [55, 56]. Although computerized tomography (CT) is the imaging technique of choice in TBI [57], magnetic resonance imaging (MRI) is more sen-

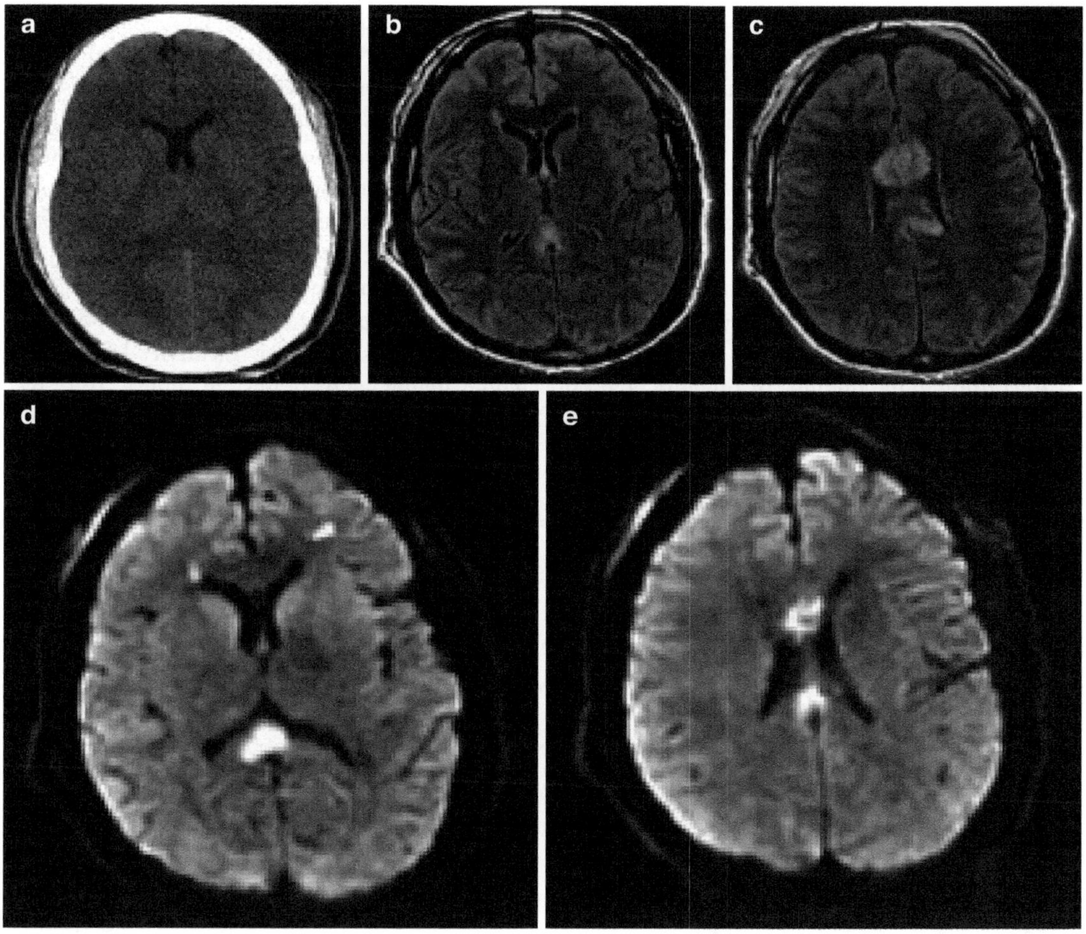

Fig. 24.2 Extensive diffuse axonal injury (DAI) involving the corpus callosum (CC). An axial CT scan demonstrates no abnormality in the genu or splenium of the CC. (**a–c**) Axial FLAIR sequences reveal extensive abnormal signal intensity involving most of the CC, consistent with DAI. (**d, e**) Axial diffusion-weighted images demonstrate abnormal signal intensity involving the majority of the CC, consistent with restricted diffusion secondary to DAI. In addition, small areas of signal abnormalities are seen in the frontal white matter bilaterally. (From Chi-Shing, Z., Maya, M., Go, J.L., Kim, P.E., Kovanlikaya, I. (2005). Head Trauma. In: Filippi, M., De Stefano, N., Dousset, V., McGowan, J.C. (eds) MR Imaging in White Matter Diseases of the Brain and Spinal Cord. Medical Radiology Diagnostic Imaging. Springer, Berlin, Heidelberg)

sitive for detecting TAI lesions [55, 58]. The relationship between CC injury and outcome in severe head trauma has been reported by many authors before [59–62], but few studies have been done with conventional MRI in subacute phase. In fact, the importance of the volume of the lesion and the location of the injury inside the CC has not been properly examined [63, 64], and only a few studies have been published in the last decade.

Unfortunately, TAI is notoriously difficult to diagnose by CT. Only 10% of TAI patients demonstrate the classic CT findings of petechial hemorrhages at the gray-white junction of the cerebral hemispheres or brainstem [65] (see Fig. 24.2).

As previously mentioned, CC lesions were classified as (1) hemorrhagic or nonhemorrhagic; (2) mild, moderate, or severe lesions (according to subjective measurement of CC lesion volume); (3) genu, splenium, or body involvement; (4) total number of lesions; and (5) total volume of the lesions [13].

A study by Cicuendez et al. [13] found that the mean volume of the lesions measured on T2 and

FLAIR sequences was 0.36 ± 1.27 cc on T2 sequences and 0.49 ± 1.59 cc on FLAIR. Hemorrhagic lesions in CC were less frequent but bigger than nonhemorrhagic lesions. The mean volume of hemorrhagic lesions on T2 sequences was 1.73 ± 2.99 and 2.34 ± 3.39 cc on FLAIR, whereas nonhemorrhagic lesions had 0.57 ± 0.78 cc on T2 and 0.79 ± 1.33 cc on FLAIR images. According to the same study, nonhemorrhagic lesions were more frequent (65%) than hemorrhagic lesions (35%) [13].

According to the findings of Figueira et al., in a patient with a closed TBI and an early CT showing no traumatic subarachnoid hemorrhage (tSAH), there is probably no severe DAI; similarly, in the absence of intraventricular hemorrhage (IVH), there is also probably no severe DAI. If there is IVH, it is possible that the patient is more severe and has a reserved prognosis, since there is, at least, CC lesion. In addition, the greater the number of ventricles affected, the more serious the clinical picture is. The presence of blood in the interpeduncular cisterns (IPC) gives a possible diagnosis of severe DAI [54].

The neuroradiological diagnosis of DAI is difficult as conventional imaging techniques such as CT and MRI underestimate the true extent of injury. In the acute phase, CT may reveal small, punctate, petechial hemorrhages, sometimes in combination with intraventricular blood (shearing of subependymal veins), and perimesencephalic SAH [66]. However, CT underestimates DAI lesions, because nonhemorrhagic lesions are difficult to identify. Indeed, the results of most admission CT scans are normal because >80% of DAIs are thought to be nonhemorrhagic [65].

MRI is far more sensitive than CT as it can detect both hemorrhagic and nonhemorrhagic lesions and has a greater sensitivity in detecting white matter and brainstem injury [65, 67]. Nonhemorrhagic DAI lesions are best appreciated on FLAIR imaging [65]. Nonhemorrhagic lesions are hyperintense on FLAIR and most probably reflect edema in the early phase and gliosis in the chronic phase [66]. Unfortunately, imaging findings on FLAIR are not specific, and it may be difficult to differentiate these TAI lesions from white matter hyperintensities caused by other premorbid conditions, for example, multiple sclerosis or chronic cerebrovascular disease, especially in elderly patients [68] (see Fig. 24.3).

MRI is the best imaging modality by which to confirm the diagnosis and to classify the lesions usually employ the grading scale of Gentry. This scale divides the findings into three groups: lesions with and without hemorrhage at the gray-white

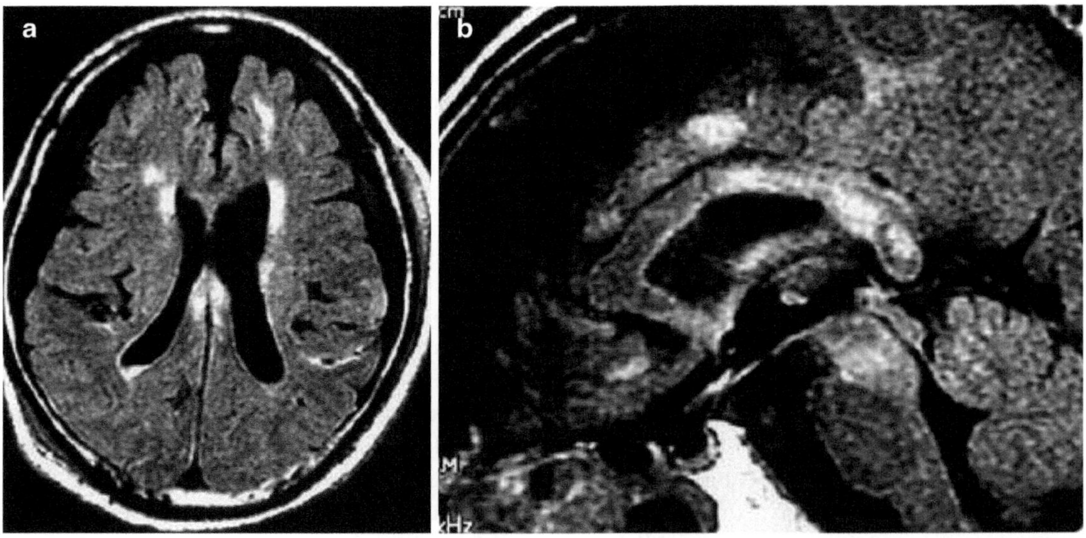

Fig. 24.3 DAI in a 79-year-old man. (**a**, **b**) FLAIR axial and midsagittal images (TR/TI/TE = 10,000/2200/125, 1.5 T) obtained 5 days after head injury show lesions in the posterior half of the CC. The midbrain, cingulate gyrus, and both frontal lobes are also involved. (From Uchino, A., Takase, Y., Nomiyama, K. et al. Acquired lesions of the corpus callosum: MR imaging. EurRadiol 16, 905–914 (2006))

matter junction, especially in the temporal and frontal areas (type 1), combined with lesions in and around the CC (type 2) and with lesions in the basal ganglia and the rostral brainstem (type 3). The scale roughly correlates with outcome [60, 69]:

Stage 1 (lobar): DAI lesions confined to the lobar white matter, especially gray-white matter junction; most common sites: parasagittal regions of frontal lobes, periventricular temporal lobes; less common sites: parietal and occipital lobes, internal and external capsules, cerebellum.

Stage 2 (callosal): DAI lesions in the CC, almost invariably in addition to the lobar white matter; most common sites: posterior body and splenium of CC; less common sites: anterior body and rostrum of CC (usually in conjunction with posterior involvement) usually unilateral and eccentric but may be bilateral and symmetric.

Stage 3 (brainstem): DAI lesions in the brainstem, almost invariably in addition to the lobar white matter and CC; most common sites: dorsolateral midbrain, upper pons, and superior cerebellar peduncles.

Hemorrhagic DAI lesions are best seen on GRE T2*-wi and SWI. Hemorrhagic DAI lesions demonstrate focal signal loss on GRE T2*-wi and SWI due to the magnetic susceptibility effects of paramagnetic blood breakdown products (deoxyhemoglobin, methemoglobin, hemosiderin) [65]. These susceptibility effects are further magnified at 3 Tesla (3 T) compared to 1.5 T, with nearly twice as many hemorrhagic lesions detectable at 3 T [70]. SWI is much more sensitive than GRE-T2 for depicting DAI [71].

Hemorrhagic DAI must be differentiated from other hypointensities on SWI or GRE T2* sequences. Vascular structures, small cavernomas, and microbleeds in small vessel disease (chronic hypertensive encephalopathy, CADASIL, and CAA), after brain radiation therapy, as well as calcifications and even fat emboli may be mistaken for DAI lesions (see schematic representation of different lesions of the CC on Fig. 24.4). Usually, they can be discriminated based on location, distribution, and morphology, as well as additional imaging abnormalities. Vascular microbleeds tend to present as rounded/spherical lesions, whereas microhemorrhagic TAI usually present as radial/elliptical or even curvilinear lesions following the perivascular spaces [72, 73]. DWI is a valuable sequence in closed head injury because it identifies additional shearing injuries that are not visible on T2/FLAIR or T2*-sequences [66] (see Fig. 24.5). Decreased ADC values can be found in DAI, indicating restricted diffusion [74]. Traumatic DAI lesions can be classified into three categories, depending on their signal characteristics on DWI and ADC maps [75]:

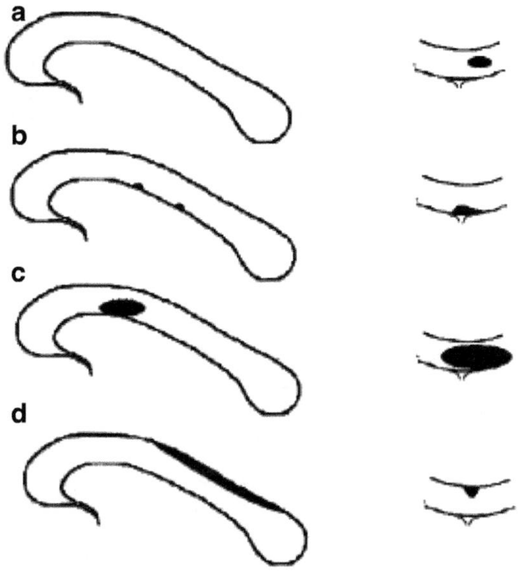

Fig. 24.4 Typical distribution of (**a**) ischemic, (**b**) inflammatory, (**c**) traumatic, and (**d**) hydrocephalus relater lesions. (From: Friese, S., Bitzer, M., Freudenstein, D. et al. Classification of acquired lesions of the corpus callosum with MRI. Neuroradiology 42, 795–802 (2000))

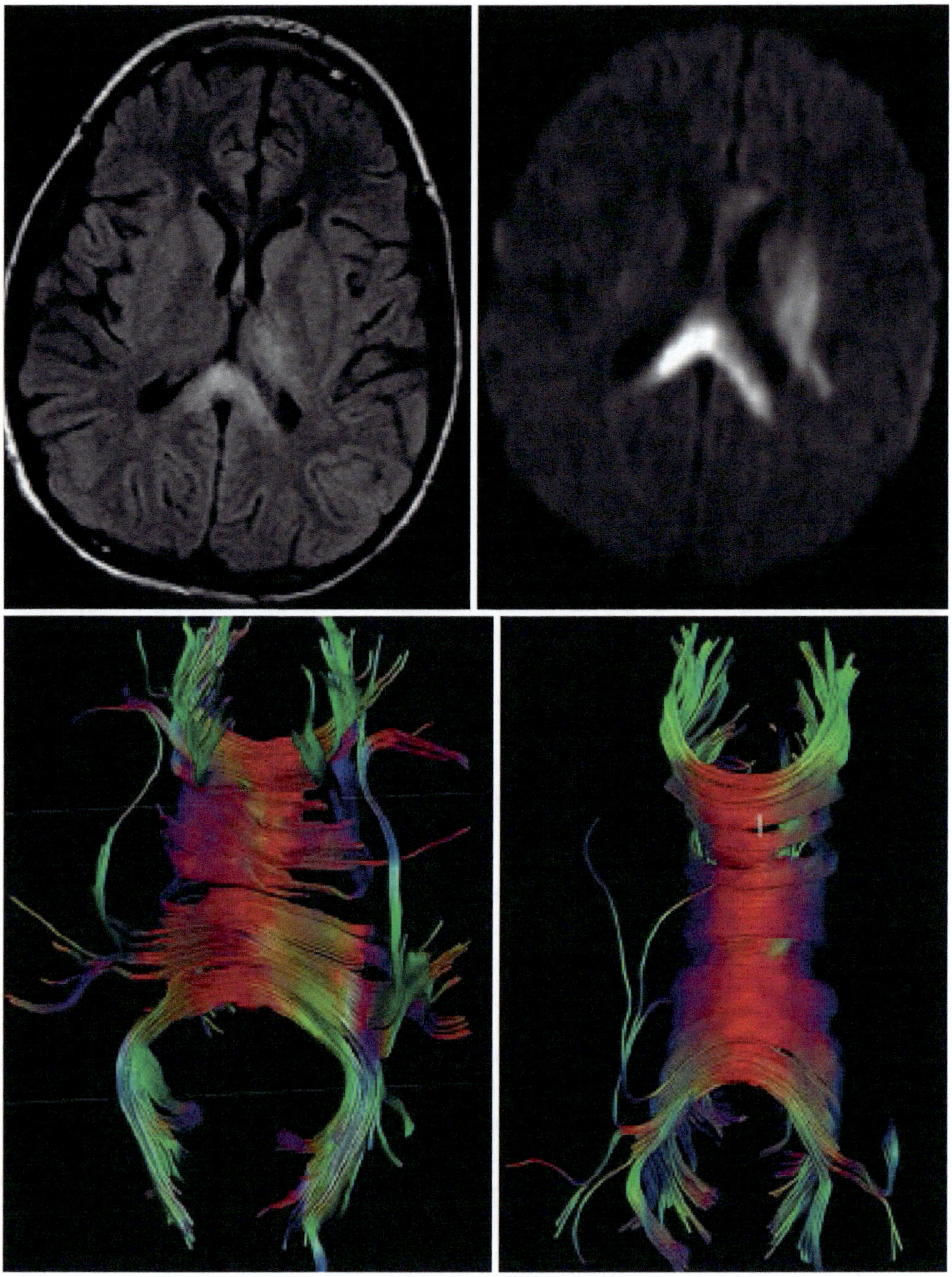

Fig. 24.5 Traumatic swelling of the splenium of the CC. (Top left) Axial FLAIR sequence demonstrates abnormal signal intensity in the splenium of the CC. (Top right) Axial diffusion-weighted image reveals abnormal signal in the splenium, consistent with restricted diffusion secondary to DAI. Some abnormal signal is also seen in the genu of the CC and left basal ganglia. (Bottom left) Three-dimensional diffusion tensor white matter tractography demonstrating alteration and irregularity of the callosal tracts, most likely indicating cytoskeletal alterations in the white matter. (Bottom right) Three-dimensional diffusion tensor white matter tractography of a normal subject for comparison. (From Chi-Shing, Z., Maya, M., Go, J.L., Kim, P.E., Kovanlikaya, I. (2005). Head Trauma. In: Filippi, M., De Stefano, N., Dousset, V., McGowan, J.C. (eds) MR Imaging in White Matter Diseases of the Brain and Spinal Cord. Medical Radiology Diagnostic Imaging. Springer, Berlin, Heidelberg)

- Type 1: DWI-hyperintense and ADC-hyperintense, most likely representing lesions with vasogenic edema.
- Type 2: DWI-hyperintense and ADC-hypointense, indicating cytotoxic edema.
- Type 3: Central hemorrhagic lesion surrounded by an area of increased diffusion [18].

However, DWI and FLAIR are less sensitive for detecting hemorrhagic lesions [66, 76]. Injury to the thalamus and basal ganglia is a rare manifestation of TBI, accounting for about 5% of parenchymal traumatic lesions. On CT, these lesions may present as small hemorrhages in the basal ganglia, postulated to be secondary to disruption of small perforating blood vessels. These lesions are also associated with a low initial GCS score [77, 78] and are more common in pediatric population [67, 79]. See Figs. 24.6 and 24.7.

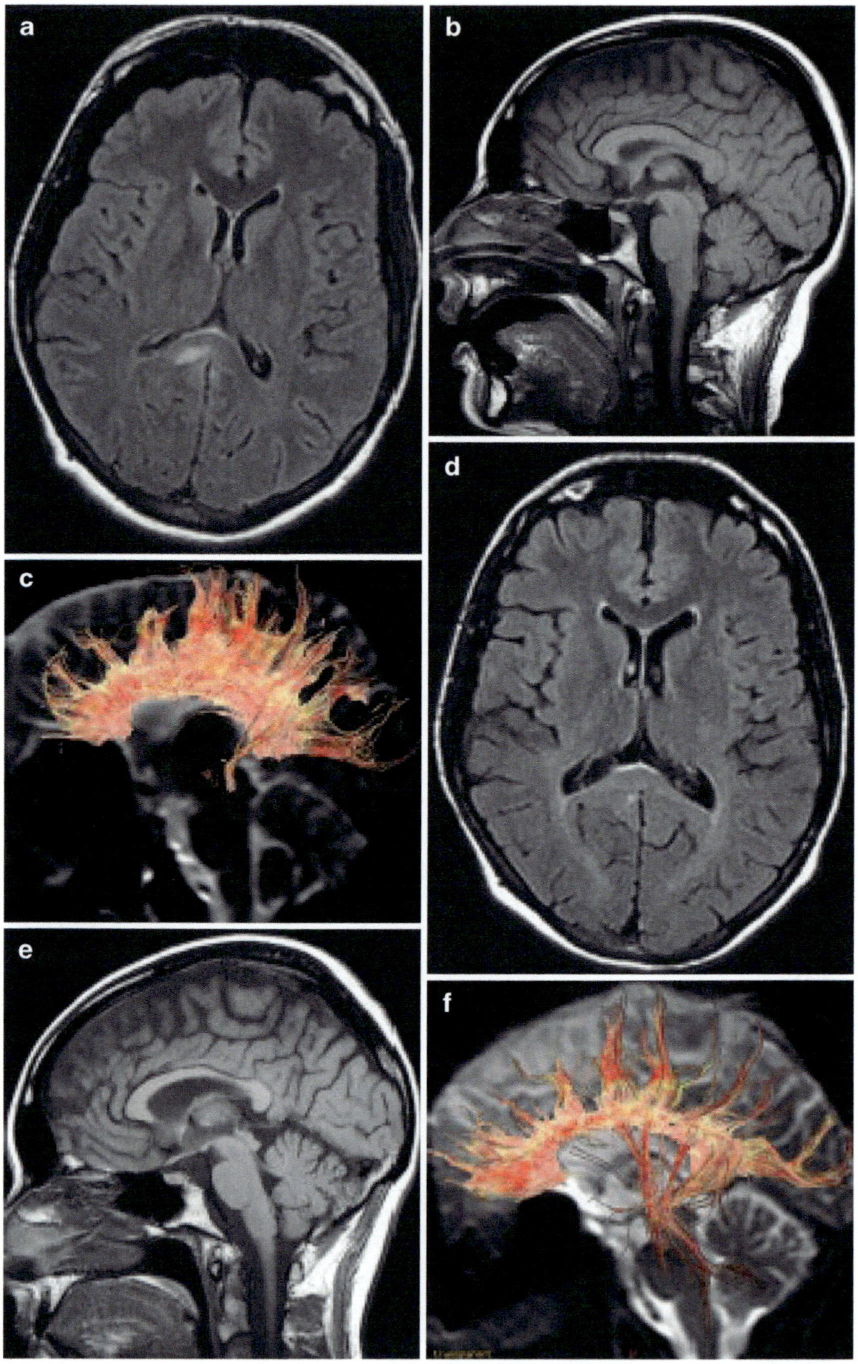

Fig. 24.6 Traffic accident. GCS = 5. Outcome, moderate disability. MRI, T2-FLAIR (**a**) shows a lesion in the splenium of the CC; T1 WI, (**b**) Tractography (**c**) in 15 days after injury does not reveal changes of the CC in a three-dimensional reconstruction. MRI in 3 years after trauma: T2-FLAIR (**d**) no marked changes, T1 WI (**e**) atrophy of the CC, and tractography (**f**) partial reduction in ascending fibers and loss of some fibers of the CC. (From Zakharova, N., Kornienko, V., Potapov, A., Pronin, I. (2014). Dynamic study of white matter fiber tracts after traumatic brain injury. In: Neuroimaging of Traumatic Brain Injury. Springer, Cham. https://doi.org/10.1007/978-3-319-04355-5_4)

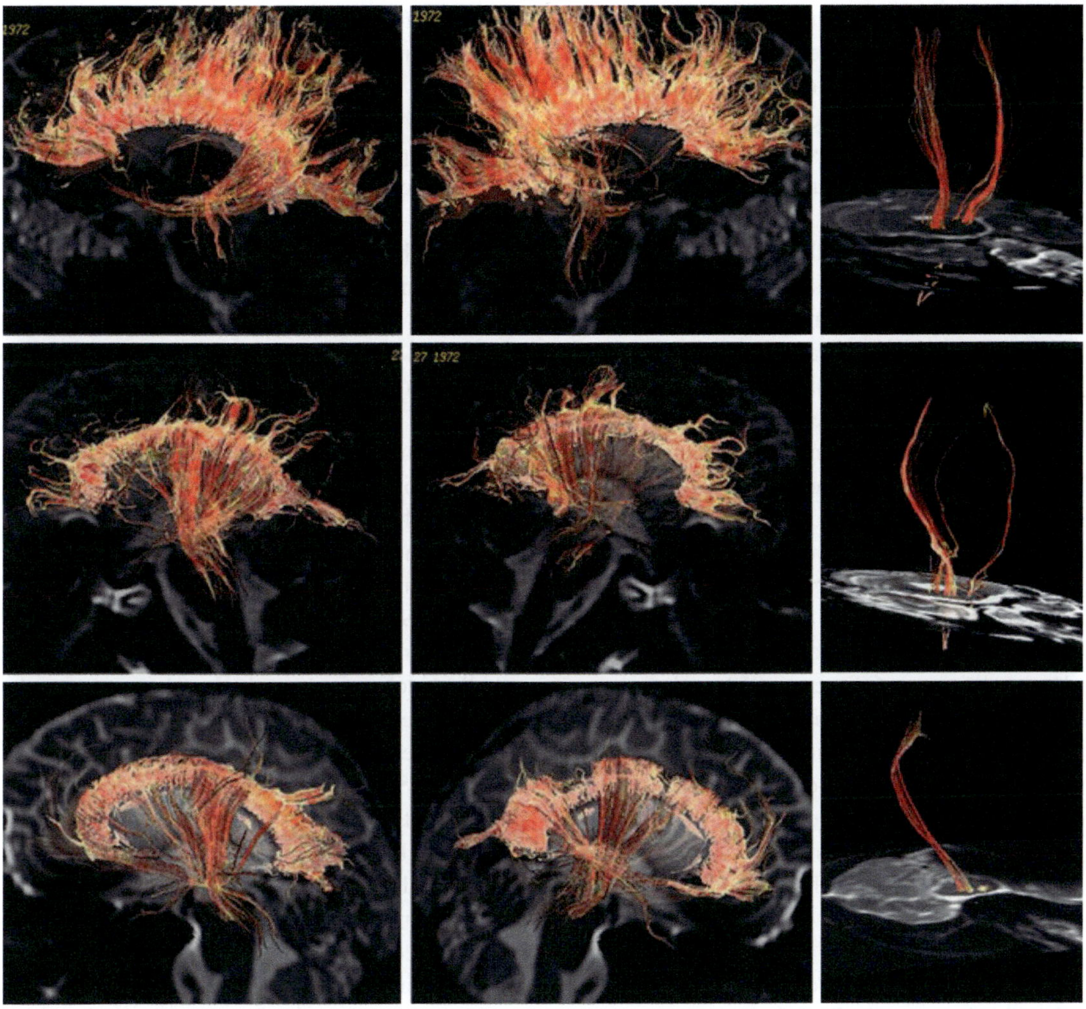

Fig. 24.7 Dynamics of MR tractography data in the same patient as in Fig. 24.6. First examination (in 5 days) reveals "defect" in the splenium of the CC, a slight asymmetry of CST (upper row). Second examination (in 22 days) (middle row), shortening and decreasing of ascending fibers on the left. Third examination (in 19 months) (lower row), practically complete disappearance of ascending fibers of the CC; left corticospinal tract is not visible. (From Zakharova, N., Kornienko, V., Potapov, A., Pronin, I. (2014). Dynamic study of white matter fiber tracts after traumatic brain injury. In: Neuroimaging of Traumatic Brain Injury. Springer, Cham. https://doi.org/10.1007/978-3-319-04355-5_4)

24.6 Clinical Presentation

The spectrum of TBI encompasses three distinct grades: mild, moderate, and severe TBI. Often there is confusion in the nosology of TBI especially in relation to mild TBI (mTBI), a term that implicitly refers to the TBI event consistent with acute concussion. TBI is categorized according to the clinical pillars of post-traumatic amnesia (PTA) and/or a disturbance of consciousness—either alteration of consciousness (AOC) or loss of consciousness (LOC). These clinical features, although correlated, allow for independent diagnosis of TBI severity [12]. See Table 24.1.

DAI is typically characterized by coma without focal lesion, on presentation, and pathologically defined by axonal damage in multiple regions of the brain parenchyma, often causing impairments in cognitive, autonomic motor, and sensory function by virtue of disrupted neuronal connectivity [11].

Head trauma, especially in motor vehicle accidents, can lead to a wide spectrum of cerebral and intracranial lesions, with usually different clinical presentations (epidural and subdural hematomas, subarachnoid and intracerebral hemorrhages, cerebral contusions, generalized cerebral edema, and secondary phenomena, such as hydrocephalus, raised intracranial pressure, and tentorial herniation) and different findings on CT and MRI.

Clinically, DAI is characterized by loss of consciousness following the accident, usually without a lucid interval, a very low score on the Glasgow Coma Scale, and discrepantly subtle abnormalities on CT scans. When lesions are seen on CT, they consist mainly of petechial hemorrhages at the cortico-subcortical junction, in the body or splenium of the CC, and/or in the basal ganglia and brainstem.

Intracranial pressure measurements in patients with DAI are, at least in the beginning, normal, in contrast to the case with most of the other brain injuries.

DAI may result in death. It is unclear how often death occurs in DAI, but one may assume that it occurs in a high percentage of the patients. When it comes to a correct estimation of the frequency of death, the problem is that numbers provided by different specialists—neurologists, traumatologists, and neuropathologists—are not consistent. In some reports (by neurologists), it is stated that DAI rarely results in death, whereas others (neuropathologists) claim that DAI is an important cause of death in acceleration-deceleration trauma. This difference is at least partly due to the different populations they encounter. About 90% of patients with a clinical diagnosis of DAI, however, will remain in a vegetative condition.

In the first episode, patients with DAI stay in the intensive care unit and in many cases artificially ventilated. During this period, it would be important for the management of these patients to have a reliable prediction of outcome, most importantly to maintain the option to discontinue treatment that will not further improve the condition of the patient. Clinical and neurophysiological data are important in that respect. EEG will reflect the seriousness of the functional disturbance of the brain and will provide important information.

24.7 Critical Care Management

Various guidelines have been presented during the last 25 years for treatment of an isolated severe traumatic brain injury in the adult.

The Lund concept [80] was presented in 1994. The US guidelines were presented in its first version in 1996 [81]. Various guidelines and their updated versions were presented over the years.

All guidelines except the Lund concept can be characterized as cerebral perfusion pressure (CPP)-targeted guidelines and show similarities with the US guidelines and are mainly based on meta-analytic surveys. The Lund concept instead is based on basal physiological principles for brain volume and perfusion control and can be characterized as an intracranial pressure (ICP) and perfusion-targeted therapy. While most traditional guidelines including US guidelines are based only on clinical studies, the Lund concept also finds support from experimental studies. None of the published guidelines has been tested in larger randomized controlled studies, but numerous investigations, some with historical controls, have been published both for the Lund concept and more conventional guidelines [82, 83].

All guidelines recommend continuous measurement of arterial pressure and ICP as well as the use of artificial ventilation. No guideline recommends treatment with steroids, except that the Lund concept accepts 1 bolus dose of Solu-Medrol (0.25–0.5 g to the adult) to reduce a life-threatening critically raised body temperature (>39.5 °C). Active cooling is an option in some conventional guidelines, but not in the Lund concept. Normal potassium and sodium concentrations are generally recommended.

A shift in the treatment paradigm has emerged recently in many "non-Lund centers" by accepting a lower CPP, indicating that the guidelines have approached the Lund concept in some respects, even though the means to reach the goal are different. Sedation should be sufficient to ameliorate the stress response. Wake-up tests are controversial but still used in some centers.

There is a general recommendation that patients with severe head injury as soon as possible should be transferred to a neurosurgical unit. The patient should be intubated and receive intensive care treatment as recommended in the guidelines. There are some differences, though, regarding the intensive care treatment, as can be seen from the different guidelines in the table above. One difference is that US guidelines recommend that ICP treatment should not start before ICP is above 20–22 mmHg, while the Lund concept recommends that it should start as soon as possible after arrival to the neurointensive care unit independent of ICP. Focus should be to prevent hypovolemia, avoid hypoxia, avoid hyper- or hypoglycemia, and avoid hyper- and hypoventilation.

Enteral feeding is the goal. The intensive care should endeavor to ensure normality in most areas of the field, such as important physiological parameters.

Dopamine, one of the catecholamines, is an important neurotransmitter in the CNS. In DAI a reduction of dopamine turnover has been found. This observation has prompted the introduction of amantadine (Symmetrel) therapy in DAI. Amantadine is a drug known from treatment of Parkinson disease. Amantadine causes release of dopamine from central neurons, facilitates dopamine release by nerve impulses, and delays the uptake of dopamine by neural cells. It may also have a profound *N*-methyl-D-aspartate receptor antagonist effect, which may contribute to the neuroprotective effects after injury, by decreasing glutamate concentrations and thus excitotoxicity. In a randomized crossover design study in DAI patients, there was a consistent trend toward more rapid functional improvement with amantadine treatment, regardless of when during the first 3 months after injury the amantadine treatment was started [84]. From these findings, it is also clear that medication that results in dopaminergic blockade is contraindicated in the early stage of recovery from DAI.

24.8 Prognosis

A study by Cicuendez et al. demonstrated that the presence of TAI at the CC was associated with poor outcome at 1 year after brain trauma ($p < 0.001$, OR 3.8, 95% CI: 2.04–7.06, sensitivity 54%, specificity 74%). A total of 62% of the patients with CC lesions had poor outcome, whereas 38% showed good prognosis. In the group of patients without CC lesions, we found that 64% of the patients experienced good outcome [13].

According to the same study, the positive predictive value (PPV) of a poor prognosis having a CC lesion was 66%. Despite similar basic characteristics, the outcome of patients suffering CC injury is worse than that of patients suffering no CC lesions. The study did not show a significant difference in outcome between hemorrhagic and nonhemorrhagic CC lesions ($p = 0.2$) nor were there significant differences among mild, moderate, and severe lesions ($p = 0.26$). The location of the CC lesion at the splenium was statistically related to worse prognosis. Comparison of patients with CC injury with the rest of the patients with TBI in the same series showed that the patients with splenium lesions presented worse outcomes ($p < 0.001$, OR = 5.93, 95% CI: 3.02–11.66; sensitivity 47%, specificity 83%, PPV 65%). Patients with lesions affecting both the splenium and body of the CC also had a worse prognosis ($p < 0.001$, OR = 7.17, 95% CI: 2.38–21.66, sensitivity 17%, specificity 96%). Splenium lesions were related to dorsal brainstem injuries ($p < 0.001$), which are also associated with poor outcome. The addition of a splenium lesion and dorsal brainstem injury was present in 16 patients of our series, and it showed a higher odds ratio for poor outcome than the splenium lesion alone ($p = 0.003$, OR 7.37, 95% CI: 1.94–28.01, sensitivity 13%, specificity 63%).

A PPV for poor prognosis of 76% was obtained when the splenium and body lesions in the CC were present in the same study. The highest PPV for poor outcome was 81% when the splenium lesion and dorsal brainstem injury were present in the same patient [13].

24.9 Conclusion

The importance of the traumatic axonal lesion of the CC comes from the frequency of these findings but also from the fact that many of these cases are often unrecognized. TAIs of the CC are often presented with coma, with varying degree of responsiveness, evaluated based on the Glasgow Coma Scale. Almost always the mechanism of injury involves acceleration/deceleration forces. The best neuroradiologic imaging modality to diagnose and objectivize these lesions is MRI. The MRI sequences used when evaluating a patient with suspected TAI are FLAIR, DWI, SWI, and T2 GRE, depending, of course, on the type of lesions present (hemorrhagic or nonhemorrhagic). Involvement of the CC is mostly associated with poor prognosis, because of impaired neural connectivity and functional sequelae but also because these injuries are rarely isolated.

References

1. Rand CW, Courville CB. Histologic changes in the brain in cases of fatal injury to the head; alterations in nerve cells. Arch Neurol Psychiatry. 1946;55:79–110.
2. Adams JH, Doyle D, Ford I, Gennarelli TA, Graham DI, McLellan DR. Diffuse axonal injury in head injury: definition, diagnosis and grading. Histopathology. 1989;15(1):49–59.
3. Gennarelli TA, Thibault LE, Adams JH, Graham DI, Thompson CJ, Marcincin RP. Diffuse axonal injury and traumatic coma in the primate. Ann Neurol. 1982;12:564–74.
4. Bruggeman GF, Haitsma IK, Dirven CMF, Volovici V. Traumatic axonal injury (TAI): definitions, pathophysiology and imaging—a narrative review. Acta Neurochir (Wien). 2021;163(1):31–44. https://doi.org/10.1007/s00701-020-04594-1. Epub 2020 Oct 2. PMID: 33006648; PMCID: PMC7778615.
5. Adams JH, Graham DI, Gennarelli TA, Maxwell WL. Diffuse axonal injury in non-missile head injury. J Neurol Neurosurg Psychiatry. 1991;54:481–3.
6. Centers for Disease Control and Prevention (CDC), National Center for Injury Prevention and Control. Report to congress on mild traumatic brain injury in the United States: steps to prevent a serious public health problem. Atlanta, GA: Centers for Disease Control and Prevention; 2003.
7. Gomez PA, Castano-Leon AM, de-la Cruz J, Lora D, Lagares A. Trends in epidemiological and clinical characteristics in severe traumatic brain injury: analysis of the past 25 years of a single centre data base. Neurocirugia. 2014;25:199–210.
8. Jennett B. Epidemiology of head injury. J Neurol Neurosurg Psychiatry. 1996;60:362–9.
9. Leibson CL, Brown AW, Ransom JE, Diehl NN, Perkins PK, Mandrekar J, Malec JF. Incidence of traumatic brain injury across the full disease spectrum: a population-based medical record review study. Epidemiology. 2011;22:836–44.
10. Moen KG, Brezova V, Skandsen T, Haberg AK, Folvik M, Vik A. Traumatic axonal injury: the prognostic value of lesion load in corpus callosum, brainstem, and thalamus in different magnetic resonance imaging sequences. J Neurotrauma. 2014;31:1486–96.
11. Su E, Bell M. Diffuse axonal injury. In: Laskowitz D, Grant G, editors. Translational research in traumatic brain injury. Boca Raton, FL: CRC Press/Taylor and Francis Group; 2016. Chapter 3. PMID: 26583181.
12. Moore DF, Jaffee M, Ling G, Radovitzky R. Overview of traumatic brain injury (TBI). In: Tsao J, editor. Traumatic brain injury. Cham: Springer; 2020. https://doi.org/10.1007/978-3-030-22436-3_1.
13. Cicuendez M, Castaño-León A, Ramos A, Hilario A, Gómez PA, Lagares A. Prognostic value of corpus callosum injuries in severe head trauma. Acta Neurochir (Wien). 2017;159(1):25–32. https://doi.org/10.1007/s00701-016-3000-4. Epub 2016 Oct 31. PMID: 27796652.
14. Vieira RCA, Paiva WS, de Oliveira DV, Teixeira MJ, de Andrade AF, de Sousa RMC. Diffuse axonal injury: epidemiology, outcome and associated risk factors. Front Neurol. 2016;7:178.
15. Department of Informatics of SUS (DATASUS). Informações de saúde, Epidemiológicas e Morbidade: banco de dados. http://datasus.saude.gov.br/informacoes-de-saude/tabnet/epidemiologicas-e-morbidade.
16. Skandsen T, Kvistad KA, Solheim O, Strand IH, Folvik M, Vik A. Prevalence and impact of diffuse axonal injury in patients with moderate and severe head injury: a cohort study of early magnetic resonance imaging findings and 1-year outcome. J Neurosurg. 2009;113(3):556–63.
17. NINDS Common Data Elements. Traumatic brain injury: data standards. 2012. https://www.commondataelements.ninds.nih.gov/TraumaticBrainInjury. Accessed 9 Dec 2020.
18. Haacke EM, Duhaime A-C, Gean AD, Riedy G, Wintermark M, et al. Common data elements in radiologic imaging of traumatic brain injury. J Magn Reson Imaging. 2010;32(3):516–43.

19. Dennis EL, Ellis MU, Marion SD, Jin Y, Moran L, Olsen A, Kernan C, Babikian T, Mink R, Babbitt C, Johnson J, Giza CC, Thompson PM, Asarnow RF. Callosal function in pediatric traumatic brain injury linked to disrupted white matter integrity. J Neurosci. 2015;35:10202–11.
20. Treble A, Hasan KM, Iftikhar A, Stuebing KK, Kramer LA, Cox CS Jr, Swank PR, Ewing-Cobbs L. Working memory and corpus callosum microstructural integrity after pediatric traumatic brain injury: a diffusion tensor tractography study. J Neurotrauma. 2013;30:1609–19.
21. Geffen G, Jones D, Geffen L. Interhemispheric control of manual motor activity. Behav Brain Res. 1994;64:131–40.
22. Bigler E. Quantitative magnetic resonance imaging in traumatic brain injury. J Head Trauma Rehabil. 2001;16:117–34.
23. Levin H, Benavidez D, Verger-Maestre K, Perachio N, Song J, Mendelsohn D, Fletcher J. Reduction of corpus callosum growth after severe traumatic brain injury in children. Neurology. 2000;54:647–53.
24. Gorrie C, Duflou J, Brown J, Gibson T, Waite P. Extent and distribution of vascular brain injury in pediatric road fatalities. J Neurotrauma. 2001;18:849–60.
25. Gennarelli TA. Mechanisms of brain injury. J Emerg Med. 1993;11(Suppl 1):5–11.
26. Oehmichen M, Meissner C, Schmidt V, Pedal I, König HG, Saternus KD. Axonal injury—a diagnostic tool in forensic neuropathology? A review. Forensic Sci Int. 1998;95:67–83.
27. Yoganandan N, Gennarelli TA, Zhang J, Pintar FA, Takhounts E, Ridella SA. Association of contact loading in diffuse axonal injuries from motor vehicle crashes. J Trauma. 2009;66:309–15.
28. Zhang J, Yoganandan N, Pintar FA, Gennarelli TA. Role of translational and rotational accelerations on brain strain in lateral head impact. Biomed Sci Instrum. 2006;42:501–6.
29. Wang HC, Duan ZX, Wu FF, Xie L, Zhang H, Ma YB. A new rat model for diffuse axonal injury using a combination of linear acceleration and angular acceleration. J Neurotrauma. 2010;27:707–19.
30. Meaney DF, Smith DH, Shreiber DI, Bain AC, Miller RT, Ross DT, Gennarelli TA. Biomechanical analysis of experimental diffuse axonal injury. J Neurotrauma. 1995;12:689–94.
31. Yoganandan N, Baisden JL, Maiman DJ, Gennarelli TA, Guan Y, Pintar FA, Laud P, Ridella SA. Severe-to-fatal head injuries in motor vehicle impacts. Accid Anal Prev. 2010;42:1370–8.
32. Gennarelli TA, Graham DI. Neuropathology of the head injuries. Semin Clin Neuropsychiatry. 1998;3(3):160–75. PMID: 10085204.
33. Gennarelli TA, Thibault LE, Graham DI. Diffuse axonal injury: an important form of traumatic brain damage. Neuroscientist. 1998;4:202–15.
34. Adams JH. The neuropathology of head injuries. In: Vinken PJ, Bruyn GW, Braakman R, editors. Handbook of clinical neurology, vol. 23. Oxford: North-Holland Publications; 1975. p. 35–65.
35. Adams JH, Graham DI, Murray LS, Scott G. Diffuse axonal injury due to nonmissile head injury in humans: an analysis of 45 cases. Ann Neurol. 1982;12:557–63.
36. Adams JH, Graham D, Scott G, Parker LS, Doyle D. Brain damage in fatal non-missile head injury. J Clin Pathol. 1980;33:1132–45.
37. Graham DI, Adams JH, Gennarelli TA. Pathology of brain damage in head injury. In: Cooper PR, editor. Head injury. Baltimore: Williams and Wilkins; 1987. p. 72–88.
38. Sahuquillo J, Poca MA. Diffuse axonal injury after head trauma. A review. Adv Tech Stand Neurosurg. 2002;27:23–86. https://doi.org/10.1007/978-3-7091-6174-6_2. PMID: 11887581.
39. Adams JH, Mitchell DE, Graham DK, Doyle D. Diffuse brain damage of immediate impact type. Brain. 1977;100:489–502.
40. Blumbergs PC, Jones NR, North JB. Diffuse axonal injury in head trauma. J Neurol Neurosurg Psychiatry. 1989;52:838–41.
41. Lindenberg R, Fisher RS, Durlacher SH, Lovitt WV Jr, Freytag E. Lesions of the corpus callosum following blunt mechanical trauma to the head. Am J Pathol. 1955;31:297–317.
42. Strich SJ. Diffuse degeneration of the cerebral white matter in severe dementia following head injury. J Neurol Neurosurg Psychiatry. 1956;19:163–85.
43. Adams JH, Graham DI, Gennarelli TA. Acceleration induced head injury in the monkey. II. Neuropathology. Acta Neuropathol Suppl. 1981;7:26–8.
44. Smith DH, Hicks R, Povlishock JT. Therapy development for diffuse axonal injury. J Neurotrauma. 2013;30:307–23.
45. Gabbita SP, Scheff SW, Menard RM, et al. Cleaved-tau: a biomarker of neuronal damage after traumatic brain injury. J Neurotrauma. 2005;22:83–94.
46. Li J, Li XY, Feng DF, et al. Biomarkers associated with diffuse traumatic axonal injury: exploring pathogenesis, early diagnosis, and prognosis. J Trauma. 2010;69:1610–8.
47. Liu S, Yin F, Zhang J, et al. The role of calpains in traumatic brain injury. Brain Inj. 2014;28:133–7.
48. McGinn MJ, Kelley BJ, Akinyi L, et al. Biochemical, structural, and biomarker evidence for calpain-mediated cytoskeletal change after diffuse brain injury uncomplicated by contusion. J Neuropathol Exp Neurol. 2009;68:241–9.
49. Neselius S, Zetterberg H, Blennow K, et al. Increased CSF levels of phosphorylated neurofilament heavy protein following bout in amateur boxers. PLoS One. 2013;8:e81249.
50. Neselius S, Brisby H, Theodorsson A, et al. CSF-biomarkers in Olympic boxing: diagnosis and effects of repetitive head trauma. PLoS One. 2012;7:e33606.
51. Papa L, Ramia MM, Edwards D, et al. Systematic review of clinical studies examining biomarkers of

brain injury in athletes after sports related concussion. J Neurotrauma. 2015;32:661–73.
52. Rubenstein R, Chang B, Davies P, et al. A novel, ultrasensitive assay for tau: potential for assessing traumatic brain injury in tissues and biofluids. J Neurotrauma. 2015;32:342–52.
53. Shahim P, Linemann T, Inekci D, et al. Serum tau fragments predict return to play in concussed professional ice hockey players. J Neurotrauma. 2016;33:1995.
54. Figueira Rodrigues Vieira G, Guedes Correa JF. Early computed tomography for acute post-traumatic diffuse axonal injury: a systematic review. Neuroradiology. 2020;62(6):653–60. https://doi.org/10.1007/s00234-020-02383-2. Epub 2020 Mar 4. PMID: 32130462; PMCID: PMC7222974.
55. Hilario A, Ramos A, Millan JM, Salvador E, Gomez PA, Cicuendez M, Diez-Lobato R, Lagares A. Severe traumatic head injury: prognostic value of brainstem injuries detected at MRI. AJNR Am J Neuroradiol. 2012;33:1925–31.
56. Murray GD, Butcher I, McHugh GS, Lu J, Mushkudiani NA, Maas AI, Marmarou A, Steyerberg EW. Multivariable prognostic analysis in traumatic brain injury: results from the IMPACT study. J Neurotrauma. 2007;24:329–37.
57. Marshall LF, Marshall SB, Klauber MR, Van Berkum CM, Eisenberg H, Jane JA, Luerssen TG, Marmarou A, Foulkes MA. The diagnosis of head injury requires a classification based on computed axial tomography. J Neurotrauma. 1992;9(Suppl 1):S287–92.
58. Lagares A, Ramos A, Alday R, Ballenilla F, Perez-Nunez A, ArreseI AJF, Pascual B, Kaen A, Gomez PA, Lobato RD. Magnetic resonance in moderate and severe head injury: comparative study of CT and MR findings. Characteristics related to the presence and location of diffuse axonal injury in MR. Neurocirugia. 2006;17:105–18.
59. Firsching R, Woischneck D, Diedrich M, Klein S, Ruckert A, Wittig H, Dohring W. Early magnetic resonance imaging of brainstem lesions after severe head injury. J Neurosurg. 1998;89:707–12.
60. Gentry LR. Imaging of closed head injury. Radiology. 1994;191(1):1–17. https://doi.org/10.1148/radiology.191.1.8134551. PMID: 8134551.
61. Gentry LR, Godersky JC, Thompson B. MR imaging of head trauma: review of the distribution and radiopathologic features of traumatic lesions. AJR Am J Roentgenol. 1988;150:663–72.
62. Paterakis K, Karantanas AH, Komnos A, Volikas Z. Outcome of patients with diffuse axonal injury: the significance and prognostic value of MRI in the acute phase. J Trauma. 2000;49:1071–5.
63. Hulkower MB, Poliak DB, Rosenbaum SB, Zimmerman ME, Lipton ML. A decade of DTI in traumatic brain injury: 10 years and 100 articles later. AJNR Am J Neuroradiol. 2013;34:2064–74.
64. Moen KG, Haberg AK, Skandsen T, Finnanger TG, Vik A. A longitudinal magnetic resonance imaging study of the apparent diffusion coefficient values in corpus callosum during the first year after traumatic brain injury. J Neurotrauma. 2014;31:56–63.
65. Kim JJ, Gean AD. Imaging for the diagnosis and management of traumatic brain injury. Neurotherapeutics. 2011;8(1):39–53.
66. Parizel PM, Van Goethem JW, Öszarlak Ö, Maes M, Phillips CD. New developments in the neuroradiological diagnosis of craniocerebral trauma. Eur Radiol. 2005;15(3):569–81.
67. Gean AD, Fischbein NJ. Head trauma. Neuroimaging Clin N Am. 2010;20(4):527–56.
68. Moen KG, Skandsen T, Folvik M, Brezova V, Kvistad KA, Rydland J, et al. A longitudinal MRI study of traumatic axonal injury in patients with moderate and severe traumatic brain injury. J Neurol Neurosurg Psychiatry. 2012;83(12):1193–200.
69. Diffuse axonal injury. In: Magnetic resonance of myelination and myelin disorders. Berlin: Springer; 2005. https://doi.org/10.1007/3-540-27660-2_104.
70. Scheid R, Ott DV, Roth H, Schroeter ML, von DY C. Comparative magnetic resonance imaging at 1.5 and 3 Tesla for the evaluation of traumatic microbleeds. J Neurotrauma. 2007;24(12):1811–6.
71. Tong KA, Ashwal S, Holshouser BA, Shutter LA, Herigault G, Haacke EM, et al. Hemorrhagic shearing lesions in children and adolescents with posttraumatic diffuse axonal injury: improved detection and initial results. Radiology. 2003;227(2):332–9.
72. Imaizumi T, Miyata K, Inamura S, Kohama I, Nyon KS, Nomura T. The difference in location between traumatic cerebral microbleeds and microangiopathic microbleeds associated with stroke. J Neuroimaging. 2011;21(4):359–64.
73. Izzy S, Mazwi NL, Martinez S, Spencer CA, Klein JP, Parikh G, Glenn MB, Greenberg SM, Greer DM, Wu O, Edlow BL. Revisiting grade 3 diffuse axonal injury: not all brainstem microbleeds are prognostically equal. Neurocrit Care. 2017;27(2):199–207.
74. Lin DD, Filippi CG, Steever AB, Zimmerman RD. Detection of intracranial hemorrhage: comparison between gradient-echo images and b(0) images obtained from diffusion-weighted echo-planar sequences. AJNR Am J Neuroradiol. 2001;22(7):1275–81.
75. Hergan K, Schaefer PW, Sorensen AG, Gonzalez RG, Huisman TAGM. Diffusion-weighted MRI in diffuse axonal injury of the brain. Eur Radiol. 2002;12(10):2536–41.
76. Huisman TAGM, Sorensen AG, Hergan K, Gonzalez RG, Schaefer PW. Diffusion-weighted imaging for the evaluation of diffuse axonal injury in closed head injury. J Comput Assist Tomogr. 2003;27(1):5–11.
77. Aiken AH, Gean AD. Imaging of head trauma. Semin Roentgenol. 2010;45(2):63–79.
78. Mutze S, Rademacher G, Matthes G, Hosten N, Stengel D. Blunt cerebrovascular injury in patients with blunt multiple trauma: diagnostic accuracy of duplex Doppler us and early CT angiography. Radiology. 2005;237(3):884–92.

79. Dekeyzer S, van den Hauwe L, VandeVyvere T, Parizel PM. Traumatic brain injury: imaging strategy. In: Barkhof F, Jäger H, Thurnher M, Rovira À, editors. Clinical neuroradiology. Cham: Springer; 2019. https://doi.org/10.1007/978-3-319-68536-6_27.
80. Asgeirsson B, Grände PO, Nordström CH. A new therapy of post-trauma brain oedema based on haemodynamic principles for brain volume regulation. Intensive Care Med. 1994;20(4):260–7.
81. Bullock R, Chesnut RM, Clifton G, Ghajar J, Marion DW, Narayan RK, Newell DW, Pitts LH, Rosner MJ, Wilberger JW. Guidelines for the management of severe head injury. Brain Trauma Foundation. Eur J Emerg Med. 1996;3(2):109–27.
82. Gerber LM, Chiu YL, Carney N, Härtl R, Ghajar J. Marked reduction in mortality in patients with severe traumatic brain injury. J Neurosurg. 2013;119:1583–90.
83. Grände PO. The "Lund concept" for the treatment of severe head trauma—physiological principles and clinical application. Intensive Care Med. 2006;32(10):1475–84.
84. Meythaler JM, Brunner RC, Johnson A, Novack TA. Amantadine to improve neurorecovery in traumatic brain injury-associated diffuse axonal injury: a pilot double-blind randomized trial. J Head Trauma Rehabil. 2002;17(4):300–13. https://doi.org/10.1097/00001199-200208000-00004. PMID: 12105999.

25 Tumoral Lesions of the Corpus Callosum

Balachandar Deivasigamani, Ved Prakash Maurya, Priyadarshi Dikshit, Vikas Dwivedi, Vipin Sahu, Kuntal Kanti Das, and Sanjay Behari

25.1 Introduction

The corpus callosum (CC) is the most extensive commissural white matter tract connecting the two cerebral hemispheres. It lies in the depth of the interhemispheric fissure between the medial surfaces of the hemispheres [1]. It connects the frontal, parietal, temporal, occipital, insular, and limbic lobes and the basal ganglia between the two sides of the cerebral cortex. It comprises more than 190 million topographically organized axons, forming homotopic or heterotopic connections, often between distant regions of the cortex. Its primary function is to integrate and transfer information from both cerebral hemispheres in order to process complex sensory, motor, and high-level cognitive functions such as language and abstract reasoning (Fig. 25.1) [2].

B. Deivasigamani · V. P. Maurya · P. Dikshit · V. Dwivedi · V. Sahu · K. K. Das · S. Behari (✉)
Department of Neurosurgery, Sanjay Gandhi Postgraduate Institute of Medical Sciences, Lucknow, India

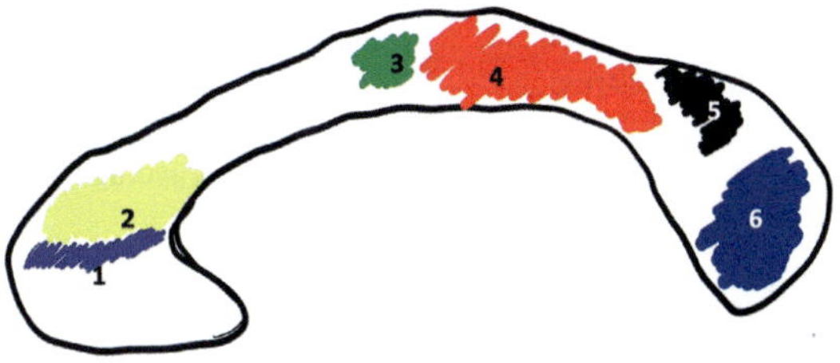

Fig. 25.1 Schematic representation of callosal activation and fiber topography. (1) Olfactory stimulation, (2) gustatory stimulation, (3) motor activation, (4) tactile stimulation, (5) auditory stimulation, and (6) visual stimulation

25.2 Structure and Function

Anatomically, from the anterior to the posterior aspect, the CC comprises four parts: the rostrum, genu, body, and splenium (Fig. 25.2). The narrow region between the body and splenium in the posterior aspect of the CC is termed the isthmus [2]. The indusium griseum, a thin layer of gray matter on the superior surface of the CC, is the supracallosal remnant of the hippocampal formation. Once this layer of gray matter is removed, the transversely running fibers of the CC can be visualized. The CC envelops the entire lateral ventricle within its callosal radiations. The rostrum forms the floor of the frontal horn, and the genu forms its anteromedial boundary. The body of the CC forms the roof of the body of the lateral ventricle.

M. Turgut et al. (eds.), *The Corpus Callosum*, https://doi.org/10.1007/978-3-031-38114-0_25

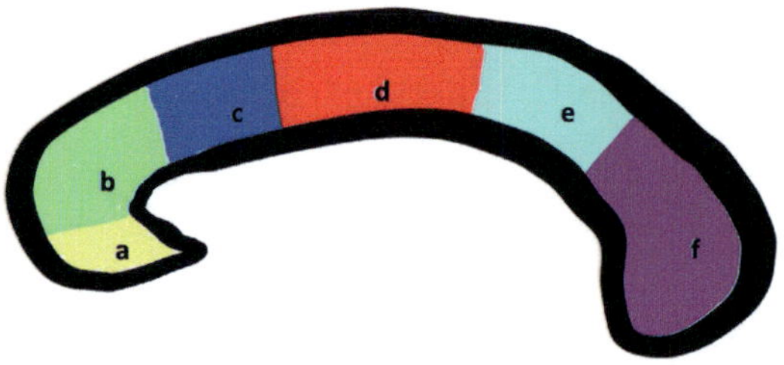

Fig. 25.2 Schematic representation of segments of the CC. (a) Rostrum, (b) genu, (c) anterior body, (d) posterior body, (e) isthmus, and (f) splenium

Table 25.1 Structural dimensions of CC

Length	Rostrum to splenium	7.5–8.2 cm (average 7.8 cm)
Breadth	Rostrum	3.0–3.5 cm (average 3.2 cm)
	Genu	3.3–3.6 cm (average 3.4 cm)
	Anterior body	3.5–3.9 cm (average 3.7 cm)
	Posterior body	3.9–4.3 cm (average 4.1 cm)
	Splenium	4.2–4.7 cm (average 4.5 cm)

The fiber connections of the CC are classified into four types: superior/dorsal, inferior/ventral, anterior, and posterior radiations (Table 25.1) [3].

1. Superior/dorsal radiation fibers traverse laterally, entering the cerebral hemisphere's medial surface and turning upward. The dorsal callosal radiation fibers are limited by the vertical fibers of the corona radiata bilaterally.
2. Inferior/ventral callosal radiations traverse laterally from the genu and the body of the CC and turn inferiorly. The ventral callosal radiations are closely related to the lateral ventricle, which they envelop with their fibers. These radiations travel laterally from the anterior portion of the body in the roof of the frontal horn to reach the caudate head. Those forming the posterior part of the body of the CC traverse laterally and inferiorly to form the lateral wall of the temporal horn, the atrium, and the occipital horn. Thus, they envelop those portions of the ventricle.

Table 25.2 Blood supply and venous drainage [8]

CC	Blood supply	Venous drainage
Rostrum, genu, and body	Pericallosal artery (a branch of the anterior cerebral artery)	Callosal and callosal cingular veins and, ultimately, the internal cerebral veins
Splenium	Vertebrobasilar system (terminal and choroidal branches of the posterior cerebral artery)	

3. Anterior radiations/forceps minor fibers begin from the genu and rostrum of the CC and turn anteriorly into the medial and basal frontal lobes. The fibers of the forceps minor curve forward and form the medial wall of the frontal horn.
4. Posterior radiations/forceps major fibers traverse posteriorly from the splenium of the CC and enter the medial aspect of the occipital lobes. The forceps major fibers form the medial wall of the occipital horn. Table 25.2 shows the blood supply and venous drainage of the CC.

25.3 Development of the Human Corpus Callosum

The human commissural plate can be anatomically subdivided into the massa commissuralis through which the CC and hippocampal commissures pass and the area septalis through which the anterior commissure crosses (Fig. 25.3) [4]. Neuroimaging studies suggest that the first axonal crossover in the commissural plate occurs at the hippocampal primordium, subsequent connections being made bidirectionally [5, 6]. In the neocortex, neurons for callosal crossover originate in layers II/III, V, and VI [7], but this is always preceded by the crossing of pioneering axons originating from the cingulate cortex [8, 9].

By 20 weeks post-conception, the final shape of the CC is complete, although exuberant axonal growth continues until 2 months after birth. This is then followed by molecular- and activity-dependent axonal pruning [2]. Although the num-

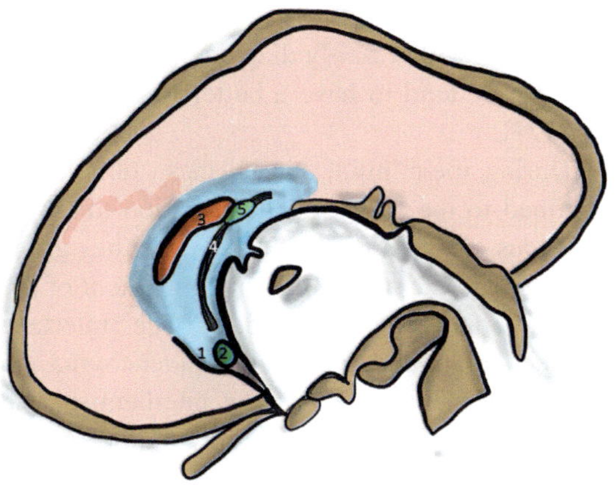

Fig. 25.3 Sagittal representation of a developing CC at 13 weeks of gestation. (1) Area septalis, (2) anterior commissure, (3) CC, (4) hippocampal commissure, and (5) fornix

ber of callosal fibers is more or less determined at birth, structural changes continue throughout postnatal development (rapid expansion of the CC due to increases in the number of axons, axon diameters, and myelin). These are most marked during childhood and adolescence [10–12]. Although the CC is completely developed by the age of 4 years, growth continues, much more slowly, until the third decade of life [1].

25.4 Hypotheses About Callosal Agenesis

25.4.1 Differential Development Hypothesis

Around weeks 13 and 14 post-conception, pioneering axons begin to cross the midline; the anterior sections begin to grow by weeks 14 and 15, whereas the posterior sections grow during weeks 18 and 19 [4, 13, 14]. The delayed development of the posterior and most anterior sections led to the assumption that early perturbation of callosal development results in complete agenesis of the CC (ACC), while later developmental disturbances result in partial agenesis, confined to the posterior CC and rostrum.

25.4.2 Dorsoventral Splenial Expansion Hypothesis

Recently, it has been found that connections are first made in two separate loci: the anterior commissure and the hippocampal commissure [15]. The early expansion of the frontal cortex results in posterior displacement of the hippocampal commissure and the associated callosal splenium while the anterior section of the CC expands. It has therefore been suggested that the absence of the posterior part of the CC in partial ACC most commonly results from failed dorsoventral expansion of the splenium [15].

25.5 Physiological Variants

Physiological variants of the CC include agenesis or hypoplasia. Agenesis of the CC is an uncommon brain malformation leading to partial or total absence of the CC. Although rare, it is one of the more frequently occurring congenital malformations, affecting 3–7 people per 1000 births [16]. It can be diagnosed using prenatal ultrasound. This anomaly can be concurrent with malformations in midline structures such as cavum septum pellucidum and cortical and posterior fossa abnormalities [16, 17].

Despite the crucial role of the CC in normal cognition, patients with isolated agenesis of the CC typically have a good prognosis. Common deficits observed in this population include difficulty with problem-solving and socializing, often placing patients within the autism spectrum [18]. These deficits, known as "disconnection syndromes," differ from those observed in commissurotomy patients, who most notably have impaired bimanual coordinated motor activity [2]. Currently, 55–70% of patients with agenesis of the CC cannot be identified on the basis of their clinical features.

Hypoplasia is most benign when the rostrum is absent; this anomaly is often not clinically significant. Therefore, it is commonly an incidental finding on a magnetic resonance imaging (MRI) scan obtained for other clinical reasons [2, 19].

25.6 Surgical Considerations

Callosal axons terminate in diverse brain regions apart from the topographically equivalent areas, so they participate in spreading synchronous corticocortical activation to multiple homotropic and heterotopic regions of the hemispheres. This synchronous activity explains the callosal propagation of an epileptic focus and the effectiveness of callosotomy in preventing the expansion of epileptic activity. Corpus callosotomy is a surgical procedure for epilepsy that remains refractory despite maximum medical management. Also, it is the preferred surgical procedure for patients with a refractory seizure disorder with drop attacks. The CC should be incised transversely, parallel to the transversely running fibers, especially when the callosotomy is performed to approach ventricular lesions. A transverse incision divides the fibers without injuring them. A longitudinal incision in the anteroposterior direction transects the transversely running fibers of the CC. Such an incision is most appropriate for a callosotomy for disconnection, as in epilepsy surgery.

For anterior interhemispheric transcallosal approaches, callosotomy is performed in the anterior portion of the body of the CC. The callosotomy site should not extend beyond 3.5 cm from the genu. During a posterior interhemispheric approach, it is preferable to perform a transverse callosotomy in the isthmus region in front of the splenium. This avoids most of the motor-sensory fibers in the posterior part of the body and the visual and auditory fibers in the splenium.

25.7 Lesions of the Corpus Callosum

25.7.1 Neoplasms

25.7.1.1 Corpus Callosal Gliomas

These belong to a diffuse group of gliomas and have a classic "butterfly" pattern of bilateral hemispherical involvement of the tumor. They spread by direct extension along white matter tracts (Table 25.3) [20]. The data reveal that 14% of patients with a newly diagnosed glioblastoma multiforme tend to have a butterfly glioma pattern [21].

During their initial stages, these tumors are confined to the subset of the radiation to which the focus of the lesion belongs. When they grow, they involve the most adjacent component of the callosal radiation (Fig. 25.4). The interface between tumor and normal brain parenchyma can be clearly identified [22]. The interhemispheric approach can be used to excise tumors in the anterior, superior, and posterior radiations.

25.7.1.2 Epidermoid Cysts

Supracallosal epidermoid cysts are very rare among the rare supratentorial epidermoid cysts (which usually occur in the basal cisternal region [0.2–1%]) [23]. They occur in the CC owing to a defect in the segregation of neuroectoderm and surface ectoderm during neurulation (third week of gestation), causing epithelial inclusion in the longitudinal cerebral fissure [24]. They present with symptoms of mass effects, seizures, and neuropsychological derangement. MRI is the investigation of choice. Complete surgical removal is the treatment of choice, as it is for epidermoid cysts in other regions of the brain (Fig. 25.5).

25.7.1.3 Lipomas

CC lipomas are congenital fat-containing brain malformations involving abnormal persistence and maldifferentiation of the primitive meninges. They develop between the eighth and tenth weeks of embryo development [25]. Among intracranial lipomas, corpus callosal lipomas are fairly common. These are primarily asymptomatic and are usually found incidentally. Asymptomatic lipomas are not operated upon as they either have no growth potential or take a long time to develop into clinically relevant entities. The risk of surgery can outweigh the benefits owing to the proximity of surrounding neurovascular structures. Surgical removal is required only when the lesion is symptomatic and is associated with a dermal sinus tract or an epidermoid tumor. CC lipomas classify as either

Table 25.3 Pattern of involvement of CC gliomas [3]

Tumor foci in the corpus callosum	Genu and forceps minor	Body (dorsal callosal radiations)	Splenium and forceps major	Ventral callosal radiations	
				Anterior	Posterior
Adjacent brain parenchyma involved	Medial orbito-basal frontal lobes	Cingulate gyrus and medial frontoparietal regions	Medial part of occipital lobes	Spreads to the genu and then to the head of caudate nucleus and all the way to the stem of temporal lobe owing to its close association with the inferior fronto-occipital fascicle (IFOF)	Spreads to tapetum and crosses to lateral parts of temporal and occipital horns
Limiting structures	Frontal horn	Projection fibers	Occipital horn	–	–

curvilinear (usually located posteriorly and usually asymptomatic) or tubulonodular (found anteriorly and occasionally associated with severe frontal lobe abnormalities) [1, 25].

25.7.1.4 Lymphoma

Primary central nervous system (CNS) lymphoma involving the CC (non-Hodgkin B-cell lymphoma) is well known, and it mostly crosses the midline [26]. Such lesions have less peritumoral edema and a lower tendency to hemorrhage. They are often multifocal with heterogeneous intensity and ring enhancement in immunocompromised patients [20]. They can also invade the periventricular regions or there can be additional lesions in those locations.

25.7.1.5 Metastasis

Metastatic lesions involving the CC are primarily due to spreading from adjacent structures (cingulate gyrus). In very rare cases, they involve the CC primarily [20]. Once the tumor involves the transverse commissural fibers, the CC provides an easy route for spread. The incidence of metastatic lesions in the CC is 11.1% (Fig. 25.6) [27].

25.7.2 Vascular Lesions

25.7.2.1 Aneurysms

Acute hemorrhages located anterior to the CC are primarily due to ruptured aneurysms of the anterior cerebral or pericallosal artery. Hence, they should be evaluated with a CT angiogram or by digital subtraction arteriography (DSA) [1].

25.7.2.2 Arteriovenous Malformations (AVMs)

Arteriovenous malformations (AVMs) are conglomerates of abnormal vessels (artery and vein) that have a congenital origin. There are no intermediate capillaries and the brain parenchyma is usually surrounded by nonfunctional tissue. The incidence of corpus callosal AVMs is 1.1–3% [28]. They form a particular category of medial cerebral hemisphere AVMs, and they are known to bleed more frequently than superficial and pial AVMs. They present at a much younger age than patients with superficial AVMs [29].

In contrast to the superficial arterial supply to most convexity malformations, CC AVMs are supplied by deep and often obscure arteries. These include the primary trunks or branches of the anterior and posterior pericallosal arteries and branches from the posterior cerebral arteries (Fig. 25.7). Care is taken to preserve the main trunks and branches of the anterior cerebral and perforating arteries of the anterior perforated space. Venous drainage of these lesions is often dual, one deep toward the ventricular region and the other superficial to the sagittal sinus. The risk of bleeding is about 4% per year [29]. The aim of treatment is to provide prophylaxis against bleeding, thereby preventing future complications and minimizing the risk of hemorrhage.

25.7.3 Metabolic Diseases

Marchiafava-Bignami disease involves acute demyelination and necrosis of the CC, most

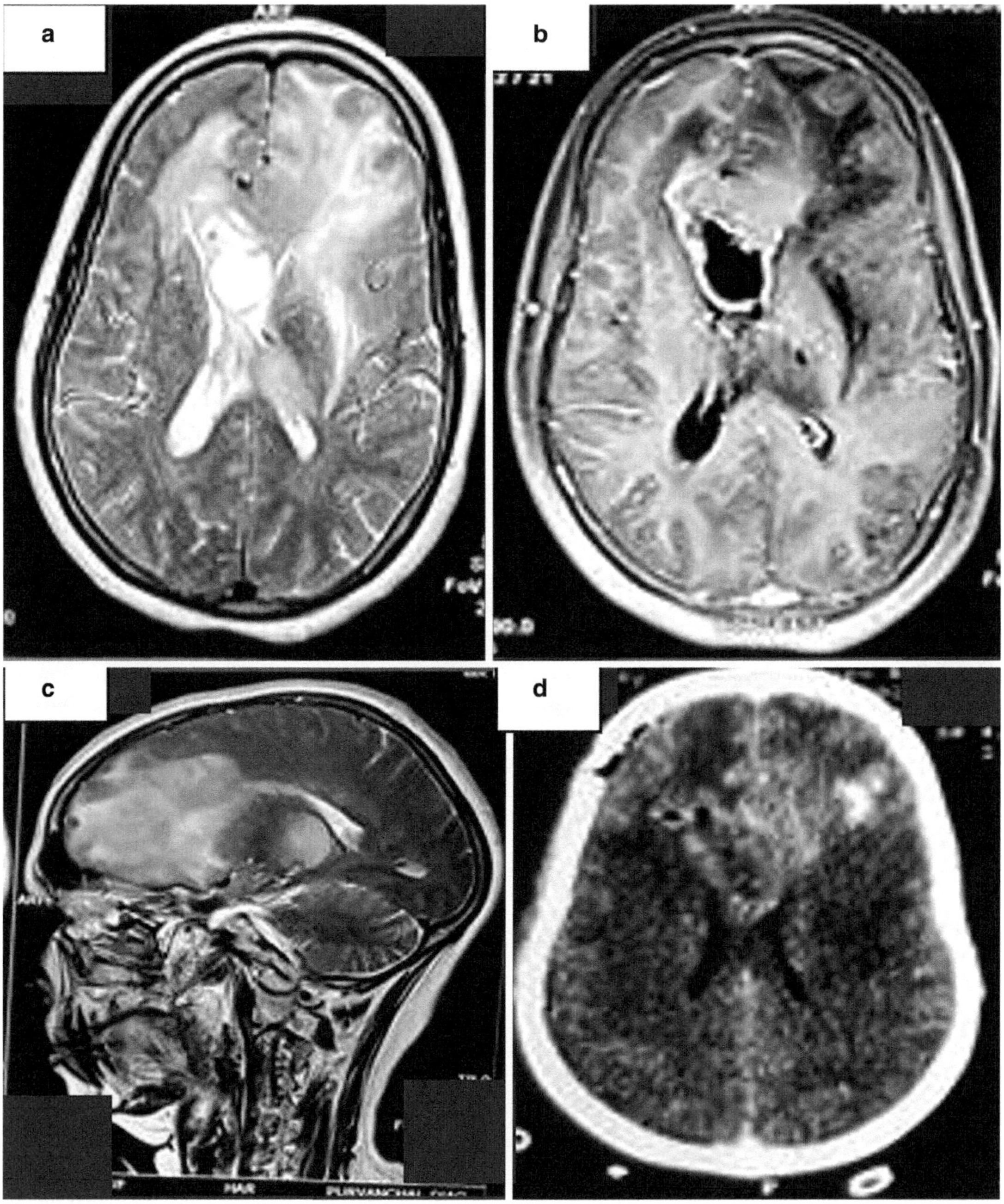

Fig. 25.4 Radiological images of a patient with a lesion in the anterior part of the body of the CC. (**a**) T2WI showing a hyperintense irregular lesion compressing the right frontal horn with extensive edema, which also extends to the left side, (**b**) T1 contrast-enhanced sequence suggests a relatively well-defined lesion involving the medial aspect of the frontal lobes (right > left), (**c**) T2WI in sagittal images show white matter edema extending up to the mid portion of the CC, and (**d**) postoperative contrast-enhanced computerized tomography scan suggests adequate decompression with a small residual lesion on the left medial aspect of the frontal lobe. This residual lesion is associated with surrounding edema

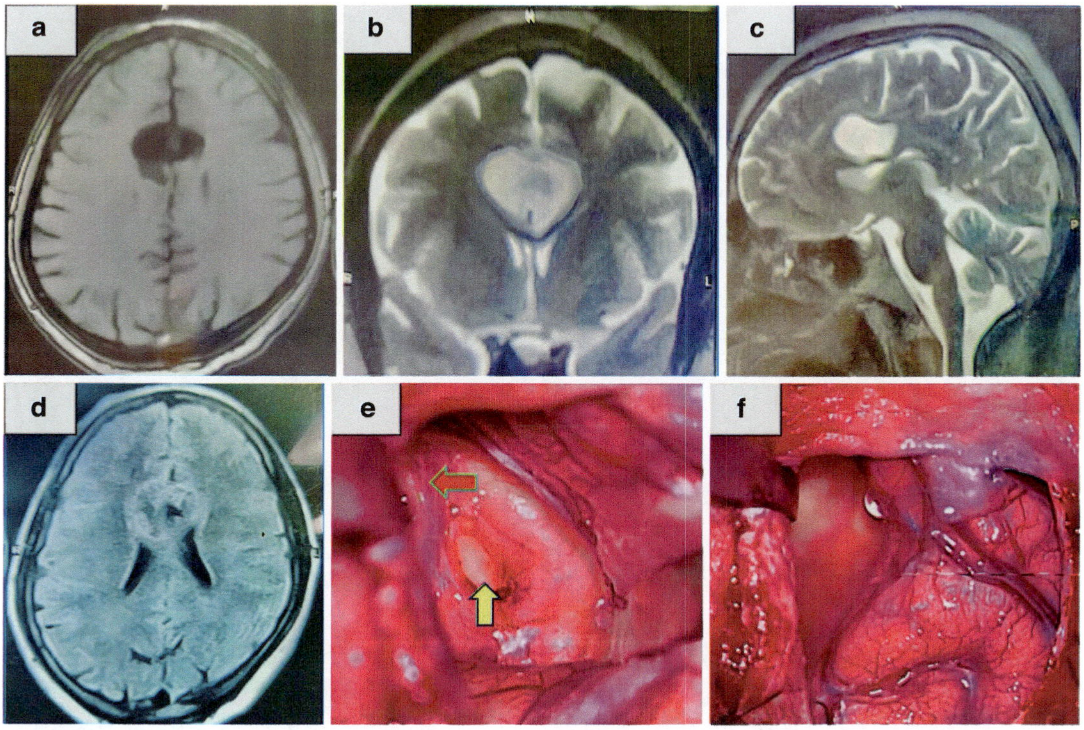

Fig. 25.5 Images of a middle-aged gentleman who presented with complaint of headache for the previous 3 months. (**a–d**) MRI brain scan suggests a lesion in the interhemispheric cistern that is hypointense on T1 and hyperintense on T2WI causing significant compression over the midportion of the body of the CC, (**e**) intraoperative picture shows evidence of a pial breach over the CC (*yellow arrow*), and the right pericallosal artery (*red arrow*) is deviated from its course by the mass effect, (**f**) image of a relaxed brain after complete removal of the epidermoid through right mini frontal craniotomy (4 × 4 cm), showing brain retractor over the falx and cottonoids over the incised dural margins

commonly observed in the body region in patients with chronic alcoholism and multiple vitamin deficiencies. There can be necrosis and cavitary lesions involving the CC, leading to a "sandwich-like" appearance on MRI images [1, 30]. The CC can also be involved in osmotic demyelinating syndrome owing to extrapontine myelinolysis [30].

25.7.4 Infective Etiologies

Although brain abscesses appear at gray-white junctions, the CC is involved in rare cases of severe *Staphylococcus aureus*, *E. coli*, and salmonella infection [20, 30]. In immunocompromised individuals, particularly in association with HIV, toxoplasmosis can rarely involve the CC. It can sometimes assume a butterfly pattern [31]. The CC is involved in viral encephalitis (cytomegalovirus infection [associated with HIV/AIDS], influenza virus, Epstein-Barr virus, varicella zoster virus, and Herpes simplex virus-6) [30].

25.7.5 Demyelinating Diseases

Multiple sclerosis affects the CC in addition to its classical involvement of periventricular white matter [20]. The inferior part of the CC (i.e., callosal-septal interface) is asymmetrically

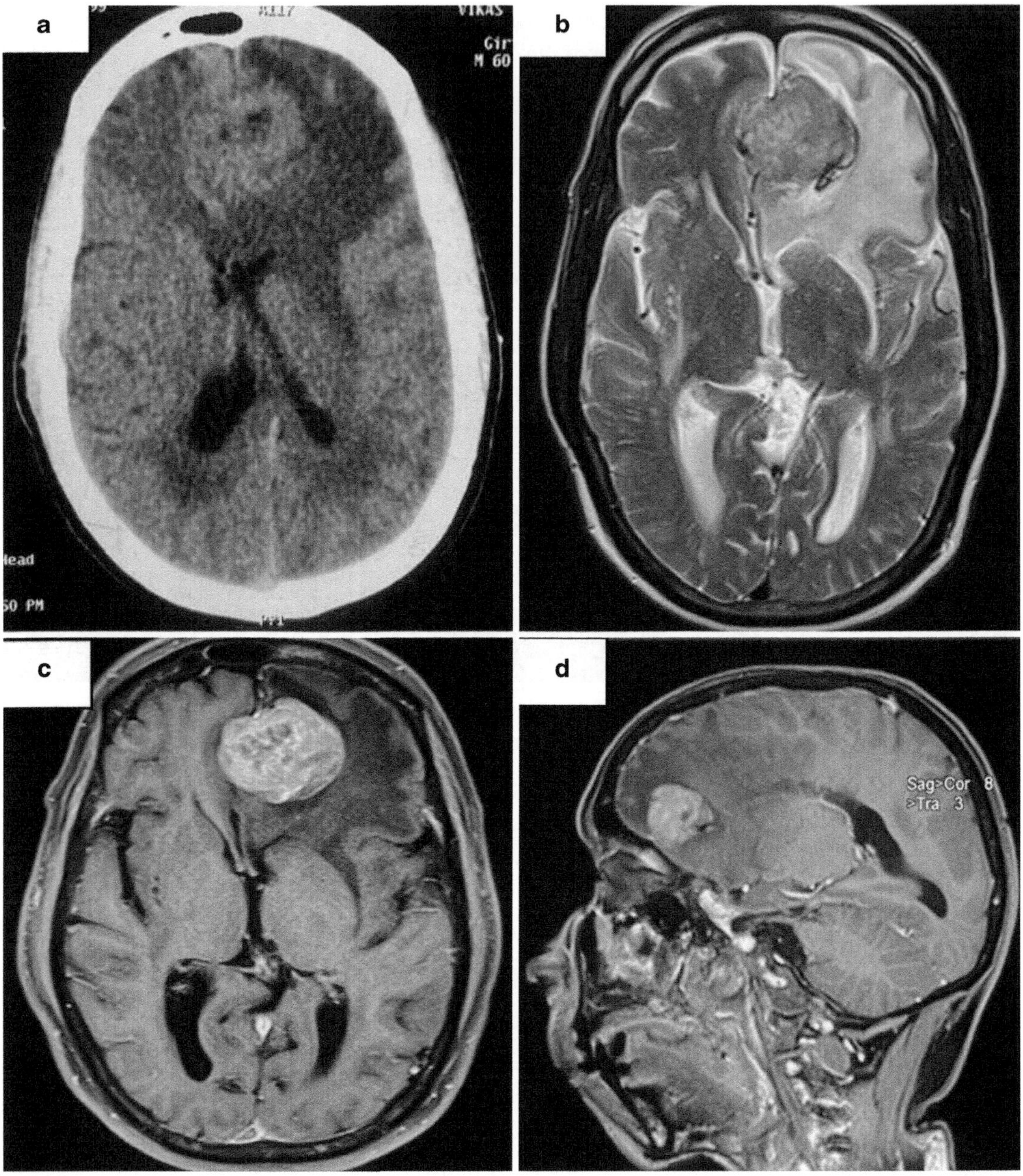

Fig. 25.6 Radiological images of an individual with right-sided lung carcinoma (*histopathology: squamous cell carcinoma*) with metastatic brain deposits. (**a**) Contrast-enhanced CT scan of the brain showing an enhancing/hyperdense lesion in the left medial frontal gyrus with significant perilesional edema; (**b–d**) brain MRI suggesting a T2 isointense lesion in the medial frontal gyrus causing a significant mass effect (left>right) and white matter edema. The lesion compresses the anterior circulation vessels toward the right side. The contrast-enhanced MRI images show significant enhancement with involvement of the genu and anterior body of the CC

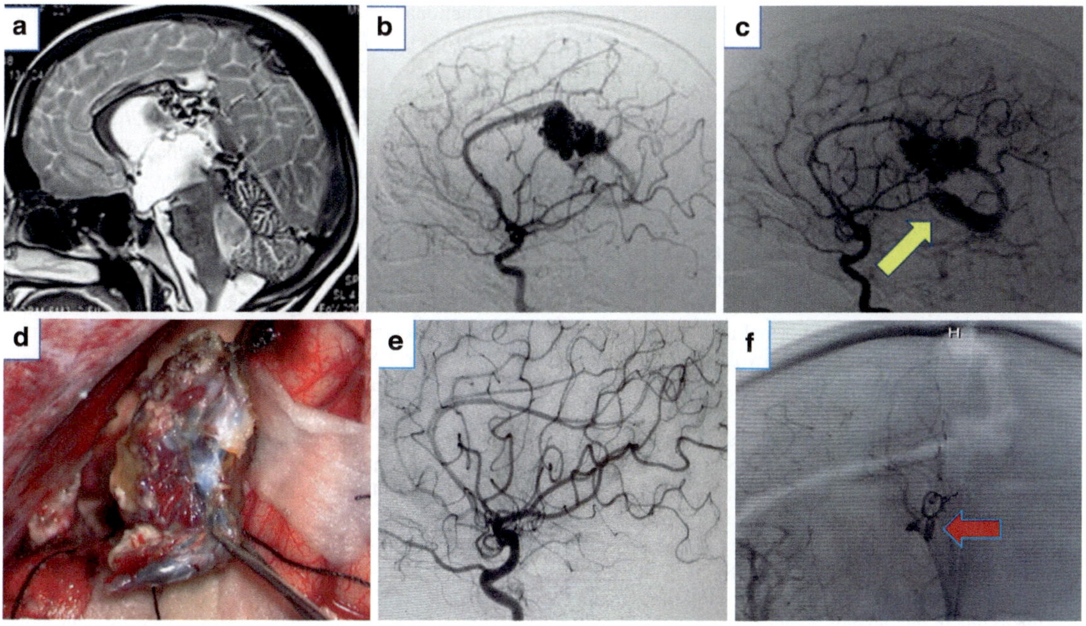

Fig. 25.7 Representative images of CC AVM are shown. (**a**) T2 sagittal section shows the AVM nidus located in the body of the CC with intraventricular extension along the septum pellucidum; (**b**) on an intraarterial angiogram, the nidus was seen as compact, receiving feeders from the pericallosal arteries with an early draining vein suggesting high flow nature of the AVM; (**c**) the draining vein (*yellow arrow*) drained into the vein of Galen, which was secondarily dilated; (**d**) intraoperative photograph showing the AVM nidus excised through an anterior interhemispheric transcallosal approach; (**e**) postoperative angiogram shows complete resection of the AVM; (**f**) image shows a permanent aneurysm clip (*red arrow*) used to block the feeding artery

involved. The CC is often involved in neuromyelitis optica (NMO) as well as its classical involvement of the periventricular white matter. The splenium of the CC is more frequently involved [30].

25.7.6 Trauma

The CC is a particularly susceptible part of the brain in traumatic injuries. It is situated in the midline and has a high incidence of secondary injuries owing to elevated intracranial pressure. The posterior part of the CC is mainly involved in diffuse axonal injury, mostly associated with other entities of head trauma. DWI and FLAIR sequence reveals these lesions as hyperintensities [30, 32].

25.8 Conclusions

The CC connects almost all parts of the cerebral hemispheres bilaterally with its rich network of white matter connections. Knowing the intricate anatomy and the pattern of white fibers in the CC elucidates the wide variety of lesions and their nature. This helps to a great extent in planning appropriate treatment and in prognosticating the disease.

References

1. Goldstein A, Covington BP, Mahabadi N, Mesfin FB. Neuroanatomy, corpus callosum. StatPearls; 2021.
2. Edwards TJ, Sherr EH, Barkovich AJ, Richards LJ. Clinical, genetic and imaging findings identify

new causes for corpus callosum development syndromes. Brain. 2014;137(6):1579–613.
3. Shah A, Jhawar S, Goel A, Goel A. Corpus callosum and its connections: a fiber dissection study. World Neurosurg. 2021;151:e1024. https://doi.org/10.1016/j.wneu.2021.05.047.
4. Rakic P, Yakovlev PI. Development of the corpus callosum and cavum septi in man. J Comp Neurol. 1968;132(1):45–72.
5. Barkovich AJ, Lyon G, Evrard P. Formation, maturation, and disorders of white matter. AJNR Am J Neuroradiol. 1992;13(2):447.
6. Kier EL, Truwit CL. The normal and abnormal genu of the corpus callosum: an evolutionary, embryologic, anatomic, and MR analysis. Am J Neuroradiol. 1996;17(9):1631–41.
7. Fame RM, MacDonald JL, Macklis JD. Development, specification, and diversity of callosal projection neurons. Trends Neurosci. 2011;34(1):41–50.
8. Koester SE, O'leary DD. Axons of early generated neurons in cingulate cortex pioneer the corpus callosum. J Neurosci. 1994;14(11):6608–20.
9. Rash BG, Richards LJ. A role for cingulate pioneering axons in the development of the corpus callosum. J Comp Neurol. 2001;434(2):147–57.
10. Garel C, Cont I, Alberti C, Josserand E, Moutard ML, le Pointe HD. Biometry of the corpus callosum in children: MR imaging reference data. Am J Neuroradiol. 2011;32(8):1436–43.
11. Luders E, Thompson PM, Toga AW. The development of the corpus callosum in the healthy human brain. J Neurosci. 2010;30(33):10985–90.
12. Luo L, O'Leary DD. Axon retraction and degeneration in development and disease. Annu Rev Neurosci. 2005;28:127–56.
13. Hewitt W. The development of the human corpus callosum. J Anat. 1962;96(Pt 3):355.
14. Ren T, Anderson A, Shen W-B, Huang H, Plachez C, Zhang J, Mori S, Kinsman SL, Richards LJ. Imaging, anatomical, and molecular analysis of callosal formation in the developing human fetal brain. Anat Rec A Discov Mol Cell Evol Biol. 2006;288(2):191–204.
15. Paul LK. Developmental malformation of the corpus callosum: a review of typical callosal development and examples of developmental disorders with callosal involvement. J Neurodev Disord. 2011;3(1):3–27.
16. Fitsiori A, Nguyen D, Karentzos A, Delavelle J, Vargas M. The corpus callosum: white matter or terra incognita. Br J Radiol. 2011;84(997):5–18.
17. Das JM, Dossani RH. Cavum septum pellucidum. StatPearls [Internet]; 2020.
18. Paul LK, Brown WS, Adolphs R, Tyszka JM, Richards LJ, Mukherjee P, Sherr EH. Agenesis of the corpus callosum: genetic, developmental and functional aspects of connectivity. Nat Rev Neurosci. 2007;8(4):287–99.
19. Das JM, Geetha R. Corpus callosum agenesis. StatPearls [Internet]; 2021.
20. Garber ST, Khoury L, Bell D, Schomer DF, Janku F, McCutcheon IE. Metastatic adenoid cystic carcinoma mimicking butterfly glioblastoma: a rare presentation in the splenium of the corpus callosum. World Neurosurg. 2016;95:621.e13–9.
21. Chaichana KL, McGirt MJ, Frazier J, Attenello F, Guerrero-Cazares H, Quinones-Hinojosa A. Relationship of glioblastoma multiforme to the lateral ventricles predicts survival following tumor resection. J Neuro-Oncol. 2008;89(2):219–24.
22. Goel A, Shah A, Vutha R, Dandpat S, Hawaldar A. Is "En Masse" tumor resection a safe surgical strategy for low-grade gliomas? Feasibility report on 74 patients treated over four years. Neurol India. 2021;69(2):406.
23. Furuhata S, Yamada F, Fukuda S, Otani M, Toya S. Epidermoid cysts of the callosal region —three case reports—. Neurol Med Chir. 1993;33(10):697–9.
24. Catala M, Pradat P-F, Cornu P, Lubetzki C. A case of epidermoid cyst arising from the rostral neuropore: case report. Neurosurgery. 1998;43(2):374–5.
25. Gezer B, Karabagli H, Koktekir E, Sahinoglu M, Karabagli P. Corpus callosum lipomas in children. Turk Neurosurg. 2019; https://doi.org/10.5137/1019-5149.JTN.27692-19.6.
26. Schlegel U. Primary CNS lymphoma. Ther Adv Neurol Disord. 2009;2(2):93–104. Received on September 12.
27. Ucmakli A. The pathological significance of corpus callosum involvement in brain scans. J Nucl Med. 1972;13(7):510–6.
28. Pabaney AH, Ali R, Kole M, Malik GM. Arteriovenous malformations of the corpus callosum: pooled analysis and systematic review of literature. Surg Neurol Int. 2016;7(Suppl 9):S228–36.
29. Gatto LAM, de Angeli Corredato R, Zanine SC, Araújo JC, Diaz FG. Intraventricular hemorrhage secondary to AVM of the septum pellucidum. Arq Bras Neurocir. 2014;33(2):151–4.
30. Renard D, Castelnovo G, Campello C, Bouly S, Floch AL, Thouvenot E, Waconge A, Taieb G. An MRI review of acquired corpus callosum lesions. J Neurol Neurosurg Psychiatry. 2014;85(9):1041–8.
31. Slone HW, Blake JJ, Shah R, Guttikonda S, Bourekas EC. CT and MRI findings of intracranial lymphoma. AJR Am J Roentgenol. 2005;184(5):1679–85.
32. Jang SH, Kim OL, Kim SH, Lee HD. Differences in corpus callosum injury between cerebral concussion and diffuse axonal injury. Medicine (Baltimore). 2019;98(41):e17467.

Iatrogenic Lesions of the Corpus Callosum

26

Mitchell W. Couldwell and R. Shane Tubbs

26.1 Introduction

Iatrogenic disease is referred to disease a result of a therapeutical or a diagnostic procedure induced on a patient [1]. The corpus callosum (CC) can be subject to iatrogenic injury via several mechanisms such as direct puncture, third ventriculostomy, callosal swelling following treatment for hydrocephalus, radiation, the trans-colossal surgical approach, and posterior reversible encephalopathy syndrome (PRES) which can include swelling of the splenium of the CC. Microsurgical anatomy and function of the CC have been an interest in the neurosurgical community due to the variety of procedures performed in close proximity or the performance of callosotomies to reach intraventricular lesions.

Clinically it is important to recognize the signs and symptoms of CC injury following procedures or treatment of conditions affecting the central nervous systems to mitigate both risks and long-term complications arising from the affected area. Potential outcomes and complications as well as patient risk factors must be evaluated prior to surgical and radiation treatment in the vicinity of the CC. Patients undergoing surgical procedures such as third ventriculostomy or transcallosal approach to the ventricles are at greater risk of iatrogenic injury compared to other procedures. Patients with the sequelae of PRES syndrome due to metabolic injury as well as radiation therapy are also at greater risk for injury to the CC.

In this chapter, iatrogenic injuries to the CC are discussed. Imaging of lesions, time frames, and neurological symptoms of procedures listed above are also discussed. Potential risk factors, mechanisms of injury, and patient outcomes are described. Surgical details and descriptions of the procedures and treatments themselves are also discussed in this chapter.

26.2 Colossal Puncture

Colossal puncture, a precursor to the third ventriculostomy in the treatment of hydrocephalus, is among the first reported outcome of surgical injury to the CC [2]. First described in 1908 by Anton and von Bramann and known as the "Balkenstich method," it was described as a CC puncture between the ventricle and subarachnoid spaces. It was indicated as a treatment for congenital hydrocephalus, obstructive hydrocephalus, and epilepsy and was postulated to offer the benefit of continuous CSF diversion without the need for a shunting device [3].

M. W. Couldwell (✉)
Tulane University School of Medicine, New Orleans, LA, USA
e-mail: mcouldwell@tulane.edu

R. S. Tubbs
Department of Neurosurgery, Tulane School of Medicine, New Orleans, LA, USA
e-mail: rtubbs@tulane.edu

M. Turgut et al. (eds.), *The Corpus Callosum*, https://doi.org/10.1007/978-3-031-38114-0_26

The procedure gained more attention after a series of patients operated on by Harvey Cushing at Johns Hopkins are among the first case series to describe a CC puncture for both obstructive and congenital hydrocephalus. Three of the seven patients operated on Cushing had postoperative left-sided weakness and one perished during surgery [2]. The procedure was utilized in the following years including treatment for a patient with obstructive hydrocephalus due to cerebrospinal meningitis by Stetten and Roberts in 1919 with full recovery of the patient following deficits in the left facial muscles and left external rectus muscle following surgery [4]. The procedure was eventually replaced by the third ventriculostomy in the treatment of hydrocephalus, and there is limited documentation of outcomes and procedural attempts.

26.3 Ventriculostomy

Ventriculostomy is among the most common neurosurgical procedures performed globally [5]. The procedure involves using an endoscope to enable CSF drainage from the floor of the third ventricle into the basal cisterns to bypass an obstruction [6]. The indications for the procedure include acute symptomatic hydrocephalus, intracranial pressure monitoring, brain relaxation during intraoperative brain edema, and an adjunct management for ventriculoperitoneal shunts. The procedure has also been used to treat hydrocephalus secondary to congenital aqueductal stenosis, cerebellar infarct, third ventricular tumors, Dandy-Walker malformation, syringomyelia with or without Chiari malformation, and posterior fossa tumors as well as many others [7].

The first third ventriculostomy was performed by William Jason Mixter in 1923, with the use of an ureteroscope in the treatment of non-communicating hydrocephalus; the procedure was further popularized by Walter Dandy. The modern approach of ventriculostomies began in the 1970s with the advent of the flexible and rigid modern endoscopes used today, and further advancement of the field has included both stereotaxic and ultrasound-guided technique to further refine the procedure [8]. Technical notes regarding the procedure have continued to evolve including the study of multiple burr hole placement and grading scales for catheter placement during bedside ventriculostomy. Modern physicians have created grading scales to attempt to standardize catheter tip placement within the brain. The Kakarla scale utilizes three grades with Grade 1 being optimal with catheter placement within the ipsilateral front horn including the tip of the third ventricle; Grade 2 being suboptimal and placed within the contralateral front horn, CC, or interhemispheric fissure; and Grade 3 being suboptimal and involving the basal ganglia, internal capsule, thalamus, brainstem, cerebellum, basal cisterns, or occipital cortex [9].

Changes in the CC have been reported in cases of successful CSF shunting following third ventriculostomy; between 5 and 17% of reported cases of sudden ventricular shunting have shown changes to the structure [10, 11]. The term "post-shunting CC swelling" has been used to describe the associated structural changes [12]. Typical changes of this condition include morphological scalloping of the dorsal CC with associated changes on magnetic resonance imaging (MRI). Patients presenting with CC changes post-shunt have been shown lesions which have predominately been in the anterior as well as the posterior body of the CC. Additionally there have been reported cases of small lesions evident upon imaging to the splenium of the CC which have been attributed to iatrogenic injury from the shunt catheter [11].

The time period of MRI changes has been documented between 3 and 75 months postoperatively [11]. MRI changes typically include decreased signal intensity on T1-weighted images and increased signal intensity on T2/FLAIR sequences. The mechanisms underlying these signal changes have not been clearly defined, but edema and ischemia associated with mechanical compression and decompression have been implicated for the development of lesions [13]. In previous studies of the CC, it has been shown that diffusor tensor imaging can be used to study the major fiber tracts within the CC [14]. Changes to structural fiber tracts including diminished fractional anisotropy as well as elevated diffusivity point toward injury involving excess of interstitial fluid buildup as opposed to demyelination, ischemic injury, or mass lesion due to the intact white matter pathways delineated upon tractography [12, 15].

Factors influencing the degree of change of the CC following ventricular shunt placement include the degree of compression of the CC against the falx cerebri due to choric moderate to severe chronic hydrocephalus followed by a drastic reduction in ventricular size over a short period of time following the shunting procedure [12]. Variation in the height of the falx cerebri dependent on individual anatomy is also thought to be a factor in degree of compression as well as following decompression injury to the CC. It is also important to note that this phenomenon is only present after the drainage of chronic hydrocephalus and is not associated with drainage of acute hydrocephalus [11, 16].

Other proposed mechanisms of injury include the superior displacement of the pericallosal artery in long-term obstructive hydrocephalus and physiologic adaptation of vessel shortening due to the smaller distance between the CC and the cingulate cortex [10]. These vessel changes may be responsible for ischemia and edema following stretching injury to the vessels after drainage of the ventricular system. MRI changes in the CC following shunt procedures indicate that these changes may also be responsible for the condition and the reason why drainage in acute hydrocephalus cases does not show in the change in comparison to chronic hydrocephalus. These physiologic vessel changes are also implicated in injury to the CC when the ventricles forcefully drained, and these procedures have been discussed in the literature as "overshunting" injuries [17].

Results of imaging as well as clinical studies have shown that many of these changes within the CC are not associated with any neuropsychological symptoms. This is a radiologically significant finding which knowledge of disease may help avoid misrepresentation of this pattern and misinterpretation of disease such as neoplasms which typically invade the CC such as glioblastoma or lymphoma or demyelination disorders such as progressive multifocal leukoencephalopathy and multiple sclerosis [11, 13].

26.4 Trans Callosal Approach

The interhemispheric transcallosal approach is used to access tumors involving the lateral as well as the third ventricles. Tumors or lesions that occur in these areas often present technical challenges due to their deep location and variable size [18]. The benefits of the procedure include a direct midline approach, with access symmetrical to the walls of the third ventricle and the lateral ventricles. Insult to the cerebral cortex and associated risks are mitigated by using a transcallosal over a transcortical approach to the ventricles [19].

The procedure itself involves a partial callosotomy, and recommendations are made to limit the incision depending upon the size and location of the lesion. The CC should be dissected down the midline between the pericallosal arteries to mitigate damage to adjacent structures or large insult to the CC [19]. Limited incisions of the colossal trunk have been associated with minimal symptoms postoperatively. Incisions up to 2.5 cm in the anterior body of the CC have historically been large enough to expose the ventricular system and avoid disconnection issues [20–23]. Longer incisions involving the anterior and mid-CC may be associated with disconnection issues and worse outcomes including the impairment of tactile data transfer [24–27].

Reported surgical complications have included patients presenting with an acute syndrome of decreased speech spontaneity to completed muteness within from hours to days lasting for several months [28, 29]. Other noted complications of the transcallosal approach included memory loss, although it has been suggested that far lateral incision of CC is favorable in avoiding this [30, 31]. Other consequences of this procedure include incontinence, lower extremity paresis, emotional disturbances, and seizures which may suggest other structures and associated changes, or damage may play a role in the etiology of potential complications [30–37].

It has been shown that if the splenium of the CC remains intact, interhemispheric information transfer should be preserved [38]. Consideration should be made to cases involving the posterior aspect of the CC (splenium) due to increased risks of cognitive dysfunction. Due to its anatomic location, the genu of the internal capsule of the CC and its proximity to the foramen of Munro must also be kept in mind when retracting the walls of the lateral ventricles [39].

26.5 Posterior Reversible Encephalopathy Syndrome

Posterior reversible encephalopathy syndrome (PRES) is a clinical-radiologic syndrome characterized by a combination of headache, altered mental status, severe hypertension, visual disturbances, and generalized seizures. The disease has been most studied in patients with eclampsia and organ transplantation with the possibility of relation to tacrolimus or cyclosporine; the disease has also been characterized in other conditions such as hypertension, post-chemotherapy, systemic lupus erythematosus, Wegener granulomatosis, and renal inflammatory conditions [40, 41]. Possible mechanisms of injury of PRES include graft-versus host disease, endothelial injury as observed in preeclampsia and eclampsia, immunosuppressive drugs, and conditioning regimens in pretransplantation preparation [42–46].

PRES can be viewed by MRI and other neurological imaging findings such as computed tomography (CT) [47]. Angiography can also be used to visualize PRES and includes focal or diffuse mixture of vasoconstriction and vasodilation and an appearance resembling a "string of beads" [48]. Criteria for conformation of PRES in previous studies has included complete or partial representation of typical PRES pattern, reversibility, and vasogenic edema [49].

The following are the three main imaging patterns that have been proposed in PRES syndrome:

- The holo-hemispheric watershed pattern in which vasogenic edema is present in a linear pattern spanning the frontal, parietal, temporal, and occipital lobes which has been reported to be in 22.8% of cases seen.
- The superior frontal sulcus pattern in which vasogenic edema was seen distinctly in the frontal lobes with minimal degrees of occipital and parietal pattern which represented 27.2% of cases studied.
- The dominant parietal-occipital pattern which represented 22.1% of cases involved the parietal and occipital cortex and minimal involvement of the temporal lobe.
- Partial or asymmetric PRES expression with traits of each aforementioned pattern have also been documented to encompass 27.9% of cases [47].

Atypical PRES lesions have also been mentioned. They can be located specifically in the white matter, basal ganglia, and brain stem have been implicated in the development of PRES, and the splenium of the CC has been identified as being affected in 9.6% of patients analyzed and have been present in the literature in other single case studies [47, 50, 51].

26.6 Radiation Injury

Radiation therapy (RT)involving the brain has been implicated in neurocognitive decline and decreases in personal relationships, quality of life prior to treatment, and difficulty in daily activities [52]. Although the mechanism of RT and its neurocognitive effects has not been clearly defined and can be considered multifactorial, thinning or damage to the CC has been associated with reduced executive function, working memory, processing speed, attention, and verbal fluency and memory, while specific deficits in attention and processing speed have been linked to the white matter tracts and projection fibers of the CC [53–60]. Brain volume loss associated with RT has been shown in multiple studies and suggest that neuronal loss may be a factor in deficits following treatment and a dose relationship of radiation may correlate with neuropsychological deficits [61–64].

The CC has previously been shown to be particularly vulnerable to dose-dependent RT, with smaller volume, higher mean diffusivity upon imaging correlating with greater amounts of white matter injury, and lower fractional anisotropy values also indicative of white matter damage. No specific region of the CC has been shown to be affected by RT, with the anterior, mid-anterior, and posterior CC all showing biomarkers of radiation therapy-induced injury [65–68].

Previous studies of pediatric brain cancer patients treated with RT have shown thinning or damage to the CC to be associated with dose-dependent radiation damage. These studies have also linked cognitive damage to the genu and splenium of the CC with reduced performance on tests of motor speed and processing, thus advocating for the potential of RT in the future to attempt to spare the genu and splenium during treatment of brain tumors to help mitigate long-term consequences of neuropsychological toxicity [66].

Apart from clinical findings, animal studies of brain irradiation have shown long-term consequences of RT leading to demyelination of the CC and behavioral changes including spatial memory loss, hyperactivity, and anxiety and depressive like symptoms. These studies have also shown imaging data associated with demyelination in the form of radial diffusivity on diffusion tensor imaging and are attempting to elucidate mechanisms of damage and potential adverse effects of RT [69, 70].

26.7 Conclusion

Iatrogenic injury to the CC is a part of the treatment of neurologic conditions whether it be surgical or medical. Recognizing the prevalence and risk factors associated with procedures and treatments as well as potential sequelae of CC injury is important for health-care professionals. By elucidating mechanisms of injury and associated findings, potential injury may be mitigated, and devastating consequences can potentially be avoided. Clinicians whose practice includes procedures such as third ventriculostomy with associated shunting, the transcallosal approach, and radiation therapy should be knowledgeable in potential complications and patient presentations associated with injury to the CC. Evaluation of patients before and after procedures may help mitigate risks associated with the procedures and provide safer outcomes.

References

1. Krishnan NR, Kasthuri AS. Iatrogenic disorders. Med J Armed Forces India. 2005;61(1):2–6. https://doi.org/10.1016/S0377-1237(05)80107-8.
2. Chesler DA, Pendleton C, Jallo GI, Quiñones-Hinojosa A. "Colossal" breakthrough: the callosal puncture as a precursor to third ventriculostomy. Minim Invasive Neurosurg. 2011;54:243–6. https://doi.org/10.1055/s-0031-1297249.
3. Archibald E. Puncture of the corpus callosum. Can Med Assoc J. 1913;3:451–62.
4. Dew S, Roberts D. Ventricular drainage by puncture of the corpus callosum: in acute obstructive hydrocephalus due to epidemic cerebrospinal meningitis. J Am Med Assoc. 1919;72:244–7. https://doi.org/10.1001/jama.1919.02610040010003.
5. Kohli G, Singh R, Herschman Y, Mammis A. Infection incidence associated with external Ventriculostomy placement: a comparison of outcomes in the emergency department, intensive care unit, and operating room. World Neurosurg. 2018;110:e135–40. https://doi.org/10.1016/j.wneu.2017.10.129.
6. Greenberg MS, Arredondo N. Handbook of neurosurgery. New York: Thieme; 2001.
7. Yadav YR, Parihar V, Pande S, et al. Endoscopic third ventriculostomy. J Neurosci Rural Pract. 2012;3:163–73. https://doi.org/10.4103/0976-3147.98222.
8. Demerdash A, Rocque BG, Johnston J, et al. Endoscopic third ventriculostomy: a historical review. Br J Neurosurg. 2017;31:28–32. https://doi.org/10.1080/02688697.2016.1245848.
9. Kakarla UK, Kim LJ, Chang SW, et al. Safety and accuracy of bedside external ventricular drain placement. Neurosurgery. 2008;63:ONS162-6; discussion ONS166-7. https://doi.org/10.1227/01.neu.0000335031.23521.d0.
10. Numaguchi Y, Kristt DA, Joy C, Robinson WL. Scalloping deformity of the corpus callosum

following ventricular shunting. Am J Neuroradiol. 1993;14:355–62.
11. Lane JI, Luetmer PH, Atkinson JL. Corpus callosal signal changes in patients with obstructive hydrocephalus after ventriculoperitoneal shunting. Am J Neuroradiol. 2001;22:158–62.
12. Ginat DT, Prabhu SP, Madsen JR. Postshunting corpus callosum swelling with depiction on tractography: case report. J Neurosurg Pediatr. 2013;11:178–80. https://doi.org/10.3171/2012.10.PEDS1242.
13. Su S, McArdle D, Gaillard F. Post-shunting corpus callosal signal change and review of the literature. J Clin Neurosci. 2020;72:466–8.
14. Huang H, Zhang J, Jiang H, et al. DTI tractography based parcellation of white matter: application to the mid-sagittal morphology of corpus callosum. NeuroImage. 2005;26:195–205. https://doi.org/10.1016/j.neuroimage.2005.01.019.
15. Bourekas EC, Varakis K, Bruns D, et al. Lesions of the corpus callosum: MR imaging and differential considerations in adults and children. AJR Am J Roentgenol. 2002;179:251–7. https://doi.org/10.2214/ajr.179.1.1790251.
16. Suh DY, Gaskill-Shipley M, Nemann MW, et al. Corpus callosal changes associated with hydrocephalus: a report of two cases. Neurosurgery. 1997;41:484–8. https://doi.org/10.1097/00006123-199708000-00037.
17. Spreer J, Ernestus RI, Lanfermann H, Lackner K. Lesions of the corpus callosum in hydrocephalic patients with ventricular drainage--a CT-study. Acta Neurochir. 1996;138:174–8. https://doi.org/10.1007/BF01411357.
18. Kasowski H, Piepmeier JM. Transcallosal approach for tumors of the lateral and third ventricles. Neurosurg Focus. 2001;10:E3. https://doi.org/10.3171/foc.2001.10.6.4.
19. Jandial R. 8 - Transcallosal approach. Core Techniques in Operative Neurosurgery. Philadelphia: Elsevier; 2020. p. 37–41.
20. Camacho A, Abernathey CD, Kelly PJ, Laws ERJ. Colloid cysts: experience with the management of 84 cases since the introduction of computed tomography. Neurosurgery. 1989;24:693–700. https://doi.org/10.1227/00006123-198905000-00006.
21. Apuzzo ML, Chikovani OK, Gott PS, et al. Transcallosal, interfornicial approaches for lesions affecting the third ventricle: surgical considerations and consequences. Neurosurgery. 1982;10:547–54. https://doi.org/10.1227/00006123-198205000-00001.
22. Apuzzo ML, Heifetz MD, Weiss MH, Kurze T. Neurosurgical endoscopy using the side-viewing telescope. J Neurosurg. 1977;46:398–400. https://doi.org/10.3171/jns.1977.46.3.0398.
23. Cabbell KL, Ross DA. Stereotactic microsurgical craniotomy for the treatment of third ventricular colloid cysts. Neurosurgery. 1996;38:301–7. https://doi.org/10.1097/00006123-199602000-00013.
24. Zaidel D, Sperry RW. Memory impairment after commissurotomy in man. Brain. 1974;97:263–72. https://doi.org/10.1093/brain/97.1.263.
25. Winston KR, Cavazzuti V, Arkins T. Absence of neurological and behavioral abnormalities after anterior transcallosal operation for third ventricular lesions. Neurosurgery. 1979;4:386–93. https://doi.org/10.1227/00006123-197905000-00003.
26. Symss NP, Ramamurthi R, Kapu R, et al. Complication avoidance in transcallosal transforaminal approach to colloid cysts of the anterior third ventricle: an analysis of 80 cases. Asian J Neurosurg. 2014;9:51–7. https://doi.org/10.4103/1793-5482.136708.
27. Jeeves MA, Simpson DA, Geffen G. Functional consequences of the transcallosal removal of intraventricular tumours. J Neurol Neurosurg Psychiatry. 1979;42:134–42. https://doi.org/10.1136/jnnp.42.2.134.
28. Nakasu Y, Isozumi T, Nioka H, Handa J. Mechanism of mutism following the transcallosal approach to the ventricles. Acta Neurochir. 1991;110:146–53. https://doi.org/10.1007/BF01400683.
29. Bellotti C, Pappadà G, Sani R, et al. The transcallosal approach for lesions affecting the lateral and third ventricles. Surgical considerations and results in a series of 42 cases. Acta Neurochir. 1991;111:103–7. https://doi.org/10.1007/BF01400496.
30. Hodges JR, Carpenter K. Anterograde amnesia with fornix damage following removal of IIIrd ventricle colloid cyst. J Neurol Neurosurg Psychiatry. 1991;54:633–8. https://doi.org/10.1136/jnnp.54.7.633.
31. Hernesniemi J, Leivo S. Management outcome in third ventricular colloid cysts in a defined population: a series of 40 patients treated mainly by transcallosal microsurgery. Surg Neurol. 1996;45:2–14. https://doi.org/10.1016/0090-3019(95)00379-7.
32. Fitsiori A, Nguyen D, Karentzos A, et al. The corpus callosum: white matter or terra incognita. Br J Radiol. 2011;84:5–18. https://doi.org/10.1259/bjr/21946513.
33. Ohnaka K, Watanabe T, Kaneko S, et al. Anterior transcallosal interforniceal approach to the third ventricle via a cavum septum pellucidum and cavum vergae. Interdiscip Neurosurg. 2022;27:101338. https://doi.org/10.1016/j.inat.2021.101338.
34. Villani R, Papagno C, Tomei G, et al. Transcallosal approach to tumors of the third ventricle. Surgical results and neuropsychological evaluation. J Neurosurg Sci. 1997;41:41–50.
35. Misra BK, Rout D, Padamadan J, Radhakrishnan VV. Transcallosal approach to anterior and mid-third ventricular tumors--a review of 62 cases. Ann Acad Med Singap. 1993;22:435–40.
36. Antunes JL, Louis KM, Ganti SR. Colloid cysts of the third ventricle. Neurosurgery. 1980;7:450–5. https://doi.org/10.1227/00006123-198011000-00004.
37. Aryan HE, Ozgur BM, Jandial R, Levy ML. Complications of interhemispheric transcallosal approach in children: review of 15 years experience.

Clin Neurol Neurosurg. 2006;108:790–3. https://doi.org/10.1016/j.clineuro.2005.10.009.
38. Bogen JE. Callosotomy without disconnection? J Neurosurg. 1994;81:328–9.
39. Timurkaynak E, Rhoton ALJ, Barry M. Microsurgical anatomy and operative approaches to the lateral ventricles. Neurosurgery. 1986;19:685–723. https://doi.org/10.1227/00006123-198611000-00001.
40. Bartynski WS. Posterior reversible encephalopathy syndrome, part 1: fundamental imaging and clinical features. AJNR Am J Neuroradiol. 2008;29:1036–42. https://doi.org/10.3174/ajnr.A0928.
41. Bartynski WS. Posterior reversible encephalopathy syndrome, part 2: controversies surrounding pathophysiology of vasogenic edema. AJNR Am J Neuroradiol. 2008;29:1043–9. https://doi.org/10.3174/ajnr.A0929.
42. Will AD, Lewis KL, Hinshaw DBJ, et al. Cerebral vasoconstriction in toxemia. Neurology. 1987;37:1555–7. https://doi.org/10.1212/wnl.37.9.1555.
43. Truwit CL, Denaro CP, Lake JR, DeMarco T. MR imaging of reversible cyclosporin A-induced neurotoxicity. Am J Neuroradiol. 1991;12:651–9.
44. Sanders TG, Clayman DA, Sanchez-Ramos L, et al. Brain in eclampsia: MR imaging with clinical correlation. Radiology. 1991;180:475–8. https://doi.org/10.1148/radiology.180.2.2068315.
45. Schwartz RB, Jones KM, Kalina P, et al. Hypertensive encephalopathy: findings on CT, MR imaging, and SPECT imaging in 14 cases. AJR Am J Roentgenol. 1992;159:379–83. https://doi.org/10.2214/ajr.159.2.1632361.
46. Rippe DJ, Edwards MK, Schrodt JF, et al. Reversible cerebral lesions associated with tiazofurin usage: MR demonstration. J Comput Assist Tomogr. 1988;12:1078–81. https://doi.org/10.1097/00004728-198811000-00037.
47. Bartynski WS, Boardman JF. Distinct imaging patterns and lesion distribution in posterior reversible encephalopathy syndrome. Am J Neuroradiol. 2007;28:1320–7. https://doi.org/10.3174/ajnr.A0549.
48. Stevens CJ, Heran MKS. The many faces of posterior reversible encephalopathy syndrome. Br J Radiol. 2012;85:1566–75. https://doi.org/10.1259/bjr/25273221.
49. Fugate JE, Claassen DO, Cloft HJ, et al. Posterior reversible encephalopathy syndrome: associated clinical and radiologic findings. Mayo Clin Proc. 2010;85:427–32. https://doi.org/10.4065/mcp.2009.0590.
50. McKinney AM, Short J, Truwit CL, et al. Posterior reversible encephalopathy syndrome: incidence of atypical regions of involvement and imaging findings. AJR Am J Roentgenol. 2007;189:904–12. https://doi.org/10.2214/AJR.07.2024.
51. Lee VH, Wijdicks EFM, Manno EM, Rabinstein AA. Clinical spectrum of reversible posterior leukoencephalopathy syndrome. Arch Neurol. 2008;65:205–10. https://doi.org/10.1001/archneurol.2007.46.
52. Saad S, Wang TJC. Neurocognitive deficits after radiation therapy for brain malignancies. Am J Clin Oncol. 2015;38:634–40. https://doi.org/10.1097/COC.0000000000000158.
53. Makale MT, McDonald CR, Hattangadi-Gluth JA, Kesari S. Mechanisms of radiotherapy-associated cognitive disability in patients with brain tumours. Nat Rev Neurol. 2017;13:52–64. https://doi.org/10.1038/nrneurol.2016.185.
54. Meyers CA, Brown PD. Role and relevance of neurocognitive assessment in clinical trials of patients with CNS tumors. J Clin Oncol Off J Am Soc Clin Oncol. 2006;24:1305–9. https://doi.org/10.1200/JCO.2005.04.6086.
55. Fling BW, Chapekis M, Reuter-Lorenz PA, et al. Age differences in callosal contributions to cognitive processes. Neuropsychologia. 2011;49:2564–9. https://doi.org/10.1016/j.neuropsychologia.2011.05.004.
56. Tuladhar AM, vanNorden AGW, de Laat KF, et al. White matter integrity in small vessel disease is related to cognition. NeuroImage Clin. 2015;7:518–24. https://doi.org/10.1016/j.nicl.2015.02.003.
57. Sorg SF, Schiehser DM, Bondi MW, et al. White matter microstructural compromise is associated with cognition but not posttraumatic stress disorder symptoms in military veterans with traumatic brain injury. J Head Trauma Rehabil. 2016;31:297–308. https://doi.org/10.1097/HTR.0000000000000189.
58. Yaldizli Ö, Penner I-K, Frontzek K, et al. The relationship between total and regional corpus callosum atrophy, cognitive impairment and fatigue in multiple sclerosis patients. Mult Scler. 2014;20:356–64. https://doi.org/10.1177/1352458513496880.
59. Wu Y, Storey P, Cohen BA, et al. Diffusion alterations in corpus callosum of patients with HIV. AJNR Am J Neuroradiol. 2006;27:656–60.
60. Agosta F, Canu E, Stefanova E, et al. Mild cognitive impairment in Parkinson's disease is associated with a distributed pattern of brain white matter damage. Hum Brain Mapp. 2014;35:1921–9. https://doi.org/10.1002/hbm.22302.
61. Asai A, Matsutani M, Kohno T, et al. Subacute brain atrophy after radiation therapy for malignant brain tumor. Cancer. 1989;63:1962–74. https://doi.org/10.1002/1097-0142(19890515)63:10<1962::aid-cncr2820631016>3.0.co;2-v.
62. Nieman BJ, de Guzman AE, Gazdzinski LM, et al. White and gray matter abnormalities after cranial radiation in children and mice. Int J Radiat Oncol Biol Phys. 2015;93:882–91. https://doi.org/10.1016/j.ijrobp.2015.07.2293.
63. Reddick WE, Russell JM, Glass JO, et al. Subtle white matter volume differences in children treated for medulloblastoma with conventional or reduced dose craniospinal irradiation. Magn Reson

Imaging. 2000;18:787–93. https://doi.org/10.1016/s0730-725x(00)00182-x.

64. Agbahiwe H, Rashid A, Horska A, et al. A prospective study of cerebral, frontal lobe, and temporal lobe volumes and neuropsychological performance in children with primary brain tumors treated with cranial radiation. Cancer. 2017;123:161–8. https://doi.org/10.1002/cncr.30313.
65. Huynh-Le M-P, Tibbs MD, Karunamuni R, et al. Microstructural injury to corpus callosum and intrahemispheric white matter tracts correlate with attention and processing speed decline after brain radiation. Int J Radiat Oncol Biol Phys. 2021;110:337–47. https://doi.org/10.1016/j.ijrobp.2020.12.046.
66. Redmond KJ, Hildreth M, Sair HI, et al. Association of neuronal injury in the genu and body of corpus callosum after cranial irradiation in children with impaired cognitive control: a prospective study. Int J Radiat Oncol Biol Phys. 2018;101:1234–42. https://doi.org/10.1016/j.ijrobp.2018.04.037.
67. Nagesh V, Tsien CI, Chenevert TL, et al. Radiation-induced changes in normal-appearing white matter in patients with cerebral tumors: a diffusion tensor imaging study. Int J Radiat Oncol Biol Phys. 2008;70:1002–10. https://doi.org/10.1016/j.ijrobp.2007.08.020.
68. Connor M, Karunamuni R, McDonald C, et al. Regional susceptibility to dose-dependent white matter damage after brain radiotherapy. Radiother Oncol. 2017;123:209–17. https://doi.org/10.1016/j.radonc.2017.04.006.
69. Song S-K, Yoshino J, Le TQ, et al. Demyelination increases radial diffusivity in corpus callosum of mouse brain. NeuroImage. 2005;26:132–40. https://doi.org/10.1016/j.neuroimage.2005.01.028.
70. Constanzo J, Midavaine É, Fouquet J, et al. Brain irradiation leads to persistent neuroinflammation and long-term neurocognitive dysfunction in a region-specific manner. Prog Neuropsychopharmacol Biol Psychiatry. 2020;102:109954. https://doi.org/10.1016/j.pnpbp.2020.109954.

Other Lesions of the Corpus Callosum

27

Alican Tahta and Mehmet Turgut

27.1 Corpus Callosum Atrophy or Tissue Loss in Chronic Alcoholism

Chronic alcohol abuse is a universal health problem. It can cause various social and economic problems. It is a condition that can be seen in the elderly and young people as well [1, 2]. It constitutes a significant proportion of hospital admissions and leads to cognitive, motor, and social behavioral consequences [3]. Alcohol abuse was found in 46% of individuals in a study evaluating patients with intellectual disability [4] and in 42% of individuals in a study evaluating patients with severe behavioral problems [5]. Psychiatric disorders, physical disabilities, and developmental handicaps have been reported to be more common in alcohol abusers [6]. Pathological findings can be detected in various parts of the brain in chronic alcohol usage.

Neuronal cell loss is usually seen in the frontal cortex in alcoholics [7]. In addition, volume loss and demyelination can also be seen in white matter [8, 9]. Demyelination, microtubule disruption, and neuronal loss in callosal, supratentorial, and infratentorial white matter have been demonstrated in animal models and postmortem studies of patients with chronic alcoholism [8, 10]. In vivo MRI studies in alcoholics have shown that volume loss can be seen in the corpus callosum (CC) and centrum semiovale in both men and women [11, 12]. The concern of which sex has a more negative impact on chronic heavy alcohol use is controversial. Some studies have shown that there is no significant difference between men and women in terms of the CC being affected [13, 14]. Some studies have shown that males are affected more [15, 16].

Various mechanisms have been proposed to explain alcoholic brain injury. These include the direct neurotoxic effect of alcohol (alterations in calcium homeostasis, degradation of components in the plasma membrane, production of reactive oxygen species and reduction of antioxidants), liver-related complications, thiamine deficiency, and homocysteine neurotoxicity [17–20] (Table 27.1).

Studies have shown that alcohol does not affect whiter matter diffusely and causes more damage to specific tracts [13]. The reason for this is thought to be the heterogeneous distribution of biochemical constituents in white matter. In the study by Kashem et al., different protein expression profiles were determined in two different white matter areas (splenium vs. BA9). They showed that 13 of the 36 proteins detected in the

A. Tahta (✉)
Department of Neurosurgery, İstanbul Medipol University Faculty of Medicine, İstanbul, Turkey
e-mail: atahta@medipol.edu.tr

M. Turgut
Department of Neurosurgery, Aydın Adnan Menderes University Faculty of Medicine, Aydın, Turkey

Department of Histology and Embryology, Aydın Adnan Menderes University Health Sciences Institute, Aydın, Turkey

M. Turgut et al. (eds.), *The Corpus Callosum*, https://doi.org/10.1007/978-3-031-38114-0_27

Table 27.1 Mechanisms in alcoholic brain injury

Direct effects	Indirect effects
Alterations in calcium homeostasis	Liver-related complications
Degradation in plasma membrane	Thiamine deficiency
Production of reactive oxygen species	Homocysteine neurotoxicity
Reduction of antioxidants	

BA9 region were expressed in a different way in the splenium. These findings suggested that these two brain areas, although they are both white matters, react differently to chronic alcohol use at the level of protein expression [21].

Deterioration in the integrity of specific fiber tracts can be detected in DTI, although no structural abnormality is detected in conventional MRI examinations in alcoholics [22, 23]. Pfefferbaum et al. found that there was a significant decrease in tissue quality along with the CC volume [14]. In addition, macrostructural and microstructural alterations were detected in the CC [14].

Alcohol primarily affects the genu of the CC and then the splenium [13, 14]. In addition, it has been reported that older alcoholics show more thinning of the genu and splenium than younger alcoholics compared to their age group [14]. This is thought to be due to myelin thinning and microtubule deletion. Therefore, the healing capacity of the brain will decrease with the normal aging process, and alcohol will cause further damage to the elderly brain compared to the young one [10, 14, 24, 25].

Fibers connecting the right and left frontal areas (prefrontal, premotor) pass through the genu, and fibers connecting the parietal, temporal, and occipital areas pass through the splenium. Therefore, it is expected that working memory will be affected due to pathologies in the genu, and visuospatial ability will be affected due to the pathologies in the splenium [14]. Working memory, visuospatial abilities, and gait and balance are typically affected in chronic alcoholics. Cognitive and motor deficits are milder in alcoholics who do not have conditions such as Korsakoff's syndrome or Marchiafava-Bignami disease and central pontine myelinolysis [26–28]. The fact that these cortical-based tasks are affected in alcoholics is due to the pathologies in white matter. Demyelination is more effective than axonal degeneration in the deterioration of white matter integrity [8]. Therefore, with the involvement of the CC, the connectivity of the cortical regions responsible for these tasks is impaired and clinical symptoms arise [29]. In individuals with impaired gait and balance, there is more global damage and thinning of the CC [30, 31].

It has been shown that the prefrontal, premotor, parietal, temporal, and occipital bundles in the CC are more affected in alcoholism than the sensory-motor bundle [16]. Pathologies related to working memory, visuospatial abilities, and balance improve with the cessation of alcohol consumption [26, 32]. These symptoms regress with increase in the volume of brain tissue and CC [33].

27.2 Corpus Callosum in High-Altitude Cerebral Edema

Low atmospheric pressure causes low partial oxygen pressure in the human body. This can cause acute mountain sickness by making it difficult to acclimatize to high altitudes. Acute mountain sickness can be seen in one-fourth of the individuals who climb to the moderate altitude (1920–2956 m) from sea level [34]. Typical clinical findings of patients with acute mountain sickness are headache, anorexia, dizziness, and insomnia [35].

High-altitude cerebral edema (HACE) is a life-threatening condition seen in individuals rapidly exposed to high altitudes [36]. It is rare in climbing below 3000 m. In the study of Hackett and Rennie, HACE was detected in 1% of the mountaineers who climbed 4243–5500 m in Nepal [37]. In the study of Bärtsch et al., HACE was found in 0.5% of mountaineers in Tibet [38].

The pathophysiology of HACE is related to vasogenic edema. Mild vasogenic edema which is related to increased cerebral perfusion can be seen in individuals rapidly climbing 3000–4000 m [39]. Hemorrhagic conversion is the most important feature that distinguishes HACE from

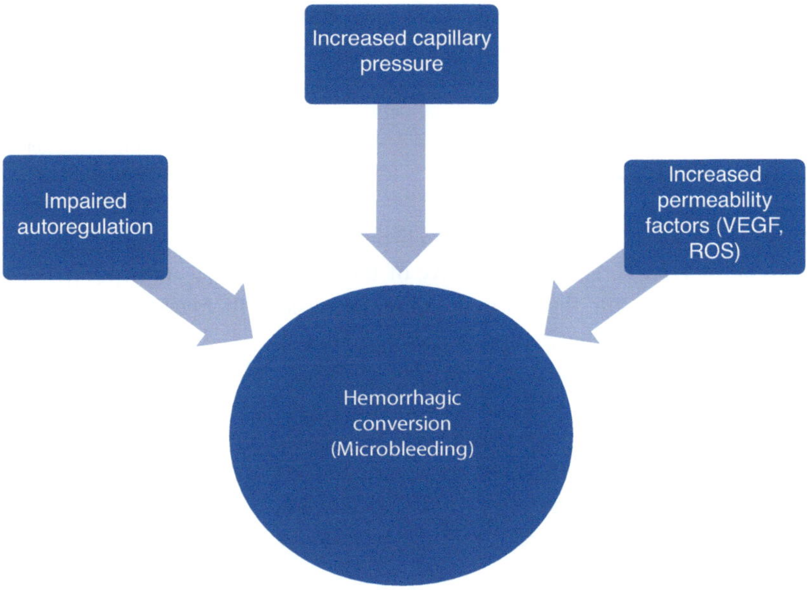

Fig. 27.1 Factors affecting hemorrhagic conversion (microbleeding) in the CC

acute mountain sickness. Vasogenic edema transforms to hemorrhagic conversion (extravasation of red blood cells leading to increased intracranial pressure) when HACE develops [40]. Impaired autoregulation, increased capillary pressure, and increased permeability factors (vascular endothelial growth factor, reactive oxygen species) cause hemorrhagic conversion by disrupting the blood-brain barrier [39, 41] (Fig. 27.1). Extravasation of red blood cells is found in high-altitude pulmonary edema due to the same pathophysiological process [42].

Microbleeding in white matter is not a typical finding for high-altitude exposure. Microbleeding is not expected unless HACE develops in patients [43]. Microbleedings appear when white matter microvascular integrity is disrupted [43]. Microbleedings in white matter and CC (especially in the splenium) are typical MRI findings of HACE [44, 45]. The CC is supplied by small and short perforating arteries, making this area more susceptible to hypoxic vasodilation and autoregulatory failure. This causes hyper perfusion that leads to microbleedings more than other parts of the brain. Also, the CC (especially the splenium) is thought to be more sensitive because it contains more glutamate and cytokine receptors [46]. Microbleedings can be seen more clearly on SWI in the early period [43]. These findings start within the first week after HACE and may remain detectable for years in MRI [43, 47].

Oxygen, dexamethasone, and hyperbaric oxygen chamber are used in the treatment of HACE. Death may occur within 24–48 h due to brain herniation despite adequate and appropriate treatment [48].

27.3 Virchow-Robin Spaces in the Corpus Callosum

Enlarged perivascular spaces, called Virchow-Robin spaces (VRS), are cavities surrounding small penetrating cerebral arterioles filled with cerebrospinal fluid [49]. VRS are seen as high-signal areas in basal ganglia and centrum semiovale on T2-weighted MRIs.

VRS can also be found in the CC [50]. They are detected as multiple, small, and ovoid-looking cystic lesions on FLAIR images [49]. Fluid accumulation is seen in VRS with the same intensity as CSF. There is no significant hyperintensity in the walls of these lesions. VRS can be detected at all ages. However, they are more common in the elderly [50]. There are studies showing that VRS are associated with lacunar ischemic stroke [51]. VRS are not expected to cause any neurological findings in individuals. VRS are more common in patients with mucopolysaccharidosis [52].

27.4 Conclusion

Lesions in the CC can be seen in chronic alcoholism or HACE. Healing capacity of the brain will decrease with the normal aging process, and alcohol will cause further damage to the elderly brain compared to the young brain. Microbleeding in the CC is a typical finding in MRI of patients suffering HACE. These lesions can be detected on MRI even years after alcohol cessation or treatment of HACE. VRS are benign lesions and more common lesions in mucopolysaccharidosis.

References

1. Dawson DA, Grant BF, Stinson FS, Chou PS. Another look at heavy episodic drinking and alcohol use disorders among college and noncollege youth. J Stud Alcohol. 2004;65:477–88. https://doi.org/10.15288/jsa.2004.65.477.
2. Grant BF, Dawson DA, Stinson FS, Chou SP, Dufour MC, Pickering RP. The 12-month prevalence and trends in DSM-IV alcohol abuse and dependence: United States, 1991-1992 and 2001-2002. Drug Alcohol Depend. 2004;74:223–34. https://doi.org/10.1016/j.drugalcdep.2004.02.004.
3. Arciniegas DB, Beresford TP. Neuropsychiatry: an introductory approach. New York: Cambridge University Press; 2001.
4. O'Phelan K, McArthur DL, Chang CWJ, Green D, Hovda DA. The impact of substance abuse on mortality in patients with severe traumatic brain injury. J Trauma. 2008;65:674–7. https://doi.org/10.1097/TA.0b013e31817db0a5.
5. Didden R, Embregts P, van der Toorn M, Laarhoven N. Substance abuse, coping strategies, adaptive skills and behavioral and emotional problems in clients with mild to borderline intellectual disability admitted to a treatment facility: a pilot study. Res Dev Disabil. 2009;30:927–32. https://doi.org/10.1016/j.ridd.2009.01.002.
6. Tyas S, Rush B. The treatment of disabled persons with alcohol and drug problems: results of a survey of addiction services. J Stud Alcohol. 1993;54:275–82. https://doi.org/10.15288/jsa.1993.54.275.
7. Courville CB. Effects of alcohol on the nervous system of man. Los Angeles: San Lucas Press; 1966.
8. Lewohl JM, Wang L, Miles MF, Zhang L, Dodd PR, Harris RA. Gene expression in human alcoholism: microarray analysis of frontal cortex. Alcohol Clin Exp Res. 2000;24:1873–82.
9. Putzke J, De Beun R, Schreiber R, De Vry J, Tölle TR, Zieglgänsberger W, Spanagel R. Long-term alcohol self-administration and alcohol withdrawal differentially modulate microtubule-associated protein 2 (MAP2) gene expression in the rat brain. Brain Res Mol Brain Res. 1998;62:196–205. https://doi.org/10.1016/s0169-328x(98)00253-8.
10. Mayfield RD, Lewohl JM, Dodd PR, Herlihy A, Liu J, Harris RA. Patterns of gene expression are altered in the frontal and motor cortices of human alcoholics. J Neurochem. 2002;81:802–13. https://doi.org/10.1046/j.1471-4159.2002.00860.x.
11. Pfefferbaum A, Sullivan EV, Hedehus M, Adalsteinsson E, Lim KO, Moseley M. In vivo detection and functional correlates of white matter microstructural disruption in chronic alcoholism. Alcohol Clin Exp Res. 2000;24:1214–21.
12. Sullivan EV, Pfefferbaum A. Alcohol use disorder: neuroimaging evidence for accelerated aging of brain morphology and hypothesized contribution to age-related dementia. Alcohol. 2022;107:44. https://doi.org/10.1016/j.alcohol.2022.06.002.
13. Hommer D, Momenan R, Kaiser E, Rawlings R. Evidence for a gender-related effect of alcoholism on brain volumes. Am J Psychiatry. 2001;158:198–204. https://doi.org/10.1176/appi.ajp.158.2.198.
14. Pfefferbaum A, Adalsteinsson E, Sullivan EV. Dysmorphology and microstructural degradation of the corpus callosum: interaction of age and alcoholism. Neurobiol Aging. 2006;27:994–1009. https://doi.org/10.1016/j.neurobiolaging.2005.05.007.
15. Pfefferbaum A, Rosenbloom M, Serventi KL, Sullivan EV. Corpus callosum, pons, and cortical white matter in alcoholic women. Alcohol Clin Exp Res. 2002;26:400–6.
16. Pfefferbaum A, Rosenbloom MJ, Fama R, Sassoon SA, Sullivan EV. Transcallosal white matter degradation detected with quantitative fiber tracking in alcoholic men and women: selective relations to dissociable functions. Alcohol Clin Exp Res. 2010;34:1201–11. https://doi.org/10.1111/j.1530-0277.2010.01197.x.
17. Azuine MA, Patel SJ, Lakshman MR. Effects of chronic ethanol administration on the activities and relative synthetic rates of myelin and synaptosomal plasma membrane-associated sialidase in the rat brain. Neurochem Int. 2006;48:67–74. https://doi.org/10.1016/j.neuint.2005.07.005.
18. Harper C, Matsumoto I. Ethanol and brain damage. Curr Opin Pharmacol. 2005;5:73–8. https://doi.org/10.1016/j.coph.2004.06.011.
19. Rouach H, Houzé P, Gentil M, Orfanelli MT, Nordmann R. Changes in some pro- and antioxidants in rat cerebellum after chronic alcohol intake. Biochem Pharmacol. 1997;53:539–45. https://doi.org/10.1016/s0006-2952(96)00770-8.

20. Zieminska E, Matyja E, Kozlowska H, Stafiej A, Lazarewicz JW. Excitotoxic neuronal injury in acute homocysteine neurotoxicity: role of calcium and mitochondrial alterations. Neurochem Int. 2006;48:491–7. https://doi.org/10.1016/j.neuint.2005.12.023.
21. Kashem MA, James G, Harper C, Wilce P, Matsumoto I. Differential protein expression in the corpus callosum (splenium) of human alcoholics: a proteomics study. Neurochem Int. 2007;50:450–9. https://doi.org/10.1016/j.neuint.2006.10.009.
22. Lehéricy S, Ducros M, Van de Moortele P-F, Francois C, Thivard L, Poupon C, Swindale N, Ugurbil K, Kim D-S. Diffusion tensor fiber tracking shows distinct corticostriatal circuits in humans. Ann Neurol. 2004;55:522–9. https://doi.org/10.1002/ana.20030.
23. Stieltjes B, Kaufmann WE, van Zijl PC, Fredericksen K, Pearlson GD, Solaiyappan M, Mori S. Diffusion tensor imaging and axonal tracking in the human brainstem. NeuroImage. 2001;14:723–35. https://doi.org/10.1006/nimg.2001.0861.
24. de la Monte SM. Disproportionate atrophy of cerebral white matter in chronic alcoholics. Arch Neurol. 1988;45:990–2. https://doi.org/10.1001/archneur.1988.00520330076013.
25. Wiggins RC, Gorman A, Rolsten C, Samorajski T, Ballinger WE, Freund G. Effects of aging and alcohol on the biochemical composition of histologically normal human brain. Metab Brain Dis. 1988;3:67–80. https://doi.org/10.1007/BF01001354.
26. Fein G, Bachman L, Fisher S, Davenport L. Cognitive impairments in abstinent alcoholics. West J Med. 1990;152:531–7.
27. Sullivan EV, Fama R, Rosenbloom MJ, Pfefferbaum A. A profile of neuropsychological deficits in alcoholic women. Neuropsychology. 2002;16:74–83. https://doi.org/10.1037//0894-4105.16.1.74.
28. Sullivan EV, Rosenbloom MJ, Pfefferbaum A. Pattern of motor and cognitive deficits in detoxified alcoholic men. Alcohol Clin Exp Res. 2000;24:611–21.
29. Cabeza R, Anderson ND, Locantore JK, McIntosh AR. Aging gracefully: compensatory brain activity in high-performing older adults. NeuroImage. 2002;17:1394–402. https://doi.org/10.1006/nimg.2002.1280.
30. Baloh RW, Ying SH, Jacobson KM. A longitudinal study of gait and balance dysfunction in normal older people. Arch Neurol. 2003;60:835–9. https://doi.org/10.1001/archneur.60.6.835.
31. Cahn DA, Malloy PF, Salloway S, Rogg J, Gillard E, Kohn R, Tung G, Richardson ED, Westlake R. Subcortical hyperintensities on MRI and activities of daily living in geriatric depression. J Neuropsychiatry Clin Neurosci. 1996;8:404–11. https://doi.org/10.1176/jnp.8.4.404.
32. Moselhy HF, Georgiou G, Kahn A. Frontal lobe changes in alcoholism: a review of the literature. Alcohol Alcohol. 2001;36:357–68. https://doi.org/10.1093/alcalc/36.5.357.
33. O'Neill J, Cardenas VA, Meyerhoff DJ. Effects of abstinence on the brain: quantitative magnetic resonance imaging and magnetic resonance spectroscopic imaging in chronic alcohol abuse. Alcohol Clin Exp Res. 2001;25:1673–82.
34. Honigman B, Theis MK, Koziol-McLain J, Roach R, Yip R, Houston C, Moore LG, Pearce P. Acute mountain sickness in a general tourist population at moderate altitudes. Ann Intern Med. 1993;118:587–92. https://doi.org/10.7326/0003-4819-118-8-199304150-00003.
35. Singh I, Khanna PK, Srivastava MC, Lal M, Roy SB, Subramanyam CS. Acute mountain sickness. N Engl J Med. 1969;280:175–84. https://doi.org/10.1056/NEJM196901232800402.
36. Hackett PH, Roach RC. High-altitude illness. N Engl J Med. 2001;345:107.
37. Hackett PH, Rennie D, Levine HD. The incidence, importance, and prophylaxis of acute mountain sickness. Lancet. 1976;2:1149–55. https://doi.org/10.1016/s0140-6736(76)91677-9.
38. Bärtsch P, Swenson ER. Acute high-altitude illnesses. N Engl J Med. 2013;369:1666–7. https://doi.org/10.1056/NEJMc1309747.
39. Wilson MH, Newman S, Imray CH. The cerebral effects of ascent to high altitudes. Lancet Neurol. 2009;8:175–91. https://doi.org/10.1016/S1474-4422(09)70014-6.
40. Stokum JA, Gerzanich V, Simard JM. Molecular pathophysiology of cerebral edema. J Cereb Blood Flow Metab. 2016;36:513–38. https://doi.org/10.1177/0271678X15617172.
41. Hackett PH, Roach RC. High altitude cerebral edema. High Alt Med Biol. 2004;5:136–46. https://doi.org/10.1089/1527029041352054.
42. Swenson ER, Maggiorini M, Mongovin S, Gibbs JSR, Greve I, Mairbäurl H, Bärtsch P. Pathogenesis of high-altitude pulmonary edema: inflammation is not an etiologic factor. JAMA. 2002;287:2228–35. https://doi.org/10.1001/jama.287.17.2228.
43. Hackett PH, Yarnell PR, Weiland DA, Reynard KB. Acute and evolving MRI of high-altitude cerebral edema: microbleeds, edema, and pathophysiology. AJNR Am J Neuroradiol. 2019;40:464–9. https://doi.org/10.3174/ajnr.A5897.
44. Hackett PH, Yarnell PR, Hill R, Reynard K, Heit J, McCormick J. High-altitude cerebral edema evaluated with magnetic resonance imaging: clinical correlation and pathophysiology. JAMA. 1998;280:1920–5. https://doi.org/10.1001/jama.280.22.1920.
45. Yarnell PR, Heit J, Hackett PH. High-altitude cerebral edema (HACE): the Denver/Front range experience. Semin Neurol. 2000;20:209–17. https://doi.org/10.1055/s-2000-9830.
46. Starkey J, Kobayashi N, Numaguchi Y, Moritani T. Cytotoxic lesions of the corpus callosum that show restricted diffusion: mechanisms, causes, and manifestations. Radiogr Rev. 2017;37:562–76. https://doi.org/10.1148/rg.2017160085.
47. Kallenberg K, Dehnert C, Dörfler A, Schellinger PD, Bailey DM, Knauth M, Bärtsch PD. Microhemorrhages in nonfatal high-altitude

cerebral edema. J Cereb Blood Flow Metab. 2008;28:1635–42. https://doi.org/10.1038/jcbfm.2008.55.
48. Luks AM, Auerbach PS, Freer L, Grissom CK, Keyes LE, McIntosh SE, Rodway GW, Schoene RB, Zafren K, Hackett PH. Wilderness medical society clinical practice guidelines for the prevention and treatment of acute altitude illness: 2019 update. Wilderness Environ Med. 2019;30:S3–S18. https://doi.org/10.1016/j.wem.2019.04.006.
49. Braffman BH, Zimmerman RA, Trojanowski JQ, Gonatas NK, Hickey WF, Schlaepfer WW. Brain MR: pathologic correlation with gross and histopathology. 1. Lacunar infarction and Virchow-Robin spaces. Am J Neuroradiol. 1988;9:621–8.
50. Heier LA, Bauer C, Schwartz L, Zimmerman R, Morgello S, Deck M. Large Virchow-Robin spaces: MR-clinical correlation. Am J Neuroradiol. 1989;10:929–36.
51. Doubal FN, MacLullich AMJ, Ferguson KJ, Dennis MS, Wardlaw JM. Enlarged perivascular spaces on MRI are a feature of cerebral small vessel disease. Stroke. 2010;41:450–4. https://doi.org/10.1161/STROKEAHA.109.564914.
52. Uchino A, Takase Y, Nomiyama K, Egashira R, Kudo S. Acquired lesions of the corpus callosum: MR imaging. Eur Radiol. 2006;16:905–14. https://doi.org/10.1007/s00330-005-0037-9.

28 Corpus Callosum in Hydrocephalus

Mehmet Saim Kazan and Ahmet Özak

28.1 Introduction

The corpus callosum (CC), situated in the center of the human brain, is the largest white matter structure in the brain, consisting of over 190 million contralateral axonal projections and the major commissural pathway connecting the cerebral hemispheres [1].

The basic structure of the CC begins to develop at about the seventh week of gestation and continues through the 20th week. At this juncture, the CC has its essential appearance; however, it continues to thicken during the rest of gestation and after birth. Other than the rostral part, the CC develops from an anterior to posterior direction. The rostrum (the most anterior segment of CC) develops between 18 and 20 weeks of gestation. A disturbance in the process can cause a range of possible congenital defects, including partial or complete agenesis of CC [2].

The word "commissure" derives its origin from the Latin word *commissura*, which means joining or connecting together. In the human brain, a "commissure" is defined as the junction where two anatomical compartments meet. The CC is the largest commissure (which comprises of transverse running fibers extending from the rostrum to the splenium) in the brain and connects the frontal, parietal, temporal, occipital, insular, and limbic lobes and the basal ganglia of the hemispheres to each other in the depth of the interhemispheric fissure [3–5].

28.2 Relationship Between Ventricular System and Corpus Callosum

The CC envelops the entire lateral ventricles within its callosal radiations. The rostrum of the CC forms the floor of the frontal horn, and the genu forms the anteromedial boundary of the frontal part the of the lateral ventricles [6]. The fibers of the forceps minor curve forward and form the medial wall of the frontal horns. The forceps major forms the medial wall of the occipital horns of the lateral ventricle.

The body of the CC forms the roof of the body of the lateral ventricles. The ventral callosal radiations of the anterior portion of the body extend laterally in the roof of the frontal horn to reach the heads of the nuclei. The ventral callosal radiations from the posterior part of the body of CC (tapetal fibers) extend laterally-inferiorly and form the lateral wall of the temporal horns, atria, and occipital horns, thus enveloping these parts of the ventricles [3].

M. S. Kazan (✉) · A. Özak
Department of Neurosurgery, Akdeniz University
School of Medicine, Antalya, Turkey

M. Turgut et al. (eds.), *The Corpus Callosum*, https://doi.org/10.1007/978-3-031-38114-0_28

28.3 Corpus Callosum in Hydrocephalus and After Shunt Surgery

Hydrocephalus is the consequence of various causes such as congenital malformations, intracerebral/subarachnoid hemorrhage, infection, traumatic brain injury, and benign/malignant brain tumors. Regardless of etiology, the most common feature among the patients is obstruction of cerebrospinal fluid (CSF) interfering with the function of the healthy brain. The symptoms are somewhat different depending on whether the patient is an infant, child, or diagnosed with the hydrocephalus at an older age. The expansion of the ventricular system influences surrounding areas such as CC and the corticospinal fiber tracts [7].

Depending on the age at which hydrocephalus develops together with the size of ventricular expansion, the neuropathological pattern has various presentations. Prolonged mechanical distortion may cause more advanced neuropathological signs such as damage to the axons and myelin. Also, when the arteries and capillaries of the cerebral circulation are substantially distorted, there is a potential for reduced regional cerebral blood flow [8–10].

The CC is a band of compact white matter composed of transversely oriented fibers, most of the axons crossing directly to homologous areas in the opposite cerebral hemisphere. A number of axons, however, travel obliquely to reach contralateral association areas [11, 12]. CC abnormalities in normal pressure hydrocephalus include changes in the morphology and the magnetic resonance imaging (MRI) signal. The most consistent findings are stretching, uniform and focal/diffuse thinning, and upward elevation of the CC (Fig. 28.1). Hydrocephalus-associated CC damage has been primarily attributed to lateral ventricle dilatation and to the impingement of the CC against the falx cerebri [13, 14].

In hydrocephalus, axonal degeneration is the major neuropathological feature. White matter destruction is due to a combination of mechanical distortion (injury), impaired regional blood flow, and accumulation of neuronal waste products in the CSF [15, 16]. Callosal damage is probably a consequence of injury in and around the CC. Direct mechanical primary CC injury is caused by ventricular compression and impingement against the falx (callosal impingement), and secondary callosal damage is a result of periventricular white matter damage that causes axonal loss and Wallerian degeneration [16, 17].

Elevation of the CC by the dilated ventricles in hydrocephalus can lead to the falx cerebri impinging especially on the middle and posterior thirds of the CC in the midline producing notch-

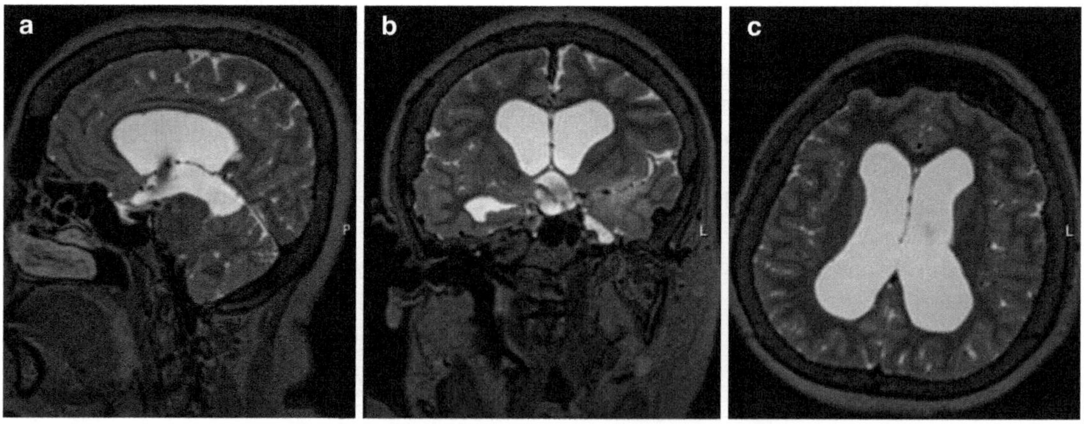

Fig. 28.1 Ventriculomegaly and upward displaced, diffuse thin CC on T2-weighted sagittal (**a**), coronal (**b**), and axial (**c**) section. Relationship between ventricular system and CC in patients with normal pressure hydrocephalus before shunt surgery. Coronal T2-weighted section shows stretching of fibers of CC over posterior falx (**b**)

like deformities [13, 18]. The mechanism is thought to be a mechanical ischemia, and an animal model has revealed a reduction in the number of capillaries [19]. Anterior scalloping deformities of CC are considered to be associated with over-shunting due to ventricle collapse and stretching of perforating branches of the pericallosal arteries by transient edema [20].

In 1995, Hofmann et al. showed the visual impression that with increasing hydrocephalus the CC extends mainly superiorly and, to a lesser degree, in the antero-posterior direction. In noncommunicating hydrocephalus (obstructive hydrocephalus), displacement of the CC was significantly more pronounced than in communicating hydrocephalus [13].

Jinkins et al. differentiated between uniform callosal and focal thinning of the posterior body of the CC in patients with symptomatic hydrocephalus (Fig. 28.2). Uniform thinning was frequently associated with asymptomatic or occult hydrocephalus; however, focal thinning and/or dorsal flattening was commonly seen with symptomatic hydrocephalus [21].

Numaguchi et al. [2] described the main causes of transient scalloping deformity of the dorsal surface of the CC in 6 of 35 patients who underwent shunting for hydrocephalus. Transient scalloping deformity exclusively involves the body of the CC and is associated with decreased signal intensity on T1-weighted MR images. Decreased signal intensity in the CC reflects edema, gliosis, and/or loss of myelinated fibers. These pathological changes are postulated to result from a combination of factors, including compression of the CC and its related structures secondary to hydrocephalus and decompression of these structures after shunting [22]. Long-standing hydrocephalus leads to thinning and softening of the CC via a process of edema and/or loss of myelinated fibers and replacement by reactive astrocytes. There is also stretching of the pericallosal artery and its branches, which tether the CC to the overlying cingulate cortex. After shunt surgery, the reduction in pressure on the ventral surface of the CC results in a collapse of the CC ventrally. The ventricular collapse is segmented between sites where rami of the pericallosal artery enter and tether the CC, thereby developing a "scalloped appearance" [22, 23].

Spreer et al. noted "hypodensities in the anterior part of the CC" on CT scans in 7 of 79 patients after ventriculoperitoneal shunting. The most common causative factor in their series was aqueductal stenosis (noncommunicating hydrocephalus). They associated the presence of

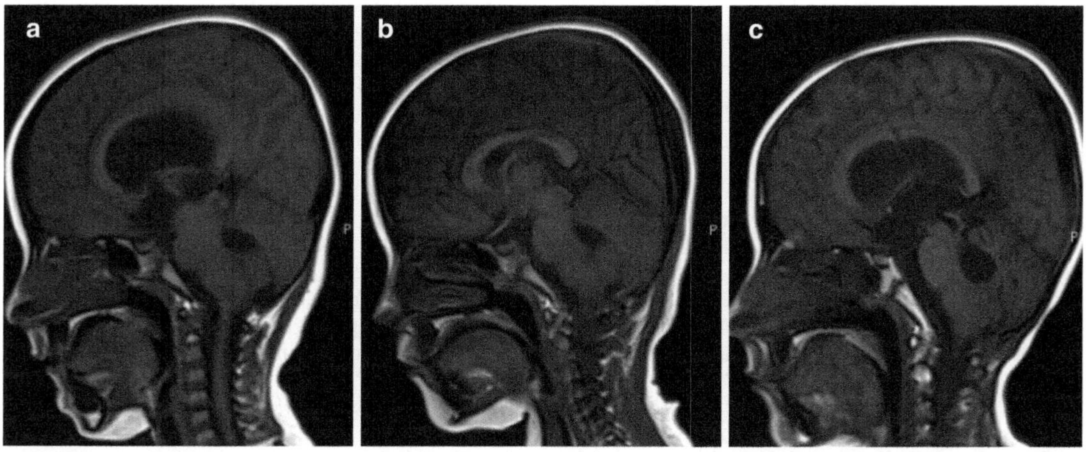

Fig. 28.2 (**a**) Preoperative sagittal T1-weighted MRI scan reveals dilatation of third and lateral ventricles, with mild curvilinear flattening and thinning of dorsal surface of the posterior body of the CC. (**b**) Early period after shunt surgery, a sagittal T1-weighted MRI scan demonstrates normalization of the lateral ventricles. (**c**) Sagittal T1-weighted MRI scans, performed 2 years after ventricular shunting, reveal persistent thickening and unchanged signal within the CC

callosal hypodensity with "forced ventricular drainage," noting that all seven patients had extreme narrowing of the ventricles and/or subdural hygromas. They invoked the theory of Numaguchi et al. that callosal scalloping was indicative of tethering by branches of the pericallosal artery but furthermore contended that this tethering, exacerbated by overshunting, caused a degree of traction upon these callosal perforators sufficient to induce postshunt ischemia [18, 22, 23].

References

1. Paul LK, Brown WS, Adolphs R, Tyszka JM, Richards LJ, Mukherjee P, et al. Agenesis of the corpus callosum: genetic, developmental and functional aspects of connectivity. Nat Rev Neurosci. 2007;8(4):287–99.
2. Klaas PA, Hannay JH, Caroselli JS, Fletcher JM. Interhemispheric transfer of visual, auditory, tactile, and visuomotor information in children with hydrocephalus and partial agenesis of the corpus callosum. J Clin Exp Neuropsychol. 1999;21(6):837–50.
3. Shah A, Jhawar S, Goel A, Goel A. Corpus callosum and its connections: a fiber dissection study. World Neurosurg. 2021;151:e1024–e35.
4. Ten Tusscher MPM, Houtman AC, De Mey J, Van Schuerbeek P. Cortical visual connections via the corpus callosum are asymmetrical in human infantile Esotropia. Strabismus. 2018;26(1):22–7.
5. Xiao G, Li H, Hu X, Niu X, Xu Q, Yang Y, et al. Interference of commissural connections through the genu of the corpus callosum specifically impairs sensorimotor gating. Behav Brain Res. 2021;411:113383.
6. Rhoton AL Jr. The lateral and third ventricles. Neurosurgery. 2002;51(4 Suppl):S207–71.
7. Caligiuri ME, Quattrone A, Mechelli A, La Torre D, Quattrone A. Semi-automated assessment of the principal diffusion direction in the corpus callosum: differentiation of idiopathic normal pressure hydrocephalus from neurodegenerative diseases. J Neurol. 2022;269(4):1978–88.
8. Momjian S, Owler BK, Czosnyka Z, Czosnyka M, Pena A, Pickard JD. Pattern of white matter regional cerebral blood flow and autoregulation in normal pressure hydrocephalus. Brain. 2004;127(Pt 5):965–72.
9. Owler BK, Pena A, Momjian S, Czosnyka Z, Czosnyka M, Harris NG, et al. Changes in cerebral blood flow during cerebrospinal fluid pressure manipulation in patients with normal pressure hydrocephalus: a methodological study. J Cereb Blood Flow Metab. 2004;24(5):579–87.
10. Owler BK, Pickard JD. Normal pressure hydrocephalus and cerebral blood flow: a review. Acta Neurol Scand. 2001;104(6):325–42.
11. Arda KN, Akay S. The relationship between corpus callosum morphometric measurements and age/gender characteristics: a comprehensive MR imaging study. J Clin Imaging Sci. 2019;9:33.
12. Georgy BA, Hesselink JR, Jernigan TL. MR imaging of the corpus callosum. AJR Am J Roentgenol. 1993;160(5):949–55.
13. Hofmann E, Becker T, Jackel M, Metzner D, Schneider M, Meixensberger J, et al. The corpus callosum in communicating and noncommunicating hydrocephalus. Neuroradiology. 1995;37(3):212–8.
14. Moody DM, Bell MA, Challa VR. The corpus callosum, a unique white-matter tract: anatomic features that may explain sparing in Binswanger disease and resistance to flow of fluid masses. AJNR Am J Neuroradiol. 1988;9(6):1051–9.
15. Del Bigio MR. Neuropathological changes caused by hydrocephalus. Acta Neuropathol. 1993;85(6):573–85.
16. Mataro M, Matarin M, Poca MA, Pueyo R, Sahuquillo J, Barrios M, et al. Functional and magnetic resonance imaging correlates of corpus callosum in normal pressure hydrocephalus before and after shunting. J Neurol Neurosurg Psychiatry. 2007;78(4):395–8.
17. Gammal TE, Allen MB Jr, Brooks BS, Mark EK. MR evaluation of hydrocephalus. AJR Am J Roentgenol. 1987;149(4):807–13.
18. Spreer J, Ernestus RI, Lanfermann H, Lackner K. Lesions of the corpus callosum in hydrocephalic patients with ventricular drainage--a CT-study. Acta Neurochir. 1996;138(2):174–8.
19. Xiong L, Rauch RA, Hagino N, Jinkins JR. An animal model of corpus callosum impingement as seen in patients with normal pressure hydrocephalus. Investig Radiol. 1993;28(1):46–50.
20. Friese SA, Bitzer M, Freudenstein D, Voigt K, Kuker W. Classification of acquired lesions of the corpus callosum with MRI. Neuroradiology. 2000;42(11):795–802.
21. Jinkins JR. Clinical manifestations of hydrocephalus caused by impingement of the corpus callosum on the falx: an MR study in 40 patients. AJNR Am J Neuroradiol. 1991;12(2):331–40.
22. Numaguchi Y, Kristt DA, Joy C, Robinson WL. Scalloping deformity of the corpus callosum following ventricular shunting. AJNR Am J Neuroradiol. 1993;14(2):355–62.
23. Suh DY, Gaskill-Shipley M, Nemann MW, Tureen RG, Warnick RE. Corpus callosal changes associated with hydrocephalus: a report of two cases. Neurosurgery. 1997;41(2):488–93; discussion 493–4.

Corpus Callosum in Dementia

29

Güneş Devrim Kicali

29.1 Introduction

During the nineteenth century, with improved microscopes and the development of methods for pathological examination, two tissue types of brain tissue, gray and white matter, were distinguished at the cellular level [1]. The main commissural pathway bundle, which contains dense pathways connecting the two hemispheres and is identified as white matter because the nerves have lipid-rich myelin sheaths, is called the corpus callosum (CC). The first clinical investigations of this anatomical structure were based on the complex clinical findings it revealed with common local lesions. With the development of imaging technologies, connections have been established between agenesis of the CC and many clinical syndromes. Agenesis has been associated with developmental delay [1, 2] and difficulties in learning and performing movements that require both hands, such as cycling and rowing. In the current psychiatric literature, its relationship with schizophrenia has been closely examined; some findings support the hypothesis that this congenital malformation is implicated in many mental disorders. The CC was originally defined as an axon bundle connecting only the primary cortical areas. However, most of our information about its physiology has come from patients who had undergone callosotomy to treat epilepsy [3–5]. In the present chapter, this transitional region, which is of great importance in areas ranging from artistic creativity to functions such as complex speech and sensory interpretation, will be discussed in relation to its pathophysiological and clinical importance in people diagnosed with cognitive disorders and dementia.

29.2 Structure and Function of the CC

The CC consists of approximately 200 million myelinated axon bundles in the form of a convex C shape between the hemispheres, visible in the midsagittal line on brain magnetic resonance imaging (MRI). It is considered to comprise four regions that cannot be distinguished by anatomically clear borders: from anterior to posterior, rostrum, genu, mid body, and splenium. The axon bundles often form heterotropic and dual-terminating cortical projections while connecting homeotropic regions in the two cortices. The CC is supplied by the main callosal arteries and the rich anastomotic network they form, via both the anterior and posterior communicating arteries [5, 6].

The CC is the principal commissural structure responsible for interhemispheric communication and data transfer. There are two main hypotheses about how it establishes such communication.

G. D. Kicali (✉)
Department of Psychiatry, Muğla Research and Training Hospital, Republic of Turkish Ministry of Health, Menteşe, Muğla, Turkey

M. Turgut et al. (eds.), *The Corpus Callosum*, https://doi.org/10.1007/978-3-031-38114-0_29

According to the first (excitatory) model, the CC provides active channels between the hemispheres; according to the second (inhibitory) model, one hemisphere suppresses the activities and messages of the other. Electrophysiological studies provide evidence supporting both hypotheses [7–9]. Axon bundles of different diameters and mostly myelinated bundles form in different regions of the CC. The commissural white matter pathways in the anterior part are rich in fine fibers that form the prefrontal cortex connection. In the middle part, there are more motor and somatosensory commissural pathways comprising larger-diameter fibers. In the posterior part, there are thick fibers connecting the occipital visual cortex and narrower axon bundles forming the parietal cortex connections [10, 11]. Advances in magnetic resonance imaging technology and the application of DW-MRI have yielded images of water molecules that have restricted diffusion owing to lipid-rich myelin barriers, providing valuable evidence for tractography studies [12].

29.3 Cognitive Functions and Pathophysiology

Studies relating the CC to cognitive functions have mostly been performed on split-brain patients who have undergone callosotomy [13]. These cases, especially patients who were disconnected to treat epilepsy, were evaluated using specialized tests for visuospatial functions, language processing, face recognition, self-awareness, abstract reasoning, and manual dexterity learning. When partial dissections were examined, incisions in the anterior parts of the CC were shown to affect executive functions and forebrain activities; incisions in the posterior part affected visuospatial functions and coordination [14]. Another source of evidence that CC injury leads directly to cognitive loss is provided by isolated vascular injuries or tumors. Cases with defects in attention, executive functions, verbal memory, alexia, dyslexia, and agraphia have been identified [15, 16]. A valuable source of information about cognitive activities comes from developmental follow-up of children with congenital CC agenesis [17]. These individuals achieve important developmental milestones and bimanual skills later than normal [17, 18], and they respond less to the sensory stimuli required for certain motor responses. It would not be appropriate to generalize these results relating the CC to cognitive functions, especially to cases of dementia and diseases causing mental deterioration. Although most of the cognitive and motor findings mentioned occur in dementia and related neuropsychiatric syndromes, they can also be direct or indirect results of cortical atrophy, isolated or diffuse gray matter lesions, and vascular or neoplastic pathologies.

Two explanations have been proposed for the relationship between CC injury and dementia and aging. Conditions secondary to gray matter lesions such as valerian degeneration or ARWMC secondary to physiological or pathological cortical atrophy can impair the callosal fibers by causing direct white matter injury. Many studies have provided evidence to explain the relationship by these two mechanisms in cases of advanced age and the diagnosis of dementia. Other settings can also indicate CC injury; transsynaptic degeneration, diaschisis, synaptic degeneration, and CC atrophy can also be indirect results of a distant lesion that causes loss of functional connectivity. Neuropathological findings such as demyelination, loss of action potentials, and asynchronous activity that cannot be detected by structural MRI can also contribute to CC dysfunction [19–21]. These changes can be detected by neuropsychological tests, neurophysiological examinations, and functional MRI (fMRI).

Many studies have examined CC neuroanatomy in dementia patients and elderly individuals with cross-sectional designs, mostly using structural neuroimaging. Measurements consistent with shrinking and reduced structural integrity of the anterior CC are often the dominant findings. Significant volume loss, demyelination, and thinning of CC anatomy, especially in the prefrontal commissural axon bundles, have also been detected in postmortem studies [22]. These neuropathological changes, which occur by the mechanisms described previously, certainly contribute to dementia clinic findings in the areas of coordination, loss of bimanual skills, and slowing of executive function, which cannot be explained by cortical atrophy alone. These specific changes, which also occur during normal aging, could be an evidence that the CC is essential for frontal brain functions [23, 24]. Jokinen et al. examined the relationship of the CC to executive functions and cognitive slowness in patients with age-related white matter hyperintensity (WMH) and investigated the effect of regional CC atrophy on these areas [25]. The 567 patients with WMH were evaluated by neuropsychological testing and quantitative MRI, and the findings indicating cognitive loss were related to linear regression to the volumes of CC regions and other measurements. Total CC atrophy was associated with poor mental performance, primarily on the Trail Making A and Stroop I/II tests. Losses related to executive functions and attention were also correlated with anterior CC atrophy, and losses related to language skills were associated with total volume and isthmus region atrophy [26, 27]. Data derived from the leukoareosis and disability study (LADIS) with many participants added to the power of the study, and correlations with quantitative imaging and neuropsychological tests, which are adequate for measuring cognitive function, were examined. However, the major limitation of this study was its cross-sectional design, and the clinical characteristics and treatments of the participants were not included. Nevertheless, the findings illuminated the importance of the CC in cognitive function and its role in dementia and the aging process [19, 25].

The role of the CC in cognitive functions and disability is also revealed by functional MRI studies. fMRI can detect signs that determine lateralizing brain functions. Studies show that lateralization is more prominent in young people, and interhemispheric asymmetry becomes more subtle in advanced age [28]. This could indicate increased activity in the nondominant hemisphere, which is trying to compensate for age-related loss of function. The loss of asymmetry in elderly individuals is associated with loss of both cognitive and motor skills. According to the dedifferentiation hypothesis, bihemispheric activity has increased because neuronal mechanism development has been inhibited secondary to degenerative changes in the CC. On the other hand, alternative and compensatory changes that occur with age could also explain the loss of lateralization [29–32].

As mentioned earlier, the importance of the CC for bimanual tasks is well known. However, in dementia patients with CC atrophy, and children and young individuals with congenital CC anomalies, the CC also acts as an important conduit for the necessary inhibition of unilateral motor activities [33].

29.4 Evidence of CC Involvement in Dementia

The relationships of CC atrophy and neurodegeneration to Alzheimer's disease (AD), Parkinson's disease (PD), vascular dementia (VaD), progressive supranuclear palsy (PSP), and frontotemporal dementia (FTD) have been investigated. Many studies comparing AD with healthy older adults give evidence of CC atrophy [34, 35]. A striking feature common to these studies is that atrophy in the posterior region predominates initially, and this atrophy becomes more widespread as the disease progresses. This could be associated with involvement of the hippocampus, parahippocampal gyrus, and mesial temporal cortex, which are fragile regions in the neuropathology of AD. There is even evidence that posterior CC atrophy is the first and earliest imaging finding. There are also imaging studies showing CC shrinkage in VaD, but the specific and significant correlation was less than in AD. The prominent observation of atrophy and degeneration in the anterior region in VaD can be explained by different pathophysiological sources. Anterior atrophy correlates significantly, especially with VaD in which cortical dysfunction predominates. For AD, in contrast, episodic memory loss correlates with posterior CC atrophy [36–38].

In the radiological studies reviewed by Di Paola et al. [39], structural MRI, DWI, and DTI were used to examine the relationships between MCI, Alzheimer's dementia, and the

CC. Although the imaging types and designs of the studies differed, all clinical features from MCI to severe AD were evaluated. Between 1999 and 2009, 52 imaging studies addressing clinical features were selected for review. Studies were examined in methodological groups such as region of interest (ROI), voxel-based morphometry, and diffusion-weighted/tensor imaging. The results common to these studies, in which CC integrity and function were explored with different techniques, are that anterior genial and posterior CC atrophy predominate in AD patients, and this can be generalized to AD. However, it is noteworthy that studies examining homogeneous groups (e.g., mild AD or amnestic MCI) gave no specific findings. The anterior CC was associated with posterior-splenium region atrophy in some studies and with AD in others. However, some studies established no significant relationship. Another result common to the studies is that the observed level of CC atrophy did not differ significantly between the pre- and early stages of the disease, especially in amnestic MCI.

In region-related evaluations, there was a significant correlation with decreased CC volume in the anterior and posterior commissural pathways. There is evidence for a significant correlation between the loss of the posterior connecting fibers in structures such as the temporal-parietal region, mesial temporal gyrus, and hippocampus, which are important in the pathophysiology of AD, and anterior fibers of the bilateral frontal cortex. In addition, there is evidence that the middle trunk part, where the motor and somatosensory commissural pathways are dense, is preserved until late in AD.

To establish a pathophysiological link between CC atrophy and AD, several studies have supported the Valerian degeneration and myelin damage hypotheses. During the early stages of AD, the progressive loss of episodic memory and the associated temporoparietal gray matter, and the loss of prefrontal tissue with its executive and attention functions, could explain why atrophy of the posterior CC begins first and the anterior CC atrophies later. In the myelin loss hypothesis, neurodevelopmental deterioration occurs first in the neurons that were the last to be myelinated. Since atrophy progresses following demyelination and loss of function, the neuronal network function acquired last is lost first. The loss of prefrontal commissural connections can be explained by atrophy of the rostrum and anterior CC [40–42].

Di Paola et al. [39] cited the cross-sectional and AD-control group design as obvious limitations common to the studies they reviewed. They concluded that serial cross-sectional studies are needed, especially in longitudinal follow-up and high-risk populations.

In 2018, Kautzky et al. [43] examined the relationship between postmortem neuropathological examinations and medical data and MRI measurements obtained during the previous follow-up period in AD patients. Evidence indicates that these data can be useful for predicting AD and cognitive impairment through machine learning. In particular, when the relationship between early changes in brain regions and AD was examined, shrinkage of the anterior CC was shown to predict cognitive impairment. The individuals studied were selected from participants of the Vienna Trans-Danube Aging (VITA) prospective study. In this study, initiated in Vienna in 2000, elderly individuals born between 1925 and 1926 were followed. The participants' medical histories, examinations, and neuropsychological and MRI tests were recorded during follow-ups throughout 2001. When the participants died (between 2001 and 2016), autopsies were conducted in Vienna's Danube state hospital, and the brains were examined neuropathologically. In total, 49 of 63 individuals diagnosed with AD were included in the study. It was concluded that the caudal and rostral anterior cingulate gyrus and the fusiform and insular gyrus regions were areas of interest as well as the CC. In view of this important finding, it could be possible in the future to evaluate the brain and perhaps prevent the onset of AD before clinical features develop, with the help of artificial intelligence and machine learning [43].

Frontotemporal dementia (FTD) entails regression of language and speech skills, loss of prefrontal executive functions, and personality changes. Losses of semantic memory and working memory are prominent, but episodic memory loss is minimal compared to AD. DTI studies examining the relationship between the CC and FTD show that white matter volume is decreased in the anterior and genu regions of the CC. The posterior CC maintains its functionality better than in AD cases [44]. There is clinical evidence for a relationship between loss of CC volume and emotional bluntness in FTD.

In studies examining CC changes in people with vascular and multi-infarct dementias, cases have often been compared with AD and healthy controls. In the first MRI study of this kind, Pantel et al. [45] compared measurements of midsagittal CC sections in 32 AD and 17 VaD patients and 13 healthy controls. VaD was associated with nonspecific atrophy, while AD was associated with significant loss of anterior and mid-trunk volume [35]. Therefore, individuals with VaD suffer more significant cortical or focal gray matter lesions, but this loss and the associated clinical features cannot be related directly to CC volume. The aforementioned LADI study of leukoaraiosis and white matter injury in the elderly examined CC atrophy with age-related WMC. The study revealed no correlation between VaD and specific CC region atrophy. In the functional imaging study by Wang et al. (2020), fractionated anisotropy values, the CC, and the subcortical structures were compared between the leukoareosis (LA) group ($n = 56$) and controls ($n = 38$). The LA group was divided into VaD+ and noncognitive (NC). All CCs in the LA group had significantly lower FA than controls. FA was significantly lower in the CC trunk and genu regions in the VaD group than in the NC group or controls. The correlation between functional images and neuropsychological tests was also investigated, revealing a positive correlation between low anterior corona radiata FA value and visuospatial skills and cognitive loss. However, there was no relationship with CC volume and FA value. The authors hypothesized that functional imaging of the CC and different brain regions can be predictive in elderly patients with LA. Access to findings supporting this hypothesis is invaluable for follow-up for dementia in the LA+ elderly.

Tomimoto et al. performed an important study examining the value of the CC in diagnosing dementia postmortem [21]. A total of 22 brains with AD (presenile onset, 9; senile onset, 13); 6 with Binswanger's disease (BD), a form of vascular dementia; and 3 with LVOD were compared with six non-neurological control brains. White matter lesions in the deep gray matter and the CC were quantified as fiber density scores by image analysis of myelin-stained sections. Atrophy of the anterior and trunk portions of the CC was significantly less than in LVOD and BD. In addition, loss of total brain volume in AD correlated with CC atrophy. In the study revealing that deep white matter lesions and CC volume are related in BD, immune histochemical and axonal examination showed that axonal injury, demyelination, and gliosis were associated with BD and LVOD, but not directly correlated with AD. This finding supports the interhemispheric disconnection hypothesis, indicating that CC volume decreases with atrophy resulting from AD [35].

29.5 Conclusion

In the history of medicine, the CC was defined as a simple anatomical structure that connects the brain hemispheres, but as medical technology has advanced, it has been subjected to detailed examination using structural and functional imaging tools. The CC is critical for motor and mental development. In addition to connecting reciprocal primary and axillary gray matter areas, it is a critical transition point for cognitive abilities, cognition, memory, computation, visuospatial functions, and attention. CC functions have mostly been studied in patients with AD and MCI, most studies being cross-sectional examinations of the correlation between neuropsychological testing and imaging results. While the focus was on region of interest (ROI) studies, a relationship was found between specific CC lesions and the results of specific structural and functional imaging studies such as DWI and DTI. Large population studies with prospective design also show that relationships between white matter injury and the CC are of potential future interest as early markers of dementia and loss of cognitive functions. Another important result is the correlation between clinical findings and the temporal order of regional losses in the CC. Pathological examination of this atrophy showed that the associated loss of CC function was proportional to the development and etiology of the dementia. It seems that the relationship between the activity of specific commissural pathways and dementia will become the focus of attention in future studies, dividing the functional neuroanatomy of the CC into more detailed regions.

References

1. Catani M, Ffytche DH. The rises and falls of disconnection syndromes. Brain. 2005;128:2224–39. https://doi.org/10.1093/BRAIN/AWH622.
2. Catani M, Mesulam M. What is a disconnection syndrome? Cortex. 2008;44:911–3. https://doi.org/10.1016/J.CORTEX.2008.05.001.
3. Geschwind N. Disconnection syndromes in animals and man. I. Brain. 1965;88:237. https://doi.org/10.1093/BRAIN/88.2.237.
4. Geschwind N. Disconnection syndromes in animals and man. II. Brain. 1965;88:585. https://doi.org/10.1093/BRAIN/88.3.585.
5. Sexton CE, Kalu UG, Filippini N, Mackay CE, Ebmeier KP. A meta-analysis of diffusion tensor imaging in mild cognitive impairment and Alzheimer's disease. Neurobiol Aging. 2011;32:2322.e5–2322.e18. https://doi.org/10.1016/J.NEUROBIOLAGING.2010.05.019.
6. Nonaka H, Akima M, Hatori T, Nagayama T, Zhang Z, Ihara F. The microvasculature of the cerebral white matter: arteries of the subcortical white matter. J Neuropathol Exp Neurol. 2003;62:154–61. https://doi.org/10.1093/JNEN/62.2.154.
7. Aboitiz F, Scheibel AB, Fisher RS, Zaidel E. Fiber composition of the human corpus callosum. Brain Res. 1992;598:143–53. https://doi.org/10.1016/0006-8993(92)90178-C.
8. Bloom JS, Hynd GW. The role of the corpus callosum in interhemispheric transfer of information: excitation or inhibition? Neuropsychol Rev. 2005;15:59–71. https://doi.org/10.1007/S11065-005-6252-Y.
9. Lee AY. Vascular dementia. Chonnam Med J. 2011;47:66. https://doi.org/10.4068/CMJ.2011.47.2.66.
10. Gazzaniga MS. Forty-five years of split-brain research and still going strong. Nat Rev Neurosci. 2005;6:653–9. https://doi.org/10.1038/NRN1723.
11. Whitford TJ, Kubicki M, Ghorashi S, Schneiderman JS, Hawley KJ, McCarley RW, Shenton ME, Spencer KM. Predicting inter-hemispheric transfer time from the diffusion properties of the corpus callosum in healthy individuals and schizophrenia patients: a combined ERP and DTI study. NeuroImage. 2011;54:2318. https://doi.org/10.1016/J.NEUROIMAGE.2010.10.048.
12. Zarei M, Johansen-Berg H, Smith S, Ciccarelli O, Thompson AJ, Matthews PM. Functional anatomy of interhemispheric cortical connections in the human brain. J Anat. 2006;209:311. https://doi.org/10.1111/J.1469-7580.2006.00615.X.
13. Heidi J-B, Behrens TEJ. Diffusion MRI: from quantitative measurement to in-vivo neuroanatomy. 2014. p. 614.
14. Aralasmak A, Ulmer JL, Kocak M, Salvan CV, Hillis AE, Yousem DM. Association, commissural, and projection pathways and their functional deficit reported in literature. J Comput Assist Tomogr. 2006;30:695–715. https://doi.org/10.1097/01.RCT.0000226397.43235.8B.
15. Devinsky O, Laff R. Callosal lesions and behavior: history and modern concepts. Epilepsy Behav. 2003;4:607–17. https://doi.org/10.1016/J.YEBEH.2003.08.029.
16. Zaidel D, Sperry RW. Memory impairment after commissurotomy in man. Brain. 1974;97:263–72. https://doi.org/10.1093/BRAIN/97.1.263.
17. Moes P, Schilmoeller K, Schilmoeller G. Physical, motor, sensory and developmental features associated with agenesis of the corpus callosum. Child Care Health Dev. 2009;35:656–72. https://doi.org/10.1111/J.1365-2214.2009.00942.X.
18. Paul LK. Developmental malformation of the corpus callosum: a review of typical callosal development and examples of developmental disorders with callosal involvement. J Neurodev Disord. 2011;3:3. https://doi.org/10.1007/S11689-010-9059-Y.
19. Ryberg C, Rostrup E, Sjöstrand K, Paulson OB, Barkhof F, Scheltens P, van Straaten ECW, Fazekas F, Schmidt R, Erkinjuntti T, Wahlund LO, Basile AM, Pantoni L, Inzitari D, Waldemar G. White matter changes contribute to corpus callosum atrophy in the elderly: the LADIS study. AJNR Am J Neuroradiol. 2008;29:1498. https://doi.org/10.3174/AJNR.A1169.
20. Teipel SJ, Bayer W, Alexander GE, Zebuhr Y, Teichberg D, Kulic L, Schapiro MB, Möller HJ, Rapoport SI, Hampel H. Progression of corpus callosum atrophy in Alzheimer disease. Arch Neurol. 2002;59:243–8. https://doi.org/10.1001/ARCHNEUR.59.2.243.
21. Teipel SJ, Hampel H, Alexander GE, Schapiro MB, Horwitz B, Teichberg D, Daley E, Hippius H, Möller HJ, Rapoport SI. Dissociation between corpus callosum atrophy and white matter pathology in Alzheimer's disease. Neurology. 1998;51:1381–5. https://doi.org/10.1212/WNL.51.5.1381.

22. Christopher F. The behavioral neurology of white matter; 2012. p. 442. Second edition (2012) ISBN: 9780199743261.
23. McLaughlin NCR, Paul RH, Grieve SM, Williams LM, Laidlaw D, DiCarlo M, Clark CR, Whelihan W, Cohen RA, Whitford TJ, Gordon E. Diffusion tensor imaging of the corpus callosum: a cross-sectional study across the lifespan. Int J Dev Neurosci. 2007;25:215–21. https://doi.org/10.1016/J.IJDEVNEU.2007.03.008.
24. Sullivan EV, Rohlfing T, Pfefferbaum A. Longitudinal study of callosal microstructure in the normal adult aging brain using quantitative DTI fiber tracking. Dev Neuropsychol. 2010;35:233–56. https://doi.org/10.1080/87565641003689556.
25. Jokinen H, Ryberg C, Kalska H, Ylikoski R, Rostrup E, Stegmann MB, Waldemar G, Madureira S, Ferro JM, van Straaten ECW, Scheltens P, Barkhof F, Fazekas F, Schmidt R, Carlucci G, Pantoni L, Inzitari D, Erkinjuntti T. Corpus callosum atrophy is associated with mental slowing and executive deficits in subjects with age-related white matter hyperintensities: the LADIS study. J Neurol Neurosurg Psychiatry. 2007;78:491–6. https://doi.org/10.1136/JNNP.2006.096792.
26. Fabri M, Polonara G, Mascioli G, Salvolini U, Manzoni T. Topographical organization of human corpus callosum: an fMRI mapping study. Brain Res. 2011;1370:99–111. https://doi.org/10.1016/J.BRAINRES.2010.11.039.
27. Ota M, Obata T, Akine Y, Ito H, Ikehira H, Asada T, Suhara T. Age-related degeneration of corpus callosum measured with diffusion tensor imaging. NeuroImage. 2006;31:1445–52. https://doi.org/10.1016/J.NEUROIMAGE.2006.02.008.
28. Dolcos F, Rice HJ, Cabeza R. Hemispheric asymmetry and aging: right hemisphere decline or asymmetry reduction. Neurosci Biobehav Rev. 2002;26:819–25. https://doi.org/10.1016/S0149-7634(02)00068-4.
29. Cabeza R, Anderson ND, Locantore JK, McIntosh AR. Aging gracefully: compensatory brain activity in high-performing older adults. NeuroImage. 2002;17:1394–402. https://doi.org/10.1006/NIMG.2002.1280.
30. Fling BW, Chapekis M, Reuter-Lorenz PA, Anguera J, Bo J, Langan J, Welsh RC, Seidler RD. Age differences in callosal contributions to cognitive processes. Neuropsychologia. 2011;49:2564. https://doi.org/10.1016/J.NEUROPSYCHOLOGIA.2011.05.004.
31. Kennedy KM, Raz N. Aging white matter and cognition: differential effects of regional variations in diffusion properties on memory, executive functions, and speed. Neuropsychologia. 2009;47:916–27. https://doi.org/10.1016/J.NEUROPSYCHOLOGIA.2009.01.001.
32. Sullivan EV, Rohlfing T, Pfefferbaum A. Quantitative fiber tracking of lateral and interhemispheric white matter systems in normal aging: relations to timed performance. Neurobiol Aging. 2010;31:464. https://doi.org/10.1016/J.NEUROBIOLAGING.2008.04.007.
33. Langan J, Peltier SJ, Bo J, Fling BW, Welsh RC, Seidler RD. Functional implications of age differences in motor system connectivity. Front Syst Neurosci. 2010;4:17. https://doi.org/10.3389/FNSYS.2010.00017.
34. Filippini N, Douaud G, MacKay CE, Knight S, Talbot K, Turner MR. Corpus callosum involvement is a consistent feature of amyotrophic lateral sclerosis. Neurology. 2010;75:1645–52. https://doi.org/10.1212/WNL.0B013E3181FB84D1.
35. Tomimoto H, Lin JX, Matsuo A, Ihara M, Ohtani R, Shibata M, Miki Y, Shibasaki H. Different mechanisms of corpus callosum atrophy in Alzheimer's disease and vascular dementia. J Neurol. 2004;251:398–406. https://doi.org/10.1007/S00415-004-0330-6.
36. Hensel A, Wolf H, Kruggel F, Riedel-Heller SG, Nikolaus C, Arendt T, Gertz HJ. Morphometry of the corpus callosum in patients with questionable and mild dementia. J Neurol Neurosurg Psychiatry. 2002;73:59–61. https://doi.org/10.1136/JNNP.73.1.59.
37. Wang PJ, Saykin AJ, Flashman LA, Wishart HA, Rabin LA, Santulli RB, McHugh TL, MacDonald JW, Mamourian AC. Regionally specific atrophy of the corpus callosum in AD. MCI and cognitive complaints Neurobiol Aging. 2006;27:1613. https://doi.org/10.1016/J.NEUROBIOLAGING.2005.09.035.
38. Wiltshire K, Concha L, Gee M, Bouchard T, Beaulieu C, Camicioli R. Corpus callosum and cingulum tractography in Parkinson's disease. Can J Neurol Sci. 2010;37:595–600. https://doi.org/10.1017/S0317167100010751.
39. di Paola M, Spalletta G, Caltagirone C. In vivo structural neuroanatomy of corpus callosum in Alzheimer's disease and mild cognitive impairment using different MRI techniques: a review. J Alzheimers Dis. 2010;20:67–95. https://doi.org/10.3233/JAD-2010-1370.
40. Lyoo IK, Satlin A, Lee CK, Renshaw PF. Regional atrophy of the corpus callosum in subjects with Alzheimer's disease and multi-infarct dementia. Psychiatry Res. 1997;74:63–72. https://doi.org/10.1016/S0925-4927(97)00009-7.
41. di Paola M, Luders E, di Iulio F, Cherubini A, Passafiume D, Thompson PM, Caltagirone C, Toga AW, Spalletta G. Callosal atrophy in mild cognitive impairment and Alzheimer's disease: different effects in different stages. NeuroImage. 2010;49:141. https://doi.org/10.1016/J.NEUROIMAGE.2009.07.050.
42. Wang PJ, Saykin AJ, Flashman LA, Wishart HA, Rabin LA, Santulli RB, McHugh TL, MacDonald JW, Mamourian AC. Regionally specific atrophy of the corpus callosum in AD, MCI and cognitive complaints. Neurobiol Aging. 2006;27:1613–7. https://doi.org/10.1016/J.NEUROBIOLAGING.2005.09.035.

43. Kautzky A, Seiger R, Hahn A, Fischer P, Krampla W, Kasper S, Kovacs GG, Lanzenberger R. Prediction of autopsy verified neuropathological change of Alzheimer's disease using machine learning and MRI. Front Aging Neurosci. 2018;10:406. https://doi.org/10.3389/FNAGI.2018.00406.
44. Yamauchi H, Fukuyama H, Nagahama Y, Katsumi Y, Hayashi T, Oyanagi C, Konishi J, Shio H. Comparison of the pattern of atrophy of the corpus callosum in frontotemporal dementia, progressive supranuclear palsy, and Alzheimer's disease. J Neurol Neurosurg Psychiatry. 2000;69:623–9. https://doi.org/10.1136/jnnp.69.5.623.
45. Pantel J, Schröder J, Essig M, Minakaran R, Schad LR, Friedlinger M, Jauss M, Knopp M, v. Corpus callosum in Alzheimer's disease and vascular dementia--a quantitative magnetic resonance study. J Neural Transm Suppl. 1998;54:129–36. https://doi.org/10.1007/978-3-7091-7508-8_12.

30 Changes in Callosal Area in Tourette Syndrome

Burcu Çalışkan and Abdullah Canbal

30.1 Introduction

Tic disorders are characterized by sudden, rapid, repetitive movements (motor tics) or vocalization (vocal or phonic tics) [1]. Tourette syndrome (TS), in which tic disorder is the main phenomenon, was first described in 1885 by the French neurologist Georges Albert Gilles de la Tourette [2]. TS is a neuropsychiatric disorder that lasts more than 1 year and begins before the age of 18 years old and is accompanied by multiple motor and vocal tics [3, 4]. The prevalence of TS in school-age children is 0.3–1% and the male/female ratio is 4:1. Clinical signs peak at 10–12 years of age and typically resolve during or after puberty. Tics in TS can be voluntarily suppressed temporarily. However, there is a "rebound" phenomenon when, after suppression, the accumulated tics erupt with a greater intensity. Tics (both qualitative and quantitative) can occur with stress, excitement, or fatigue. Conversely, tic expressions regress when the patient is engaged in a mental or physical activity that requires concentration [5].

Psychiatric disorders such as attention deficit hyperactivity disorder (ADHD), obsessive-compulsive disorder (OCD), self-harming behavior, depression, and specific learning disability may often accompany TS [6, 7]. TS patients with ADHD and/or OCD have impaired social function and a poor quality of life [5, 8].

The etiology of TS is not fully understood, yet genetic, environmental, and neuroanatomical factors may play a role. Recent studies have mentioned the importance of genetic factors that predispose individuals to develop TS [9].

30.2 Pathophysiology of Tourette Syndrome

TS's pathophysiological mechanisms have not yet been elucidated. Neuroanatomical studies in TS focus on the interaction of different regions. Morphological studies have focused on the basal ganglia as an important part of the cortico-striato-thalamo-cortical (CSTC) circuits [10, 11].

The relationship between dysfunction of CSTC circuits and tic symptoms in TS has been demonstrated by studying advanced neuroimaging modalities. In addition to this intrahemispheric dysfunction, morphological and diffusion tensor imaging (DTI) studies have also shown abnormal interhemispheric connectivity [12]. Abnormal corpus callosum (CC) microstructure revealed by reduced fractional anisotropy (FA) using DTI has also been detected in children with TS [13]. Resting-state functional magnetic resonance imaging (fMRI) is a functional neuroimaging method used to evaluate regional interactions

B. Çalışkan (✉) · A. Canbal
Department of Pediatric Neurology, Faculty of Medicine, Necmettin Erbakan University, Konya, Turkey
e-mail: abdullahcanbal@yaani.com

M. Turgut et al. (eds.), *The Corpus Callosum*, https://doi.org/10.1007/978-3-031-38114-0_30

when a person cannot perform a desired task. One study investigated changes in spontaneous neural activity in children with TS. Abnormal spontaneous neural activity was observed in children with TS with decreased spontaneous amplitude of low-frequency fluctuations (ALFF) and fractional ALFF in the superior parietal lobe, bilateral middle and superior frontal cortices, anterior cingulated cortex, posterior cingulated gyrus, precuneus, and bilateral parietal gyrus [14]. Magnetoencephalography (MEG) is a functional neuroimaging method for mapping brain activity using sensitive magnetometers. Brain magnetic resonance spectroscopy (MRS) is a noninvasive technique that demonstrates metabolic changes in the brain. In a study by Puts et al., with this technique, it has been shown that GABA concentrations in the primary sensorimotor cortex are decreased in children with TS compared to controls. These findings may explain the motor tics and sensory impairments in children with TS [15]. Near-infrared spectroscopy (NIRS) is a spectroscopic method that uses the near-infrared region of the electromagnetic spectrum. In one study, a decreased hemoglobin concentration was detected in the prefrontal cortex of children with TS compared to healthy controls by the NIRS method [16]. Positron emission tomography (PET) scan has shown a correlation between the presence of tics and hypermetabolism in various parts of the brain. Studies show that serotonergic pathways play an important role in the pathophysiology of TS. According to another hypothesis, the imbalance in the noradrenergic, glutamatergic, serotonergic, opioid, cholinergic, and GABA-ergic systems is emphasized [17, 18]. Anatomical abnormalities of the basal ganglia in childhood have been found to be associated with the severity of tics in early adulthood [19, 20]. Cortical regions are predicted to play an important role in the modulation and suppression of tics. Additionally, connectivity problems involving parietal, temporal, and occipital lobes, and interhemispheric brain abnormalities have been demonstrated in patients with TS that may be associated with mental and behavioral changes [7, 12, 21]. Children and young adults with TS have been found to have larger prefrontal regions and smaller CC area compared to healthy controls, and these anatomical features appear to reflect neural plasticity that helps reduce the severity of tics [22]. In a study using volumetric brain MRI, gray matter volumes in the left frontal lobes were found to be smaller in patients with TS than in controls [23]. In another study using 3-Tesla MRI, 52 adult patients with TS and 52 control patients were compared, and it was shown that the gray matter thickness and depth of the patients with TS decreased in the pre-post central region [24].

30.3 Corpus Callosum in Tourette Syndrome

The CC is the large commissural pathway consisting of myelinated and unmyelinated axons that provide axonal connection between the cerebral hemispheres. Changes in CC size are largely related to the number and size of axons. The morphology of CC can be affected by hand preference and cerebral lateralization pattern for the tongue and sex [25]. In a study conducted by Mostofsky et al. on 19 girls with TS and 22 control groups, the mean CC measurements were 6.1 cm^2 in patients with TS, 6.44 cm^2 in patients with TS and ADHD, and 6.27 cm^2 in the control group [26]. In a study by Peterson et al., the midsagittal cross-sectional areas of the CC were compared in 14 untreated TS patients and 14 healthy controls. While the mean CC midsagittal cross-sectional area was 6.16 cm^2 in patients with TS, it was 7.84 cm^2 in the control group. Accordingly, the CC area was smaller in patients with TS [24]. In another study by Baumgardner et al., 77 children aged 6–16 years were evaluated. Of these, 16 were had TS, 21 had TS + ADHD, 13 had ADHD, and 27 were healthy controls. The total CC area was 6.90 cm^2 in the TS group, 5.84 cm^2 in the ADHD group, 6.53 cm^2 in the TS + ADHD group, and 6.19 cm^2 in the control group. It has been stated that the higher total CC area in the TS group may be related to an increase in the number of axons or an increase in axonal myelination [25]. An increase in regional CC areas was reported in two studies in children with TS [27, 28]. In a study involving 37 children with TS and 27 control groups, it was determined that there was a significant increase in four parts of the CC (splenium, isthmus/posterior body, mid-body, and rostral body) in children with TS. It was thought that this increase in CC size might be due to the central pathological process or the compensatory response of the atypical brain in TS. Increase in CC size carrying excitatory axons causes greater spread of neural activation to the opposite hemisphere, resulting in decreased behavioral control [25]. In a study comparing 158 patients with TS and 121 healthy controls, the size of the midsagittal CC was smaller in children with TS but larger in adults with TS compared with healthy controls of the same age. The size of the CC correlated positively with the severity of motor tics. This suggests that the smaller CC in children may have increased the severity of the tics. Furthermore, CC size was inversely proportional to prefrontal cortex volumes in both the TS and control groups, although the magnitudes of these inverse correlations were significantly higher in the TS group. This exaggerated correlation may indicate the presence of an adaptive or compensatory process in the brains of people with TS, given that smaller CCs and larger prefrontal cortices are both associated with less severe tic symptoms [29]. Synapses in the prefrontal regions are slowed in individuals who successfully adapt to the presence of tics; possibly these brain regions are markedly activated by children's need to suppress tics in social situations [30]. Furthermore, another study reported that 20 boys with TS had lower measures of anisotropy in their CC as measured by DTI compared to 20 controls, suggesting the presence of reduced interhemispheric connectivity in people with TS [13]. The cavum septum pellucidum is a brain structure closely associated with the development of the CC and limbic system during the early development of the brain. A significantly smaller cavum septum has been demonstrated in children with TS [31]. Because the development of the cavum septum ends during fetal or early postnatal life, it has been suggested that a smaller cavum in children and adults with TS is a marker for a persistent early developmental vulnerability in people with TS [32].

As a result, CC is a large commissural pathway consisting of myelinated and unmyelinated axons that provide axonal connection between cerebral hemispheres. Changes in CC size are largely related to the number and size of axons. The studies show that changes in CC size are associated with TS.

References

1. American Psychiatric Association. Diagnostic and statistical manual of mental disorders (DSM-5®). Washington, DC: American Psychiatric Publishing; 2013.
2. Gunduz A, Okun MS. A review and update on Tourette syndrome: where is the field headed? Curr Neurol Neurosci Rep. 2016;16(4):1–13.
3. Foltynie T. Vocal tics in Tourette's syndrome. Lancet Neurol. 2016;15(3):e1.
4. Ueda K, Kim S, Greene DJ, Black KJ. Correlates and clinical implications of tic suppressibility. Curr Dev Disord Rep. 2021;8:112–20.
5. Robertson MM, Eapen V, Cavanna AE. The international prevalence, epidemiology, and clinical phenomenology of Tourette syndrome: a cross-cultural perspective. J Psychosom Res. 2009;67:475–83.
6. Bloch MH, Peterson BS, Scahill L. Adulthood outcome of tic and obsessive compulsive symptom severity in children with Tourette syndrome. Arch Pediatr Adolesc Med. 2006;160:65–9.
7. Zykov VP, Begasheva OI. Cognitive disturbances in patients with tics and Tourette's syndrome and their correction with encephabol. Zh Nevrol Psikhiatr Im S S Korsakova. 2003;103:10–4.
8. Cavanna AE, Seri S. Tourette's syndrome. BMJ. 2013;347:f4964.
9. Debes NM, Preel M, Skov L. Functional neuroimaging in Tourette syndrome: recent perspectives. Neurosci Neuroecon. 2017;6:1–13.
10. Augustine F, Singer HS. Merging the pathophysiology and pharmacotherapy of tics. Tremor Other Hyperkinet Mov. 2020;8:595.
11. Peterson BS, Thomas P, Kane MJ, Scahill L, Zhang H, Bronen R, King RA, Leckman JF, Staib L. Basal ganglia volumes in patients with Gilles de la Tourette syndrome. Arch Gen Psychiatry. 2003;60:415–24.
12. Felling RJ, Singer HS. Neurobiology of Tourette syndrome: current status and need for further investigation. J Neurosci. 2011;31:12387–95.
13. Plessen KJ, Grüner R, Lundervold A, Hirsch JG, Xu D, Bansal R. Reduced white matter connectivity in the

corpus callosum of children with Tourette syndrome. J Child Psychol Psychiatry. 2006;47:1013–22.
14. Cui Y, Jin Z, Chen X, He Y, Liang X, Zheng Y. Abnormal baseline brain activity in drug-naïve patients with Tourette syndrome: a resting-state fMRI study. Front Hum Neurosci. 2014;7:913.
15. Puts NA, Harris AD, Crocetti D, et al. Reduced GABAergic inhibition and abnormal sensory symptoms in children with Tourette syndrome. J Neurophysiol. 2015;114(2):808–17.
16. Yamamuro K, Ota T, Iida J. Prefrontal dysfunction in pediatric Tourette's disorder as measured by near-infrared spectroscopy. BMC Psychiatry. 2015;15:102.
17. Saporta ASD, Chugani HT, Juhász C. Multimodality neuroimaging in Tourette syndrome: alpha-[11C] methyl-L-tryptophan positron emission tomography and diffusion tensor imaging studies. J Child Neurol. 2010;25:336–42.
18. Wong DF, Brasic JR, Singer HS. Mechanisms of dopaminergic and serotonergic neurotransmission in Tourette syndrome: clues from an in vivo neurochemistry study with PET. Neuropsychopharmacology. 2007;33:1239–51.
19. Ganos C. Tics and Tourette's: update on pathophysiology and tic control. Curr Opin Neurol. 2016;29:513–8.
20. Martino D, Ganos C, Worbe Y. Neuroimaging applications in Tourette's syndrome. Int Rev Neurobiol. 2018;143:65–108.
21. Cavanna AE, Stecco A, Rickards H, Servo S, Terazzi E, Peterson B. Corpus callosum abnormalities in Tourette syndrome: an MRI-DTI study of monozygotic twins. J Neurol Neurosurg Psychiatry. 2010;81:533–5.
22. Pascual-Leone A, Amedi A, Fregni F, Merabet LB. The plastic human brain cortex. Annu Rev Neurosci. 2005;28:377–401.
23. Fredericksen KA, Cutting LE, Kates WR. Disproportionate increases of white matter in right frontal lobe in Tourette syndrome. Neurology. 2002;58:85.
24. Muellner J, Delmaire C, Valabrégue R. Altered structure of cortical sulci in gilles de la Tourette syndrome: further support for abnormal brain development. Mov Disord. 2015;30:655.
25. Baumgardner TL, Singer HS, Denckla MB, Rubin MA, Abrams MT, Colli MJ. Corpus callosum morphology in children with Tourette syndrome and attention deficit hyperactivity disorder. Neurology. 1996;47:477–82.
26. Mostofsky SH, Wendlandt J, Cutting L, Denckla MB, Singer HS. Corpus callosum measurements in girls with Tourette syndrome. Neurology. 1999;53(6):1345–5.
27. Reutens DC, Berkovic SF, Macdonell RAL, Bladin PF. Magnetic stimulation of the brain in generalized epilepsy: reversal of cortical hyperexcitability by anticonvulsants. Ann Neurol. 1993;34:351–5.
28. Ziemann U, Steinhoff BJ, Tergau F, Paulus W. Transcranial magnetic stimulation: its current role in epilepsy research. Epilepsy Res. 1998;30:11–30.
29. Plessen KJ, Wentzel-Larsen T, Hugdahl K, Feineigle P, Klein J, Staib LH. Altered interhemispheric connectivity in individuals with Tourette's disorder. Am J Psychiatry. 2004;161:2028–37.
30. Peterson BS, Leckman J, Duncan J, Wetzles R, Riddle M, Hardin M. Corpus callosum morphology from magnetic resonance images in Tourette's syndrome. Psychiatry Res. 1994;55:85–99.
31. Kim KJ, Peterson BS. Cavum septi pellucidi in Tourette syndrome. Biol Psychiatry. 2003;54:76–85.
32. Mol DN. Co-morbid disorders in Tourette syndrome. Behav Neurol. 2013;27:7–14.

Callosal Abnormalities in Schizophrenia

31

Damla Balkan Erdoğan, Nilfer Şahin, Öykü Özçelik, and Osman Vırıt

31.1 Introduction

Schizophrenia is a severe mental illness that impairs social interaction, self-care abilities, and the capacity to discriminate between reality and hallucinations. Perception, emotion, cognition, thoughts, and behaviors are all abnormal [1]. According to research, 1% of people develop schizophrenia during their lives. Schizophrenia affects men and women equally; however, its onset is earlier in men [2].

Numerous genetic and environmental variables are involved in the etiology of schizophrenia. Previous studies have shown that the multifactorial neurodevelopmental model explains this etiology best [3]. Numerous genes on different chromosomes have been potentially linked to schizophrenia. Also, several hypotheses propose that neurotransmitters such as nicotine, glutamate, acetylcholine, serotonin, norepinephrine, GABA, and dopamine contribute to its etiology. Neuropathology has been identified in the limbic system, basal ganglia, cerebellum, prefrontal cortex, and cerebral ventricles by structural brain imaging. Research on schizophrenia patients has also revealed decreased corpus callosum (CC) size and fiber density as well as impaired symmetry in numerous brain areas, including the temporal, frontal, and occipital lobes [4, 5].

The CC, a cortico-cortical white matter tract, links the cortex of one hemisphere to that of the other. During the second century CE, Galen of Pergamon named this structure "callus" [6]. It extends from the anterior commissure in the front to the hippocampal commissure in the back and covers most of the frontal and parietal lobes. It is the most obvious forebrain commissure in mature animals. There are anatomical divisions in the human CC: the genu, trunk/body, isthmus, and spleen [7].

The lamina rostralis is present from the anterior commissure to the posterior-inferior part (the genu). It is typically considered the final callosal segment to mature, but it is already present in the a 14-week-old fetus [8]. Although they have not been precisely studied, the fibers it contains are most likely to connect the frontobasal cortex [9, 10].

The genu is a thicker section of the CC. It forms the anterior boundary of the septum pellucidum and links the prefrontal cortex and the anterior cingulate cortex genu [9]. The ventral genu houses the fibers of the ventromedial pre-

D. B. Erdoğan
Kocaeli City Hospital, İzmit, Turkey

N. Şahin
Department of Child and Adolescent Psychiatry, Muğla Sıtkı Koçman University School of Medicine, Muğla, Turkey

Ö. Özçelik · O. Vırıt (✉)
Department of Psychiatry, Muğla Sıtkı Koçman University School of Medicine, Muğla, Turkey

M. Turgut et al. (eds.), *The Corpus Callosum*, https://doi.org/10.1007/978-3-031-38114-0_31

frontal cortex, while the dorsal genu houses those of the dorsolateral prefrontal cortex [10].

The horizontal segment that runs from the genu to the point where the fornix sits next to the lower surface of the CC is called the trunk or body. Its superior border is the septum pellucidum. The fibers of the callosal body constitute the roofs of the lateral ventricular bodies. It spreads along the anterior radiations of the thalamus and between the cingulum bundle superiorly and the occipitofrontal fascicle inferiorly. The trunk connects the superiorly located part of the insula, the precentral cortex (premotor region, supplementary motor area), and frequently the cingulate gyrus [9, 10].

The CC and fornix meet at the isthmus, which transmits commissural fibers of the motor and somatosensory strips from the precentral and postcentral gyrus [9, 10] and fibers from the primary auditory area [11].

The thickest section of the CC is the spleen. It comprises three parts: superior, posterior, and inferior [9–11]. The superior group contains commissural fibers from the posterior parietal cortex. The posterior group contains commissural fibers from the medial occipital cortex.

Between the homologous portions of the two hemispheres, the CC is essential for sensory integration, inhibition, and attention as well as memory and language [12, 13]. The CC is also presumed to be fundamental to cognition [14]. There is evidence that its integrity is directly implicated in the deterioration in cognitive performance as people age [15], whereas the typically greater callosal thickness in children during development is linked to intelligence [16], processing speed [17], and problem-solving skills [18]. Owing to its proximity, the CC is also thought to be indirectly linked to the limbic system [19].

Structural abnormalities in the CC frequently accompany psychiatric and developmental conditions such as schizophrenia, autism, mental retardation, Down syndrome, attention deficit hyperactivity disorder, and developmental language difficulties. Changed interhemispheric connections brought on by abnormalities in CC morphology could be involved in the pathogenesis of cognitive disorders [20]. Studies of the CC have accelerated because it has been discovered that abnormalities in it could be implicated in the early stages of schizophrenia [21].

The notion that schizophrenia is a "disconnection syndrome" is supported by several neuroimaging investigations of schizophrenia patients that have revealed atypical structural and functional brain connections [22, 23]. Investigations into the etiopathology of schizophrenia have drawn attention to disorders of the CC [24, 25].

Postmortem brain examinations of schizophrenia patients have revealed such CC abnormalities. Highley et al. found decreased fiber count in all regions of the CC except the rostrum [4]. Later studies using magnetic resonance imaging (MRI) gave different results. MRI studies on first-episode schizophrenia patients have yielded inconsistent findings. Some researchers have highlighted a lower two-dimensional midsagittal cross-sectional area and lower three-dimensional CC volume in schizophrenia patients than healthy controls. Collinson et al. found these differences in first-episode schizophrenia patients, with no significant difference between the sexes [26]. Bachmann et al. found that the CC and its subregions were smaller among first-episode schizophrenia patients than controls and the CC area was larger in women than in men [27]. Kesavan et al. observed that the genu, trunk, and splenium in particular were smaller among first-episode schizophrenia patients than the control group [28]. However, some studies have reported no significant difference in three-dimensional CC volume or two-dimensional CC area between first-episode schizophrenia patients and control groups [29]. These contradictory findings imply that CC morphology is highly correlated with the duration of first-episode schizophrenia disease; CC volume and area are not necessarily smaller than normal in first-episode schizophrenia patients with disease duration less than 1 year [30].

Some researchers have specifically examined the longitudinal course of CC abnormalities among adults with schizophrenia. Mitelman et al. found that patients with chronic schizophrenia showed a more rapid decrease in absolute CC size at a four-year follow-up than healthy controls, with a stronger effect among patients with poor prognosis [31]. Another longitudinal study investigated patients with first-episode schizophrenia-related psychosis and found no significant difference from healthy controls in the rate of CC volume change at one-year follow-up [32].

Little information is available about the effect of long-term antipsychotic use on the CC in schizophrenic patients. One study compared individuals undergoing long-term antipsychotic treatment with healthy volunteers and found larger posterior CC volumes in those who had been exposed to these medications for longer [33]. In a different study, posttreatment patients with a poor prognosis, patients with a good prognosis, and healthy controls were contrasted. The authors concluded that patients with poor prognosis displayed a more significant reduction in CC size than the other two groups [31]. These studies did not include patients whose use of the drugs was of shorter duration, so whether the findings relate to the neuropathology or the drugs remains unclear [34]. Another study compared schizophrenia patients treated with risperidone, those treated with clozapine, those who had never received treatment, and healthy controls. The untreated schizophrenia patients had significantly smaller volumes of the CC and its subregions and significantly lower white matter integrity than the healthy controls. The clozapine and risperidone groups showed higher fractional anisotropy (FA) in the anterior CC region than the untreated patients. The volume of the middle CC region was dramatically enlarged only in the risperidone group, but the mid-anterior CC volume was lower in the clozapine-treated patients than the healthy controls [34].

Nearly 30% of schizophrenia patients do not respond to antipsychotic medications. To be considered treatment-resistant, they must have failed to respond to at least two distinct antipsychotic medications at the right doses and timings [35]. Treatment resistance is a major issue since it significantly reduces functionality in these patients. Therefore, research has focused on elucidating it. One study compared treatment-resistant schizophrenia patients and healthy controls and showed that the patients had lower FA and increased radial diffusion (RD) in the genu, trunk, and splenium of the CC [36]. Another study showed significantly reduced FA in the splenium of the CC among treatment-resistant than nontreatment-resistant schizophrenia patients [37].

In areas of the brain where there differences are sex , such as the CC, schizophrenia patients sometimes exhibit sex differences in brain morphology [38]. There are gender disparities in the macrostructure and FA of the CC in both schizophrenia patients and healthy people [39–41]. According to one meta-analysis, seven studies reported lower FA among women than men, regardless of diagnosis group. Three studies showed no sex difference, though female patients had lower FA than female controls. Two studies showed that female patients had lower FA than male patients. Thus, the meta-analysis showed a significant difference in FA between patients and controls for both sexes, particularly in the genu [42].

Following the introduction of the "bicameral mentality model" in 1976, more hypotheses concerning schizophrenia were formulated. According to the bicameral mentality model, the right hemisphere is the creative, poetic, "god-brain," and the place where decisions are made. The left hemisphere is the analytical, rational, "human-brain" part, which follows the decisions made by the right hemisphere. In schizophrenia patients, however, the right temporal lobe produces hallucinations and the left lobes obeys them. Hence, interhemispheric connections (the CC and others) were believed to have a key role in the pathogenesis of schizophrenia [43]. Researchers have argued that anterior corpus callosotomy could help to reduce auditory and visual hallucinations among treatment-resistant schizophrenia patients in light of the bicameral mentality perspective and of information from neuroimaging investigations. This intervention could prevent the transmission of abnormal perception and stimuli from the right to the left brain and the left brain's obedience to them. However, relevant data are limited. Reports from previous trials seem promising, but the findings remain anecdotal. Therefore, well-designed, controlled research with reversible approaches such as "deep brain stimulation" is needed to determine the effectiveness of anterior callosotomy in schizophrenia patients [44, 45].

31.2 Conclusion

Earlier research showed a significant connection between schizophrenia and the CC. A clearer understanding of the etiopathogenesis of schizophrenia, better assessments of treatment efficacy, earlier diagnosis in at-risk populations, and better management of treatment-resistant schizophrenia patients could all emanate from further research to elucidate this relationship.

References

1. Tandon R, Gaebel W, Barch DM, Bustillo J, Gur RE, Heckers S, et al. Definition and description of schizophrenia in the DSM-5. Schizophr Res. 2013;150(1):3–10.
2. Hany M, Rehman B, Azhar Y, Chapman J. Continuing education activity. 2021.
3. Kuniyoshi JS, McClellan JM. Early-onset schizophrenia. Washington, DC: American Psychiatric Publishing; 2010.
4. Highley JR, Esiri MM, McDonald B, Cortina-Borja M, Herron BM, Crow TJ. The size and fibre composition of the corpus callosum with respect to gender and schizophrenia: a post-mortem study. Brain. 1999;122(1):99–110.
5. Hoff AL, Neal C, Kushner M, DeLisi LE. Gender differences in corpus callosum size in first-episode schizophrenics. Biol Psychiatry. 1994;35(12):913–9.
6. Apak S. "Korpus Kallozum" Beynin Merkezindeki Gizemli Bölge-Derleme. 2009.
7. Raybaud C. The corpus callosum, the other great forebrain commissures, and the septum pellucidum: anatomy, development, and malformation. Neuroradiology. 2010;52(6):447–77.
8. Kier EL, Truwit CL. The lamina rostralis: modification of concepts concerning the anatomy, embryology, and MR appearance of the rostrum of the corpus callosum. Am J Neuroradiol. 1997;18(4):715–22.
9. Hofer S, Frahm J. Topography of the human corpus callosum revisited—comprehensive fiber tractography using diffusion tensor magnetic resonance imaging. NeuroImage. 2006;32(3):989–94.
10. Velut S, Destrieux C, Kakou M. Anatomie morphologique du corps calleux. Neurochir Suppl. 1998;44(1):17–30.
11. Aboitiz F, Montiel J. One hundred million years of interhemispheric communication: the history of the corpus callosum. Braz J Med Biol Res. 2003;36:409–20.
12. Banich MT. The missing link: the role of interhemispheric interaction in attentional processing. Brain Cogn. 1998;36(2):128–57.
13. Giedd JN, Rumsey JM, Castellanos FX, Rajapakse JC, Kaysen D, Vaituzis AC, et al. A quantitative MRI study of the corpus callosum in children and adolescents. Dev Brain Res. 1996;91(2):274–80.
14. Brown WS, Paul LK, Symington M, Dietrich R. Comprehension of humor in primary agenesis of the corpus callosum. Neuropsychologia. 2005;43(6):906–16.
15. Voineskos AN, Rajji TK, Lobaugh NJ, Miranda D, Shenton ME, Kennedy JL, et al. Age-related decline in white matter tract integrity and cognitive performance: a DTI tractography and structural equation modeling study. Neurobiol Aging. 2012;33(1):21–34.
16. Luders E, Thompson PM, Narr KL, Zamanyan A, Chou Y-Y, Gutman B, et al. The link between callosal thickness and intelligence in healthy children and adolescents. NeuroImage. 2011;54(3):1823–30.
17. Penke L, Maniega SM, Murray C, Gow AJ, Hernández MCV, Clayden JD, et al. A general factor of brain white matter integrity predicts information processing speed in healthy older people. J Neurosci. 2010;30(22):7569–74.
18. van Eimeren L, Niogi SN, McCandliss BD, Holloway ID, Ansari D. White matter microstructures underlying mathematical abilities in children. Neuroreport. 2008;19(11):1117–21.
19. de Sonneville L, Njiokiktjien C. Pediatric behavioural neurology. Amsterdam: Suyi Publications; 1988.
20. Bloom JS, Hynd GW. The role of the corpus callosum in interhemispheric transfer of information: excitation or inhibition? Neuropsychol Rev. 2005;15(2):59–71.
21. Henze R, Brunner R, Thiemann U, Parzer P, Klein J, Resch F, et al. White matter alterations in the corpus callosum of adolescents with first-admission schizophrenia. Neurosci Lett. 2012;513(2):178–82.
22. Lui S, Yao L, Xiao Y, Keedy S, Reilly J, Keefe R, et al. Resting-state brain function in schizophrenia and psychotic bipolar probands and their first-degree relatives. Psychol Med. 2015;45(1):97–108.
23. Zhang W, Lei D, Keedy SK, Ivleva EI, Eum S, Yao L, et al. Brain gray matter network organization in psychotic disorders. Neuropsychopharmacology. 2020;45(4):666–74.
24. Francis AN, Mothi SS, Mathew IT, Tandon N, Clementz B, Pearlson GD, et al. Callosal abnormalities across the psychosis dimension: bipolar schizophrenia network on intermediate phenotypes. Biol Psychiatry. 2016;80(8):627–35.

25. Walterfang M, Velakoulis D. Callosal morphology in schizophrenia: what can shape tell us about function and illness? Br J Psychiatry. 2014;204(1):9–11.
26. Collinson SL, Gan SC, San Woon P, Kuswanto C, Sum MY, Yang GL, et al. Corpus callosum morphology in first-episode and chronic schizophrenia: combined magnetic resonance and diffusion tensor imaging study of Chinese Singaporean patients. Br J Psychiatry. 2014;204(1):55–60.
27. Bachmann S, Pantel J, Flender A, Bottmer C, Essig M, Schröder J. Corpus callosum in first-episode patients with schizophrenia–a magnetic resonance imaging study. Psychol Med. 2003;33(6):1019–27.
28. Keshavan M, Diwadkar V, Harenski K, Rosenberg D, Sweeney J, Pettegrew JW. Abnormalities of the corpus callosum in first episode, treatment naive schizophrenia. J Neurol Neurosurg Psychiatry. 2002;72(6):757–60.
29. Walterfang M, Wood AG, Reutens DC, Wood SJ, Chen J, Velakoulis D, et al. Morphology of the corpus callosum at different stages of schizophrenia: crosssectional study in first-episode and chronic illness. Br J Psychiatry. 2008;192:429–34. https://doi.org/10.1192/bjp.bp.107.041251.
30. Takahashi M, Matsui M, Nakashima M, Takahashi T, Suzuki M. Callosal size in first-episode schizophrenia patients with illness duration of less than one year: a cross-sectional MRI study. Asian J Psychiatr. 2017;25:197–202. https://doi.org/10.1016/j.ajp.2016.10.031.
31. Mitelman SA, Nikiforova YK, Canfield EL, Hazlett EA, Brickman AM, Shihabuddin L, et al. A longitudinal study of the corpus callosum in chronic schizophrenia. Schizophr Res. 2009;114(1–3):144–53.
32. Del Re EC, Bouix S, Fitzsimmons J, Blokland GA, Mesholam-Gately R, Wojcik J, et al. Diffusion abnormalities in the corpus callosum in first episode schizophrenia: associated with enlarged lateral ventricles and symptomatology. Psychiatry Res. 2019;277:45–51.
33. de Moura MT, Zanetti MV, Duran FL, Schaufelberger MS, Menezes PR, Scazufca M, et al. Corpus callosum volumes in the 5 years following the first-episode of schizophrenia: effects of antipsychotics, chronicity and maturation. Neuroimage Clin. 2018;18:932–42.
34. Tao B, Xiao Y, Cao H, Zhang W, Yang C, Lencer R, et al. Characteristics of the corpus callosum in chronic schizophrenia treated with clozapine or risperidone and those never-treated. BMC Psychiatry. 2021;21(1):1–10.
35. Howes OD, McCutcheon R, Agid O, De Bartolomeis A, Van Beveren NJ, Birnbaum ML, et al. Treatment-resistant schizophrenia: treatment response and resistance in psychosis (TRRIP) working group consensus guidelines on diagnosis and terminology. Am J Psychiatr. 2017;174(3):216–29.
36. Holleran L, Ahmed M, Anderson-Schmidt H, McFarland J, Emsell L, Leemans A, et al. Altered interhemispheric and temporal lobe white matter microstructural organization in severe chronic schizophrenia. Neuropsychopharmacology. 2014;39(4):944–54.
37. Assunção-Leme IB, Zugman A, De Moura LM, Sato JR, Higuchi C, Ortiz BB, et al. Is treatment-resistant schizophrenia associated with distinct neurobiological callosal connectivity abnormalities? CNS Spectr. 2021;26(5):545–9.
38. Abel KM, Drake R, Goldstein JM. Sex differences in schizophrenia. Int Rev Psychiatry. 2010;22:417–28.
39. Liu F, Vidarsson L, Winter JD, Tran H, Kassner A. Sex differences in the human corpus callosum microstructure: a combined T2 myelin-water and diffusion tensor magnetic resonance imaging study. Brain Res. 2010;1343:37–45.
40. Westerhausen R, Kreuder F, Sequeira SDS, Walter C, Woerner W, Wittling RA, et al. Effects of handedness and gender on macro-and microstructure of the corpus callosum and its subregions: a combined high-resolution and diffusion-tensor MRI study. Cogn Brain Res. 2004;21(3):418–26.
41. Witelson SF. Hand and sex differences in the isthmus and genu of the human corpus callosum: a postmortem morphological study. Brain. 1989;112(3):799–835.
42. Shahab S, Stefanik L, Foussias G, Lai M-C, Anderson KK, Voineskos AN. Sex and diffusion tensor imaging of white matter in schizophrenia: a systematic review plus meta-analysis of the corpus callosum. Schizophr Bull. 2018;44(1):203–21.
43. Jaynes J. The origin of consciousness in the breakdown of the bicameral mind. Boston, MA: Houghton Mifflin Harcourt; 2000.
44. Neumaier F, Paterno M, Alpdogan S, Tevoufouet EE, Schneider T, Hescheler J, et al. In reply to corpus Callosotomy for drug-resistant schizophrenia; novel treatment based on pathophysiology. World Neurosurg. 2018;116:485–6.
45. Taghipour M, Ghaffarpasand F. Corpus callosotomy for drug-resistant schizophrenia; novel treatment based on pathophysiology. World Neurosurg. 2018;116:483–4.

Corpus Callosum in Autism Spectrum Disorder

32

William Smith, Cuong C. J. Bui, and R. Shane Tubbs

32.1 Introduction

Autism spectrum disorder (ASD) is a neurodevelopmental disorder with a worldwide population prevalence of about 1%. The brain etiology is currently unknown, but genetics, developmentally early environmental factors, large-effect rare mutations, and small-effect common mutations all contribute to the risk. The corpus callosum (CC) is the main white matter bundle connecting the two brain hemispheres, as detailed previously in this book, and there have been repeated findings of abnormal CC structure in ASD. The correlation between clinical manifestations of ASD and CC characteristics has been examined in relation to brain connectivity, and the findings indicate a correlation between CC volume and those clinical manifestations [1]. In addition, findings suggest that age and sex significantly affect morphometric measurements of the CC [2].

Clinically, ASD is characterized by early-onset symptoms that fall into two atypical cognitive profiles that include persistent difficulties in social communication and social interaction and restricted, repetitive interests, patterns of behaviors, and activities [3]. ASD is associated with intellectual disability in up to half of all cases, and more than 70% have concurrent medical, developmental, or psychiatric conditions [4–8]. It is four or five times more prevalent in males than females and can be classified into three levels. Level 1 is the mildest form and requires the least support; level 3 is the most severe level of impairment and requires the most support [9]. Generally, the greater the individual's level of impairment, the more co-occurring conditions they have [10].

This chapter will compare the structural and functional differences in the CC between ASD cases and typically developing (TD) controls generally and as they relate to the age and sex differences in those differences.

32.2 Structural and Functional Changes in ASD

Knowledge of structural brain development in ASD comes mainly from cross-sectional studies, so there is limited understanding of the changes across a lifetime. The structure of the CC in ASD varies according to the age at which subjects are examined. Studies of young children with ASD revealed an atypical CC growth trajectory and an

W. Smith (✉)
Tulane University School of Medicine, New Orleans, LA, USA
e-mail: wsmith15@tulane.edu

C. C. J. Bui
Department of Neurosurgery, Ochsner Medical Center, New Orleans, LA, USA
e-mail: cjbui@ochsner.org

R. S. Tubbs
Department of Neurosurgery, Tulane School of Medicine, New Orleans, LA, USA
e-mail: rtubbs@tulane.edu

M. Turgut et al. (eds.), *The Corpus Callosum*, https://doi.org/10.1007/978-3-031-38114-0_32

increase in total CC volume as well as in several of its subregions as early as the first year of life [11–14]. At the earliest age of clinical onset of ASD (1–2.5 years), there are already multiple structural differences from TD controls in the callosal tracts connecting the temporal lobes.

The extent of abnormality in the temporal callosal pathway at the age of clinical onset is correlated with the severity of clinical symptoms measured by the Vineland test later in life [11].

The diffusion pattern in the callosal neurons was explored using mean diffusivity (MD) or the average total diffusion, axial diffusivity (AD), or diffusion parallel to the principal diffusion direction, radial diffusivity (RD), or diffusion perpendicular to the principal diffusion direction, and fractional anisotropy (FA), comparing the scalar value of the normalized standard deviation of the three previous values. Larger FA and smaller MD, AD, and RD values were apparent in the CC but were only noted when the clinical symptoms of ASD emerged very early. The elevated FA level gradually disappears or is reversed from abnormally high to abnormally low during development [11, 15–17] and is only apparent in toddlers who present with symptoms when younger than 2.5 years [11, 17–19]. Differences from TD controls in the callosal growth trajectory are no longer apparent in those who present with clinical symptoms between 3 and 5 years of age [20], and accompanying the increase in CC volume is an age-related increase in ventricle volume in early childhood. It seems that the atypical growth trajectory in ASD individuals reflects their total brain volume growth trajectory and is characterized by rapid early development followed by slower age-related growth, leading to smaller CCs in adolescents and adults with ASD than in age-matched TD controls [11, 15–17].

Generally, CCs are smaller in patients with ASD than TD controls [1]. The global nature of the CC abnormality suggests a disturbance of heterotopic and homotopic callosal connections [21], and this is supported by the fact that patients with agenesis of the CC exhibit social and behavioral problems typically associated with ASD [22, 23].

The disrupted connectivity model derived from resting-state and task-dependent fMRI [24–28] asserts that individuals with ASD show that functional connectivity is stronger between closer brain regions and weaker between regions that are farther apart [26, 29, 30]. This is supported by evidence of a bias in ASD favoring short connections such as arcuate fibers at the expense of longer cortical connections such as commissural fibers [31, 32]. Changes in the superficial cortical layer (II/III) or the deep cortical region layer (V/VI) have been implicated in ASD [33, 34]; roughly 80% of callosal projection neurons are located in the superficial layer and roughly 20% in the deep layer [35].

There is also variability in the allocation of cortical regions involved in visuo-motor tasks in ASD cases. Activation of visual areas on presentation of visual stimuli is lower than in TD individuals. Additionally, there is greater functional connectivity between bilateral visual areas than motor areas in ASD; the opposite obtains in TD controls. Furthermore, there is impaired interhemispheric connectivity across the temporal lobe and greater involvement of posterior brain areas in interhemispheric transfer in ASD [11, 36]. Impaired interhemispheric connectivity across the temporal lobes could have specific consequences for the development of lateralized functions such as language and social communication [11].

32.2.1 Sex Differences

ASD is much more common in males than females. High-functioning females are diagnosed later than males and must typically present with more concurrent behavioral or cognitive issues than males to be diagnosed [23–26]. Because of the difference in incidence, little is known about the neuropathology in females, but there is evidence that they present with a different neuroanatomical profile from males.

There is no significant sex difference in volume of the whole CC or its subregions among ASD subjects when total brain volume and age are considered. In preschool children (3–5 years), both males and females with ASD have smaller areas of the CC with fibers projecting to the superior frontal cortex than TD children.

Measurements of cortical diffusivity were more altered in females with ASD than males. Females with ASD show greater AD, RD, and MD than TD females, while males with ASD show no differences. There are no diagnostic or sex differences in growth rate or diffusion measurements in the CC from three- to five-year-olds.

32.3 Conclusion

ASD is a neurodevelopmental disorder characterized by persistent difficulties in social communication and social interaction and restricted, repetitive interests, patterns of behaviors, and activities. The disorder is accompanied by abnormalities in the CC that have been shown to evolve throughout development and result in a smaller CC during adulthood. When clinical symptoms are manifest, diffusion of callosal neurons is atypical, and this is thought to contribute to the deficits noted in ASD. The atypical patterning differs between male and female children, but because of the limited number of females diagnosed with ASD, more research into these differences is needed. Ultimately, familiarity with the structural changes as they begin and progress longitudinally will help with identifying ASD earlier in life and therefore contribute to earlier intervention.

References

1. Frazier TW, Hardan AY. A meta-analysis of the corpus callosum in autism. Biol Psychiatry. 2009;66(10):935–41.
2. Allouh MZ, Al Barbarawi MM, Ali HA, Mustafa AG, Alomari SO. Morphometric analysis of the corpus callosum according to age and sex in middle eastern Arabs: racial comparisons and clinical correlations to autism Spectrum disorder. Front Syst Neurosci. 2020;14:30.
3. Lai M-C, Lombardo MV, Baron-Cohen S. Autism. Lancet. 2014;383(9920):896–910.
4. Hofvander B, Delorme R, Chaste P, Nydén A, Wentz E, Ståhlberg O, et al. Psychiatric and psychosocial problems in adults with normal-intelligence autism spectrum disorders. BMC Psychiatry. 2009;9:35.
5. Lugnegård T, Hallerbäck MU, Gillberg C. Psychiatric comorbidity in young adults with a clinical diagnosis of Asperger syndrome. Res Dev Disabil. 2011;32(5):1910–7.
6. Lugnegard T, Hallerback MU, Gillberg C. Personality disorders and autism spectrum disorders: what are the connections? Compr Psychiatry. 2012;53(4):333–40.
7. Mattila ML, Hurtig T, Haapsamo H, Jussila K, Kuusikko-Gauffin S, Kielinen M, et al. Comorbid psychiatric disorders associated with Asperger syndrome/high-functioning autism: a community- and clinic-based study. J Autism Dev Disord. 2010;40(9):1080–93.
8. Simonoff E, Pickles A, Charman T, Chandler S, Loucas T, Baird G. Psychiatric disorders in children with autism spectrum disorders: prevalence, comorbidity, and associated factors in a population-derived sample. J Am Acad Child Adolesc Psychiatry. 2008;47(8):921–9.
9. Association AP. Diagnostic and statistical manual of mental disorders, vol. 5. Washington, DC: American Psychiatric Publishing; 2013.
10. Woolfenden S, Sarkozy V, Ridley G, Coory M, Williams K. A systematic review of two outcomes in autism spectrum disorder–epilepsy and mortality. Dev Med Child Neurol. 2012;54(4):306–12.
11. Fingher N, Dinstein I, Ben-Shachar M, Haar S, Dale AM, Eyler L, et al. Toddlers later diagnosed with autism exhibit multiple structural abnormalities in temporal corpus callosum fibers. Cortex. 2017;97:291–305.
12. Wolff JJ, Gerig G, Lewis JD, Soda T, Styner MA, Vachet C, et al. Altered corpus callosum morphology associated with autism over the first 2 years of life. Brain. 2015;138(Pt 7):2046–58.
13. Courchesne E, Redcay E, Morgan JT, Kennedy DP. Autism at the beginning: microstructural and growth abnormalities underlying the cognitive and behavioral phenotype of autism. Dev Psychopathol. 2005;17(3):577–97.
14. Zhang Y, Qin B, Wang L, Zhang K, Song C, Chen J, et al. Corpus callosum volumes in children with autism Spectrum disorders: sex-associated differences. J Autism Dev Disord. 2022;53:2421.
15. Travers BG, do PM T, Adluru N, Lange N, Destiche D, Ennis C, et al. Atypical development of white matter microstructure of the corpus callosum in males with autism: a longitudinal investigation. Mol Autism. 2015;6:15.
16. Ben Bashat D, Kronfeld-Duenias V, Zachor DA, Ekstein PM, Hendler T, Tarrasch R, et al. Accelerated maturation of white matter in young children with autism: a high b value DWI study. NeuroImage. 2007;37(1):40–7.
17. Weinstein M, Ben-Sira L, Levy Y, Zachor DA, Ben Itzhak E, Artzi M, et al. Abnormal white matter integ-

rity in young children with autism. Hum Brain Mapp. 2011;32(4):534–43.
18. Solso S, Xu R, Proudfoot J, Hagler DJ Jr, Campbell K, Venkatraman V, et al. Diffusion tensor imaging provides evidence of possible axonal overconnectivity in frontal lobes in autism spectrum disorder toddlers. Biol Psychiatry. 2016;79(8):676–84.
19. Wolff JJ, Gu H, Gerig G, Elison JT, Styner M, Gouttard S, et al. Differences in white matter fiber tract development present from 6 to 24 months in infants with autism. Am J Psychiatry. 2012;169(6):589–600.
20. Nordahl CW, Iosif AM, Young GS, Perry LM, Dougherty R, Lee A, et al. Sex differences in the corpus callosum in preschool-aged children with autism spectrum disorder. Mol Autism. 2015;6:26.
21. Mihrshahi R. The corpus callosum as an evolutionary innovation. J Exp Zool B Mol Dev Evol. 2006;306(1):8–17.
22. Badaruddin DH, Andrews GL, Bölte S, Schilmoeller KJ, Schilmoeller G, Paul LK, et al. Social and behavioral problems of children with agenesis of the corpus callosum. Child Psychiatry Hum Dev. 2007;38(4):287–302.
23. Casanova MF, El-Baz A, Elnakib A, Switala AE, Williams EL, Williams DL, et al. Quantitative analysis of the shape of the corpus callosum in patients with autism and comparison individuals. Autism. 2011;15(2):223–38.
24. Hull JV, Dokovna LB, Jacokes ZJ, Torgerson CM, Irimia A, Van Horn JD. Resting-state functional connectivity in autism Spectrum disorders: a review. Front Psych. 2016;7:205.
25. Just MA, Cherkassky VL, Keller TA, Minshew NJ. Cortical activation and synchronization during sentence comprehension in high-functioning autism: evidence of underconnectivity. Brain. 2004;127(Pt 8):1811–21.
26. Kana RK, Libero LE, Moore MS. Disrupted cortical connectivity theory as an explanatory model for autism spectrum disorders. Phys Life Rev. 2011;8(4):410–37.
27. Kana RK, Keller TA, Cherkassky VL, Minshew NJ, Just MA. Sentence comprehension in autism: thinking in pictures with decreased functional connectivity. Brain. 2006;129(9):2484–93.
28. Just MA, Keller TA, Malave VL, Kana RK, Varma S. Autism as a neural systems disorder: a theory of frontal-posterior underconnectivity. Neurosci Biobehav Rev. 2012;36(4):1292–313.
29. Maximo JO, Cadena EJ, Kana RK. The implications of brain connectivity in the neuropsychology of autism. Neuropsychol Rev. 2014;24(1):16–31.
30. Vasa RA, Mostofsky SH, Ewen JB. The disrupted connectivity hypothesis of autism Spectrum disorders: time for the next phase in research. Biol Psychiatry Cogn Neurosci Neuroimaging. 2016;1(3):245–52.
31. Belmonte MK, Allen G, Beckel-Mitchener A, Boulanger LM, Carper RA, Webb SJ. Autism and abnormal development of brain connectivity. J Neurosci. 2004;24(42):9228–31.
32. Mottron L, Dawson M, Soulieres I, Hubert B, Burack J. Enhanced perceptual functioning in autism: an update, and eight principles of autistic perception. J Autism Dev Disord. 2006;36(1):27–43.
33. Parikshak Neelroop N, Luo R, Zhang A, Won H, Lowe Jennifer K, Chandran V, et al. Integrative functional genomic analyses implicate specific molecular pathways and circuits in autism. Cell. 2013;155(5):1008–21.
34. Willsey AJ, Sanders Stephan J, Li M, Dong S, Tebbenkamp Andrew T, Muhle Rebecca A, et al. Coexpression networks implicate human Midfetal deep cortical projection neurons in the pathogenesis of autism. Cell. 2013;155(5):997–1007.
35. Li J, Shi M, Ma Z, Zhao S, Euskirchen G, Ziskin J, et al. Integrated systems analysis reveals a molecular network underlying autism spectrum disorders. Mol Syst Biol. 2014;10(12):774.
36. Contarino VE, Bulgheroni S, Annunziata S, Erbetta A, Riva D. Widespread focal cortical alterations in autism spectrum disorder with intellectual disability detected by threshold-free cluster enhancement. AJNR Am J Neuroradiol. 2016;37(9):1721–6.

Part IV

Surgery of the Corpus Callosum

Editor's Summary

Located in the deep interior of the brain, historically, lesions of the corpus callosum (CC) were approached with caution. However, today, neurosurgeons, with a greater knowledge of the anatomy and function of this part of the human neuroanatomy, have learned to approach pathology in and around this structure. Knowledge of the blood supply and relationship of the CC to surrounding structures is of paramount importance to the neurosurgeon. Additionally, our understanding of the results of disconnecting this largest of the commissural tracts has grown. Therefore, callosotomy procedures are now a regular procedure offered to patients with medically recalcitrant epilepsy. This part offers the reader the topics of partial and complete callosotomy, endoscopic-assisted microsurgery of the CC, and various commissurotomy procedures.

33 Partial and Complete Callosotomy of the Corpus Callosum

Mohammed Benzagmout, Meryem Himmiche, Zouhayr Souirti, and Abad Cherif El Asri

33.1 Introduction

The corpus callosum (CC) represents the largest white commissure that crosses the longitudinal cerebral fissure and interconnects the homologous areas of the two hemispheres. The total length of the adult CC is about 6.5 cm, with a thickness that varies from 0.5 cm in the body to 1 cm in the splenium, generally the thickest segment [1]. Moreover, the CC is almost considered as the major anatomical substratum for seizure bilateralization and bisynchronization [2].

Corpus callosotomy consists of partial or total disconnection of the CC. The procedure was described to treat intractable atonic seizures and complex epileptic syndromes and to block the interhemispheric spread of secondary generalized seizures. Certainly palliative, not curative, the corpus callosotomy is one of the disconnection surgical procedures employed to medically treat intractable seizures in patients who are not amenable to focal resection [3]. The goal is to achieve a potential reduction in seizure frequency or intensity or to eliminate certain seizure types that deeply impact patient's life, the family, and caregivers.

In addition, anterior and posterior callosotomy can be used as a surgical pathway to approach lesions in the lateral and third ventricle, the thalamus, and the pineal region.

In epilepsy surgery, the extent of corpus callosotomy varies according to surgical centers and surgeons' preferences. To minimize complications, especially disconnection syndrome, anterior or two-thirds CC split, sparing the splenium, might be advantageous over complete callosotomy [4, 5]. However, complete callosotomy could be indicated if the partial one failed.

In this chapter, we will focus on the role of callosotomy in epilepsy surgery, its indications, techniques, functional results, and complications.

M. Benzagmout (✉)
Department of Neurosurgery, University Hospital of Fez, Fez, Morocco

Clinical Neurosciences Laboratory, Faculty of Medicine, Pharmacy and Dentistry, University Sidi Mohamed Ben Abdellah, Fez, Morocco

M. Himmiche
Department of Neurosurgery, University Hospital of Tangiers, Tangiers, Morocco

Z. Souirti
Clinical Neurosciences Laboratory, Faculty of Medicine, Pharmacy and Dentistry, University Sidi Mohamed Ben Abdellah, Fez, Morocco

Department of Neurology, University Hospital of Fez, Fez, Morocco

A. C. El Asri
Department of Neurosurgery, Mohamed V Military Hospital, Rabat, Morocco

M. Turgut et al. (eds.), *The Corpus Callosum*, https://doi.org/10.1007/978-3-031-38114-0_33

33.2 History

Corpus callosotomy was firstly described in 1940 by Van Wagenen and York Herren as a palliative surgical treatment for intractable seizures [5]. They reported a small sample of epileptic patients with various disruptive lesions of the CC [5]. The authors hypothesized that surgical division of the CC might confine the epileptogenic foci to a single hemisphere, stop bihemispheric seizures spread, and block the secondary generalization. They ligated the anterior one-third of the superior sagittal sinus, divided the CC, and sectioned the anterior commissure and in some cases the fornix [6]. However, this approach had a significant morbidity rate and was for years a contentious procedure [7].

Twenty years later, Bogen and colleagues revisited the technique with total and partial disconnection of the interhemispheric commissures [7]. They performed two approaches: (1) complete disconnection by dividing the entire CC, anterior commissure, unilateral fornix, and hippocampal commissure in some patients. In contrast, (2) partial disconnection was performed by dividing the anterior third of the CC, unilateral fornix, and anterior commissure in patients with bilateral independent seizure discharge limited to frontal and temporal lobes [8]. Later, the corpus callosotomy procedure has gradually gained popularity and started to be frequently used in patients with epilepsy. In the 1970s, Wilson applied the microscope to the procedure [9, 10], and in the 1980s, Maxwell introduced the concept of staged callosotomy. The anterior two-thirds to three-quarter callosotomy was performed in the first stage and posterior callosotomy later on if patients did not respond to the first treatment [7].

Since then, the procedure has obtained wider acceptance and undergone many improvements of its technique. Bahuleyan et al. described endoscopic hemispherotomy in cadaveric dissection, using anterior and posterior burr holes [11]. Although not commonly used, this conventional one-handed technique was not practical and did not offer perioperative safety. Chandra et al. recently used an alternative three-handed technique with a large craniotomy. In this technique, the endoscope is held by the assistant or a holder, which leaves the surgeon's hands free to easily handle the instruments [12, 13]. Since 2011, Sandeep Sood has improved the technique by holding the endoscope with suction in the left hand and the cautery system in the right hand to obtain safe control and bimanual microsurgical dissection via a small entry. A neuronavigation guide is attached to the endoscope to guide corpus callosotomy [14]. For a less invasive and more effective work, the use of endoscope tends to be more common in the field of epilepsy surgery [13–16]. Recent techniques including radiosurgical callosotomy with gamma knife [4, 17–19] and corpus callosotomy performed with laser interstitial thermal therapy [20] have also been reported.

33.3 Indications

The CC is conventionally considered responsible for bilateral seizure synchronization in generalized epilepsy and contributes to rapid spread of seizure activity from one hemisphere to another.

Corpus callosotomy is usually performed as a palliative procedure to mitigate seizures in patients with drug refractory epilepsy, especially in children [21].

A well-defined surgical planning with careful patient's selection, good preoperative assessment, and definition of the extent of callosotomy has to be clearly established beforehand in order to ensure the efficacy of the treatment and limit its side effects.

Roughly, patients for corpus callosotomy might generally, but not always, meet these main criteria [3, 22–24]:

- Multidrug-resistant epilepsy (including major antiepileptic medications).
- Generalized seizures without identifiable and resectable brain foci.
- Epileptic syndromes: Lennox-Gastaut syndrome, West syndrome, infantile hemiplegia syndrome, and drop attack seizures.

- Recurrent episodes of status epilepticus or complex partial seizures with rapid secondary generalization.
- Disabling types of seizures affecting the quality of life with high risk of injury.
- Nonresponse to a ketogenic diet [25].

To date, the debate on the extent of callosotomy persists and requires further discussion. Historically, callosotomy has evolved from total to partial/staged corpus callosotomy including anterior one-half, anterior two-thirds, or posterior callosotomy to avoid brain split syndrome (disconnection syndrome).

Calculating the extent of the callosotomy should take into consideration the patient's age, seizures type, etiology, extent of cerebral lesions, ictal electroencephalogram pattern, epilepsy syndrome, and the surgical risks and complications [26, 27]. Partial or total section of the CC is suitable for generalized major motor seizures but is less beneficial for simple or complex epilepsy [28].

In some series, seizure control is better with total callosotomy than with partial callosotomy, especially in children [29, 30]. On the other hand, staged anterior callosotomy causes fewer complications [31]. Complete corpus callosotomy should be considered as the initial choice in lower-functioning children affected by absence, atonic, or myoclonic seizures [32]. However, anterior callosotomy may be effective in severely affected higher-functioning children, failing that complete posterior callosotomy might be discussed [31].

In medically refractory epilepsy where rapid secondary bisynchrony is suspected but the EEG is nonlocalizing, callosotomy should be considered to treat generalized seizures and may also help identify potentially operable epileptogenic foci [33].

33.4 Contraindications

The presence of a progressive neurological or medical disease or severe psychiatric problems might represent absolute or relative contraindication to corpus callosotomy.

Major behavioral and/or language deficits may occur even with partial splits in patients where language and dominant handedness are located in opposite hemispheres. Thus, a Wada test is recommended in all left-handed patients.

33.5 Techniques

Corpus callosotomy is a palliative disconnection procedure mostly indicated in patients with intractable epileptic seizures. It can be performed via an open surgery or alternatively [15]. Total versus partial CC disconnection varies according to preoperative evaluation and patient's selection [29].

33.5.1 Preoperative Evaluation

Preoperative management of patients considered for corpus callosotomy has to adhere to three fundamental principles [9]: (1) intractable seizures with multiple antiepileptic drugs are used in combination, (2) resectable epileptogenic foci must be excluded, and (3) a patient's expectations from the surgery must be clearly defined and informed consent of the patient with details regarding the procedure alternatives and possible complications must be obtained [24].

33.5.2 General Recommendations

- Patients undergoing surgery should maintain all antiepileptic drugs at the same preoperative doses [3].
- Image-guided navigation system is very useful and helps determine the correct trajectory to map the CC margins, to minimize the size of the bone flap, to determine the anterior-posterior extent of callosotomy, and to distinguish the CC from the cingulate gyri [34, 35].
- Venography with contrast CT scan or magnetic resonance imaging (MRI) is useful to determine the side of entry and to identify the bridging veins to the superior sagittal sinus [24].

– Intraoperative electroencephalogram (EEG) monitoring is used in some institutions when bisynchronous epileptiform discharges are present under general anesthesia. The use of EEG during surgery can guide the extent of CC disconnection. However, it did not effectively predict long-term results after callosotomy [36, 37].

33.5.3 Microsurgical Technique

The patient is placed in a supine lounge chair position. This is important to allow both Trendelenburg and reverse Trendelenburg positioning during surgery. The angle of the surgical approach should help visualize both the genu and the splenium of the CC with minimal brain retraction.

The skin incision depends on the surgeon's preference and the hairline. A bicoronal incision is generally performed and typically biased to the nondominant side. A zig-zag incision might be used to give better cosmetic results.

The bone flap is large, approximately 10–12 cm in anteroposterior dimension, depending on the extent of the planned callosotomy. The lateral dimension of the bone flap is approximately 6 cm biased to the side of the approach. Typically, four burr holes are made at 1 cm to either side of the sagittal sinus. The flap is generally biased to the nondominant hemisphere, and the most critical craniotome cut is made last across the sagittal sinus.

The dural opening is hinged at the sinus. A 1.5–2 cm window will be needed between bridging veins. All efforts should be made to preserve the draining veins to minimize the risk of venous infarction. Self-retaining retractors can be used during the initial dissection of the interhemispheric fissure, but they are usually replaced deeply by cotton balls or pledgets to ensure soft brain retraction.

The interhemispheric dissection requires a slow and gentle manipulation as the pia-arachnoid layers are often adherent or fused below an incompetent falx. A meticulous dissecting technique is recommended to prevent damage to the pia mater and cingulate parenchyma. This permits early visualization of callosomarginal and pericallosal arteries just below the falx cerebri. The extent of exposure depends on whether a complete or partial callosotomy is planned. Generally, it is preferable to complete arachnoid dissection whenever it is possible, to the entire interhemispheric CC area.

Between and under the two pericallosal arteries, the body of the CC is visualized. The pericallosal arteries are dissected carefully. Then, a small entry incision of the CC is created with microaspiration, bipolar cautery, sharp knife, microscissors, or ultrasonic aspiration device and deepened to expose the midline cavum of the septum pellucidum. It is imperative to extend the resection through all commissural fibers to ensure the efficacy of the procedure.

The CC is generally dissected until the ependyma becomes visible. It is mandatory to remain extraependymal by respecting the edges of the lining ependymal membrane (Fig. 33.1). The goal is to reduce the risk of intraventricular bleeding and to avoid cerebrospinal fluid leakage with eventual postoperative hydrocephalus. In case of inadvertent ventricle opening, it is important to seal the breach with large Surgicel pieces to prevent bleeding.

Prior to initiating callosal section, some authors advocate using electrical stimulation of the interhemispheric cortex to identify lower limb motor cortex which in turn defines the mini-

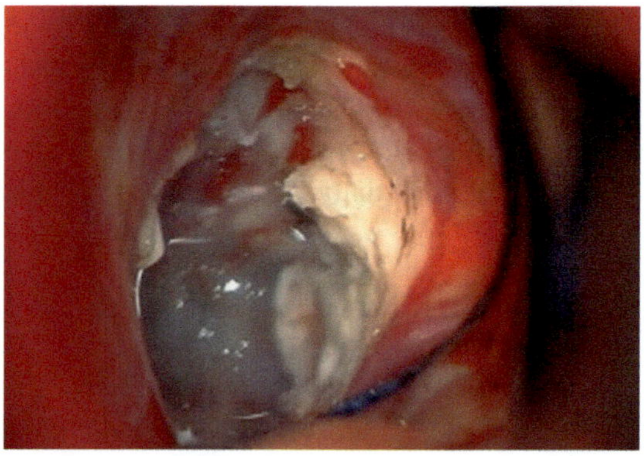

Fig. 33.1 Preoperative microscopic view of a corpus callosotomy. The black arrow indicates the thin ependymal membrane covering the ventricle

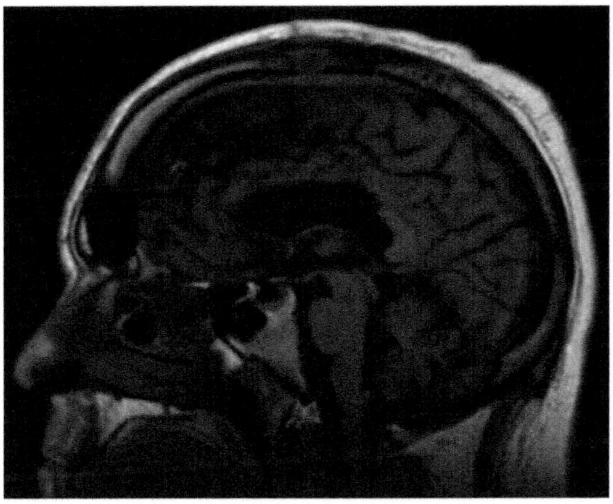

Fig. 33.2 MRI in sagittal T1-weighted view demonstrating a two-thirds anterior callosotomy

mal posterior extent of callosotomy [3]. The latter can also be defined using image-guided navigation. Typically, it is about 5 cm from the anterior genu of the callosum.

Callosotomy should start posteriorly and be continued toward the genu of the CC. The anterior edge of resection is limited by the dorsal surface of the anterior cerebral arteries as well as the anterior commissure which is commonly used as a stopping point to avoid the anterior communicating artery perforators (Fig. 33.2). In case of compete callosotomy, the posterior resection stops upon seeing the arachnoid covering the vein of Galen and the internal cerebral veins.

Papaverine can be used intraoperatively to prevent pericallosal arteries' spasm. After callosotomy, copious irrigation and meticulous hemostasis are routinely performed before closure.

33.5.4 Endoscopic Corpus Callosotomy

The use of endoscopy for corpus callosotomy in epilepsy surgery has recently been reported in several cadaveric studies and case series [11, 14, 16].

The patient is placed in the supine position with the head fixed to a Mayfield head holder in mild extension. A scalp linear frontal parasagittal incision, 1.5 cm lateral to the midline, is made. A D-shaped craniotomy is afterward done with the straight line of "D" run over the midline. The dura is then opened with the flap based medially on the sagittal superior sinus.

For the interhemispheric dissection step, the combined use of the microscope and endoscope seems to be more effective, especially for anterior or complete corpus callosotomy. The microscope is used to meticulously dissect the interhemispheric fissure, whereas the endoscope is used to divide the CC. However, the endoscope can be used alone to perform posterior corpus callosotomy [38] where the endoscope guides the cavitron ultrasonic surgical aspirator.

The advantages of the technique lie in the possibility to reduce the length of the skin incision and that of the craniotomy and hence the shorter surgical time, shorter hospital stay, and lower complication rates [38]. Nonetheless, these advantages should be proved by larger multicentric studies.

It is important to underline that patients who already underwent cranial surgery are not eligible for endoscopic callosotomy. In addition, patients with distorted anatomy and significant pial adhesions are predisposed to perioperative difficulties with more morbidity and mortality [15].

33.5.5 Corpus Callosotomy Without Craniotomy

Minimally invasive approaches to corpus callosotomy have been described to ensure more effectiveness and decreased complication rate of the procedure. Both stereotactic radiosurgery [19] and laser interstitial thermal therapy (LITT) [20, 39–41] have recently been developed to perform corpus callosotomy.

Nowadays, radiosurgery appears as an effective alternative treatment option to open callosotomy based on its effect on reducing epilepsy and significantly improving patients' quality of life [42]. In 1999, Pendl et al. firstly used radiosurgical callosotomy in three adult patients with multifocal refractory epilepsy [19]. Over the past two decades, several reports demonstrated the efficacy of gamma knife callosotomy as a safe and effective noninvasive approach [17, 18, 43] with up to 84% of improvement of drop attacks and generalized tonic-clonic seizures during the follow-up [43]. Conversely, previous studies on callosotomy showed that such control of seizure tends to decrease after 2 years [44].

MRI-guided laser interstitial thermal therapy is a safe and effective approach for corpus callosotomy. It represents an emerging technique that offers significant advantages over open callosotomy, including its minimally invasive nature, shorter hospitalization and operative time, decreased total blood loss, and comparable seizure control rates in both adult [41, 45–47] and pediatric patients [20, 45, 48]. The technique consists roughly in stereotactic insertion of ablation probes in the CC through the guidance bolt along the planned trajectory already created using stereotactic and navigation systems. LITT is then performed under direct MR thermography. MRI images are subsequently obtained to assess tissue changes and successful ablation of fibers. Afterward, the ablation bolt is removed and the holes are sutured. Recently, Singh et al. have performed LITT callosotomy using robotic system [47].

The effectiveness of these techniques is still under study and has to be evaluated in larger series to define the patient's selective criteria as well as functional results.

33.6 Functional Results

Generally, atonic seizure is the seizure prototype that has the best response to callosal section, whereas myoclonic seizure has the least [49].

In children under 16 years old, atonic seizures respond much better than any other generalized seizure type to anterior callosotomy. Total callosotomy has also improved response rates, but much more surgical benefits were seen with tonic, tonic-clonic, and absence seizures [50]. On the other hand, complete arrest of drop attacks (atonic and tonic seizures) is observed in up to 91% of pediatric cases after total callosotomy and in 67% after partial callosotomy [51].

In adults, total callosotomy offers approximately 10% higher response rate for all seizure types compared with anterior callosotomy alone [50].

In a recent meta-analysis, Chan et al. found that patients were more likely to be completely seizure free after callosotomy if they suffered from infantile spasms and their preoperative MRI was normal or had epilepsy for less than 15 years [52]. In addition, patients were more likely to be free of drop attacks if the procedure was a single-stage complete rather than partial callosotomy [52].

Anterior callosotomy is an effective treatment for patients with frontal complex partial seizures, particularly when EEG demonstrates rapid propagation to contralateral frontal regions with or without generalization. At least half of these patients achieve considerable improvement after surgery. Similarly, good surgical results of callosotomy were obtained in patients with bilateral diffuse cortical dysplasia and lissencephaly [51]. In contrast, temporal lobe complex partial seizures did not respond to anterior callosotomy [53].

Finally, it is important to note that only 35% of patients remain drop attacks-free in the long-term follow-up (>5 years) [54].

33.7 Complications

33.7.1 General Complications

Even with refinement of microsurgical techniques, surgical complications following corpus callosotomy are common and join the general brain surgery complications. The combined morbidity and mortality in microsurgical corpus callosotomy are less than 11% [55].

These complications include epidural/subdural hematomas, subdural cerebrospinal fluid accumulation, hydrocephalus, infections (e.g., meningitis, empyema, osteomyelitis), and deep vein thrombosis [51, 56–58].

Cerebrospinal fluid leaks might occur after corpus callosotomy, especially if there was brain atrophy or accidental injury to brain ventricle wall. This can cause subdural fluid collection or even hydrocephalus.

Subdural hematoma occurs secondarily to bridging vein damage or aggressive brain retraction. This can be managed by preoperative venography to choose the entry zone and by careful arachnoid dissection to limit brain retraction.

Early complications such as hemispheric edema or brain infarction in the territory of the anterior cerebral artery can result from damage to callosomarginal or pericallosal arteries and lead to death or severe morbidity.

33.7.2 Disconnection Syndrome

Disconnection syndrome is more common with total section than with anterior callosotomy [56, 59]. The syndrome features are left tactile anomia, left-sided dyspraxia (resembling hemiparesis), and incontinence. Briefly, stimulations reaching solely the nondominant hemisphere may not be verbally reported by the patient as the dominant hemisphere has no access to the information. Besides, the nondominant hand no longer responds reliably to verbal commands, because the dominant hemisphere may not readily transfer information to the nondominant motor cortex [59, 60].

Most patients are unaware of the deficit [59]. Transient postoperative apathy is sometimes observed and is probably related to mesial and convexity frontal lobe disconnections. Moreover, right heminegligence is usually observed and could be attributed to the posterior callosotomy [49].

This complication generally resolves over several months, but mild to moderate residual symptoms may persist [61, 62].

33.7.3 Alien Hand Syndrome (AHS)

AHS appears more commonly after total callosotomy, particularly in patients with significant unilateral cerebral injury prior to surgery [3].

It refers to a stressful condition in which a patient's nondominant hand acts without apparent guidance by the patient's own motivation. Sometimes, the hand performs disagreeable or conflicting actions, and patients must use their dominant hand to restrain the alien hand actions [63]. This creates the scene of intermanual conflict. The ipsilateral lower extremity is often involved as well [3].

These symptoms usually resolve with time, albeit are lasting and may fluctuate over the years [3].

33.7.4 Chemical Meningitis

In 1993, Burchiel reported that the frequency of chemical meningitis following corpus callosotomy in children was higher than 33% [64]. This complication was also mentioned in other reports [12, 65].

Chemical meningitis is probably due to intraventricular inflammation induced by the iatrogenic breach of the ependymal membrane. Interestingly, Fujimoto et al. found that post-callosotomy pneumocephalus was a risk factor for chemical meningitis in those patients [66].

Many authors advocate the use of postoperative short-term corticosteroids to prevent post-callosotomy chemical meningitis [3, 24, 59].

33.7.5 Transient Neurological Deficits

These include hemiparesis or apraxia of variable degrees, disinhibited forced grasp in the nondominant hand (usually left), gait difficulty, and urinary incontinence. These deficits are likely related to acute parasagittal disconnection and/or traction effect on the nondominant frontal paracentral cortex. They usually completely resolve within few weeks to months [24, 49, 57, 58, 60].

33.7.6 Language Impairment

Different language disturbances might appear after corpus callosotomy depending on the hemisphere dominance of the patient. They include isolated speech difficulty, speech and writing difficulties, and/or dysgraphia with intact speech [67].

These impairments are related to the supplementary motor area traction during surgery and/or suboptimal integration of cortical responses after corpus callosotomy [3, 24, 60].

These troubles usually recover within 4 weeks [56, 59]. However, cases with prolonged mutism have been reported [62].

33.7.7 Cognitive and Neuropsychological Deficits

Memory dysfunction has been reported in a minority of patients following callosotomy. This may be explained by the section of the hippocampal commissure and/or the posterior part of CC [3].

In addition, a variety of neuropsychological signs might occur in patients after corpus callosotomy such as depression, excessive self-insistence, slight stammering, or schizophrenic symptoms [56, 59]. These personality changes require careful neuropsychological evaluation after surgery.

33.8 Conclusion

Over the past two decades, improvements in surgical technique, seizure monitoring, and postoperative care have established corpus callosotomy as an effective treatment for patients with idiopathic intractable generalized epilepsy. The procedure involves partial or total disconnection of the CC to block the interhemispheric spread of secondary generalized seizures.

Corpus callosotomy can be performed using basic microsurgical techniques or via endoscopy. Furthermore, stereotactic laser interstitial thermal therapy and radiosurgical corpus callosotomy have recently been developed to reduce the invasiveness of the procedure. However, data available on the long-term follow-up of these techniques are not enough to evidence their effectiveness.

Debate is still open regarding the extent of callosotomy. While many authors have suggested that only anterior callosotomy suffices to achieve clinical efficacy, others advocate for a complete callosotomy [68].

Several complications might occur in the postoperative stage. Thus, surgeons should make the maximum effort to minimize the complication rate by the good selection of patients and strict adherence to the different surgical steps of callosotomy.

References

1. Musiek FE. Neuroanatomy, neurophysiology, and central auditory assessment: part III. Corpus callosum and efferent pathways. Ear Hear. 1986;7:349–58.
2. Wada JA. Midline subcortical structures for transhemispheric ictal and interictal transmission. In: Reeves AG, Roberts DW, editors. Epilepsy and the corpus callosum 2. advances in behavioral biology, vol. 45. New York: Plenum; 1995. p. 61–78.
3. Asadi-Pooya AA, Sharan A, Nei M, Sperling MR. Corpus callosotomy. Epilepsy Behav. 2008;13(2):271–8.
4. Maxwell RE, Gates JR, Gumnit RJ. Corpus callosotomy at the University of Minnesota. In: Engel Jr J, editor. Surgical treatment of the epilepsies. New York: Raven Press; 1986. p. 659–66.
5. Van Wagenen WP, Herren RY. Surgical divisions of commissural pathways in the corpus callosum: relation to spread of an epileptic attack. Arch Neurol Psychiatr. 1940;44:740–59.
6. Mathews MS, Linskey ME, Binder DK. Wiliam P. van Wagenen and the first corpus callosotomies for epilepsy. J Neurosurg. 2008;108(3):608–13.
7. Vaddiparti A, Huang R, Blihar D, Du Pleissis M, Montalbano MJ, Tubbs S, Loukas M. The evolution of the corpus callosotomy for epilepsy management. World Neurosurg. 2021;145:455–61.

8. Bogen JE, Fisher ED, Vogel PJ. Cerebral commissurotomy. A second case report. JAMA. 1965;194(12):1328–9.
9. Park EK, Kim DS. Corpus callosotomy in pediatric intractable epilepsy: microsurgical technique implication and variation. In: Gadze ZP, editor. Epilepsy in children—clinical and social aspects. Chicago, IL: InTech; 2011.
10. Wilson DH, Reeves A, Gazzaniga M. Division of the corpus callosum for uncontrollable epilepsy. Neurology. 1978;28(7):649–53.
11. Bahuleyan B, Vogel TW, Robinson S, Cohen AR. Endoscopic total corpus callosotomy: cadaveric demonstration of a new approach. Pediatr Neurosurg. 2011;47(6):455–60.
12. Chandra SP, Kurwale NS, Chibber SS, Banerji J, Dwivedi R, Garg A, Bal C, Tripathi M, Sarkar C, Tripathi M. Endoscopic-assisted (through a mini-craniotomy) corpus callosotomy combined with anterior, hippocampal, and posterior commissurotomy in Lennox-Gastaut syndrome: a pilot study to establish its safety and efficacy. Neurosurgery. 2016;78(5):743–51.
13. Sood S, Rajah GB. Endoscopic callosotomy and hemispherotomy: bimanual endoscopic technique. In: Villemure JG, Baltuch GH, Cukiert A, editors. Operative techniques in epilepsy surgery. Stuttgart: Thieme; 2020.
14. Sood S, Marupudi NI, Asano E, Haridas A, Ham SD. Endoscopic corpus callosotomy and hemispherotomy. J Neurosurg Pediatr. 2015;16(6):681–6.
15. Smyth MD, Vellimana AK, Asano E, Sood S. Corpus callosotomy-open and endoscopic surgical techniques. Epilepsia. 2017;58(S1):73–9.
16. Sood S, Asano E, Altinok D, Luat A. Endoscopic posterior interhemispheric complete corpus callosotomy. J Neurosurg Pediatr. 2016;25:689–92.
17. Eder HG, Feichtinger M, Pieper T, Kurschel S, Schroettner O. Gamma knife radiosurgery for callosotomy in children with drug-resistant epilepsy. Childs Nerv Syst. 2006;22(8):1012–7.
18. Feichtinger M, Schröttner O, Eder H, Holthausen H, Pieper T, Unger F, Holl A, Gruber L, Körner E, Trinka E, Fazekas F, Ott E. Efficacy and safety of radiosurgical callosotomy: a retrospective analysis. Epilepsia. 2006;47(7):1184–91.
19. Pendl G, Eder HG, Schroettner O, Leber KA. Corpus callosotomy with radiosurgery. Neurosurgery. 1999;45:303–8.
20. Palma AE, Wicks RT, Popli G, Couture DE. Corpus callosotomy via laser interstitial thermal therapy: a case series. J Neurosurg Pediatr. 2018;23(3):303–7.
21. Abou-khalil BW. When should corpus callosotomy be offered as palliative therapy. Epilepsy Curr. 2010;10(1):9–10.
22. Pinard JM, Delalande O, Chiron C, Soufflet C, Plouin P, Kim Y, Dulac O. Callosotomy for epilepsy after West syndrome. Epilepsia. 1999;40(12):1727–34.
23. Stigsdotter-Broman L, Olsson I, Flink R, Rydenhag B, Malmren K. Long term follow-up after callosotomy-a prospective population based, observational study. Epilepsia. 2014;55(2):316–21.
24. Wong TT, Kwan SY, Chang KP, Hsiu-Mei W, Yang TF, Chen YS, Yi-Yen L. Corpus callosotomy in children. Childs Nerv Syst. 2006;22(8):999–1011.
25. Lemmon ME, Terao NN, Ng Y, Reisig W, Rubenstein JE, Kossoff EH. Efficacy of the ketogenic diet in Lennox-gastaut syndrome: a retrospective review of one institution's experience and summary of the literature. Dev Med Child Neurol. 2012;54(5):464–8.
26. Luat AF, Asano E, Kumar A, Chugani HT, Sood S. Corpus callosotomy for intractable epilepsy revisited: the children's hospital of Michigan series. J Child Neurol. 2017;32(7):624–9.
27. Schaller K, Cabrilo I. Corpus callosotomy. Acta Neurochir. 2016;158(1):155–60.
28. Chen PC, Messina SA, Castillo E, Baumgartner J, Seo JH, Skinner H, Gireesh ED, Lee KH. Altered integrity of corpus callosum in generalized epilepsy in relation to seizure lateralization after corpus callosotmy. Neurosurg Focus. 2020;48(4):E15.
29. Arifin MT, Muttaqin Z, Bakhtiar Y, Andar E, Priambada D, Kurnia H, Risdianto A, Tsaniadi K, Kusnarto G, Bunyamin J. Seizure outcomes in patients with complete versus anterior corpus callosotomy: analysis of outcome. Int J Gen Med. 2020;13:105–10.
30. Rathore C, Abraham M, Rao RM, George A, Sarma PS, Radhakrishnan K. Outcome after corpus callosotomy in children with injurious drop attacks and severe mental retardation. Brain and Development. 2007;29(9):577–85.
31. Pilcher WH, Rusyniak WG. Complications of epilepsy surgery. Neurosurg Clin N Am. 1993;4(2):311–25.
32. Jalilian L, Limbrick DD, Sterger-may K, Johnston J, Powers AK, Smyth MD. Complete versus anterior two-third corpus callosotomy in children: analysis of outcome. J Neurosurg. 2010;6(3):257–66.
33. Clarke DF, Wheless JW, Chacon MM, et al. Corpus callosotomy: a palliative therapeutic technique may help identify resectable epileptogenic foci. Seizure. 2007;16:545–53.
34. Hashizume K, Tanaka T, Kunimuto M, Yonemasu Y. Image-guided surgery for epilepsy. Neurol Surg. 1997;25(4):329–35.
35. Hodaie M, Musharbash A, Otsubo H, Snad OC, Chitoku S, Holowka S, Hoffman HJ, Rutka JT. Image-guided, frameless stereotactic sectioning of the corpus callosum in children with intractable epilepsy. Pediatr Neurosurg. 2001;34(6):286–94.
36. Fiol ME, Gates JR, Mireles R, Maxwell RE, Erickson DM. Value of intraoperative EEG changes during corpus callosotomy in prediction surgical results. Epilepsia. 1993;34(1):74–8.
37. Kwan SY, Lin JH, Wong TT, Chang KP, Yiu CH. Prognostic value of electrocorticography finding during callosotomy in children with Lennox-gastaut syndrome. Seizure. 2005;14(7):470–5.
38. Uda T, Kunihiro N, Umaba R, Koh S, Kawashima T, Ikeda S, Ishimoto K, Goto T. Surgical aspects of corpus callosotomy. Brain Sci. 2021;11(12):1608.

39. Ho AL, Miller KJ, Cartmell S, Inoyama K, Fisher RS, Halpern CH. Stereotactic laser ablation of the splenium for intractable epilepsy. Epilep Behav Case Rep. 2016;5:23–6.
40. Lehner KR, Yeagle EM, Argyelan M, Klimaj Z, Du V, Megevand P, Hwang ST, Mehta AD. Validation of corpus callosotomy after laser interstitial thermal therapy: a multimodal approach. J Neurosurg. 2018:1–11.
41. Pruitt R, Gamble A, Black K, Schulder M, Mehta AD. Complication avoidance in laser interstitial thermal therapy: lessons learned. J Neurosurg. 2017;126:1238–45.
42. Bodaghabadi M, Bitaraf MA, Aran S, Alikhani M, Ashrafian H, Zahiri A, Alahverdi M. Corpus callosotomy with gamma knife radiosurgery for a case of intractable generalised epilepsy. Epileptic Disord. 2011;13(2):202–8.
43. Celis MA, Moreno-Jiménez S, Lárraga-Gutiérrez JM, Alonso-Vanegas MA, García-Garduño OA, Martínez-Juárez IE, Fernández-Gónzalez MC. Corpus callosotomy using conformal stereotactic radiosurgery. Childs Nerv Syst. 2007;23(8):917–20.
44. Pressler RM, Binnie CD, Elwes RD, Polkey CE. Return of generalized seizures and discharges after callosotomy. Adv Neurol. 1999;81:171–82.
45. Huang Y, Yecies D, Bruckert L, Parker JJ, Ho AL, Kim LH, Fornoff L, Wintermark M, Porter B, Yeom KW, Halpern CH, Grant GA. Stereotactic laser ablation for completion corpus callosotomy. J Neurosurg Pediatr. 2019;2:1–9.
46. Rich CW, Fasano RE, Isbaine F, Saindane AM, Qiu D, Curry DJ, Gross RE, Willie JT. MRI-guided stereotactic laser corpus callosotomy for epilepsy: distinct methods and outcomes. J Neurosurg. 2021;22:1–13.
47. Singh H, Essayed WI, Deb S, Hoffman C, Schwartz TH. Minimally invasive robotic laser corpus callosotomy: a proof of concept. Cureus. 2017;9(2):e1021.
48. Caruso JP, Janjua MB, Dolce A, Price AV. Retrospective analysis of open surgical versus laser interstitial thermal therapy callosotomy in pediatric patients with refractory epilepsy. J Neurosurg Pediatr. 2021;27(4):420–8.
49. Cukiert A, Burattini JA, Mariani PP, Câmara RB, Seda L, Baldauf CM, Argentoni M, Baise-Zung C, Forster CR, Mello VA. Extended, one-stage callosal section for treatment of refractory secondarily generalized epilepsy in patients with Lennox-Gastaut and Lennox-like syndromes. Epilepsia. 2006;47(2):371–4.
50. Spencer SS, Spencer DD. Seizure types: results of partial and complete callosotomy in adults. In: Reeves AG, Roberts DW, editors. Epilepsy and the corpus callosum 2. Advances in behavioral biology, vol. 45. New York: Plenum; 1995. p. 145–52.
51. Shimizu H. Our experience with pediatric epilepsy surgery focusing on corpus callosotomy and hemispherotomy. Epilepsia. 2005;46(Suppl 1):30–1.
52. Chan AY, Rolston JD, Lee B, Vadera S, Englot DJ. Rates and predictors of seizure outcome after corpus callosotomy for drug-resistant epilepsy: a meta-analysis. J Neurosurg. 2018;1:1–10.
53. Purves SJ, Wada JA, Woodhurst WB. Corpus callosum section for complex partial seizures. In: Reeves AG, Roberts DW, editors. Epilepsy and the corpus callosum 2. Advances in behavioral biology, vol. 45. New York: Plenum; 1995. p. 175–82.
54. Téllez-Zenteno JF, Dhar R, Wiebe S. Long-term seizure outcomes following epilepsy surgery: a systematic review and meta-analysis. Brain. 2005;128(Pt 5):1188–98.
55. Petrozza PH. Brain cortex resection (for epilepsy). In: Fleisher LA, Roizen MF, editors. Essence of anesthesia practice. Third ed. Amsterdam: Elsevier; 2011. p. 414–5.
56. Maehara T, Shimizu H. Surgical outcome of corpus callosotomy in patients with drop attacks. Epilepsia. 2001;42(1):67–71.
57. Maehara T, Shimizu H, Oda M, Arai N. Surgical treatment of children with medically intractable epilepsy—outcome of various surgical procedures. Neurol Med Chir (Tokyo). 1996;36(5):305–9.
58. Nei M, O'Connor M, Liporace J, Sperling MR. Refractory generalized seizures: response to corpus callosotomy and vagal nerve stimulation. Epilepsia. 2006;47(1):115–22.
59. Kim DS, Yang KH, Kim TG, Chang JH, Chang JW, Choi JU, Lee BI. The surgical effect of callosotomy in the treatment of intractable seizure. Yonsei Med J. 2004;45(2):233–40.
60. Reeves AG, Risse G. Neurological effects of callosotomy. In: Reeves AG, Roberts DW, editors. Epilepsy and the corpus callosum 2. Advances in behavioral biology, vol. 45. New York: Plenum; 1995. p. 241–51.
61. Andersen B, Rogvi-Hansen B, Kruse-Larsen C, Dam M. Corpus callosotomy: seizure and psychosocial outcome. A 39-month follow-up of 20 patients. Epilepsy Res. 1996;23(1):77–85.
62. Jenssen S, Sperling MR, Tracy JI, Nei M, Joyce L, David G, O'Connor M. Corpus callosotomy in refractory idiopathic generalized epilepsy. Seizure. 2006;15(8):621–9.
63. Assal F, Schwartz S, Vuilleumier P. Moving with or without will: functional neural correlates of alien hand syndrome. Ann Neurol. 2007;62:301–6.
64. Burchiel KJ. Corpus callosotomy for intractable epilepsy in children. West J Med. 1993;158:66–7.
65. Hader W, Bezchlibnyk I, Pillay N, Wiebe S. Corpus callosotomy, risks and benefits: a systematic review of the evidence. Epilepsy Curr. 2012;12:1.
66. Fujimoto A, Hatano K, Nozaki T, Sato K, Enoki H, Okanishi T. Postoperative pneumocephalus on computed tomography might predict post-corpus callosotomy chemical meningitis. Brain Sci. 2021;11(5):638.
67. Sass KJ, Novelly RA, Spencer DD, Spencer SS. Postcallosotomy language impairments in patients with crossed cerebral dominance. J Neurosurg. 1990;72(1):85–90.
68. Malmgren K, Rydenhag B, Hallbook T. Reappraisal of corpus callosotomy. Curr Opin Neurol. 2015;28:175–81.

Commissurotomy of the Corpus Callosum

34

Mengzhao Feng, Yuchao Zuo, and Fuyou Guo

34.1 Introduction

The corpus callosum (CC) is a group of white matter tracts that span the left and right cerebral hemispheres. It is approximately 6.5 cm long and 0.5–1.0 cm thick. From cranial to caudal regions, it forms a characteristic "C" shape. It is segmented into rostrum, genu, body, isthmus, and splenium. The CC is the largest forebrain commissure, with approximately 190 million axons crossing the midline. It has been reported that up to two-thirds of patients with epilepsy have complete or partial CC agenesis [1]. Corpus callosotomy, first used in the management of epilepsy by William P. van Wagenen in 1940, has been chosen for patients with secondary generalized seizures originating from a unilateral focus not well controlled by medication [2]. It is usually indicated if patients demonstrate a resectable epileptic focus that cannot be identified and are not candidates for resection surgery. However, some data indicate no effect on or worsening of seizures after commissurotomy [3]. In view of the complexity of many categories and potential complications caused by improper management, it is necessary to investigate optimal selection criteria for corpus callosotomy to achieve excellent outcomes following surgery.

Commissurotomy of the CC is a major disconnection procedure that functionally isolates the cerebral hemispheres, interrupting the spread of epileptic activity from one hemisphere to the other. It is extremely useful in children suffering from nonlocalized drug refractory epilepsy, especially drop attacks [4]. However, commissurotomy of the CC is rarely curative and is only adopted when the epileptogenic focus cannot be identified after multimodality neuroimaging and electroencephalography. Although callosotomy is considered a palliative surgical method, thorough evaluation of patient histories and seizure activities is still required to identify appropriate candidates. In addition, patients who have undergone CC can experience disconnection syndromes, which can be temporary or persist long after surgery [5]. These complications predominantly develop if patients undergo total corpus callosotomies. Consequently, this chapter focuses on patient selection for surgery and surgical techniques and on surgical outcomes as revealed in previous literature and by our institutional practices.

M. Feng · Y. Zuo · F. Guo (✉)
Department of Neurosurgery, First Affiliated Hospital of Zhengzhou University,
Zhengzhou, Henan Province, China

International Joint Laboratory of Nervous System Malformations, First Affiliated Hospital of Zhengzhou University,
Zhengzhou, Henan Province, China

M. Turgut et al. (eds.), *The Corpus Callosum*, https://doi.org/10.1007/978-3-031-38114-0_34

34.2 Indication for Surgery

Every candidate patient being assessed for epilepsy surgery should undergo detailed examinations, and preoperative evaluations should be used in the following multidisciplinary approach. They should include (1) neuroimaging examinations, such as computed tomography (CT) or magnetic resonance imaging (MRI); (2) neuropsychology and functional imaging, including ictal single-photon emission computed tomography (SPECT) or magnetoencephalography (MEG); (3) neurophysiological (both invasive and noninvasive) assessment, including routine electroencephalography (EEG) or video EEG monitoring (Figs. 34.1 and 34.2); and (4) clinical assessment consisting of a history of seizures and general/neurological examination.

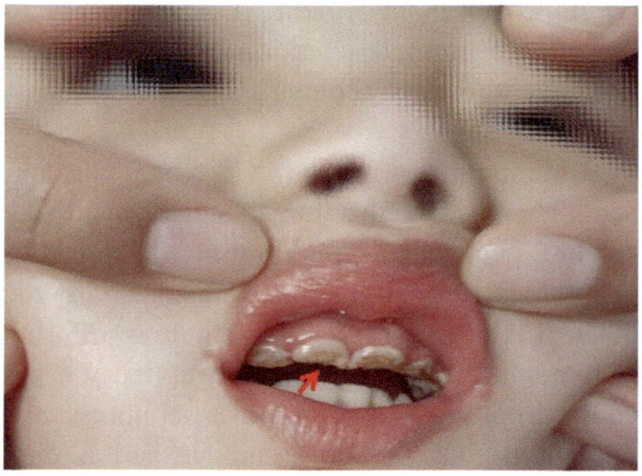

Fig. 34.1 Female, 3.5 years of age, seizure onset for 1 year. Typical generalized seizures characterized by sudden drop attacks. The red arrow indicates a broken tooth due to long-term sudden drop attacks

The aim of surgical intervention is either to resect a structural abnormality or to modify brain function to improve the control of epilepsy. Resection surgery can eliminate seizures in up to 80% of patients; in contrast, functional procedures lead to significant seizure control in approximately 30–80%. Hence, resection is the first choice for optimal epilepsy surgery, which consists of identifying the seizure focus and resecting it safely. The primary goal of commissurotomy of the CC is to prevent the rapid spread of epileptic activity from one hemisphere to the other, so that patients experience less severe seizures and injuries. Callosotomy is obviously indicated for patients with medically intractable epilepsy involving severe and frequent atonic or tonic seizures, especially in the absence of localized resectable pathology.

Generally, according to the views of Roberts and Siegel [6], there are three selection criteria: (1) medical intractability of at least two and usually 4 years' duration, with exhaustive anticonvulsant regimens and documented adequate serum anticonvulsant levels; (2) generalized seizures, other types including atonic or akinetic seizures generally characterized by sudden drop attacks; and (3) potential functional benefit if the seizure disorder improves. In addition, absence of demonstrable focal seizure onset and bilateral independent EEG abnormalities are recommended for commissurotomy.

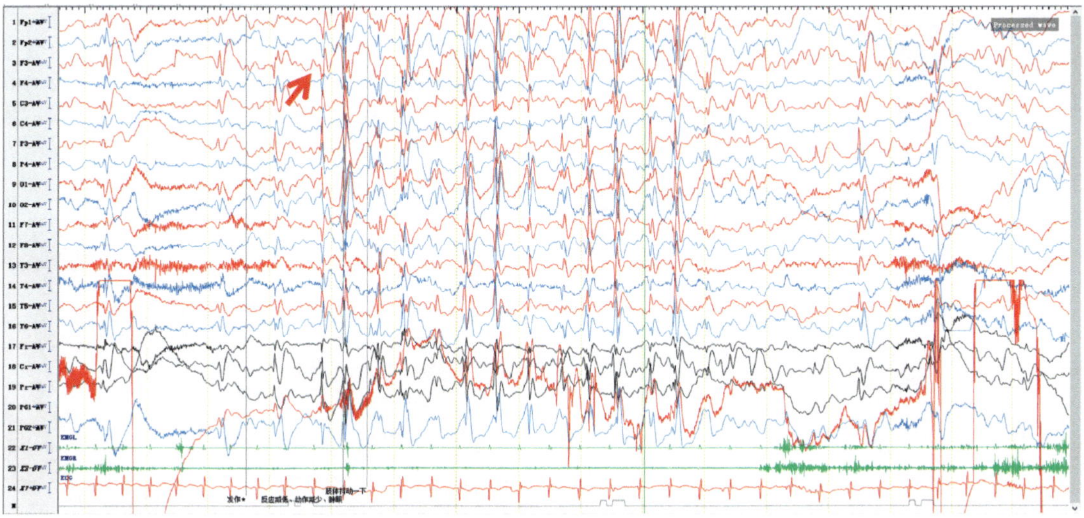

Fig. 34.2 Sleep EEG before callosotomy demonstrating generalized spikes and waves, 2 s^{-1} sharp and slow waves, bilateral runs of rapid spikes, and awake EEG revealing myoclonic seizures. The 24-h video EEG monitored (red arrow) several atypical distraction seizures (sometimes mixed myoclonic seizures), myoclonic-tone seizures, and several myoclonic seizures during the waking period

34.3 Surgical Procedures and Techniques

34.3.1 Standard Surgical Procedures

Microsurgery was performed under general anesthesia with the patient in the supine position with the head fixed in a Mayfield holder. A limited bicoronal incision was made 2 cm anterior to the coronal suture and extended more toward the nondominant hemisphere. Craniotomy was usually performed in the right parasagittal approach, and a medically based dural flap was reflected on the sagittal sinus. Intraoperative EEG recording was then placed for real-time monitoring. The interhemispheric fissure must be opened; fixed brain retractors proved helpful for exposure. The operating microscope was used to divide the adhesions and carefully protect the pericallosal artery. The glistening white appearance of the CC distinguishes it from the more superficial cingulate gyrus. It is vital to expose along the length of the callosum before entering the commissure. The callosal fibers are gently divided using the microseptal and a small sucker or the cavitation ultrasonic surgical aspirator (CUSA). Intraoperative photographs in microsurgical total corpus callosotomy are shown in Fig. 34.3.

Further attention should be given to the extent of resection during surgery. On the basis of previous literature [7], total corpus callosotomy is superior to anterior two-thirds callosotomy in eliminating seizures. However, anterior two-thirds callosotomy preserves the splenium of the corpus, which could prevent disconnection syndrome in older patients (over 15 years). Total corpus callosotomy was advocated if the patient was under 10. For patients between 10 and 15 years old, the extent of callosotomy was determined on the basis of activities of daily living, cognitive functions, and electroencephalographic findings. Additional posterior one-third callosotomy was performed at least three months later. Of course, neuronavigation facilitates accurate orientation and easily identifies the four segments, the rostrum, genu, body, and splenium, to ensure the precision of surgery [8].

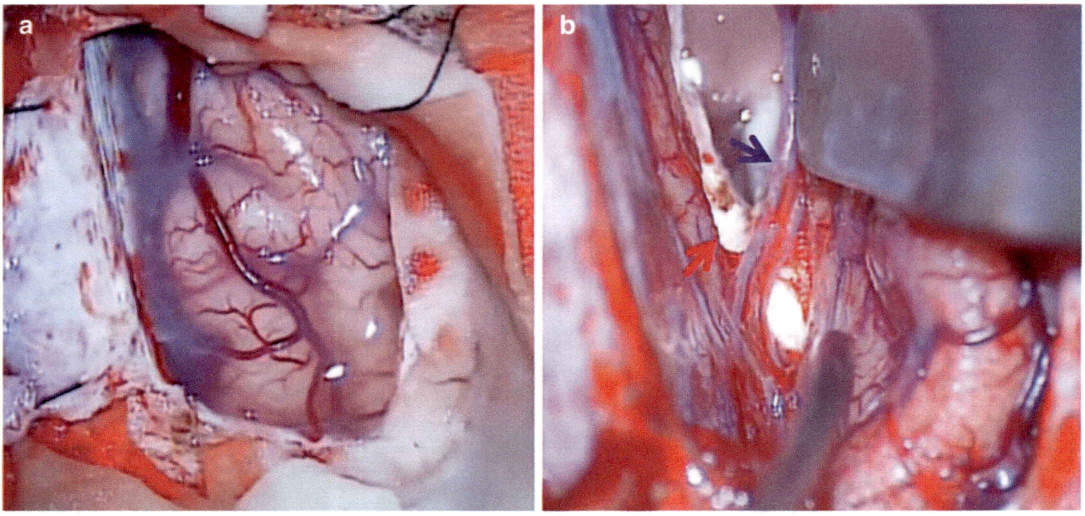

Fig. 34.3 Intraoperative photographs in microsurgical total corpus callosotomy. Dural flap was reflected on the sagittal sinus (**a**); anterior two-thirds callosotomy was performed. Purple arrow shows pericallosal arteries; red arrow shows corpus callosotomy resorption (**b**)

Several key surgical skills contribute to the success of corpus callosotomy. First, vessel protection is important. Not only should major bridging veins be preserved to prevent venous infarction, care should also be taken to avoid injuring deep-seated veins such as the internal cerebral veins and vein of Galen; the pericallosal arteries also need to be protected from injury. Second, the pial integrity of the cingulate gyri should be preserved to reduce the likelihood of postoperative mutism. Third, no attempt was made to divide the anterior commissure blindly; posterior callosotomy is more challenging as the angle of the splenium falls away from the surgeon under the microscope. It is wise to adopt a second-stage procedure through a separate posterior incision, and craniotomy provides more direct access to the splenium [9].

34.3.2 Surgical Techniques

The evolution of corpus callosotomy has focused on yielding better outcomes with minimal invasion. Newer technical instruments and navigational tools are being introduced to provide better visualization, lessen injury, and reduce side effects. In addition, endoscope-assisted callosotomy provides for a smaller incision and craniotomy and can decrease operative morbidity [10]. Endoscope-assisted surgical approaches can help to reduce morbidity related to traditional open surgical approaches while allowing for good visualization of surgical targets [11]. Sood [12] reported four patients with drop attacks and two with intractable seizures related to a neonatal stroke who underwent endoscopic complete corpus callosotomy and hemispherotomy. All operations were accomplished successfully in all patients owing to the excellent visualization. A short operating time and minimal blood loss and a median hospital stay of 4 days were found in the present series with endoscopic corpus callosotomy. Satisfactory outcomes, including complete resolution of drop attacks and seizure-free status, were achieved after corpus callosotomy and hemispherotomy, respectively. Posterior quadrantectomy and corpus callosotomy were similarly shown to be effective and helpful for endoscope-assisted surgery to curb medically refractory epilepsy [13].

Although microsurgical callosotomy is regarded as the gold standard, radiosurgical callosotomy can be selected as a supplementary or alternative technique for microsurgical callosotomy if there is an unfavorable microsurgical risk and subsequent complications. Radiosurgical corpus callosotomy offers a noninvasive technique for division of the CC, which can confer improved seizure control on patients suffering from frequent atonic seizures due to rapid interhemispheric generalization. There is increasing evidence that stereotactic radiosurgery methods such as the gamma knife approach are alternative treatments if open anterior corpus callosotomy fails [14]. In addition, drop attacks were reduced after gamma knife surgery for posterior extension of callosotomy when patients suffered from unsatisfactory open anterior callosotomy and vagal nerve stimulation [15].

Laser interstitial thermal therapy (LITT) is another less invasive procedure that has been widely adopted for various intracranial lesions including refractory epilepsy. Roland [16] described 10 patients who underwent 11 LITT procedures; the functional outcomes of LITT were similar to those of open callosotomy. Most importantly, the maximal benefits from LITT are drop attacks in their seizure semiology. Likewise, LITT ablation of the CC was proven to be a safe, feasible, and minimally invasive treatment option for patients with refractory epilepsy; it can also be an attractive alternative for patients unwilling or unable to undergo open surgery [17]. Huang et al. [18] recently reported a retrospective study of a case series of six patients (three children and three adults) who underwent callosotomy using LITT. Engel outcomes for complete corpus callosotomy by LITT were similar to reported outcomes of open complete callosotomy. Magnetic resonance-guided laser interstitial thermal therapy (MR-gLiTT) is a novel minimally invasive treatment approach for drug-resistant focal epilepsy. The indications for MR-gLiTT in pediatric epilepsy surgery include hypothalamic hamartoma, tuberous sclerosis complex, and cavernoma-related epilepsy. No patients experienced new neurological morbidity or endocrine dysfunction after the operation [19, 20].

34.4 Surgical Outcomes and Complications

Callosotomy remains an effective palliative procedure for intractable epilepsy whether it is anterior or complete. Luat et al. [21] described the ideal outcomes in 20 patients with various seizure types; complete callosotomy was performed on fifteen patients and anterior 2/3 callosotomy on three, and two underwent anterior 2/3 followed later by complete callosotomy. An endoscopic approach and microsurgical approach were adopted for ten patients. Satisfactory outcomes were obtained after the operation; 65% of all patients had ≥ 50% reduction of generalized seizures (atonic, tonic, and myoclonic) leading to falls; 35% became seizure-free; and seven subjects who failed vagus nerve stimulator therapy responded with ≥50% seizure reduction after callosotomy. Another convincing excellent outcome was obtained from the largest pediatric case series of corpus callosotomy outcomes; 26 (26/55, 47%) patients achieved a good outcome for drop attacks at the last follow-up evaluation. Twenty-six of the 29 (90%) had a return of drop attacks within 12 months of surgery. Patients with drop attacks plus one other seizure type before surgery were significantly more likely to achieve a good outcome for other seizures than patients with two other seizure types or three or more other seizure types (100% vs. 23% [p<0.01]; 100% vs. 10% [p<0.01]); moreover, there was a significant decrease in the number of patients who regularly sustained injuries caused by their drop attacks (77%vs. 21% p<0.01) [22]. Consequently, corpus callosotomy is a safe and effective procedure for drop attacks in pediatric or adult patients. Moreover, obvious benefits from reduced seizure frequency and decreased antiepileptic drug administration were achieved after microsurgical commissurotomy. In our practice, the excellent outcome from corpus callosotomy was also confirmed by postoperative EEG and MRI examination (Figs. 34.4 and 34.5). During postoperative follow-up, the child recovered uneventfully with a right frontal surgical incision class I/class A (Fig. 34.6).

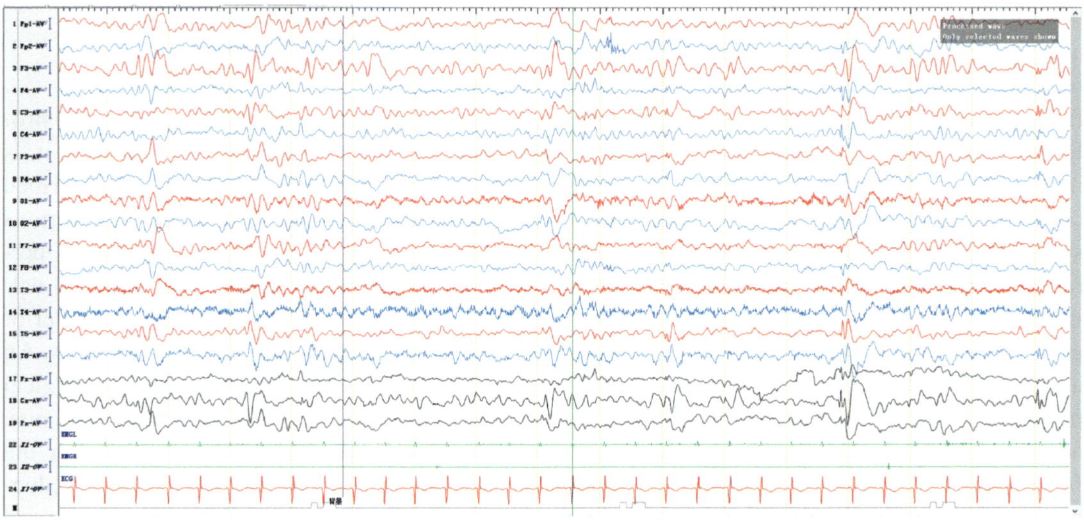

Fig. 34.4 EEG 6 months after partial callosotomy. EEG during sleep. Spike and wave focus over C4. No bilateral epileptiform activities were observed

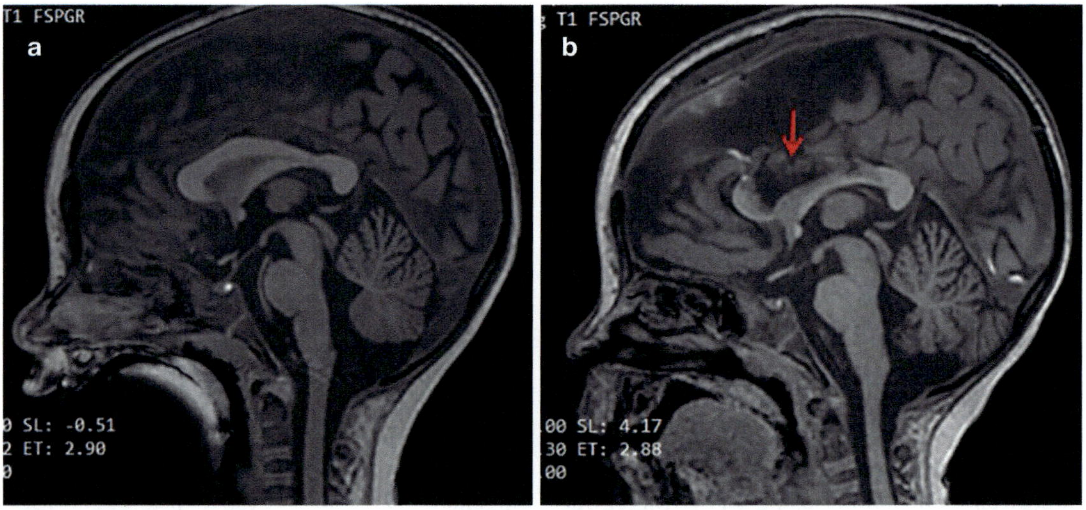

Fig. 34.5 Preoperative (**a**) and two-week postoperative (**b**) T1 FSPGR sequence MRIs, red arrow showing the result of callosotomy in the sagittal planes

Recently, Chan [23] identified 1742 eligible patients with corpus callosotomy from 58 study centres to explore the predictors of seizure outcome. The 12 variables were: (1) age at surgery, (2) sex, 3) corpus callosotomy extent, (4) MRI findings, (5) EEG findings, (6) seizure frequency, (7) epilepsy duration, (8) presence of generalized seizures, (9) presence of focal seizures, 10) presence or history of infantile spasms, (11) presence of absence seizures, and (12) epilepsy etiology. The incidences of complete seizure freedom and drop attack freedom after corpus callosotomy were 18.8% and 55.3%, respectively. In addition, infantile spasms and normal MRI findings as well as shorter epilepsy duration could predict complete seizure freedom, and freedom from drop

Fig. 34.6 Postoperative follow-up showing no seizure attack and right frontal surgical incision (red arrow) class I/class A

attacks was predicted by complete over partial callosotomy and idiopathic over known epilepsy etiology. A favorable outcome was positively associated with complete callosotomy. However, the extent of corpus callosotomy was also separately related to the presence or absence of disconnection syndrome symptoms after surgery, which included hemispatial neglect, tachistoscopic visual suppression, non-dominant hand apraxia, alexia without agraphia, tactile aphasia, dichotic listening suppression, or alien hand syndrome. Consequently, it is important to maintain a balance between obtaining maximal seizure freedom and minimizing postoperative complications. Individual surgery Should be recommended on the basis of Precise Preoperative assessment of multiple factors

Rare complications such as venous ischemia, or even thrombosis, when unilateral, can manifest as hemiparesis with the possible addition of focal seizures. There can be transient paresis owing to retraction on the medial surface of the hemisphere. A more serious complication is akinetic mutism, probably due to bilateral anterior cerebral artery spasm. Overall, the risk of death from callosotomy seems to be between 0% and 6%, of permanent neurological deficit less than 5%, and of transient deficit up to 20% [24]. Other major surgical complications occasionally developed, including hydrocephalus or deep infections, and minor complications such as CSF leakage, aseptic meningitis, and pneumonia were also observed after callosotomy [25]. Once potential complications occur in practice, early accurate diagnosis should be made, and prompt management should be considered to prevent severe outcomes and poor prognosis. Although there are complications in a minority of patients, corpus callosotomy is still vital for patients with drug-resistant epilepsy when appropriate in low-to-middle-income countries where resources are limited [26].

34.5 Conclusion

Corpus callosotomy remains a safe and efficacious procedure, and successful surgery depends on the following factors: precise preoperative assessment for appropriate surgical indication, meticulous surgical procedures during operation, and prompt management for potential complications. Microsurgical corpus callosotomy is important for treating patients with drug-resistant epilepsy after appropriate selection; however, endoscope-assisted surgical approaches can help to lower the morbidity related to traditional open surgical approaches. In addition, radiosurgical corpus callosotomy approaches, including gamma knife or LITT, offer a noninvasive technique for dividing the CC, and excellent outcomes are similarly achieved after using the foregoing surgical techniques.

References

1. Unterberger I, Bauer R, Walser G, Bauer G. Corpus callosum and epilepsies. Seizure. 2016;37:55–60.
2. Vaddiparti A, Huang R, Blihar D, Du Plessis M, Montalbano MJ, Tubbs RS, Loukas M. The evolution of corpus Callosotomy for epilepsy management. World Neurosurg. 2021;145:455–61.
3. Passamonti C, Zamponi N, Foschi N, Trignani R, Luzi M, Cesaroni E, Provinciali L, Scerrati M. Long-term seizure and behavioral outcomes after corpus callosotomy. Epilepsy Behav. 2014;41:23–9.
4. Doddamani R, Kota RC, Ahemad N, Chandra SP, Tripathi M. Endoscopic Total corpus Callosotomy and pan Commissurotomy for Lennox-Gastaut syndrome. Neurol India. 2022;70(1):63–7.
5. Hussain I, Kocharian G, Tosi U, Schwartz TH, Hoffman CE. Foundations of the diagnosis and surgical treatment of epilepsy. World Neurosurg. 2020;139:750–61.
6. Schmidek HH. Chapter 111. In: Operative neurosurgical techniques, indications, methods and results, vol. 2. 4th ed. Amsterdam: Elsevier; 2002. p. 1490–8.
7. Uda T, Kunihiro N, Umaba R, et al. Surgical aspects of corpus Callosotomy. Brain Sci. 2021;11(12):1608.
8. Schaller K, Cabrilo I. Corpus callosotomy. Acta Neurochir. 2016;158(1):155–60.
9. Smyth MD, Vellimana AK, Asano E, Sood S. Corpus callosotomy-open and endoscopic surgical techniques. Epilepsia. 2017;58(Suppl 1):73–9.
10. Chandra SP, Kurwale NS, Chibber SS, Banerji J, Dwivedi R, Garg A, Bal C, Tripathi M, Sarkar C, Tripathi M. Endoscopic-assisted (through a mini craniotomy) corpus Callosotomy combined with anterior, hippocampal, and posterior Commissurotomy in Lennox-Gastaut syndrome: a pilot study to establish its safety and efficacy. Neurosurgery. 2016;78(5):743–51.
11. Sufianov AA, Cossu G, Iakimov IA, Sufianov RA, Markin ES, Stefanov SZ, Zemmoura I, Messerer M, Daniel RT. Endoscopic interhemispheric disconnection for intractable multifocal epilepsy: surgical technique and functional neuroanatomy. Oper Neurosurg (Hagerstown). 2020;18(2):145–57.
12. Sood S, Marupudi NI, Asano E, Haridas A, Ham SD. Endoscopic corpus callosotomy and hemispherotomy. J Neurosurg Pediatr. 2015;16(6):681–6.
13. LoPresti MA, Wagner K, Lam S. Endoscope-assisted posterior quadrant disconnection plus corpus callosotomy: case report. J Neurosurg Pediatr. 2021;27(4):406–10.
14. Sachdev S, Sita TL, Shlobin NA, Gopalakrishnan M, Sucholeiki R, Régis J, Bandt SK. Completion corpus callosotomy with stereotactic radiosurgery for drug-resistant, intractable epilepsy. World Neurosurg. 2020;143:440–4.
15. Hamdi H, Boissonneau S, Hadidane S, Guenot M, Bartolomei F, Regis J. Effective posterior extension of callosotomy by gamma knife surgery. Epileptic Disord. 2020;22(3):342–8.
16. Roland JL, Akbari SHA, Salehi A, Smyth MD. Corpus callosotomy performed with laser interstitial thermal therapy. J Neurosurg. 2019;13:1–9.
17. Ball T, Sharma M, White AC, Neimat JS. Anterior corpus callosotomy using laser interstitial thermal therapy for refractory epilepsy. Stereotact Funct Neurosurg. 2018;96(6):406–11.
18. Huang Y, Yecies D, Bruckert L, Parker JJ, Ho AL, Kim LH, Fornoff L, Wintermark M, Porter B, Yeom KW, Halpern CH, Grant GA. Stereotactic laser ablation for completion corpus callosotomy. J Neurosurg Pediatr. 2019:1–9.
19. Consales A, Cognolato E, Pacetti M, Mancardi MM, Tortora D, Di Perna G, Piatelli G, Nobili L. Magnetic resonance-guided laser interstitial thermal therapy (MR-gLiTT) in pediatric epilepsy surgery: state of the art and presentation of Giannina Gaslini Children's hospital (Genoa, Italy) series. Front Neurol. 2021;12:739034.
20. Rich CW, Fasano RE, Isbaine F, Saindane AM, Qiu D, Curry DJ, Gross RE, Willie JT. MRI-guided stereotactic laser corpus callosotomy for epilepsy: distinct methods and outcomes. J Neurosurg. 2021;135:1–13.
21. Luat AF, Asano E, Kumar A, Chugani HT, Sood S. Corpus Callosotomy for intractable epilepsy revisited: the Children's Hospital of Michigan series. J Child Neurol. 2017;32(7):624–9.
22. Graham D, Gill D, Dale RC, Tisdall MM. Corpus callosotomy outcomes study group. Seizure outcome after corpus callosotomy in a large paediatric series. Dev Med Child Neurol. 2018;60(2):199–206.

23. Chan AY, Rolston JD, Lee B, Vadera S, Englot DJ. Rates and predictors of seizure outcome after corpus callosotomy for drug-resistant epilepsy: a meta-analysis. J Neurosurg. 2018;1:1–10.
24. Moore AJ, Newell DW. Chapter 34. In: Neurosurgery principles and practice. London: Springer; 2005. p. 591–606.
25. Graham D, Tisdall MM, Gill D. Corpus callosotomy outcomes in pediatric patients: a systematic review. Epilepsia. 2016;57(7):1053–68.
26. Markosian C, Patel S, Kosach S, Goodman RR, Tomycz LD. Corpus Callosotomy in the modern era: origins, efficacy, technical variations, complications, and indications. World Neurosurg. 2022;159:146–55.

Endoscope-Assisted Microsurgery of the Corpus Callosum

35

Oreste de Divitiis and Vincenzo Meglio

35.1 Corpus Callosum: An Unusual Neuroanatomical Structure

The corpus callosum (CC) is a thick transverse bundle of commissural fibers traveling between the medial surfaces of the cerebral hemispheres. It is the largest white matter structure in mammalian brains [1] and also the largest commissure in the brain [2]. Aside from its function, it has particular anatomical features that distinguish it from other neuroanatomical structures.

As first reported by Galen, the CC is unusually thick for a structure composed by nervous tissue [3]. This led many important anatomists including Galen, Vesalius, and others to consider its significance as purely mechanical not neurological [4]. They named it "callosum" (from the Greek word *tyloeides*) because of its unusual thickness, thus referring to it as the "callus of the brain" [1]. It participates in forming the two lateral ventricles and the third ventricle, supporting the septum pellucidum and the fornix [3]. It occupies a strategic position at the center of the cerebrum, providing a passage for many surgical routes directed at the ventricles. However, as first noticed by Vic d'Azyr, the father of comparative anatomy, in 1784, and better described by later authors such as Sir Victor Horsley, the CC is not merely a supportive anatomical structure but is fundamental in uniting the halves of the brain [5, 6]. Unfortunately, this task is performed under both physiological and pathological conditions, making it the main pathway, for example, for the spread of epileptic seizures. As shown first by William P. van Wagenen, chair of neurosurgery at the University of Rochester, it is actually the critical link for generalizing epileptic seizures [7, 8]. In an epilepsy syndrome with generalized seizures that are intractable and/or originate from a nonfocal area of the cortex, "interrupting the main diffusion pathway" is the only therapeutic option available. This involves the surgical splitting of the CC as a recognized target for the surgical management of epilepsy. This splitting effectively reduces the frequency of seizures, improving the patient's quality of life [9].

Because of its "double soul" (structural supportive central structure and neurological route of diffusion), the CC is intrinsically implicated in surgical approaches and techniques applied to the brain that conceptually and technically reflect this dualism. There are two main categories of approaches involving the CC: those directed to the CC itself (as a target) or to its parts and those targeted to neighboring parts, involving the CC only indirectly as a gateway [10].

O. de Divitiis (✉) · V. Meglio
Division of Neurosurgery, Department of Neurosciences and Reproductive and Odontostomatological Sciences, Università degli Studi di Napoli Federico II, Naples, Italy

M. Turgut et al. (eds.), *The Corpus Callosum*, https://doi.org/10.1007/978-3-031-38114-0_35

35.2 Newest Conceptions and Surgical Applications: The Debut of Endoscopy

Following the latest technological innovations in neurosurgery, and surgery in general, minimally invasive procedures are often considered [11]. In this setting, many new tools have been implemented, especially in neurosurgical practice [11, 12].

More specifically, in relation to surgical approaches to the CC, this technical and conceptual evolution has helped neurosurgeons to grasp the importance of removing the least possible amount of CC tissue without undermining the effectiveness of the approach or the surgical outcomes, leading to the concept of a *tailored corpus callosotomy* [13]. The delicacy required for dissections in the pericallosal area is well described in the literature [3]. In this setting, the endoscope, with its intrinsic aspects and technical properties, is valuable for reducing cranial traumatization and improving the postoperative course in both corpus capsulotomies and transcallosal approaches [11]. Apart from surgical planning and tissue removal, neurosurgical operations also involve a further unavoidable challenge: the opportunity to optimize visualization of anatomical landmarks, surgical targets, and neighboring structures with the smallest possible approach and the least possible neurovascular manipulation and retraction. After the introduction of the microscope, the second groundbreaking revolution in neurosurgery has surely been the introduction of the endoscope.

From its conceptual introduction in 1773 by Philipp Bozzini and its first introduction in surgery by the Chicago urologist Victor Darwin Lespinasse [14], and in neurosurgery by Guiot in 1963 [15], endoscopy has been used in more and more neurosurgical indications. Nowadays, neuroendoscopy can be applied to a broad range of procedures: transnasal approaches for pituitary and other skull-base tumors, third ventriculostomy, resection of intraventricular tumors, and pediatric neurosurgical pathologies. This broad range of applications continues to expand, continually finding new indications, most recently in nerve and spine surgery [12, 16]. In this expansion, CC surgery has been affected by technological advances. While its application in procedures with CC as a gateway to other neurological structures is well described and accepted for some approaches, its use for approaches to the CC as a target is very new, with only a few publications in the literature, mostly cadaveric studies.

Before proceeding with the discussion, it is important to emphasize that there is no single method of application of the endoscope to cranial surgery. It can be used for inspection and for endoscope-assisted and endoscope-controlled procedures, though the main tool for visualization during those interventions remains the operating microscope [12]. In 1998, Hopf and Perneczky provided a clinical classification of the different modalities by which the endoscope is implemented in cranial surgery. They distinguished four such modalities:

1. Pure neuroendoscopy (NE), when the endoscope is used alone, surgical manipulations being performed with special endoscopic instruments through the endoscope itself.
2. Endoscope-controlled microsurgery (ECM), when conventional instruments are used for microsurgical manipulations but are guided by the video image from the endoscope (video surgery).
3. Endoscopic inspection (EI), when the endoscope is only used as adjunct to a surgical procedure, exclusively for inspection.
4. Endoscope-assisted microsurgery (EAM), when the endoscope is coupled with the microscope but dissection is performed under the operating microscope.

In this chapter, we will focus on the last modality and on the implementation of endoscope-assisted techniques in CC surgery, describing its advantages and disadvantages. We will describe the different approaches from a technical point of view [17].

35.3 Endoscope-Assisted Corpus Callosotomy

When the CC is the target, resection of it in whole or in part is the main surgical modality. This is called "corpus callosotomy" (CCy). Although the term "callosotomy" literally means resection of parts of the CC, it also includes cases where the structure is simply divided, a development of the old commissurotomy [18].

The main indication for CCy is the palliative treatment of medically refractory epilepsy, when no focal area of genesis is detectable, so a focal surgical resection is ineffective [18, 19]. It is also valuable for secondary generalized seizures originating from a unilateral focus but not well controlled by medications [20, 21]. It is more effective for treating drop attacks caused by tonic/atonic seizures and childhood organic encephalopathic epilepsies (Lennox-Gastaut syndrome, West syndrome, and severe epilepsy with multiple independent spike foci) [22].

The first application of this procedure was by Dendy in 1936 for resecting a pineal tumor. It was proposed for treating epilepsy by van Wagen and Herren in 1940 and applied effectively for the first time by Joseph Bogen and Philip Vogel in 1962 [23]. The surgical technique has undergone numerous improvements and refinements since its inception, leading to its current conception and acceptance [18, 24]. This continual methodological evolution has produced different versions, aimed at reducing the risk of complications and optimizing clinical outcomes. Anterior callosotomy, posterior callosotomy, partial callosotomy, and staged callosotomy are the best described. In view of the satisfactory results attained over the years, the procedure, initially considered contentious, is nowadays a widely accepted surgical solution for appropriately selected epilepsy patients.

Along with the different extensions of CC resection or division, the modalities for surgical management of the CC have evolved substantially, starting with "simple" open surgical resection and leading to microsurgical and radiosurgical techniques. The gradual achievement of minimal invasiveness with maximal efficiency has inevitably made the endoscope a valuable tool for corpus callosotomy, reducing the traumatization of the patient during various surgical approaches [18].

Although the properties of this tool are evident, its application in corpus callosotomy is much more recent than endoscope-assisted transcallosal approaches, as indicated by the relatively sparse literature, limited to a few papers describing cadaveric dissections and a few case reports, mostly of pediatric patients.

The endoscope was first proposed as an adjunct to the operating microscope in corpus callosotomy by Guerrero and Cohen in 2003. On the basis of their cadaveric study, they proposed its use for enhancing surgical visualization of the CC [25]. In 2004, Tubbs et al. performed the first endoscopic anterior callosotomies on human cadavers and swine models, opting for supraorbital trephination through a hole 2 cm above the nasion on the midline, avoiding the frontal sinus but causing an unacceptably unesthetic result for the patient, despite greater safety [26]. In 2011, Bahuleyan et al. proposed a more suitable endoscopic solution, using two parasagittal burr holes on silicon-injected cadavers, comparing their surgical results with microsurgical ones [27]. Considering these promising preliminary results on cadavers, the application of endoscopy in corpus callosotomy could have been more rapid. In 2016, Scerrati et al. published the first paper describing endoscopic corpus callosotomy on an adult patient with acceptable results; there are more papers describing its use for pediatric patients [11]. To the best of our knowledge, the largest adult patient population treated by this surgical modality was reported by Sood et al. They described the surgical issues and clinical outcomes of ten adult patients successfully cured using this approach [24].

35.3.1 Surgical Technique

Considering the novelty of introducing endoscopy to this well-accepted and theorized microsurgical approach, there are only a few descriptions of endoscope-assisted callosotomy in the literature, especially in relation to living patients. Despite this very limited number, the experiences reported differ among patient groups. We will offer a concise synthesis of the surgical steps, including some variations adopted by the authors who have provided a more complete description, also integrating our personal experience and preferences.

35.3.1.1 Preoperative Stage

As in all surgical approaches, indication and patient selection are the first fundamental and critical steps. It is important to assess clinical histories; patients who have been previously undergone cranial operations or have anatomical abnormalities can be seriously challenging for the surgeon, even to the point of being ineligible for such surgery [24]. An extensive preoperative clinical evaluation of the patient is needed, comprising electroencephalography, MRI, and a PET study, before an endoscope-assisted disconnective surgical procedure is undertaken [28].

After patient selection, it is mandatory to obtain a preoperative neuronavigation MRI scan and an additional time-resolved imaging of contrast kinetics sequence with magnetic resonance venography to determine the location of bridging veins. According to Sood et al., scalp electroencephalography electrodes should be placed for EEG during the procedure [24].

Most surgeons including our group consider a rigid endoscope with a 4 mm diameter zero-degree straightforward lens to be optimal for this type of approach. A video monitor is necessary to project the image recurring to a three chip camera [25]. A two-dimensional or a three-dimensional endoscope is suitable for this procedure. There are more reports in the literature involving the former, but Sood et al. clearly recommend a three-dimensional endoscope. If a two-dimensional endoscope is chosen, they recommend stereotactic attachment [9, 28].

35.3.1.2 Surgical Mindset and Patient Positioning

The earliest reports about this approach, mostly involving cadavers (see above), described a one-handed technique [25]. However, as all surgical choices, this modality can have advantages and disadvantages. It can limit the ability of the surgeon to control hemorrhage and perform bimanual dissection, against the advantage of a smaller craniotomy [24]. For these reasons, variations of the "surgical mindset" have been developed according to the surgeon's preference. Chandra et al., for example, described a three-handed technique, the endoscope being held by a second operator, so the first operator has both hands free. Guerrero et al. reported that having the lens mounted on a holder is an acceptable solution. In both cases, the first operator gains increased freedom, though at the price of a bigger craniotomy and more frequent repositioning of the endoscope (owing to the wider surgical field) [24, 29]. In order to overcome those disadvantages, Sood et al. proposed a third option: a two-handed technique with the endoscope fixed to the suction, both held in the left hand of the first operator, the other hand being free to accommodate another tool, for example, for dissection or coagulation [30]. According to these authors, this type of surgery could then be performed with better bleeding control, bimanual dissection ability, and less need for endoscope repositioning [24].

The patient's body is positioned supine. Some authors recommend a slight flexion, which coupled with the extension of the neck allows the face to look straight upward. Appropriate securing of the body is desirable because Trendelenburg and reverse Trendelenburg can be needed during the procedure to ensure the right orientation, maintaining the proper trajectory of the surgical corridor. Some authors also prefer a lateral position with the entry side located downward to exploit the effect of gravity on the brain, facilitating interhemispheric dissection with less brain retraction [31, 32]. However, this position has disadvantages so it is less often used than the supine one. Mayfield three pin fixation of the head is obviously necessary, with a neutral head orientation [29].

35.3.1.3 Approach

The type of skin incision depends on the type of craniotomy chosen; there are some variations among groups. A pre-coronal craniotomy is advisable because it allows both the splenium and the genu of the CC to be accessed. It is important to place it anterior to the major bridging veins but behind the hairline, with the assistance of intraoperative neuronavigation; the same applies after craniotomy. Craniotomies around the central sulcus should be avoided since the vein of Trolard, the biggest bridging vein in the superior sagittal sinus, runs nears it. A paramedian precoronal skin incision 1.5 cm lateral to the midline and 1–1.5 cm long on the sagittal plane is sufficient for this. After the skin is incised, small skin retractors can be used without risk of interference with surgical tools or the endoscope.

After subgaleal dissection, craniotomy is started with a burr hole on the sagittal sinus. Almost all the literature approves a semicircular D-shaped craniotomy with the base located medially on the superior sagittal sinus itself. Interestingly, Sood et al. provided further important details about the bone opening shape. They suggested drawing the craniotomy bevel medially in an outward direction; anteriorly and posteriorly, it must be inward. This meticulous craniotomy planning ensures maximization of the angulation needed to reach the anterior and posterior limits of the resection of the CC.

In a series of 11 patients treated by endoscope-assisted microsurgical corpus callosotomy, Manjari Tripathi et al. opted for a slightly bigger mini-craniotomy (4 × 3 cm) but always drawn in a semicircle with the base located medially on the superior sagittal sinus and anteriorly to the coronal suture. This type of craniotomy sufficed to yield satisfactory results [29]. Even the first groups describing endoscope-assisted corpus callosotomy on cadavers reported that the approach using a paramedian craniotomy was optimal [9, 25]. Guerrero and Cohen, like Manjari et al., described a craniotomy of at least 4 cm diameter. Moreover, they specified that it should be made on the nondominant side. However, Uda et al. recommended that the craniotomy side should be decided according to the distribution of bridging veins so these can be respected [32]. The dural opening mirrors the craniotomy shape. It has to be made semicircular (or more horseshoe-shaped according to other authors [25]), with the base medially sited, corresponding to the superior sagittal sinus.

35.3.1.4 Intradural Phase

Once the dura is opened, the interhemispheric fissure is targeted. At this point, as for the endoscope-assisted technique, CSF must be removed to relax the brain. If ventricular drainage has not already been placed for hypertension pathologies or has not been placed before surgery, it can be placed during this phase, in the frontal horn of the lateral ventricle. Intraoperative mannitol administration can help to achieve the same result.

Some slight lateral retraction can be applied to the brain to expose the deeper portion of the falx, allowing the cingulate gyrus to be visualized. The interhemispheric fissure is opened under the microscope. During these stages, large bridging veins from the frontal lobe to the superior sagittal sinus should be preserved.

All these steps can be made without the endoscope. Thereafter, the endoscope is introduced into the surgical field to complement the microscope. Following the lateral face of the falx, it allows the cingulate gyrus to be dissected precisely, exposing the CC, the dissection moving anteriorly and posteriorly. The cingulate gyrus is located at the end of the falx cerebri. In the cadaveric description by Guerrero and Cohen, the endoscope is introduced only once the CC is visualized [25]. We suggest introducing it earlier, because visualizing the CC without the endoscope can involve excessive and dangerous retraction of the brain parenchyma. Importantly, the pericallosal arteries are the lateral boundaries during the dissection and CC exposure; they must be precisely recognized and carefully preserved. For this reason and because of the callosomarginal arteries in the interhemispheric fissure, it can be safer to have the endoscope already in place.

The most important moment of the approach, the corpus callosotomy, is now possible. Through microscopic endoscope-assisted dissection, the cavum of septum pellucidum (CSP) can be found and isolated, a fundamental landmark for starting the dissection of the CC [32]. To split the CC, a 1.5–2 cm exposure usually suffices through a cavitron ultrasonic surgical aspirator [24]. Microscissors, forceps, and suction tube can also be used during this phase. Splitting should begin from the anterior portion of the CC, starting from the CSP and proceeding forward, involving the genu and the rostrum.

How the corpus callosotomy continues thereafter depends on the variant of the procedure chosen. For an anterior callosotomy, the division should be posterior to the CSP up to the point where the body passes to the isthmus of the CC. According to various authors, this point represents the posterior limit of the anterior corpus callosotomy. Neuronavigation can be very useful for determining where to stop. For total corpus callosotomy, dissection has to proceed further posteriorly and include the splenium of the CC. There is a new landmark to recognize to guide this phase of the dissection: the upper wall of the cavum velum interpositum [32]. Again, endoscope assistance is very useful here. Finally, visualization of the vein of Galen through the arachnoid space confirms the end of the posterior dissection.

Any bleeding must be promptly and carefully coagulated. Immediately before beginning the corpus callosotomy, small vessels on the surface of the CC must be coagulated and cut.

35.3.1.5 Closure

After callosal resection is completed, closure can be accomplished simply. The endoscope can be useful at the start of this final phase for ensuring precise and effective hemostasis. The dura is then closed and made watertight and the bone flap is replaced using plates and screws. The scalp is closed in standard fashion.

Sood et al. suggest postoperative MRI with diffusion tensor imaging (DTI) to confirm that disconnection has been satisfactory [24].

35.4 Endoscope-Assisted Interhemispheric Transcallosal Approaches

Because of its deep location in the midline of the brain, the CC is an unavoidable gateway to many neuronal elements. It forms an important landmark for surgical approaches to the pericallosal vessels, ventricular system, and basal ganglia [33]. It is especially involved in surgical routes aimed at reaching and removing median or paramedian lesions, notably those inside the supratentorial ventricles. Although there are two main surgical options for reaching the intraventricular space, transcortical transventricular (TCo-TV) and transcallosal transventricular (TCa-TV) approaches, the latter have become more widely used, especially for managing third ventricle pathology [12, 34]. They offer advantages over the alternatives, particularly the benefit of avoiding cerebral cortex violation, reducing the risks of seizures, and also the opportunity to gain a more direct midline orientation, with symmetrical access to both lateral ventricles and third ventricle walls [35].

In 1930, Dandy presented a series of three patients with intraventricular choroid plexus tumors, which were removed by inserting a cystoscope into the third ventricle cavity [36]. That was the first description of an intraventricular procedure. Subsequently, the techniques for reaching those intracerebral spaces have been slowly but continually refined and have exploited the most recent technological advances.

Nevertheless, a major concern about this technique is the surgical management and splitting of the CC, which must be carried out in a deep and delicate region of the brain. Neurosurgeons therefore attempt to be as noninvasive as possible, using the smallest callosotomy obtainable, and seeking to use the endoscope for intraventricular transcallosal surgery [17].

Implementation of the endoscope as a support in microsurgical removal of a tumor with a complex location was first described during the 1970s by Apuzzo et al. [37] and by Prott [38]. They described endoscope assistance for posterior fossa microsurgery, especially for tumors in the cerebellopontine angle extending into the internal auditory meatus. Since then, application of the endoscope for deeply located tumors has stood the test of time, becoming a standard route for accessing deep midline intra- and paraventricular lesions.

There are two modalities by which endoscopy is applied to ventricular surgery [18, 25]. The first is the pure neuroendoscopic modality. As described earlier, this involves using this tool alone with endoscopic instruments, without the microscope. It can be used for ventricular pathology with different indications, such as third ventriculostomy for obstructive hydrocephalus patients and biopsy or resection of ventricle tumors [39–41]. The second modality is endoscope-assisted microsurgical resection of purely interventricular lesions, taking advantage of the natural cavities of the cerebral ventricles and thus reducing the need for brain retraction [12, 42–44]. This modality requires a combination of microscopic and endoscopic techniques.

In the following section, we will describe the indications and surgical technique of endoscope-assisted transcallosal approaches to the ventricles, emphasizing third ventricle approaches and discussing their limits and the opportunities they afford. Transcortical approaches do not involve the CC and are therefore outside the scope of our discussion. Since endoscope-assisted resection of lateral ventricular tumors is preferable a TCa-TV approach, we will focus on the management of third ventricle lesions [45].

35.4.1 Indications

Indications for endoscope-assisted transcallosal approaches have changed over the years as techniques and instruments have evolved. According to Cappabianca et al., it is possible nowadays to remove some intraventricular tumors endoscopically, subject to the following strict criteria: moderate vascularity of the lesion, soft consistency, small diameter (2–3 cm), associated hydrocephalus, low histological grade, and location inside the lateral ventricle [45, 46]. All other interventricular lesions that are not amenable to completely pure endoscopic resection are virtually indicated for intraventricular endoscope-assisted microsurgical resection.

35.4.2 Surgical Technique

35.4.2.1 Preoperative Stage, Mind Setting, and Patient Positioning

Preoperative preparation and the surgical steps preceding dural opening are almost the same as for endoscope-assisted callosotomy. However, it is important on preoperative MRI to focus particularly on any adhesions of the upper forniceal surface to the CC and their points of anterior and posterior insertion because they can interfere with surgical route planning, especially during the post-callosotomy steps [47].

Moreover, since the CC is not the target in this approach but a gateway to other areas, another important difference is that variations of the TCa-TV approach can be tailored according to the location and extension of the lesion [48]. The most important variants are the anterior, middle, and posterior TCa-TV approaches. The choice of variant, determined from CT or MRI scans, dictates the differences in subsequent surgical steps, starting from patient positioning. Although some authors prefer to use a supine position for all variants, others suggest the prone position for the posterior TCa-TV approach [45]. Although the supine position is most widely used for the anterior and middle TCa-TV approaches, lateral decubitus with the body rotated toward the side of the lesion can be an option [48]. However, this is less often preferred because it has important disadvantages such as discomfort for the surgeon, septum pellucidum collapse into the surgical field, and greater time demand [49].

All authors agree on a neutral position of the head with three pin Mayfield fixation. Some neck flexion (15°) can be applied in both the anterior and middle variations of the approach, according to surgeon's preference [13, 45]. Some authors suggest 30° flexion [48]. For posterior third ventricle or pineal regions lesions, where a posterior TCa-TV approach is advised and patient is placed in prone position, the head must be in a neutral position [49]. However, regardless of the choice of body and head positioning, the anatomical peculiarities of this approach, and the depth of the lesions, neuronavigation is mandatory and must be used to check the position and decide the best locations for skin incision and craniotomy to optimize the surgical trajectory [13, 45, 49].

35.4.2.2 Exposure

The steps in exposure of the intraventricular lesion form two stages: the pre-endoscope stage, which comprises the skin incision, craniotomy, and dural opening, and the endoscope-assisted stage (Fig. 35.1).

Pre-Endoscope Stage

For the anterior and middle variants of the approach, the pre-endoscope stage broadly echoes the steps in endoscope-assisted callosotomy as far as exposure of the interhemispheric fissure. Skin incision, dissection, craniotomy, and dural opening do not require magnification or visualization tools and are performed without microscope or endoscope.

A linear incision usually suffices; the main advantage of the endoscope-assisted TCa-TV approach is the use of a small incision. However, following descriptions of the standard non-endoscopic technique for this approach, as provided by Delfini et al., an L-shaped skin incision should be made with its vertical portion on the midline and its horizontal one parallel to the coronal suture line (Fig. 35.2) [49]. For the endoscope-assisted modality, we consider that a

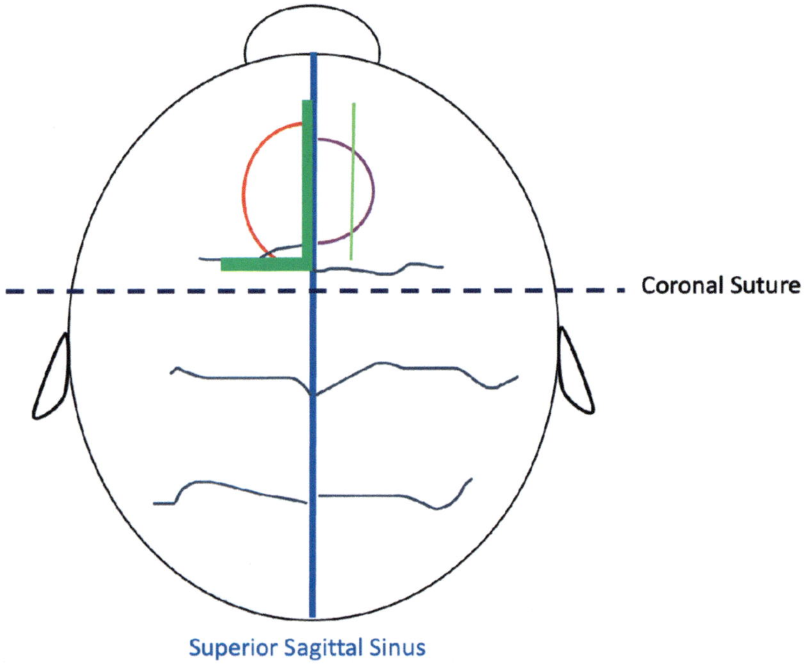

Fig. 35.1 Schematic drawing of the exposures by transcallosal approaches. On the left, exposure by the microscopic approach. An L-shaped skin incision (in green) is depicted, the vertical portion on the midline and the horizontal portion parallel to the coronal suture. A "D-shaped" craniotomy (in red) is also shown, its base directed toward the midline. On the right, the same exposure but for an endoscope-assisted approach, with a linear skin incision (light green) and a smaller craniotomy (purple). It should be noted that besides the skin incision and craniotomy type, the approach must be placed so as to preserve the superior sagittal sinus tributary veins. The smallest required dimension of the endoscope-assisted craniotomy makes this preservation much easier

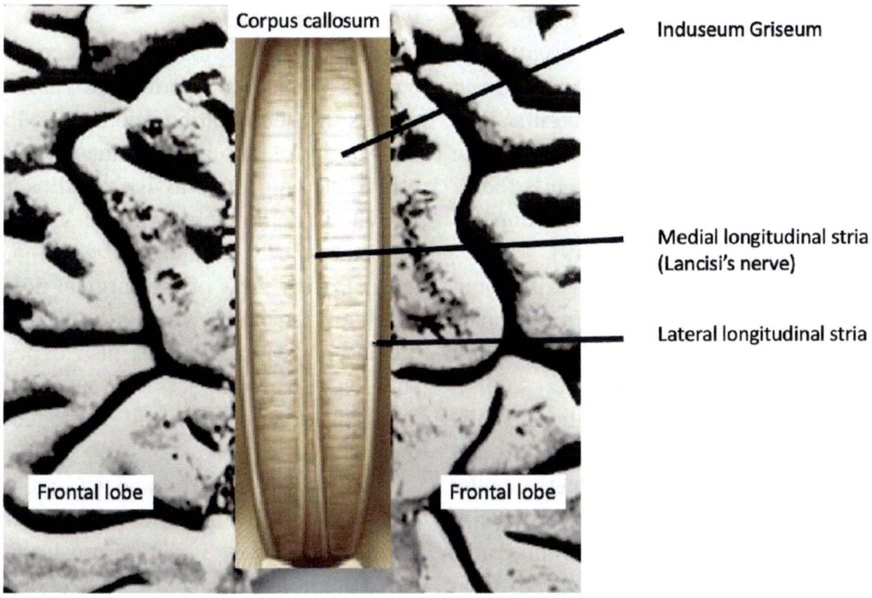

Fig. 35.2 Schematic representation of the superior surface of the CC, as it appears from an anatomical exposure. On both sides (in gray), the frontal lobes can be appreciated. On the midline is the CC with medial and lateral striae, important surgical landmarks for placing the corpus callosotomy

smaller linear incision slightly paramedian and parallel to the midline is sufficient.

Regardless of the shape of the skin incision (and of the subsequent craniotomy), the placement of the approach depends on the location of the lesion to be treated. For lesions in the anterior portion of the third ventricle, anterior to the epithalamus and the foramen of Monro, the anterior variant of the TCa-TV approach is the best option, so the skin incision is located anterior to the coronal suture (Fig. 35.1) with just the most posterior part moved slightly over the coronal suture line. However, for lesions in the middle portion of the third ventricle, behind the foramen of Monroe, or in the lateral ventricles, when a TCo-TV approach cannot be used, the middle variant of the TCa-TV approach is indicated, and the skin incision (and craniotomy) must be placed more posteriorly, one-third of the incision anterior to the coronal suture and two-thirds posterior to it [49].

Totally different planning is required for lesions manageable by a posterior TCa-TV approach. The patient is positioned in a completely different way. Although some authors such as Charalampaki et al. still prefer a supine position, we believe that prone positioning is more suitable [45]. Nevertheless, in either case, the skin incision (and obviously the craniotomy) must be placed totally posterior to the coronal suture, 4–5 cm anterior to the torcular Herophili, as Defini et al. specified in their chapter [49].

Following the skin incision, as described above, the craniotomy is placed according to the anatomical aspects of the lesion. However, as well as the anteroposterior position of the skin incision and craniotomy, the side should be specified. According to Delfini et al., the craniotomy should be placed on the nondominant side because the TCa-TV approach provides good visualization of both lateral ventricles and even better visualization for third ventricle exposure [49]. Nevertheless, other authors prefer to place the craniotomy on the same side as the lesion. For instance, Charalampaki et al. in their series of 35 patients with differently located intraventricular tumors made at least one left-sided craniotomy for a left-sided transcallosal approach directed to the contralateral ventricle [45]. It is important to remark that the sides of the craniotomy and of the approach in general needs not to depend on the location of the lesion alone, because vascular anatomy, and in particular the location of the parasagittal vein in reaching the superior sagittal sinus, must be the principal factor in side selection. It is therefore mandatory to check their location preoperatively by venous MR angiography [49].

The literature also provides few details about the best modality for dura opening. In our opinion, the best way is semicircular with the base of the dural flap parallel to the superior sagittal sinus (SSS), as indicated by other authors. When the dural flap is rotated over the SSS, it is a good practice to place long cotton sponges between the two, avoiding SSS strangulation. However, with endoscope assistance, the skin incision and craniotomy are smaller and so is the required dural opening, so this risk is reduced [48, 49].

Endoscope-Assisted Stage

Once the dural opening is completed and the small dural flap fixed, the truly endoscope-assisted procedure begins. Interestingly, Charalampaki et al. used a three-dimensional ultrasound device to locate and visualize the tumors in patients undergoing the TCa-TV approach before they introduced the endoscope [AP 68]. We consider this a valid way of making the approach less invasive. Nevertheless, this intradural phase corresponds to the endoscope-assisted phase, during which different surgical stages can be distinguished:

(a) Pre-callosal phase.
(b) Callosal phase.
(c) Post-callosal phase.

It is important to clarify that, in our opinion, the use of the endoscope as adjunct confers advantages during all these stages. The intradural stage of the TCa-TV approach, despite the opportunity to use neuronavigation systems for guidance, remains a matter of steering by sight, with careful identification of important landmarks. In this scenario, the endoscope is a great help for the neurosurgeon, analogous to the telescope for the sailor.

Pre-Callosal Phase

Introduction of the endoscope is considered during this phase. It is inserted and guided into position using the operating microscope [50]. Again, the surgical steps are similar to the intradural phase of endoscope-assisted callosotomy until the CC is completely exposed. Once the endoscope has reached the desired position, providing a satisfactory view of the surgical field, it can be fixed and secured, freeing both the surgeon's hands [17]. However, this also depends on the surgeon's preference; some authors prefer an assistant surgeon specifically assigned to manage the endoscope [40].

The cingulate gyrus and CC are endoscopically visualized and distinguished. Instruments for microsurgical dissection are introduced into the craniotomy adjacent to the endoscope lens. Exposure can stop when the two main anatomical landmarks, the nerves of Lancisi (also called medial longitudinal striae), are completely exposed and visualizable. These are important guides to the next surgical step, the callosotomy [49, 50]. They form the border of the longitudinal raphe, which corresponds to the midline of the CC [51] (Fig. 35.2).

Callosal Phase

The focus of the corpus callosotomy approach involves the main surgical stage, the callosotomy. This is the final step; however, when the TCa-TV approach is used, it is only a preparation for reaching deeper brain regions. According to some authors describing the "classical" microscopic transcallosal approach, a 1 cm longitudinal callosotomy suffices to expose the ventricles. Nevertheless, they recommend 2 cm callosotomy. However, thanks to endoscope assistance, we believe that a 1 cm callosotomy is more than enough to access the lateral and third ventricles.

Moreover, the location of the callosotomy is not standard but depends on the location of the lesion and therefore on the choice of TCa-TV variation. The callosotomy is more safely located in the anterior half of the CC. It can be placed strictly median, between the two nerves of Lancisi, to access to the cavum septi pellucidi when an interforniceal approach to the third ventricle is considered. If the lateral ventricles are the target, then a paramedian incision of the CC is possible, at the cost of limited injury to the indusium griseum [49]. Esposito et al. and others have described a modified anterior transcallosal approach, opting for an interhemispheric transgenual route with good clinical results [52].

However, although the anterior portion of the CC is safe and suitable for the anterior and middle TCa-TV approaches, callosotomy via a posterior transcallosal route must be placed more posteriorly. In this case, the callosotomy cannot be median. It should be paramedian because the posterior portion of the CC adheres to the midline of the choroid plexus and to the roof of the third ventricle, making direct contact with the fornixes. A paramedian incision is therefore required.

Moreover, since the posterior part of the CC is rich in commissural fibers and sectioning it can cause severe clinical symptoms, callosotomy should be even smaller in this portion. Here, implementing the endoscope is very useful. The rostrum and splenium should be preserved to preclude neurological deficits [10, 45, 53].

Post-Callosal Phase

Once callosotomy has been completed, the endoscope is passed through the CC, providing the surgeon with a panoramic view of the cavum septum pellucidum. Again, the subsequent steps depend on the target.

If the lesion is in the lateral ventricles, then moving the endoscope forward with microscope magnification after the paramedian callosotomy has been completed enables several landmarks to be identified easily: the septum pellucidum, frontal horn, middle cell (both containing the choroid plexus), foramen of Monroe, thalamostriate veins, anterior septal vein, thalamus, and caudate nucleus [49].

If access to the third ventricle is required, different routes are possible after the callosotomy: paraforniceal, interforniceal, suprachoroidal, and subchoroidal routes. The first of these takes advantage of the communication between the third and lateral ventricles via the foramen of Monroe. The second considers dissection and separation of the bodies of the fornixes on the midline, together with the internal cerebral veins and posteromedial choroidal arteries. The suprachoroidal route crosses the tenia fornicis between the fornix and the choroid plexus, while the subchoroidal route gains surgical access to the third ventricle by displacing the choroid plexus medially to the internal cerebral vein and the posterior medial choroidal artery [54, 55].

Further descriptions of these approaches would be tedious and go beyond the remit of our chapter, which is CC endoscope-assisted surgery. However, a general principle is that all post-callosal approaches are very deep and increasingly narrow. Endoscope assistance is therefore necessary to complete them without injuring the patient.

Closure

For closure, the same procedure as in endoscope-assisted corpus callosotomy is valid.

35.5 Discussion

In the past, CC surgery required much larger craniotomies than today's techniques, making them traumatic for the patient and leading to worse surgical outcomes [23]. There were various reasons for this, such as the lack of advanced diagnostic tools and intraoperative field illumination and magnification tools and of microscopic instruments dedicated to neurosurgery. Fortunately, those times have passed.

The introduction of the microscope with its magnification and illumination capacities reduced the space necessary for illuminating the surgical field and also for the surgeon to visualize and explore the surgical route. This allowed neurosurgeons to reach anatomical structures that were previously difficult to observe, such as the CC [56]. However, despite the remarkable reduction of craniotomies in approaches to the CC, some limitations still had to be overcome to make them as noninvasive as possible. The microscope is restricted by the need for a minimum craniotomy diameter; also, it can only visualize what is directly opposite the approach entrance. This limited the reduction of craniotomy size using the microscope because the space provided for adequate view angles was insufficient, there was unacceptable lowering of light intensity inside the surgical field, and it was impossible to maintain correct coaxial control of the microsurgical instruments [56].

The introduction of the endoscope, especially in the endoscope-assisted modality, has provided the perfect fit, synergistically combining the strengths of two different sets of tools: microsurgical ones that make difficult and dangerous manipulations possible and those from endoscopy that affect visualization dramatically [17]. This combination is especially valuable in surgical approaches to deep-seated lesions or deeply located anatomical structures, such as the CC or lesions located beside it, which require narrow surgical routes. However, as for all surgical tools, the advantages are coupled with drawbacks. We will analyze both in more detail, focusing on the effects on CC surgery.

35.5.1 Advantages

According to Perneczky and Fries [56], the three main advantages that the development of endoscope-assisted surgery has brought to neurosurgery are the following:

1. Providing intense illumination of the surgical field.
2. Allowing for more detailed observation of the anatomical structures involved.
3. Widening of viewing angles.

All these advantages are especially valuable in CC surgery. Although the CC seems easy to access, it is covered by other neurovascular formations that need to be preserved in order to guarantee the best clinical outcome for the patient. For this reason, the surgeon's first requirement in accessing this structure is to ensure the maximum possible illumination of the fundus of the surgical field. This is mainly because the anatomical structures covering the CC on both sides cause steric hindrance, for example, the cerebral hemispheres with the frontal lobe and the cingulate gyrus. The direct illuminating capacity of the endoscope provides crucial assistance for the surgeon. The microscope light (and lens) has to be completely outside the surgical field, but the endoscope can be placed directly inside that field despite the narrowed hole approach. The craniotomy should only be large enough to see the tip of the endoscope (Fig. 35.3). The possibility of having the light source inside the surgical field, adjusting its position according to the surgical phase, opens the possibility of illuminating deeper structures covered and overshadowed by anatomical protuberances, even with a very small cranial opening such as a burr hole [26, 57]. This advantage does not obtain when a straight frontal visualization trajectory is mandatory, as with the microscope alone.

As reported previously, another distinct advantage of supplementing the microscope with the endoscope is the surgeon's enhanced ability to capture anatomical details, even of the deepest structures. The microscope lens is constrained by necessarily being outside the surgical field, especially when the surgical route to the lesion is narrow and deep, as during CC surgery. This can mean that the magnifying power of the microscope cannot be fully exploited. In contrast, the endoscope offers the opportunity of bringing the lens directly into the center of the surgical field. Endoscope assistance therefore helps the surgeon to transcend the natural limits of his senses, "putting his eyes directly into the surgical field." This is of the utmost importance in both TCa-TV and callosotomy approaches because it is hard to reach the final target in those procedures, implying the progressive lowering of the surgical plane, relying only on recognition of the anatomical landmarks that gradually appear during the dissection. These landmarks consist of both neuronal and vascular formations, so they need to be recognized precisely not only to ensure that the surgical target is reached correctly but also to preclude iatrogenic injury. Even small visual details can make a big difference to the result. For example, a common pitfall for surgeons in this type of procedure happens during the intradural

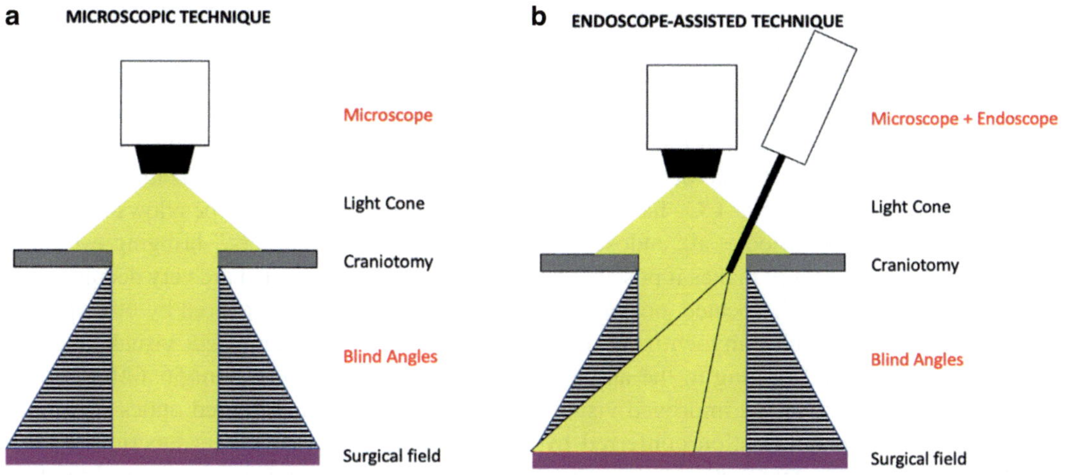

Fig. 35.3 Schematic representation of the illumination provided when the endoscope complements the microscope, especially for deep approaches and small craniotomies. (**a**) On the left, some angles of the surgical field cannot be reached by the microscope light so they are blind. (**b**) On the right, the endoscope has the ability to introduce a cone of light directly into the surgical field, overcoming the limitations of the small craniotomy and also illuminating more lateral aspects

pre-callosal phase of the approach, when the gyrus of the cingulum must be distinguished from the CC itself. As noted by Delfini et al., the two cerebral hemispheres can be so attached to the gyrus of the cingulum in some patients that the right and the left ones together can resemble the CC [49]. Since incision of the CC is a delicate and essential phase in the approaches we are discussing, such mistakes must evidently be avoided. Because only visualization criteria are relied upon (the gray matter of the gyrus has a different color from the pale hypovascular CC), an instrument such as the endoscope tangibly enhances the likelihood of success of the surgical intervention. With its slim fit, impressive maneuverability, and the possibility of applying differently angled lenses, the endoscope dramatically increases the amount of visual detail that the surgeon can capture during the procedure; the endoscope lens can be moved very close to the brain tissue [58]. This increases the surgeon's capacity to distinguish among anatomical structures (e.g., CC or gyrus of cingulum), improving his/her ability to follow the main landmarks and making the entire procedure safer, with lower risk of nerve or vessel injury (e.g., to the pericallosal arteries). Moreover, exact placement of the callosotomy is crucial. As described in the "surgical technique" section, the posterior portion of the CC has to be preserved to preclude iatrogenic neurological deficits in the patient. The endoscope makes it easier to recognize the different portions of the CC, allowing a safer callosotomy to be performed. In particular, beyond the corpus callosotomy approach, a closer view of the CC surface is important for transcallosal approaches. As reported earlier, when the anterior or middle transcallosal routes are used, a small callosotomy must be placed after the longitudinal raphe bordered by the two nerves of Lancisi is recognized. All these anatomical structures mentioned are microscopic and are covered by others, making it difficult to distinguish them without the endoscope. Furthermore, the lateral borders of the callosotomy are represented by the two pericallosal arteries. Injury to them is a concrete risk during a callosotomy. Having the opportunity to see them much more closely is an invaluable advantage of the endoscope [47, 59, 60].

Besides the recognition of small and deep anatomical landmarks, endoscope implementation is also useful for evaluating the surgical efficacy of the approach. Sood et al. noted that the improved visualization and magnification of the endoscope enable the completeness of lateral disconnection of the CC to be confirmed [28].

Thanks to this improvement of the surgeon's capacity to observe anatomical details, some surgical approaches to the CC have been rethought. According to Sood et al., endoscope-assisted corpus callosotomy makes it possible to split the splenium of CC, located posteriorly, through an anterior craniotomy with minimal brain retraction [28]. According to the authors, this is possible because of the improved visualization and greater magnification conferred by the endoscope, as also demonstrated in cadaveric studies [9]. This is important because avoiding a posterior approach means reducing the risk of disrupting the parasagittal draining veins, which are more often encountered in the posterior portion of the frontal bone, during posterior craniotomies. Up to 70% of the venous tributaries reach the superior sagittal sinus within a region of 2 cm posterior to the coronal suture [61].

Additionally, better and closer anatomical visualization facilitates more effective hemostasis, not only during the closure steps but also during the different intradural phases of the approach. Needless to say, this ensures better clinical outcomes for patients.

Finally, thanks to its well-known maneuverability, the endoscope also enables the lens to be moved with a wider range of angulations, following the surgical route dynamically without harming the brain parenchyma [17]. This makes the endoscope particularly powerful since the microscope provides only a rectilinear field of vision, so structures that are not directly in that line of vision are overlooked. A suggestive but effective metaphor by Cohen et al. compares using the microscope for exploring the surgical field to "peering into a room through a keyhole in the door" [62]. Structures that are located straight ahead of the surgeon's eyes are seen adequately, but structures that are more lateral or placed on the same side as the craniotomy cannot be visualized. The surgeon can move his or her head to one side, looking obliquely through the keyhole, but there are sides of the room that cannot be seen. Analogously, the operating microscope can be moved in several directions, but parts that are deep, lateral, or opposite to the surgical field will always be invisible to the surgeon. However, coupling the simultaneous visualization viewpoints of microscope and endoscope allows one to look around the "blind corners," bringing even those anatomical structures that are very deeply located in the surgical field or covered by other portions of the brain that hinder direct visualization into the surgeon's range of vision [56, 62]. Moreover, the opportunity to use angled optics with different angulation grades confers superior visualization quality on portions of the surgical field [63] (Fig. 35.4).

Having the capacity to obtain newer visualization angles that are impossible with the microscope alone is very desirable for surgeons conducting procedures directed to deep regions of the brain, as in CC surgery. This is one of the main reasons why endoscope-assisted surgery is particularly indicated for surgical management of lesions that are hard to reach, such as intraventricular lesions. Guerrero and Cohen claim that thanks to support from the endoscope, the surgeon can clearly visualize even the most anterior and posterior parts of the CC, and the adjacent vascular structures, through a smaller craniotomy. They add that this ability to see the length of the CC from stem to stern can provide information about local pathology and anatomical variations [25].

In addition, as reported by Fries and Perneczky in a series of 380 patients, endoscope-assisted procedures (some of which were TCa-TV endoscope-assisted approaches) applied to microsurgical dissection of lesions also ensure better control of their posterior and lateral margins than the standard microsurgical approach [42]. This is paramount because the tumors are vascularized from their lateral borders, where detailed visualization of the lesion-parenchyma interface is often difficult. The endoscope-assisted technique therefore not only makes resection safer, with the possibility of seeing and coagulating even very small bleeds, but also leads to more complete total and supratotal resections.

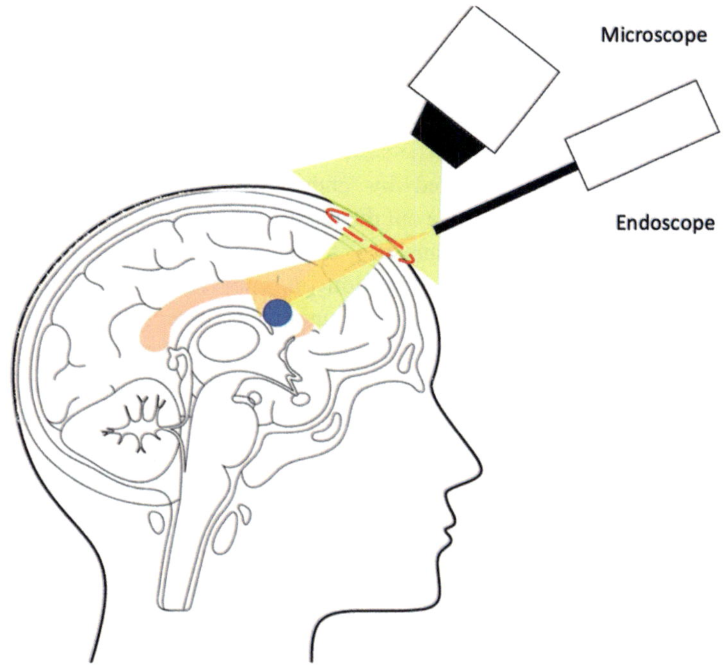

Fig. 35.4 The picture illustrates how the endoscope unlocks more visual angles in the surgical field. Importantly, this tool improves the surgeon's ability to visualize the posterior and lateral aspects of a deep intraventricular lesion, making its removal safer and more effective

Although they can be implemented alone, all these additions provided by endoscope microsurgical assistance can lead synergistically to further advantages. A superior detailed visualization capacity, coupled with wider visual angles and more efficient surgical field illumination, enables the surgeon to rule out more invasive surgery. As a result, smaller craniotomies and dural openings are allowed without impairing the surgical outcome [42].

It is common knowledge that in order to avoid excessive brain retraction in the absence of endoscope support, wider resection of bone and dural borders is required to control the operating field adequately. The literature demonstrates that the introduction of endoscope-assisted transcallosal and corpus callosotomy approaches has reduced the dimensions of both craniotomies and dural openings. For the "classical" open microscopic approaches to the CC, 6–8 cm craniotomies are deemed necessary [49, 64]; in the endoscope-assisted techniques, these minimum diameters are almost halved to 2–4 × 3 cm [24, 25].

A smaller craniotomy has important positive implications for surgical planning, especially in CC surgical midline approaches. One of the first issues that a neurosurgeon must face in applying these approaches to a patient is to decide the right location for the craniotomy in relation to the major superior sagittal sinus parasagittal draining veins, the anatomy of which is a major determinant of the surgical approach entry point. Craniotomies intended for open CC surgery can be impeded by the venous anatomy of the patient, making it difficult to obtain the frontal straight line of view required for microscope visualization without interrupting those vessels. In contrast, with the endoscope-assisted technique, not only is this straightness not required, but also the craniotomies are smaller, unlocking more cranial areas suitable for bone overture placement. This is particularly true with the posterior corpus callosotomy approach, where the craniotomy is particularly challenging to place because most of the bridging veins are posterior to the coronal suture. Sood et al. described a technique that allows posterior portions of the CC to be reached with a small anterior craniotomy, anterior to the most important bridging veins. These authors state that the same results can be obtained thereby with less risk to the patient [28].

Another improvement conferred by endoscope assistance is the ability to take out the most from a given approach. This translates to a reduced need for manipulation and traction on both neuronal and vascular structures. The main driver of neuroendoscopy is to obtain the maximum resection in a minimally invasive way, minimizing the need for brain retraction. In describing the open variant of TCa-TV approaches, Delfini et al. suggest brain retraction of at least 3–4 cm in order to gain the "optimal cone of vision and working angles" with the microscope [LAC]. However, according to Guerrero and Cohen, dissecting under endoscopic guidance allows the CC to be exposed without significant retraction of the brain, making all interventions through a corridor not more than 10 mm wide [25]. This seems very important, especially in CC surgery, where taking great care of arteries and veins is mandatory in order to avoid ischemic injury [65]. The reduction of retraction is even more appreciated for posterior CC approaches because retraction of the posterior frontal lobe risks injury to the bridging veins. Applying less retraction can reduce this injury risk.

Apart from the vessels, manipulation and retraction of neuronal tissue is a big issue in CC surgery, more than in other procedures. This is mainly because of the extreme vulnerability and functional importance of several delicate structures surrounding the CC. Galen understood this concept many centuries ago, underlining the friability of the septum pellucidum, emphasizing that sectioning the CC should be precise, skillful, and meticulous [3]. In an overview of endoscope application in neurosurgery, Rigante et al. note the risk of injury to the fornix resulting from excessive and reckless neural manipulation and retraction, leading to a high risk of memory deficits in such approaches [12]. Applying less pressure on the neuronal tissue during CC approaches, by working in a narrower corridor, reduces local cerebral tissue compression and the decrease of regional cerebral blood flow, lowering the risk of ischemic complications with a double modality [66, 67]. These advances in CC surgery are more appreciable by neurosurgeons since excessive retraction and manipulation lead to complications that are even less acceptable for the patient because they are due to the surgical procedure not the pathology itself, impairing the postoperative outcome.

The list of benefits from endoscope assistance in CC surgery would not be complete without mentioning the indirect benefits that arise from them.

First, smaller craniotomies, smaller dural openings (with smaller closures), and more contained corpus callosotomies inevitably shorten the duration of CC surgical procedures [42]. In a prospective series of 11 patients affected by drug-resistant epilepsy, treated by endoscope-assisted corpus callosotomy and only microscopic surgery, Chandra et al. reported a shorter duration in endoscope-assisted patients than the microsurgery group (mean duration of microscope surgery was 153 (627) minutes vs. 124 (129) minutes for endoscopic-assisted surgery) [29].

Secondly, a smaller craniotomy reduces blood loss and postoperative pain [62]. In one word, the introduction of the endoscope as an assistance tool in CC surgery has led to less traumatic approaches, with better surgical outcomes and shorter hospital stay [28].

Importantly, synergizing the stronger features of two instruments, endoscope-assisted CC surgery has advantages over purely endoscopic techniques. For example, commonly available microsurgical instruments can be used, so there is no need to buy or produce specifically endoscopic tools. This means there are no additional costs for hospitals and no need for extra training for the surgeon. Also, endoscope-assisted CC surgery, in contrast to pure neuroendoscopy, offers a simultaneous dual view of the surgical field, so both ipsilateral and contralateral aspects of an intraventricular lesion, or of the CC itself, can be visualized. Purely endoscopic approaches to the CC do not have this advantage [42].

35.5.2 Disadvantages

As with every newly developed technique or a new use of a well-known instrument, the balance of advantages versus disadvantages must be analyzed. The application of endoscope assistance to CC surgery has drawbacks as well as its many benefits.

In our opinion, the main disadvantage is that neurosurgeons can lack confidence in using the endoscope. As Sood et al. reported, experience or familiarity with this surgical modality is a prerequisite for practicing endoscope-assisted CC surgery [28]. This involves a steep learning curve. This is a well-known problem with neuroendoscopy in general, for example, with endoscopic approaches to pituitary adenomas. In the comparison between endoscope-assisted techniques and purely endoscopic surgery, the problem can be even more acute. The endoscope-assistant technique requires not only familiarity with a certain visualization tool, the endoscope, but also mastery of another tool, the microscope. Most of all, both instruments must be maneuvered together simultaneously. In our opinion, this last requirement can be especially challenging, causing enough frustration to make surgeons quit this effective surgical modality in favor of pure microscopic dissection, with which they are more confident. This confidence problem is even more significant when endoscope-assisted techniques are applied to CC surgery, which requires a deep surgical field and approaches to lesions that are intimately associated with critical neurovascular structures. Moreover, since the use of endoscopes requires an endoscope-adapted knowledge of neuroanatomy, it is important to emphasize that microscope visualization of interhemispheric callosal anatomy is already complex and becomes even more difficult to master when the endoscope is applied. The earliest reports on this technique, mostly applied to cadavers, state that it is "simple but adds time" [25].

Obviously, practice mitigates this problem. Although few studies have analyzed the learning curve specific for endoscope-assisted surgical approaches to the CC, several studies assessing the learning curve for skull base endoscopic techniques agree that after 30–50 procedures, surgeons acquire great confidence with the endoscope [68–70]. The learning curve for endoscope-assisted CC surgery was specifically examined by Chandra et al. in 2015. In reporting the results of 16 patients treated by endoscope-assisted callosal or transcallosal approaches, these authors noted the duration of the procedure. The time spent on the first three endoscopic surgeries was comparable to that spent on microscopic surgeries, but the last two were significantly quicker; the duration of the final case was almost half that of case 1. Most of the time in the earlier procedures was expended on setting up the endoscopic equipment rather than the actual surgery [29]. These data show that although the learning curve constitutes a disadvantage for the endoscope technique compared with the purely microscopic one, practice minimizes that disadvantage. However, some could argue that the rarity of the pathologies and the indications for such approaches makes it difficult for a surgeon in a single center to gain sufficient experience to acquire the required confidence.

Another possible disadvantage of endoscope-assisted techniques for CC surgery is the difficulty of placing two visualization tools in the same strict visual line offered by a minimally invasive surgical corridor. Although there is little literature concerning the endoscope-assisted technique for CC surgery, there is debate about the best modalities for managing the microscope and the endoscope simultaneously in the same surgical field for the same approach. While Chandra et al. prefer a three-handed technique [53] that corresponds to the "classical" operative setting of endoscopic pituitary surgery, Sood et al. implemented a two-handed technique with the endoscope fixed to the suction, both placed on the left hand of the surgeon, freeing the right hand for delicate microscopic dissection [28]. There were two main reasons why the latter authors adopted this different technique. First, the three-handed technique requires another component to handle the endoscope, different from the first operator, a second operator, or a holding device. Since the first operator decides what to see, the second element must move the endoscope to a new position every time the field of surgery has to change, making long operations tedious and frustrating. Second, with three-handed techniques, small craniotomies and surgical corridors can cause crowding of the instruments through the narrow opening. This can be inconvenient, leading to irritating instrument “scissoring,” which can be especially dangerous in hindering the control of bleeding [17]. The three-handed technique therefore requires bigger craniotomies and major brain retraction, eliminating two of the main advantages of the endoscope-assisted technique.

Besides the solutions for these problems proposed by different authors, it seems evident to us that this debate highlights an important issue that can make the surgical technique uncomfortable, especially for skilled surgeons who have practiced open microscopic CC surgery for several years.

The availability of instrument settings is another possible disadvantage of the endoscope-assisted technique, which requires the simultaneous use of two visualization tools with different points of view and image capturing systems. The main difficulty is how to make these two different image settings available to the surgeon in real time in a user-friendly way. The neurosurgeon must mentally combine information about the surgical field from those two sources, which can be problematic for two reasons. First, applying this information integration skill in real time throughout the procedure is very time-demanding for the surgeon. Second, an operating theater is required with enough room to accommodate a microscope and an endoscopic visualization setting simultaneously, and that implies a center where both instruments have been bought and updated. These considerations can make the technique unsuitable for smaller neurosurgical centers. Nevertheless, that could be advantageous in relation to the learning curve: centers that have this technology provide more cases for the surgeon to use this type of approach.

Many technical solutions have been proposed to facilitate the surgeon’s need to combine the information sources. The two main ones are alternation of microscopic and endoscopic viewing by the surgeon and fusion of the two pictures provided. The former is the less expensive way but requires regular interruptions for the surgeon, who has to switch from viewing the microscope to the endoscope and vice versa. Alternatively, the endoscopic image can be mounted on the microscope, reducing the need to move the eyes and neck [71]. This ergonomic advantage can be enhanced by a picture-in-picture mode for visualizing endoscopic images, as with the twin-video system from Storz, for example. However, fusing the two pictures from microscope and endoscope in the microscope eyepiece or in a head-mounted LCD screen is the best way of combining them [56].

However, taking stock of all the advantages and disadvantages of this technique, it seems clear that despite a few small and almost logistical issues, its wide range of advantages and high efficiency seem to us to constitute a revolution in the surgical management of callosal and transcallosal pathologies. We recommend choosing this surgical modality whenever possible.

35.6 Perspectives: A Glimpse of the Future

Although the introduction of the endoscope into CC surgery has obviously reduced invasiveness and improved clinical results, progress must continue. Some limitations of the endoscope-assisted approach to the CC have been described above, and these need to be overcome, encouraging further development of these operative techniques.

In our opinion, the exoscope holds promise for achieving the required improvement in neurosurgery. This instrument has provided the most recent innovations in surgeries in general, where blind spots can be hard to observe. Moreover, it has interesting features that, used in association with the endoscope, could overcome some problems of the endoscope-assisted modality. First, the exoscope frees the surgeon from the binocular vision of the microscope, providing a so-called head-up surgery system [72]. This can have many positive consequences, such as the opportunity to see the endoscopic and exoscopic images on the same screen without the surgeon having to move his/her head continually from microscope to monitor and vice versa. Some exoscopes, such as the ORBEYE exoscope, already offer a picture-in-picture view for capturing navigation and also endoscopic images [73].

Moreover, the exoscope is very much smaller than the microscope. Therefore, keeping it side by side with the endoscope in the same visual line offered by a minimally invasive surgical approach is simpler and more effective than with the microscope [74]. We therefore see a possible future for CC surgery in the exoscope-endoscope-assisted surgical approach.

35.7 Conclusion

CC approaches are surgically very challenging. Despite their minor disadvantages, we consider that endoscope-assisted techniques represent a real breakthrough for the neurosurgeon, reducing risks for the patient and allowing better clinical outcomes to be obtained, with the advantages outweighing the disadvantages. We believe that the endoscope-assisted technique for CC surgery will improve in the future, helped by new technological tools that can contribute to promoting its wider use.

References

1. Pearce JM. Corpus callosum. Eur Neurol. 2007;57:249–50. https://doi.org/10.1159/000101293.
2. Shah A, Jhawar S, Goel A. Corpus callosum and its connections: a fiber dissection study. World Neurosurg. 2021;151:e1024–35. https://doi.org/10.1016/j.wneu.2021.05.047.
3. Rocca J. Galen on the brain: anatomical knowledge and physiological speculation in the second century AD. Stud Anc Med. 2003;26:1–313.
4. Vesalius A. De humani corporis fabrica. Padua: School of Medicine; 1543.
5. F VdA Mémoires sur l ' Anatomie Humaine et Comparée. 13th ed. 1847 [in French].
6. Uff C, Frith D, Harrison C, Powell M, Kitchen N. Sir Victor Horsley's 19th century operations at the National Hospital for neurology and neurosurgery, Queen Square. J Neurosurg. 2011;114:534–42. https://doi.org/10.3171/2010.9.JNS09731.
7. Jea A, Vachhrajani S, Widjaja E, Nilsson D, Raybaud C, Shroff M, Rutka JT. Corpus callosotomy in children and the disconnection syndromes: a review. Childs Nerv Syst. 2008;24:685–92. https://doi.org/10.1007/s00381-008-0626-4.
8. Mathews MS, Linskey ME, Binder DK. William P. van Wagenen and the first corpus callosotomies for epilepsy. J Neurosurg. 2008;108:608–13. https://doi.org/10.3171/JNS/2008/108/3/0608.
9. Bahuleyan B, Vogel TW, Robinson S, Cohen AR. Endoscopic total corpus callosotomy: cadaveric demonstration of a new approach. Pediatr Neurosurg. 2011;47:455–60. https://doi.org/10.1159/000338984.
10. de Divitiis E, Zona G. Tumours of the callosal area: problems and surgical strategies. J Neurosurg Sci. 1995;39:147–52.
11. Nasi D, Iacoangeli M, Di Somma L, Dobran M, Di Rienzo A, Gladi M, Benigni R, Passamonti C, Zamponi N, Scerrati M. Microsurgical endoscopy-assisted anterior corpus callosotomy for drug-resistant epilepsy in an adult unresponsive to vagus nerve stimulation. Epilepsy Behav Case Rep. 2016;5:27–30. https://doi.org/10.1016/j.ebcr.2016.01.001.
12. Rigante L, Borghei-Razavi H, Recinos PF, Roser F. An overview of endoscopy in neurologic surgery. Cleve Clin J Med. 2019;86:16ME–24ME. https://doi.org/10.3949/ccjm.86.me.18142.
13. Shoakazemi A, Evins AI, Burrell JC, Stieg PE, Bernardo A. A 3D endoscopic transtubular transcallosal approach to the third ventricle. J Neurosurg. 2015;122:564–73. https://doi.org/10.3171/2014.11.JNS14341.
14. Longatti P, Fiorindi A, Bruscella S, Cappabianca P. Development of cerebral endoscopy in Italy: background, narration, and legacy. World Neurosurg. 2020;134:396–401. https://doi.org/10.1016/j.wneu.2019.10.165.
15. Abbott R. History of neuroendoscopy. Neurosurg Clin N Am. 2004;15(1):1–7.
16. de Divitiis O, d'Avella E, de Notaris M, Di Somma A, De Rosa A, Solari D, Cappabianca P. The (R)evolution of anatomy. World Neurosurg. 2019;127:710–35. https://doi.org/10.1016/j.wneu.2019.03.050.
17. Hopf NJ, Perneczky A. Endoscopic neurosurgery and endoscope-assisted microneurosurgery for the treatment of intracranial cysts. Neurosurgery. 1998;43:1330–6. discussion 1336–1337. https://doi.org/10.1097/00006123-199812000-00037.
18. Vaddiparti A, Huang R, Blihar D, Du Plessis M, Montalbano MJ, Tubbs RS, Loukas M. The evolution of corpus callosotomy for epilepsy management. World Neurosurg. 2021;145:455–61. https://doi.org/10.1016/j.wneu.2020.08.178.
19. Jalilian L, Limbrick DD, Steger-May K, Johnston J, Powers AK, Smyth MD. Complete versus anterior two-thirds corpus callosotomy in children: analysis of outcome. J Neurosurg Pediatr. 2010;6:257–66. https://doi.org/10.3171/2010.5.PEDS1029.
20. Chan AY, Rolston JD, Lee B, Vadera S, Englot DJ. Rates and predictors of seizure outcome after corpus callosotomy for drug-resistant epilepsy: a meta-

analysis. J Neurosurg. 2018;130:1193. https://doi.org/10.3171/2017.12.JNS172331.
21. Rougier A, Claverie B, Pedespan JM, Marchal C, Loiseau P. Callosotomy for intractable epilepsy: overall outcome. J Neurosurg Sci. 1997;41:51–7.
22. Bogen J. The neurosurgeon ' s interest in the corpus callosum. In: Greenblatt SH, Dagi TF, Epstein MH, editors. A history of neurosurgery: in it scientific and professional contexts. Stuttgart: Thieme; 1997. p. 489–98.
23. Dandy WE. Operative experience in cases of pineal tumor. Arch Surg. 1936;33:19–46. https://doi.org/10.1001/archsurg.1936.01190010022002.
24. Smyth MD, Vellimana AK, Asano E, Sood S. Corpus callosotomy-open and endoscopic surgical techniques. Epilepsia. 2017;58(Suppl 1):73–9. https://doi.org/10.1111/epi.13681.
25. Guerrero MH, Cohen AR. Endoscope-assisted microsurgery of the corpus callosum. Minim Invasive Neurosurg. 2003;46:54–6. https://doi.org/10.1055/s-2003-37968.
26. Tubbs RS, Smyth MD, Salter G, Doughty K, Blount JP. Eyebrow incision with supraorbital trephination for endoscopic corpus callosotomy: a feasibility study. Childs Nerv Syst. 2004;20:188–91. https://doi.org/10.1007/s00381-003-0894-y.
27. Bahuleyan B, Manjila S, Robinson S, Cohen AR. Minimally invasive endoscopic transventricular hemispherotomy for medically intractable epilepsy: a new approach and cadaveric demonstration. J Neurosurg Pediatr. 2010;6:536–40. https://doi.org/10.3171/2010.9.PEDS10267.
28. Sood S, Marupudi NI, Asano E, Haridas A, Ham SD. Endoscopic corpus callosotomy and hemispherotomy. J Neurosurg Pediatr. 2015;16:681–6. https://doi.org/10.3171/2015.5.PEDS1531.
29. Chandra SP, Kurwale NS, Chibber SS, Banerji J, Dwivedi R, Garg A, Bal C, Tripathi M, Sarkar C. Endoscopic-assisted (through a mini craniotomy) corpus callosotomy combined with anterior, hippocampal, and posterior commissurotomy in Lennox-Gastaut syndrome: a pilot study to establish its safety and efficacy. Neurosurgery. 2016;78:743–51. https://doi.org/10.1227/NEU.0000000000001060.
30. Sood S, Hoeprich M, Ham SD. Pure endoscopic removal of pineal region tumors. Childs Nerv Syst. 2011;27:1489–92. https://doi.org/10.1007/s00381-011-1490-1.
31. Liu JK, Majmundar N. Retractorless interforniceal approach for microsurgical resection of a papillary tumor of the pineal region: operative video and technical nuances. Neurosurg Focus Video. 2021;5:V12. https://doi.org/10.3171/2021.4.FOCVID2139.
32. Uda T, Kunihiro N, Umaba R, Koh S, Kawashima T, Ikeda S, Ishimoto K, Goto T. Surgical aspects of corpus callosotomy. Brain Sci. 2021;11:1608. https://doi.org/10.3390/brainsci11121608.
33. Fandiño-Franky J, Torres M, Nariño D, Fandiño J. Corpus callosotomy in Colombia and some reflections on care and research among the poor in developing countries. Epilepsia. 2000;41(Suppl 4):S22–7. https://doi.org/10.1111/j.1528-1157.2000.tb01541.x.
34. Yano S, Hide T, Shinojima N, Ueda Y, Kuratsu J. A flexible endoscope-assisted interhemispheric transcallosal approach through the contralateral ventricle for the removal of a third ventricle craniopharyngioma: a technical report. Surg Neurol Int. 2015;6:S113–6. https://doi.org/10.4103/2152-7806.153653.
35. Liu JK. Interhemispheric transcallosal approach for resection of intraventricular central neurocytoma. Neurosurg Focus. 2013;34:Video 3. https://doi.org/10.3171/2013.V1.FOCUS12353.
36. Anon. Benign tumors in the third ventricle of the brain: diagnosis and treatment. Arch Neurol Psychiatr. 1935;33:242–2. https://doi.org/10.1001/archneurpsyc.1935.02250130248020.
37. Apuzzo ML, Heifetz MD, Weiss MH, Kurze T. Neurosurgical endoscopy using the side-viewing telescope. J Neurosurg. 1977;46:398–400. https://doi.org/10.3171/jns.1977.46.3.0398.
38. Prott W. Cisternoscopy–endoscopy of the cerebellopontine angle. Acta Neurochir (Wien). 1974;31:105.
39. Bouras T, Sgouros S. Complications of endoscopic third ventriculostomy. World Neurosurg. 2013;79:S22.e29–12. https://doi.org/10.1016/j.wneu.2012.02.014.
40. Cappabianca P, Cinalli G, Gangemi M, Brunori A, Cavallo LM, de Divitiis E, Decq P, Delitala A, Di Rocco F, Frazee J, Godano U, Grotenhuis A, Longatti P, Mascari C, Nishihara T, Oi S, Rekate H, Schroeder HW, Souweidane MM, Spennato P, Tamburrini G, Teo C, Warf B, Zymberg ST. Application of neuroendoscopy to intraventricular lesions. Neurosurgery. 2008;62(Suppl 2):575–97. discussion 597–578. https://doi.org/10.1227/01.neu.0000316262.74843.dd.
41. Kulkarni AV, Riva-Cambrin J, Holubkov R, Browd SR, Cochrane DD, Drake JM, Limbrick DD, Rozzelle CJ, Simon TD, Tamber MS, Wellons JC, Whitehead WE, Kestle JR, Network HCR. Endoscopic third ventriculostomy in children: prospective, multicenter results from the hydrocephalus clinical research network. J Neurosurg Pediatr. 2016;18:423–9. https://doi.org/10.3171/2016.4.PEDS163.
42. Fries G, Perneczky A. Endoscope-assisted brain surgery: part 2—analysis of 380 procedures. Neurosurgery. 1998;42:226–31. discussion 231–222. https://doi.org/10.1097/00006123-199802000-00008.
43. King WA, Wackym PA. Endoscope-assisted surgery for acoustic neuromas (vestibular schwannomas): early experience using the rigid Hopkins telescope. Neurosurgery. 1999;44:1095–100. discussion 1100-1092. https://doi.org/10.1097/00006123-199905000-00084.
44. Matula C, Tschabitscher M, Day JD, Reinprecht A, Koos WT. Endoscopically assisted microneurosurgery. Acta Neurochir. 1995;134:190–5. https://doi.org/10.1007/BF01417688.

45. Charalampaki P, Filippi R, Welschehold S, Conrad J, Perneczky A. Tumors of the lateral and third ventricle: removal under endoscope-assisted keyhole conditions. Neurosurgery. 2008;62:1049–58. https://doi.org/10.1227/01.neu.0000333770.58693.ed.
46. Gaab MR, Schroeder HW. Neuroendoscopic approach to intraventricular lesions. J Neurosurg. 1998;88:496–505. https://doi.org/10.3171/jns.1998.88.3.0496.
47. de Divitiis O, Cavallo LM, Dal Fabbro M, Elefante A, Cappabianca P. Freehand dynamic endoscopic resection of an epidermoid tumor of the cerebellopontine angle: technical case report. Neurosurgery. 2007;61:E239. https://doi.org/10.1227/01.neu.0000303222.40145.1a.
48. Gagliardi F, Gragnaniello C, Mortini P. Operative cranial neurosurgical anatomy. 1st ed. Stuttgart: Thieme; 2019.
49. Delfini R, Pichierri A. Transcallosal approaches to intraventricular tumors. In: Cappabianca P, Iaconetta G, Califano L, editors. Cranial, craniofacial and skull base surgery. 1st ed. Milano: Springer; 2010. https://doi.org/10.1007/978-88-470-1167-0.
50. Grotenhuis JA. Endoscope-assisted craniotomy. In: Techniques in neurosurgery, vol. 3. Philadelphia, PA: Lippincott-Raven; 1996. p. 201–12.
51. Di Ieva A, Tschabitscher M, Rodriguez Y, Baena R. Lancisi's nerves and the seat of the soul. Neurosurgery. 2007;60:563–8. discussion 568. https://doi.org/10.1227/01.NEU.0000249283.46514.93.
52. Esposito V, di Russo P, Del Maestro M, Ciavarro M, Vangelista T, De Angelis M, Fava A, Morace R. The interhemispheric Transgenual approach for microsurgical removal of third ventricle colloid cysts. Technical note. World Neurosurg. 2020;142:197–205. https://doi.org/10.1016/j.wneu.2020.06.222.
53. Chandra PS, Kurwale N, Garg A, Dwivedi R, Malviya SV, Tripathi M. Endoscopy-assisted interhemispheric transcallosal hemispherotomy: preliminary description of a novel technique. Neurosurgery. 2015;76:485–94. discussion 494–485. https://doi.org/10.1227/NEU.0000000000000675.
54. Kasowski HJ, Nahed BV, Piepmeier JM. Transcallosal transchoroidal approach to tumors of the third ventricle. Neurosurgery. 2005;57:361–6. discussion 361–366. https://doi.org/10.1227/01.neu.0000176652.59747.40.
55. Wen HT, Rhoton AL, de Oliveira E. Transchoroidal approach to the third ventricle: an anatomic study of the choroidal fissure and its clinical application. Neurosurgery. 1998;42:1205–17. discussion 1217–1209. https://doi.org/10.1097/00006123-199806000-00001.
56. Perneczky A, Fries G. Endoscope-assisted brain surgery: part 1—evolution, basic concept, and current technique. Neurosurgery. 1998;42:219–24 . discussion 224–215. https://doi.org/10.1097/00006123-199802000-00001.
57. de Divitiis E, de Divitiis O, Elefante A. Supraorbital craniotomy: pro and cons of endoscopic assistance. World Neurosurg. 2014;82:e93–6. https://doi.org/10.1016/j.wneu.2013.03.027.
58. Perneczky A. MTKR endoscopic anatomy for neurosurgery. 1st ed. Stuttgart: George Thieme Verlag; 1993.
59. de Divitiis O, Conti A, Somma T, Angileri F, Cappabianca P. Ventral brainstem anatomy: an endoscopic transoral perspective. Acta Neurochir Suppl. 2019;125:45–50. https://doi.org/10.1007/978-3-319-62515-7_7.
60. de Divitiis O, d'Avella E, Sacco M, Somma T, Baro V, Denaro L. The role of vitom-3D in the management of spinal meningiomas: review of the literature and illustrative case. Mini-invasive Surg. 2020;4:75. https://doi.org/10.20517/2574-1225.2020.66.
61. Wen HT. Surgical approach to lesions located in the lateral, third and fourth ventricles. In: Sekhar L, Fessler R, editors. Atlas of neurosurgical techniques. Stuttgart: Thieme; 2006.
62. Cohen AR, Perneczky A, Rodziewicz GS, Gingold SI. Endoscope-assisted craniotomy: approach to the rostral brain stem. Neurosurgery. 1995;36:1128–9. discussion 1129-1130. https://doi.org/10.1227/00006123-199506000-00009.
63. Lewis AI, Crone KR, Taha J, van Loveren HR, Yeh HS, Tew JM. Surgical resection of third ventricle colloid cysts. Preliminary results comparing transcallosal microsurgery with endoscopy. J Neurosurg. 1994;81:174–8. https://doi.org/10.3171/jns.1994.81.2.0174.
64. Kasasbeh AS, Smyth MD, Steger-May K, Jalilian L, Bertrand M, Limbrick DD. Outcomes after anterior or complete corpus callosotomy in children. Neurosurgery. 2014;74:17–28. discussion 28. https://doi.org/10.1227/NEU.0000000000000197.
65. Jan M, Kakou M, Velut S. Surgical approaches to the corpus callosum. Neurochirurgie. 1998;44:133–7.
66. Arbit E, DiResta GR, Bedford RF, Shah NK, Galicich JH. Intraoperative measurement of cerebral and tumor blood flow with laser-Doppler flowmetry. Neurosurgery. 1989;24:166–70. https://doi.org/10.1227/00006123-198902000-00003.
67. Jakobsson KE, Löfgren J, Zwetnow NN. Determinants of intracranial pressure rebound after brain compression. In: Miller JD, Teasdale GM, Rowan JO, Galbraith SL, Mendelow AD, editors. Intracranial pressure VI. Berlin: Springer; 1986. p. 161–5.
68. Kshettry VR, Do H, Elshazly K, Farrell CJ, Nyquist G, Rosen M, Evans JJ. The learning curve in endoscopic endonasal resection of craniopharyngiomas. Neurosurg Focus. 2016;41:E9. https://doi.org/10.3171/2016.9.FOCUS16292.
69. Leach P, Abou-Zeid AH, Kearney T, Davis J, Trainer PJ, Gnanalingham KK. Endoscopic transsphenoidal pituitary surgery: evidence of an operative learning curve. Neurosurgery. 2010;67:1205–12. https://doi.org/10.1227/NEU.0b013e3181ef25c5.
70. Smith SJ, Eralil G, Woon K, Sama A, Dow G, Robertson I. Light at the end of the tunnel: the learn-

ing curve associated with endoscopic transsphenoidal skull base surgery. Skull Base. 2010;20:69–74. https://doi.org/10.1055/s-0029-1238214.

71. Taneda M, Kato A, Yoshimine T, Hayakawa T. Endoscopic-image display system mounted on the surgical microscope. Minim Invasive Neurosurg. 1995;38:85–6. https://doi.org/10.1055/s-2008-1053463.
72. Yano S, Hiraoka F, Morita H, Kawano H, Kuwajima T, Yoshida SI, Hama Y, Tashiro N, Hamaguchi S, Aikawa H, Go Y, Kazekawa K. Usefulness of endoscope-assisted surgery under exoscopic view in skull base surgery: a technical note. Surg Neurol Int. 2022;13:30. https://doi.org/10.25259/SNI_965_2021.
73. Broggi M, Acerbi F, Ferroli P, Tringali G, Schiariti M, Broggi G. Microvascular decompression for neurovascular conflicts in the cerebello-pontine angle: which role for endoscopy? Acta Neurochir. 2013;155:1709–16. https://doi.org/10.1007/s00701-013-1824-8.
74. El Refaee E, Langner S, Marx S, Rosenstengel C, Baldauf J, Schroeder HWS. Endoscope-assisted microvascular decompression for the management of hemifacial spasm caused by vertebrobasilar dolichoectasia. World Neurosurg. 2019;121:e566–75. https://doi.org/10.1016/j.wneu.2018.09.166.

Part V

Cognitive Neuroscience

Editor's Summary

The field of cognitive neuroscience has grown vastly since its inception. Moreover, our current understanding of this science is based on many studies focused on the corpus callosum (CC). For example, the role of the CC in behavior, intelligence, listening, and cognition is all becoming clearer. Additionally, the neuroscience of the CC in regard to the various hyperactivity disorders and attention deficit disorders has received recent attention. Finally, the CC as an evolutionary innovation will be reviewed in this part of the textbook.

The Role of the Corpus Callosum in Human Cognition and Anesthesia

36

Pelin Dilsiz and Sinem Sarı

36.1 Association Between the Corpus Callosum and Neurocognition

The corpus callosum (CC) is the largest connecting structure in the brain with more than 190 million axons. It has been stated that this structure provides cognitive performance by transmitting and integrating the information received from the cortex between the hemispheres [1, 2]. In terms of the gross anatomy, it can be divided into seven subregions: (1) rostrum, (2) genu, (3) rostral body, (4)_ anterior mid-body, (5) posterior mid-body, (6) isthmus, and (7) splenium. The splenium connects the occipital areas [3].

The earliest studies on the neurocognitive function of the CC were in split-brain cases. Roger Sperry and Gazzaniga continued their CC research, which began with animal studies in the 1950s, by examining patients whose callosal connections were cut for the treatment of epilepsy [4]. In the following years, Gordon, Gazzaniga, and Freedman presented their observations on patients with partial callosal lesions and have stated that callosal lesions lead to integration defects at varying levels [5, 6]. Recent studies [7] have supported the fact that the CC has a key role in the connection between the two hemispheres, revealed its functional specificity, and explained its topographic organization.

For a clearer determination of the significance of the CC in neurocognition, its relationship with cognitive disorders was evaluated in various studies. Hinkley et al., in their review, citing Voineskos et al. [8], have stated that the deterioration of the integrity and function of CC contributes to cognitive decline in adults [9]. In the same review, it has been also stated that changing CC thickness in childhood has effects on intelligence [10, 11], processing speed [12], and problem-solving ability [13]. Also, there is increasing literature suggesting that the changes in the callosal structure can be associated with neurodevelopmental cognitive and behavioral disorders including autism [14], attention deficit, hyperactivity disorder [15], and dyslexia [16].

In addition to examining the changes in the CC in cognitive disorders, the effect of disorders in the CC on cognition also adds a different perspective to make clear inferences. Examining individuals with complete (cAgCC) and partial (pAgCC), agenesis can yield valuable results to understand the relationship between the CC and cognitive functions. The clinical course in individuals with AgCC ranges from asymptomatic to severe neurodevelopmental disorders. Due to various factors including genetic defects and metabolic diseases, the development of CC can be impaired in many ways, resulting in complete

P. Dilsiz
Anesthesiology and Reanimation Clinic, Menteşe State Hospital, Menteşe, Muğla, Turkey

S. Sarı (✉)
Department of Anesthesiology and Reanimation, Aydın Adnan Menderes University Faculty of Medicine, Aydın, Turkey

M. Turgut et al. (eds.), *The Corpus Callosum*, https://doi.org/10.1007/978-3-031-38114-0_36

or partial AgCC. This creates various clinical cases depending on the specificity of the malformation and other accompanying defects in the patients. Epilepsy, motor impairment, and mental retardation are common clinical manifestations. The most common form of the disease, however, is the asymptomatic course [17].

Valenti et al. [18], in their review, stated that partial or complete agenesis of CC was associated with delayed and hindered interhemispheric integration [19] and slowing of execution processes [20]. Individuals with AgCC showed a decrease in both their problem-solving abilities [21] and processing speed [22]. This decrease can be explained by impairments in areas such as abstract reasoning [23], perception of paralinguistic information [24], and social cognition [9, 25, 26].

36.2 The Effects of Anesthesia Exposure on the Corpus Callosum and Neurocognition

The changes in cognitive functions in the postoperative period have been researched in numerous studies. Evered et al., in their review, evaluated this subject and provided updates [27]. In 1887, Savage defined "chronic weak-mindedness" after anesthesia and focused on delirium, rather than cognitive function [28]. In the following years, the studies that revealed cognitive decline after both general and local anesthesia have reported cognitive losses at a level that met the criteria for dementia [29]. Subsequently, it was emphasized that the changes in cognitive functions could be revealed more clearly with pre- and postoperative evaluations and objective tests were presented. The widespread use of MMSE (Mini-Mental State Examination) facilitated a clearer definition of postoperative cognitive dysfunction (POCD). POCD is a clinical picture that can emerge in a period from the recovery room until the fifth postoperative day [30]. POCD is a clinical condition that may present with different subtypes (hyperactive, hypoactive, mixed type). It has been stated that the prognosis of hypoactive delirium among clinical subtypes may be worse related to delay in diagnosis and treatment. Also, advanced age seems to be a predisposing factor for hypoactive delirium [31].

It was estimated that postoperative cognitive dysfunction is seen in 10% of the non-cardiac operations [27]. POCD is seen in different age and treatment groups; however, advanced age and cardiac and orthopedic surgeries have been known to be well-known risk factors. The frequency of POCD after specific surgeries has been reported to be 10–15% in patients undergoing general surgery, 30% after open heart surgery, and more than 50% in patients who underwent surgery for hip fractures [31].

The evidence indicates that anesthesia and surgical exposure result in postoperative neurocognitive dysfunction and long-term cognitive decline, especially in advanced age patient groups [32]. This early cognitive decline, typically called postoperative neurocognitive dysfunction, has also been associated with long-term cognitive decline after 5 years [33]. On the other hand, it remains unclear to what extent this long-term cognitive dysfunction reflects the impact of anesthesia and surgery, in addition to underlying surgical disease processes and medical comorbidities [34, 35].

POCD is a common complication; however, its pathophysiology remains unclear. Subramaniyan et al. have reported that (1) systemic inflammation, (2) changes in the neurovascular unit/blood-brain barrier, and (3) postsurgical neuroinflammation have an evident role in the pathophysiology of POCD [36].

To explain the pathogenesis of POCD, imaging studies of both neurovascular changes and neuroinflammation have been underway. Kant et al., using structural MRI, have reported a consistent relationship between postoperative neurocognitive dysfunction and neurovascular changes [37].

Other than imaging the neurovascular structure, it is difficult to demonstrate neuroinflammation. TSPO (PET with the second-generation translator protein) is one of the methods evaluated in the last few decades to show neuroinflammation in neurodegenerative diseases that develop related to neuroinflammation [38]. In the first study that adopted this method to assess the changes in the postoperative period, Forsberg et al. [39] evaluated eight patients who underwent prostatectomy surgery under general anesthesia with brain [11 C] PBR28 PET imaging. The researchers have reported that the downregulation of glial cells developing with the immunosuppressive response in the acute postoperative period was followed by the normalization of glial activity with the disappearance of immunosuppression in the third-month controls. The researchers also stated that this process could explain postoperative cognitive dysfunction [39].

Other imaging-supported studies in the literature showed that there were changes in specific brain regions of patients after anesthesia exposure. Studies on changes in the cortex, hippocampus, and CC to explain the pathogenesis of cognitive dysfunction are underway.

Kline et al. have reported that, after exposure to anesthesia, there was significant cortical area atrophy and hippocampal volume loss compared to that in the control group. The researchers have argued that this change was not associated with a permanent cognitive impairment during follow-up but rather a finding in the favor of POCD [40].

Sprung et al. [41] have reported a cortical thinning in the areas accepted as the signature region for Alzheimer's disease (AD) in the postoperative period. The study aimed to detect cortical thinning with structural MRI and hyperdense (WMH) areas in white matter with FLAIR MRI and to determine ischemic areas in the postoperative period. In patients who received general anesthesia 20 years prior, cortical thinning areas were detected in the cortical signature regions of AD, and WMH areas were not observed. Also, hippocampal involvement is expected in the early phase of AD; however, no significant changes were found in the hippocampus volume in the study. The measured CC volume was found to be similar in both groups. The researchers associated the thinning in the cortical signature regions of AD with anesthesia exposure [41].

Although studies on this subject are mainly on general anesthesia, Sprung et al., in another study in the form of a letter to the editor, emphasized that regional anesthesia may have similar results and lead to cortical thinning [42].

Tang et al. [43] adopted a different method than cortical thickness measurement and made a molecular-level examination by diffusion tensor imaging (DTI). This is the first study in the literature examining the micromolecular structure under anesthesia. Axonal degeneration and edema lead to enlargement of the extracellular space between the axons [44]. In the study, the measurements were made consecutively, as postanesthetic 40 min, postanesthetic 100 min, postanesthetic 1 day, and postanesthetic 7 days. The results of the study showed that the interaxonal space was enlarged and this enlargement was temporary. A global decrease in white matter volume and an increase in CSF were observed. The shrinkage of glial cells under anesthesia and the fact that sevoflurane causes vasogenic swelling were thought to be responsible for these changes. A temporary decrease was detected in CC volume in line with the global white matter measurements. The cited study is important as it is the first study conducted at the micromolecular level. It has been thought that the detected changes will be useful to explain the pathogenesis of POCD if supported by further studies [43].

Banerjee et al. [45] presented remarkable results in their study in which they examined the long-term outcomes of the effects of multiple anesthesia exposures and the consequent neurocognitive effects in patients with childhood acute lymphoblastic leukemia (ALL). Higher doses of propofol and a longer duration of anesthesia exposure were associated with impaired white matter integrity in the CC and consequent worse performance on measures of attention and processing speed. These results suggested that high-dose and long-term anesthesia is associated with myelin dysfunction patterns [45].

In addition to clinical studies, animal experiments also provide significant results on the effects on CC after anesthesia. In the examinations carried out after exposure to isoflurane in mice, animals exposed to repeated isoflurane were found to exhibit significant motor deficits in the balance beam and increased anxiety-like behaviors. It was determined that animals exposed to a single isoflurane administration showed poor performance in the foot failure test. With diffusion tensor imaging, repeated exposure to isoflurane was shown to result in impaired long-term diffusion in the CC white matter. Also, differences in the microstructural organization of CC white matter were demonstrated in animals with single and repeated exposure to isoflurane. These results suggest that the behavioral changes observed up to 90 days after repeated isoflurane exposure were due to altered CC white matter microstructural integrity. Furthermore, it was found that especially the anterior CC portion was more sensitive to isoflurane and caused behavioral changes in animals [46].

Zhang et al. administered propofol to newborn rats and observed that high-dose propofol administration increased apoptosis in oligodendrocytes whereas decreased MBP (myelin basic protein) expression. It was determined that these changes reduced myelin formation in the internal capsule and CC [47].

36.3 Anesthesia Management in Corpus Callosum Pathologies

Agenesis of the CC (AgCC) is an intrauterine developmental disorder. Genetic factors, metabolic disorders, maternal alcohol dependence, intrauterine infections, and neuronal developmental defects may have a role in this developmental process. The clinical picture ranges from asymptomatic to severe neurodevelopmental defects [17].

Clinical outcomes that can be seen in patients with CC agenesis or CC hypoplasia comprise microcephaly, macrocephaly, epileptic seizures, developmental delay or retardation, hypotonia, spasticity, and multiple somatic anomalies including arthrogryposis, eye anomalies, syringomyelia, and facial dysmorphisms [48]. Some syndromes that often include AgCC include Aicardi syndrome, Andermann syndrome, Shapiro syndrome, acrocallosal syndrome, septooptic dysplasia (optic nerve hypoplasia), Mowat-Wilson syndrome, and Menkes syndrome [17].

Kumar et al. have reported their anesthesia management in four cases of CC agenesis syndrome who were operated on with sevoflurane under general anesthesia [49]. Considering CC agenesis and accompanying syndromic conditions, common anesthetic concerns in this patient group have been reported as difficult airway, altered respiratory mechanics, myopathy, gastric reflux, epileptic seizures, the interaction of polypharma therapy for anesthetic agents to be used, and related cardiovascular disorders.

Considering the present risks in the reported cases, a few suggestions by Hitosugi et al. [50] that may be useful in terms of anesthesia management are listed below:

- Patients may have difficulty in venous cannulation due to syndromic conditions that may accompany it. In such cases, preoperative central venous catheterization may be considered.
- The possibility of difficult intubation due to micrognathia, spasticity, and short neck should always be considered. Equipment that can be used in difficult intubation should be readily available and the option of awake intubation should be considered.
- The use of succinylcholine as a muscle relaxant should be avoided due to the possibility of myopathy.
- Intraoperative cervical positions must be handled carefully due to cervical instability.
- In order not to trigger seizure activity in patients with epileptic seizures, normocarbia should be provided and intracranial pressure balance should be maintained.
- Nasogastric tube placement may be considered for gastric reflux and to prevent aspiration.

- There is also data that can be considered regarding agent selection:
- The use of ketamine should not be preferred. Due to being a dissociative agent, the use of ketamine is limited in cognitive disorders. Also, a case has been reported that developed rigidity after intramuscular ketamine administration.
- Another case reported the safe use of dexmedetomidine in terms of cardiovascular stability.
- Another case reported no problems as a result of sevoflurane use in induction and maintenance anesthesia [51].

36.4 Conclusions

The CC has an important role in receiving and integrating information from the cortex for neurocognitive functions. Considering both the changes in neurocognitive diseases and the data in cases of impaired structural integrity (AgCC), its role and significance in neurocognition are more clearly understood. Assessing the postanesthetic neurocognitive functions in light of the current studies, it can be argued that there is cortical influence, especially in advanced age groups; the duration of anesthesia and repetitive exposures have an effect on cortical thinning. There are different data on volume changes in the CC after anesthesia exposure; however, there is strong evidence for microstructural changes. Advanced age is regarded as a risk factor for cortical involvement. Considering the effects in the developmental period, it can be argued that CC effects in childhood exposures have more clinically noticeable results. It is thought that the data of future studies at the microstructural level will bring clarity to the subject. As another aspect of the relationship between CC and anesthesia exposure, AgCC cases were examined to evaluate anesthesia management in CC pathologies. It was concluded that the most important point in terms of anesthesia management in AgCC cases is the identification of accompanying anomalies. After the identification of accompanying anomalies, anesthetic concerns were presented, and the necessity of precautions was emphasized.

References

1. Paul LK, Brown WS, Adolphs R, Tyszka JM, Richards LJ, Mukherjee P, et al. Agenesis of the corpus callosum: genetic, developmental and functional aspects of connectivity. Nat Rev Neurosci. 2007;8:287–99.
2. Tomasch J. Size, distribution, and number of fibres in the human corpus callosum. Anat Rec. 1954;119:119–35.
3. Frazier JA, Breeze JL, Papadimitriou G, Kennedy DN, Hodge SM, Moore CM, et al. White matter abnormalities in children with and at risk for bipolar disorder. Bipolar Disord. 2007;9:799–809.
4. Wolman D. A tale of two halves. Nature. 2012;483:260–3.
5. Gazzaniga MS, Freedman H. Observations on visual processes after posterior callosal section. Neurology. 1973;23:1126–30.
6. Gordon H, Bogen JE, Sperry RW. Absence of deconnexion syndrome in two patients with partial section of the neocommissures. Brain. 1971;94:327–36.
7. Fabri M, Pierpaoli C, Barbaresi P, Polonara G. Functional topography of the corpus callosum investigated by DTI and fMRI. World J Radiol. 2014;6:895.
8. Voineskos AN, Lobaugh NJ, Bouix S, Rajji TK, Miranda D, Kennedy JL, et al. Diffusion tensor tractography findings in schizophrenia across the adult lifespan. Brain. 2010;133:1494–504.
9. Hinkley LB, Marco EJ, Findlay AM, Honma S, Jeremy RJ, Strominger Z, et al. The role of corpus callosum development in functional connectivity and cognitive processing. PLOS One. 2012;7:e39804.
10. Hutchinson AD, Mathias J, Jacobson B, Ruzic L, Bond A, Banich MT. Relationship between intelligence and the size and composition of the corpus callosum. Exp Brain Res. 2009;192:455–64.
11. Luders E, Thompson PM, Narr KL, Zamanyan A, Chou Y-Y, Gutman B, et al. The link between callosal thickness and intelligence in healthy children and adolescents. NeuroImage. 2011;54:1823–30.
12. Penke L, Maniega SM, Murray C, Gow AJ, Hernández MCV, Clayden JD, et al. A general factor of brain white matter integrity predicts information processing speed in healthy older people. J Neurosci. 2010;30:7569–74.
13. van Eimeren L, Niogi SN, McCandliss BD, Holloway ID, Ansari D. White matter microstructures underlying mathematical abilities in children. Neuroreport. 2008;19:1117–21.
14. Paul LK, Corsello C, Kennedy DP, Adolphs R. Agenesis of the corpus callosum and autism: a comprehensive comparison. Brain. 2014;137:1813–29.

15. Dramsdahl M, Westerhausen R, Haavik J, Hugdahl K, Plessen KJ. Adults with attention-deficit/hyperactivity disorder—a diffusion-tensor imaging study of the corpus callosum. Psychiatry Res Neuroimaging. 2012;201:168–73.
16. Westerhausen R, Fjell AM, Krogsrud SK, Rohani DA, Skranes JS, Håberg AK, et al. Selective increase in posterior corpus callosum thickness between the age of 4 and 11 years. NeuroImage. 2016;139:17–25.
17. Hofman J, Hutny M, Sztuba K, Paprocka J. Corpus callosum agenesis: an insight into the etiology and spectrum of symptoms. Brain Sci. 2020;10:625.
18. Valenti M, Pino MC, Mazza M, Panzarino G, Di Paolantonio C, Verrotti A. Abnormal structural and functional connectivity of the corpus callosum in autism spectrum disorders: a review. Rev J Autism Dev Disord. 2020;7:46–62.
19. Ocklenburg S, Ball A, Wolf CC, Genç E, Güntürkün O. Functional cerebral lateralization and interhemispheric interaction in patients with callosal agenesis. Neuropsychology. 2015;29:806.
20. Marco EJ, Harrell KM, Brown WS, Hill SS, Jeremy RJ, Kramer JH, et al. Processing speed delays contribute to executive function deficits in individuals with agenesis of the corpus callosum. J Int Neuropsychol Soc. 2012;18:521–9.
21. Imamura T, Yamadori A, Shiga Y, Sahara M, Abiko H. Is disturbed transfer of learning in callosal agenesis due to a disconnection syndrome? Behav Neurol. 1994;7:43–8.
22. Harrell KMME, Porter A, Sherr E, Paul LK, et al. Color-word inhibition in agenesis of the corpus callosum. Philadelphia, PA: International Meeting for Autism Research; 2010.
23. Brown WS, Paul LK. Cognitive and psychosocial deficits in agenesis of the corpus callosum with normal intelligence. Cogn Neuropsychiatry. 2000;5:135–57.
24. Paul LK, Van Lancker-Sidtis D, Schieffer B, Dietrich R, Brown WS. Communicative deficits in agenesis of the corpus callosum: nonliteral language and affective prosody. Brain Lang. 2003;85:313–24.
25. Symington SH, Paul LK, Symington MF, Ono M, Brown WS. Social cognition in individuals with agenesis of the corpus callosum. Soc Neurosci. 2010;5:296–308.
26. Turk AA, Brown WS, Symington M, Paul LK. Social narratives in agenesis of the corpus callosum: linguistic analysis of the thematic apperception test. Neuropsychologia. 2010;48:43–50.
27. Evered LA, Silbert BS. Postoperative cognitive dysfunction and noncardiac surgery. Anesth Analg. 2018;127:496–505.
28. Savage GH. Insanity following the use of anaesthetics in operations. Br Med J. 1887;2:1199.
29. Bedford P. Adverse cerebral effects of anaesthesia on old people. Lancet. 1955;266:259–64.
30. Radtke F, Franck M, Schneider M, Luetz A, Seeling M, Heinz A, et al. Comparison of three scores to screen for delirium in the recovery room. Br J Anaesth. 2008;101:338–43.
31. Benjamin J, Sadock VAS, Ruiz P. Kaplan and Sadock's comprehensive textbook of psychiatry. 10th ed. Philadelphia, PA: Lippincott Williams & Wilkins; 2017. p. 4997.
32. Berger M, Browndyke JN, Cooter Wright M, Nobuhara C, Reese M, Acker L, et al. Postoperative changes in cognition and cerebrospinal fluid neurodegenerative disease biomarkers. Ann Clin Trans Neurol. 2022;9:155–70.
33. Newman MF, Kirchner JL, Phillips-Bute B, Gaver V, Grocott H, Jones RH, et al. Longitudinal assessment of neurocognitive function after coronary-artery bypass surgery. N Engl J Med. 2001;344:395–402.
34. Selnes OA, Grega MA, Bailey MM, Pham LD, Zeger SL, Baumgartner WA, et al. Cognition 6 years after surgical or medical therapy for coronary artery disease. Ann Neurol. 2008;63:581–90.
35. Selnes OA, Grega MA, Bailey MM, Pham LD, Zeger SL, Baumgartner WA, et al. Do management strategies for coronary artery disease influence 6-year cognitive outcomes? Ann Thorac Surg. 2009;88(445–54):e3.
36. Subramaniyan S, Terrando N. Narrative review article: neuroinflammation and perioperative neurocognitive disorders. Anesth Analg. 2019;128:781.
37. Kant IM, de Bresser J, van Montfort SJ, Slooter AJ, Hendrikse J. MRI markers of neurodegenerative and neurovascular changes in relation to postoperative delirium and postoperative cognitive decline. Am J Geriatr Psychiatry. 2017;25:1048–61.
38. Alam M, Lee J, Lee S-Y. Recent progress in the development of TSPO PET ligands for neuroinflammation imaging in neurological diseases. Nucl Med Mol Imaging. 2017;51:283–96.
39. Forsberg A, Cervenka S, Jonsson Fagerlund M, Rasmussen LS, Zetterberg H, Erlandsson Harris H, et al. The immune response of the human brain to abdominal surgery. Ann Neurol. 2017;81:572–82.
40. Kline RP, Pirraglia E, Cheng H, De Santi S, Li Y, Haile M, et al. Surgery and brain atrophy in cognitively normal elderly subjects and subjects diagnosed with mild cognitive impairment. J Am Soc Anesthesiol. 2012;116:603–12.
41. Sprung J, Kruthiventi SC, Warner DO, Knopman DS, Petersen RC, Mielke MM, et al. Exposure to surgery under general anaesthesia and brain magnetic resonance imaging changes in older adults. Br J Anaesth. 2019;123:808–17.
42. Sprung J, Knopman DS, Petersen RC, Vemuri P, Mielke MM, Weingarten TN, et al. Exposure to surgery with regional anesthesia and cortical thickness in older adults. Alzheimer's Dement. 2020;6:6.
43. Tang CY, Wang VX, Lun MY, Mincer JS, Ng JC, Brallier JW, et al. Transient changes in white matter microstructure during general anesthesia. PLoS One. 2021;16:e0247678.
44. Bennett IJ, Madden DJ, Vaidya CJ, Howard DV, Howard JH Jr. Age-related differences in multiple measures of white matter integrity: a diffusion tensor imaging study of healthy aging. Hum Brain Mapp. 2010;31:378–90.

45. Banerjee P, Rossi MG, Anghelescu DL, Liu W, Breazeale AM, Reddick WE, et al. Association between anesthesia exposure and neurocognitive and neuroimaging outcomes in long-term survivors of childhood acute lymphoblastic leukemia. JAMA Oncol. 2019;5:1456–63.
46. Bajwa NM, Lee JB, Halavi S, Hartman RE, Obenaus A. Repeated isoflurane in adult male mice leads to acute and persistent motor decrements with long-term modifications in corpus callosum microstructural integrity. J Neurosci Res. 2019;97:332–45.
47. Zhang X, Lin C, Guo P, Qin J, Qin X, Liang W. Effect of propofol on myelin basic protein expression and myelination of oligodendrocytes in neonatal SD rats. Nan Fang Yi Ke Da Xue Xue Bao. 2019;39:950–6.
48. Aminoff MJSF, Sarnath DF, Curatolo HB. Malformations of nervous system. Amsterdam: Elsevier; 2008. p. 83–5.
49. Kumar N. Sevoflurane in child with carpus callosum agenesis syndrome, a series of four cases. Pediatr Anesth Crit Care J. 2014;2:48–51.
50. Hitosugi T, Tsukamoto M, Ishii K, Kadowaki M, Fujiwara S, Yokoyama T. Repeated anesthesia management in a patient with Aicardi syndrome. Masui. 2016;65:78–81.
51. Kılıç ET, Akçay ME, Akdemir MS, Atağ E, Gürsu O. Anesthetic management in corpus callosum agenesis. Anesth Essays Res. 2018;12:758.

Corpus Callosum as Anatomical Marker of Intelligence

37

Kyle Biggins and R. Shane Tubbs

37.1 Introduction

The notion that brain size or total intracranial volume (TIV) is predictive of intelligence was first proposed nearly 200 years ago by Fredrik Tiderman in 1836. Tiderman described "an indisputable connection between the size of the brain and the mental energy displayed by the individual" [1]. Despite the drastic changes and advancements in medical knowledge since the time of these initial claims, that can be attributed to modern research and technology, contemporary scientific literature continues to support Tiderman's claims that TIV exists as a legitimate, yet nuanced, anatomical predictor of human intelligence [2–5].

Moreover, the literature supports a positive correlation between TIV and midsagittal corpus callosal thickness, which brings to question the relation between callosal thickness and human intelligence [6–8].

Adding additional complexity to the relation between TIV, callosal thickness and intelligence scores are sex-specific discrepancies between average measurements. In a 2018 publication, Ritchie et al. documented a significant discrepancy between measurements taken from male and female cohorts. Despite these discrepancies, Ritchie et al. reported no discrepancies between mean intelligence scores among male and female cohorts [9]. The group's claims regarding equal intelligence between cohorts are corroborated by several other twenty-first-century publications by numerous groups [10–12]. Thus, it is important to note that callosal thickness must be considered a sex-specific anatomical marker of intelligence.

K. Biggins (✉)
Tulane University School of Medicine, New Orleans, LA, USA
e-mail: kbiggins@tulane.edu

R. S. Tubbs
Department of Neurosurgery, Tulane School of Medicine, New Orleans, LA, USA
e-mail: rtubbs@tulane.edu

37.2 Adults vs. Children

There is a consensus within the literature that callosal thickness significantly correlates with intelligence scores; however, the nature of this correlation varies between publications [13, 14].

37.2.1 Adults

In a 2007 NIH-funded study, Lauders et al. proposed a positive correlation between callosal thickness and human intelligence. In this study, researchers analyzed the brains of 26 right-handed males (mean age 28.1 ± 7.3 years) and 33 right-handed females (mean age 28.8 ± 7.4 years). The group measured callosal thickness along with performance, verbal, and overall intelligence quotient (IQ) scores, finding positive rela-

M. Turgut et al. (eds.), *The Corpus Callosum*, https://doi.org/10.1007/978-3-031-38114-0_37

tionships between colossal thickness and all IQ scores in both cohorts with the strongest correlation found in male subjects [15].

In a similar 2014 study, researchers analyzed 63 participants with two cohorts consisting of 25 males (mean age 33.46 ± 11.52 years) and 38 females (mean age 30.33 ± 11.14 years). Results from this study showed a significant positive correlation between callosal thickness and IQ scores in the male cohort with equivocal results in the female cohort [16]. A larger study involving a group of 3722 males (mean age 25.0 ± 10.2 years) and 3207 females (mean age 25.9 ± 11.0 years) was conducted in 2010 by Tang et al. Data from this study suggested a positive correlation between callosal thickness and intelligence test scores only among female participants [17].

While all three of the aforementioned analyses support a positive relationship between callosal thickness and intelligence in adults, a similar study using a large pool of older individuals (mean age 40+ years) would be an interesting area for future research that may provide more insight into the nature of this correlation.

37.2.2 Children and Adolescents

In a follow-up study to their 2007 publication, Lauders et al. employed the same methods with 200 children and adolescents ages 6–17 years. In this iteration, researchers divided subjects into cohorts based on sex and age (see Table 37.1). Contrary to the findings in adult subjects, the data from Lauders et al. (2011) suggest a significant negative correlation between callosal thickness measurements and all observed intelligence test scores [18]. Westerhausen et al. found a similar

Table 37.1 This table outlines the various cohorts examined in Lauders et al. [18]

Ages (years)	Male	Female	Total
6–8	25	25	50
9–11	25	25	50
12–14	25	25	50
15–17	25	25	50
Total	100	100	200

negative relation between callosal thickness and intelligence scores in their analysis of datasets from 495 participants (245 male and 250 female) ages 6.4–21.9 years [19].

Upon further analysis of their data, Westerhausen et al. found that "in addition to the main effect, both analyses also revealed a significant interaction of test score and sex indicating differences in the slope of the association between the sexes" with the most pronounced correlations found in female subjects [19]. Moreover, while both Lauders et al. and Westerhausen et al. analyzed the brains of children as young as 6 years, a 2009 Hutchinson et al.'s project provides evidence that the same trends hold true in older adolescents (mean age 19.2 years) [20].

37.3 Case Study Analyzing Albert Einstein's Brain

Albert Einstein is widely regarded as one of human history's greatest physicists and one of mankind's most intelligent people. The following section will explore a case study analyzing Einstein's corpus callosum (CC).

To examine whether there are regional callosal differences between the brain of Einstein and those of ordinary people, and to minimize potential differences in CC morphology due to cause of death, brain atrophy, age, and sex, in vivo *magnetic resonance imaging data sets from two different age groups were used. The high-resolution photographs of Einstein's left and right hemispheres were supplied by Dean Falk with permission from the National Museum of Health and Medicine* (Fig. 37.1). *Because Einstein was right-handed and died at the age of 76, our first control group consisted of 15 elderly, healthy right-handed males, aged 70–80 years (mean: 74.20 ± 2.60 years). All participants were college graduates or beyond college, and non-demented (Clinical Dementia Rating = 0, Mini-Mental State Examination was from 28 to 30, mean ± SD: 29.53 ± 0.64)* [21–23].

In this case study, measurements revealed Einstein had a callosal thickness of 7.88 ± 0.05 cm

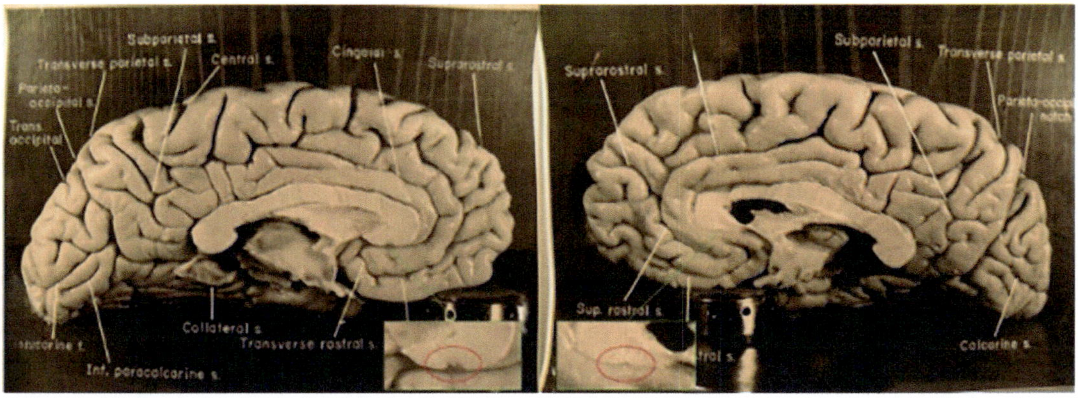

Fig. 37.1 Photographs of left and right midsagittal sections of Einstein's brain [21]

while the elderly control group exhibited a mean thickness of 6.11 cm SD 0.56 cm. These measurements would give Einstein a *Z*-score of 3.16 putting him in the 100th percentile for callosal thickness. Although this case does not provide any statistically significant data supporting a positive correlation between callosal thickness and intelligence, it provides interesting anecdotal support for the claims made in Lauders et al. [15] and Dunst et al. [16].

37.4 Conclusion

Based on the literature, callosal thickness is significantly correlated with human intelligence; however, the nature of this correlation is largely age-related. Studies involving adult participants generally support the notion that there is a positive correlation between callosal thickness and intelligence scores [15–17]. Conversely, research focusing on children and adolescents suggests a negative correlation [18–20]. A potential explanation for these age-related discrepancies is the fact that the CC is not fully matured until young adulthood [24–26]. A potential area for future research on the topic that may help explain the age discrepancies would be a longitudinal study in which participants were screened for callosal thickness and intelligence scores at three points during childhood, adolescence, and adulthood.

References

1. Hamilton JA. The association between brain size and maze ability in the white rat. Ph.D. Dissertation, University of California at Berkeley; 1935.
2. Gignac GE, Bates TC. Brain volume and intelligence: the moderating role of intelligence measurement quality. Intelligence. 2017;64:18–29. https://doi.org/10.1016/j.intell.2017.06.004.
3. Pietschnig J, Penke L, Wicherts JM, Zeiler M, Voracek M. Meta-analysis of associations between human brain volume and intelligence differences: how strong are they and what do they mean? Neurosci Biobehav Rev. 2015;57(2015):411–32. https://doi.org/10.1016/j.neubiorev.2015.09.017.
4. McDaniel MA. Big-brained people are smarter: a meta-analysis of the relationship between in vivo brain Vol. and intelligence. Intelligence. 2005;33(4):337–46. https://doi.org/10.1016/j.intell.2004.11.005.
5. Cox SR, Ritchie SJ, Fawns-Ritchie C, Tucker-Drob EM, Deary IJ. Structural brain imaging correlates of general intelligence in UK biobank. Intelligence, Article. 2019;76:101376. https://doi.org/10.1016/j.intell.2019.101376.
6. Jäncke L, Staiger JF, Schlaug G, Huang Y, Steinmetz H. The relationship between corpus callosum size and forebrain volume. Cereb Cortex. 1997;7:48–56. https://doi.org/10.1093/cercor/7.1.48. PMID: 9023431.
7. Rauch RA, Jinkins JR. Analysis of cross-sectional area measurements of the corpus callosum adjusted for brain size in male and female subjects from childhood to adulthood. Behav Brain Res. 1994;64:65–78. https://doi.org/10.1016/0166-4328(94)90119-8. PMID: 7840893.
8. Westerhausen R, Fjell AM, Krogsrud SK, Rohani DA, Skranes JS, Håberg AK, Walhovd KB. Selective increase in posterior corpus callosum thickness between the age of 4 and 11years. NeuroImage.

2016;139:17–25. https://doi.org/10.1016/j.neuroimage.2016.06.008. Epub 2016 Jun 7. PMID: 27288321.
9. Ritchie SJ, Cox SR, Shen X, Lombardo MV, Reus LM, Alloza C, Harris MA, Alderson HL, Hunter S, Neilson E, Liewald DCM, Auyeung B, Whalley HC, Lawrie SM, Gale CR, Bastin ME, McIntosh AM, Deary IJ. Sex differences in the adult human brain: evidence from 5216 UK biobank participants. Cereb Cortex. 2018;28:2959–75. https://doi.org/10.1093/cercor/bhy109. PMID: 29771288; PMCID: PMC6041980.
10. Deary IJ, Strand S, Smith P, Fernandes C. Intelligence and educational achievement. Intelligence. 2007;35(1):13–21. https://doi.org/10.1016/j.intell.2006.02.001.
11. Johnson W, Carothers A, Deary IJ. Sex differences in variability in general intelligence: a new look at the old question. Perspect Psychol Sci. 2008;3:518–31. https://doi.org/10.1111/j.1745-6924.2008.00096.x. PMID: 26158978.
12. Lakin J, Gambrell J. Sex differences in fluid reasoning: manifest and latent estimates from the cognitive abilities test. J Intelligence. 2014;2:36–55. https://doi.org/10.3390/jintelligence2020036.
13. Hulshoff Pol HE, Schnack HG, Posthuma D, Mandl RC, Baaré WF, van Oel C, van Haren NE, Collins DL, Evans AC, Amunts K, Bürgel U, Zilles K, de Geus E, Boomsma DI, Kahn RS. Genetic contributions to human brain morphology and intelligence. J Neurosci. 2006;26:10235–42. https://doi.org/10.1523/JNEUROSCI.1312-06.2006. PMID: 17021179; PMCID: PMC6674628.
14. Peterson BS, Feineigle PA, Staib LH, Gore JC. Automated measurement of latent morphological features in the human corpus callosum. Hum Brain Mapp. 2001;12:232–45. https://doi.org/10.1002/1097-0193(200104)12:4<232::AID-HBM1018>3.0.CO;2-J. PMID: 11241874; PMCID: PMC6871880.
15. Luders E, Narr KL, Bilder RM, Thompson PM, Szeszko PR, Hamilton L, Toga AW. Positive correlations between corpus callosum thickness and intelligence. Neuroimage. 2007;37:1457–64. https://doi.org/10.1016/j.neuroimage.2007.06.028. Epub 2007 Jul 12. PMID: 17689267; PMCID: PMC2754582.
16. Dunst B, Benedek M, Koschutnig K, Jauk E, Neubauer AC. Sex differences in the IQ-white matter microstructure relationship: a DTI study. Brain Cogn. 2014;91:71–8. https://doi.org/10.1016/j.bandc.2014.08.006. Epub 2014 Sep 29. PMID: 25238623; PMCID: PMC4245721.
17. Tang CY, Eaves EL, Ng JC, Carpenter DM, Mai X, Schroeder DH, Condon CA, Colom R, Haier RJ. Brain networks for working memory and factors of intelligence assessed in males and females with fMRI and DTI. Intelligence. 2010;38(3):293–303. https://doi.org/10.1016/j.intell.2010.03.003.
18. Luders E, Thompson PM, Narr KL, Zamanyan A, Chou YY, Gutman B, Dinov ID, Toga AW. The link between callosal thickness and intelligence in healthy children and adolescents. Neuroimage. 2011;54:1823–30. https://doi.org/10.1016/j.neuroimage.2010.09.083. Epub 2010 Oct 13. PMID: 20932920; PMCID: PMC3110814.
19. Westerhausen R, Friesen CM, Rohani DA, Krogsrud SK, Tamnes CK, Skranes JS, Håberg AK, Fjell AM, Walhovd KB. The corpus callosum as anatomical marker of intelligence? A critical examination in a large-scale developmental study. Brain Struct Funct. 2018;223:285–96. https://doi.org/10.1007/s00429-017-1493-0. Epub 2017 Aug 11. PMID: 28801753; PMCID: PMC5772147.
20. Hutchinson AD, Mathias JL, Jacobson BL, Ruzic L, Bond AN, Banich MT. Relationship between intelligence and the size and composition of the corpus callosum. Exp Brain Res. 2009;192:455–64. https://doi.org/10.1007/s00221-008-1604-5. Epub 2008 Oct 24. PMID: 18949469.
21. Men W, Falk D, Sun T, Chen W, Li J, Yin D, Zang L, Fan M. The corpus callosum of Albert Einstein's brain: another clue to his high intelligence? Brain. 2014;137:268. https://doi.org/10.1093/brain/awt252. Epub 2013 Sep 24. PMID: 24065724; PMCID: PMC3959548.
22. Marcus DS, Fotenos AF, Csernansky JG, Morris JC, Buckner RL. Open access series of imaging studies: longitudinal MRI data in nondemented and demented older adults. J Cogn Neurosci. 2010;22:2677–84. https://doi.org/10.1162/jocn.2009.21407. PMID: 19929323; PMCID: PMC2895005.
23. Marcus DS, Wang TH, Parker J, Csernansky JG, Morris JC, Buckner RL. Open access series of imaging studies (OASIS): cross-sectional MRI data in young, middle aged, nondemented, and demented older adults. J Cogn Neurosci. 2007;19:1498–507. https://doi.org/10.1162/jocn.2007.19.9.1498. PMID: 17714011.
24. Barnea-Goraly N, Menon V, Eckert M, Tamm L, Bammer R, Karchemskiy A, Dant CC, Reiss AL. White matter development during childhood and adolescence: a cross-sectional diffusion tensor imaging study. Cereb Cortex. 2005;15:1848–54.
25. Giedd JN, Blumenthal J, Jeffries NO, Rajapakse JC, Vaituzis AC, Liu H, Berry YC, Tobin M, Nelson J, Castellanos FX. Development of the human corpus callosum during childhood and adolescence: a longitudinal MRI study. Prog Neuro-Psychopharmacol Biol Psychiatry. 1999;23:571–88.
26. Thompson PM, Giedd JN, Woods RP, MacDonald D, Evans AC, Toga AW. Growth patterns in the developing brain detected by using continuum mechanical tensor maps. Nature. 2000;404:190–3.

Corpus Callosum in Dichotic Listening

38

Müge Ayanoğlu and Huriye Berra Ertuğrul

38.1 Introduction

The Corpus Callosum (CC) comprises most of the white matter fibers connecting the right and left hemispheres [1]. It has been proposed that despite the increase in brain size, the structure of the CC has been relatively preserved during the course of human evolution [2]. The CC consists of a set of axons and connects the left and right hemispheres. Its size is thought to be related to the increased content of myelin in the fibers. Different parts are responsible for transmitting different types of signals [3], the fibers in the posterior region of the CC being responsible for interhemispheric auditory transmission [4]. However, details of the auditory connection are not well known. There is a "slowdown" in transmission between the hemispheres [5], which could account for hemispheric specialization [6, 7]. Some authors have proposed that callosal fibers play an active role via inhibition or excitation of one hemisphere by the other [8]. The structure of the CC and callosal connections is established between the second trimester and the end of gestation. Irrelevant fibers are pruned during the first year of postnatal life [9], and myelination continues until the end of adolescence. Interhemispheric transportation accelerates during these processes [10]. Callosal fibers can be affected, and there can be partial or complete agenesis due to genetic disorders or environmental factors [11, 12]. Language lateralization is reduced in patients with CC agenesis [13, 14]. Auditory fibers mediate the interaction between right and left speech-related brain parts. Moreover, interhemispheric connections in the CC are essential for normal speech processing. According to the callosal relay model, information flows from the right to the left hemisphere.

38.2 Dichotic Listening

Verbal dichotic-listening paradigmas ensure association between hemispheric differences and language processing to be evaluated [15, 16]. Verbal dichotic-listening is also essential for diagnosing auditory processing disorders [17–19]. The main advantage of the test is that it can easily be performed by both younger and older children and also by patients with cognitive disorders [20–23]. The procedure consists of presenting verbal stimuli (words or syllables) via a headphone. One stimulus is presented to the left ear and simultaneously one to the right ear [24]. The participant is then asked to report the stimulus that was heard better. The right-ear stimulus is reported more frequently, more rapidly, and more accurately than

M. Ayanoğlu (✉)
Department of Pediatric Neurology, Aydın Adnan Menderes University Faculty of Medicine, Aydın, Turkey

H. B. Ertuğrul
Department of Pediatrics, Aydın Adnan Menderes University Faculty of Medicine, Aydın, Turkey

M. Turgut et al. (eds.), *The Corpus Callosum*, https://doi.org/10.1007/978-3-031-38114-0_38

the left. This right-ear dominance, also known as right-ear advantage (REA), is related to left hemisphere dominance in speech processing and also interhemispheric auditory integration [16, 25–27]. However, there is left-ear dominance in approximately 20% of right-handed individuals; this is considered an atypical right-hemispheric specialization for language processing [24]. Important factors that can affect dichotic-listening test results include stimulus choices, stimulus composition, stimulus order, sound intensity in the two auditory channels, and the instructions given to the participants. Disregarding these factors may result in a "reduced right-ear advantage" or "left-ear advantage" [28]. Numerous different verbal dichotic-listening paradigmas have been suggested. However, since the test-retest reliability of the paradigmas is not optimal, the measurements can differ, which can be considered a disadvantage [29–31].

38.3 Maturation of Corpus Callosum and Improvement of Dichotic-Listening

The CC continues to mature during the first two decades of life, with increased myelination and axonal growth and shrinkage [32]. Dichotic-listening processing ability improves with maturation of the callosal fibers [33, 34]. Children who have a delay in callosal fiber maturation show reduced dichotic processing scores, described as central auditory processing disorder. Children with this disorder frequently have problems with hearing in noisy environments and learning disabilities and usually show lower left-ear scores than their healthy peers [33, 35, 36].

Dichotic deficits can occur in healthy people as a result of aging owing to reduced CC integrity [37, 38]. Left-ear performance is significantly poorer in older people. This decrease in left-ear performance emerges between 55 and 60 years of age [39, 40]. Also, age-related declines in the CC begin around these ages, and men tend to develop this left-ear deficit earlier than women.

38.4 The Effects of Corpus Callosum Lesions on Dichotic-Listening Performance

Most studies regarding dichotic-listening have been performed on patients with injury/surgery or agenesis of the CC [41, 42]. Commissurotomy is a surgical procedure to control the spread of intractable seizures. Studies on patients who have undergone commissurotomy have contributed in this area since the 1940s. These studies have shown that after total section of the CC, also known as split-brain, the responses of the left ear are reduced while right-ear performance is normal or better [43, 44]. On the other hand, there is no or minimal suppression in the left ear in subjects with anterior callosotomy [45, 46]. Presurgical brain insults in early life, prior to callosotomy, are also important in dichotic-listening. If the left hemisphere is injured during early life, some functions of the left hemisphere are taken over by the right hemisphere owing to neuronal plasticity. Therefore, dichotic-listening results can differ after callosotomy in such patients. Dichotic-listening scores in patients with CC agenesis are generally within the normal range, or there can be milder asymmetry than it occurs to split-brain subjects [47]. Results from studies on neurological disorders involving the CC such as multiple sclerosis, vascular and mass lesions, and traumatic brain injury also yielded information about the association between the CC and dichotic-listening. The CC is a heavily myelinated part of the brain, and demyelinating lesions are common in patients with multiple sclerosis [42]. The left-ear dichotic test was reduced in cases with demyelinating lesions in the CC. Results from neuroimaging studies supported this observation [48–51]. Moreover, these deficits seemed to be associated with lesions in the auditory regions of the CC [52]. Dichotic-listening test results were also reduced in patients with vascular and mass lesions of the CC [45]; lesions located deep in the parietal cortex especially tend to involve the CC. When the mass lesion is located in the left hemisphere without involving the CC, a right-ear deficit is observed. However, if the CC is involved, a left-ear deficit is also observed [44]. Fibers in the CC can be injured during abrupt hemispheric movements. An abnormal left-ear deficit can be observed in patients with such head injuries [53–55].

A relationship between genetic factors and dichotic-listening has been identified. Morell et al. demonstrated that monozygotic twins have more strongly related dichotic-listening results than dizygotic twins [56]. Also, dichotic-listening (left-ear deficit) performance was reduced in patients with PAX6 gene mutations and X-linked adrenoleukodystrophy, probably because the CC structure was abnormal [41, 42, 56].

38.5 Conclusion

In summary, the CC is responsible for connecting the hemispheres. The auditory fibers are located in the posterior region of the CC and are essential for listening and speech processing. The dichotic-listening paradigmas, which enable hemispheric differences and language processing to be evaluated, are strongly associated with the structure and size of the CC. Lesions in the posterior region of the CC entail abnormal dichotic-listening test results.

References

1. Aboitiz F, Scheibel AB, Fisher RS, Zaidel E. Fiber composition of the human corpus callosum. Brain Res. 1992;598(1–2):143–53.
2. Caminiti R, Ghaziri H, Galuske R, Hof PR, Innocenti GM. Evolution amplified processing with temporally dispersed slow neuronal connectivity in primates. Proc Natl Acad Sci. 2009;106(46):19551–6.
3. Musiek FE, Baran JA. The auditory system: anatomy, physiology, and clinical correlates. San Diego: Plural Publishing; 2018.
4. Bloom JS, Hynd GW. The role of the corpus callosum in interhemispheric transfer of information: excitation or inhibition? Neuropsychol Rev. 2005;15(2):59–71.
5. Jäncke L, Steinmetz H. Interhemispheric transfer time and corpus callosum size. Neuroreport. 1994;5(17):2385–8. https://doi.org/10.1097/00001756-199411000-00043. PMID: 7881064.
6. Cook ND. Homotopic callosal inhibition. Brain Lang. 1984;23(1):116–25.
7. Dennis M. Language in a congenitally acallosal brain. Brain Lang. 1981;12(1):33–53.
8. Yazgan MY, Wexler BE, Kinsbourne M, Peterson B, Leckman JF. Functional significance of individual variations in callosal area. Neuropsychologia. 1995;33(6):769–79.
9. Kostović I, Jovanov-Milošević N, editors. The development of cerebral connections during the first 20–45 weeks' gestation. In: Seminars in fetal and neonatal medicine. Amsterdam: Elsevier; 2006.
10. Brody BA, Kinney HC, Kloman AS, Gilles FH. Sequence of central nervous system myelination in human infancy. I. An autopsy study of myelination. J Neuropathol Exp Neurol. 1987;46(3):283–301.
11. Bedeschi MF, Bonaglia MC, Grasso R, Pellegri A, Garghentino RR, Battaglia MA, et al. Agenesis of the corpus callosum: clinical and genetic study in 63 young patients. Pediatr Neurol. 2006;34(3):186–93.
12. Bonneau D, Toutain A, Laquerriere A, Marret S, Saugier-Veber P, Barthez MA, et al. X-linked lissencephaly with absent corpus callosum and ambiguous genitalia (XLAG): clinical, magnetic resonance imaging, and neuropathological findings. Ann Neurol. 2002;51(3):340–9.
13. Hinkley LB, Marco EJ, Brown EG, Bukshpun P, Gold J, Hill S, et al. The contribution of the corpus callosum to language lateralization. J Neurosci. 2016;36(16):4522–33.
14. Komaba Y, Senda M, Mori T, Ishii K, Mishina M, Kitamura S, et al. Bilateral representation of language function. J Neuroimaging. 1998;8(4):246–9.
15. Lezak MD, Howieson DB, Loring DW, Fischer JS. Neuropsychological assessment. Oxford: Oxford University Press; 2004.
16. Ocklenburg S, Gunturkun O. The lateralized brain: the neuroscience and evolution of hemispheric asymmetries. London: Academic Press; 2017.
17. Keith RW. Development and standardization of SCAN-A: test of auditory processing disorders in adolescents and adults. J Am Acad Audiol. 1995;6(4):286–92.
18. Moncrieff D, Miller E, Hill E. Screening tests reveal high risk among adjudicated adolescents of auditory processing and language disorders. J Speech Lang Hear Res. 2018;61(4):924–35.
19. Musiek FE. Assessment of central auditory dysfunction: the dichotic digit test revisited. Ear Hear. 1983;4(2):79–83.
20. Gootjes L, Van Strien JW, Bouma A. Age effects in identifying and localising dichotic stimuli: a corpus callosum deficit. J Clin Exp Neuropsychol. 2004;26(6):826–37.

21. Helland T, Morken F, Bless JJ, Valderhaug HV, Eiken M, Helland WA, et al. Auditive training effects from a dichotic listening app in children with dyslexia. Dyslexia. 2018;24(4):336–56.
22. Passow S, Westerhausen R, Wartenburger I, Hugdahl K, Heekeren HR, Lindenberger U, et al. Human aging compromises attentional control of auditory perception. Psychol Aging. 2012;27(1):99.
23. Westerhausen R, Helland T, Ofte S, Hugdahl K. A longitudinal study of the effect of voicing on the dichotic listening ear advantage in boys and girls at age 5 to 8. Dev Neuropsychol. 2010;35(6):752–61.
24. Bryden MP. An overview of the dichotic listening procedure and its relation to cerebral organization. 1988.
25. Carey DP, Johnstone LT. Quantifying cerebral asymmetries for language in dextrals and adextrals with random-effects meta analysis. Front Psychol. 2014;5:1128.
26. Hugdahl K, Westerhausen R. Speech processing asymmetry revealed by dichotic listening and functional brain imaging. Neuropsychologia. 2016;93:466–81.
27. Westerhausen R, Hugdahl K. The corpus callosum in dichotic listening studies of hemispheric asymmetry: a review of clinical and experimental evidence. Neurosci Biobehav Rev. 2008;32(5):1044–54.
28. Westerhausen R, Passow S, Kompus K. Reactive cognitive-control processes in free-report consonant–vowel dichotic listening. Brain Cogn. 2013;83(3):288–96.
29. Kelley KS, Littenberg B. Dichotic listening test–retest reliability in children. J Speech Lang Hear Res. 2019;62(1):169–76.
30. Pedhazur EJ, Schmelkin LP. Measurement, design, and analysis: an integrated approach: psychology press; 2013.
31. Voyer D. On the reliability and validity of noninvasive laterality measures. Brain Cogn. 1998;36(2):209–36.
32. Luders E, Thompson PM, Toga AW. The development of the corpus callosum in the healthy human brain. J Neurosci. 2010;30(33):10985–90.
33. Berrick JM, Shubow GF, Schultz MC, Freed H, Fournier SR, Hughes JP. Auditory processing tests for children: normative and clinical results on the SSW test. J Speech Hear Disord. 1984;49(3):318–25.
34. Bissell JC, Clark F. Dichotic listening performance in normal children and adults. Am J Occup Ther. 1984;38(3):176–83.
35. Musiek FE, Geurkink NA. Auditory perceptual problems in children: considerations for the otolaryngologist and audiologist. Laryngoscope. 1980;90(6):962–71.
36. Sharma M, Purdy SC, Kelly AS. Comorbidity of auditory processing, language, and reading disorders. J Speech Lang Hear Res. 2009;52:706.
37. Bastin ME, Maniega SM, Ferguson KJ, Brown LJ, Wardlaw JM, MacLullich AM, et al. Quantifying the effects of normal ageing on white matter structure using unsupervised tract shape modelling. NeuroImage. 2010;51(1):1–10.
38. Sullivan EV, Rohlfing T, Pfefferbaum A. Longitudinal study of callosal microstructure in the normal adult aging brain using quantitative DTI fiber tracking. Dev Neuropsychol. 2010;35(3):233–56.
39. Bellis TJ, Wilber LA. Effects of aging and gender on interhemispheric function. J Speech Lang Hear Res. 2001;44:246.
40. Gootjes L, Bouma A, Van Strien J, Van Schijndel R, Barkhof F, Scheltens P. Corpus callosum size correlates with asymmetric performance on a dichotic listening task in healthy aging but not in Alzheimer's disease. Neuropsychologia. 2006;44(2):208–17.
41. Bamiou D-E, Davies R, Jones S, Musiek FE, Rudge P, Stevens J, et al. An unusual case of X-linked adrenoleukodystrophy with auditory processing difficulties as the first and sole clinical manifestation. J Am Acad Audiol. 2004;15(02):152–60.
42. Bamiou D-E, Free SL, Sisodiya SM, Chong WK, Musiek F, Williamson KA, et al. Auditory interhemispheric transfer deficits, hearing difficulties, and brain magnetic resonance imaging abnormalities in children with congenital aniridia due to PAX6 mutations. Arch Pediatr Adolesc Med. 2007;161(5):463–9.
43. Kimura D. Some effects of temporal-lobe damage on auditory perception. Can J Psychol. 1961;15(3):156.
44. Musiek FE, Reeves AG, Baran JA. Release from central auditory competition in the split-brain patient. Neurology. 1985;35(7):983.
45. Musiek FE, Baran JA, Pinheiro ML. Neuroaudiology: case studies. San Diego: Singular Publishing Group; 1994.
46. Pandya D, Karol E, Heilbronn D. The topographical distribution of interhemispheric projections in the corpus callosum of the rhesus monkey. Brain Res. 1971;32(1):31–43.
47. Lassonde MC, Sauerwein HC, Lepore F. Agenesis of the corpus callosum. 2003.
48. Barkhof F, Tas M, Valk J, Elton M, Lindeboom J, Schmidt W, et al. Functional correlates of callosal atrophy in relapsing-remitting multiple sclerosis patients. A preliminary MRI study. J Neurol. 1998;245(3):153–8.
49. Gadea M, Marti-Bonmatí L, Arana E, Espert R, Casanova V, Pascual A. Dichotic listening and corpus callosum magnetic resonance imaging in relapsing-remitting multiple sclerosis with emphasis on sex differences. Neuropsychology. 2002;16(2):275.
50. Pelletier J, Suchet L, Witjas T, Habib M, Guttmann C, Salamon G, et al. A longitudinal study of callosal atrophy and interhemispheric dysfunction in relapsing-remitting multiple sclerosis. Arch Neurol. 2001;58(1):105–11.
51. Rao SM, Bernardin L, Leo GJ, Ellington L, Ryan SB, Burg LS. Cerebral disconnection in multiple sclerosis: relationship to atrophy of the corpus callosum. Arch Neurol. 1989;46(8):918–20.
52. Pelletier J, Habib M, Lyon-Caen O, Salamon G, Poncet M, Khalil R. Functional and magnetic resonance

imaging correlates of callosal involvement in multiple sclerosis. Arch Neurol. 1993;50(10):1077–82.
53. Benavidez DA, Fletcher JM, Hannay HJ, Bland ST, Caudle SE, Mendelsohn DB, et al. Corpus callosum damage and interhemispheric transfer of information following closed head injury in children. Cortex. 1999;35(3):315–36.
54. Meyers JE, Roberts RJ, Bayless JD, Volkert K, Evitts PE. Dichotic listening: expanded norms and clinical application. Arch Clin Neuropsychol. 2002;17(1):79–90.
55. Wilde EA, Chu Z, Bigler ED, Hunter JV, Fearing MA, Hanten G, et al. Diffusion tensor imaging in the corpus callosum in children after moderate to severe traumatic brain injury. J Neurotrauma. 2006;23(10):1412–26.
56. Morell RJ, Brewer CC, Ge D, Snieder H, Zalewski CK, King KA, et al. A twin study of auditory processing indicates that dichotic listening ability is a strongly heritable trait. Hum Genet. 2007;122(1):103–11.

Corpus Callosum in Attention Deficit Hyperactivity Disorder

39

Graham Dupont and R. Shane Tubbs

39.1 Introduction

Attention deficit hyperactivity disorder (ADHD) is a complex neurobehavioral disorder featuring impulsivity, emotionality, varying degrees of inattention, issues with psychosocial adjustment and social cohesion, and hyperactivity. Many children with ADHD have issues with focusing on and completing tasks, causing learning issues in early education [1]. The focus of this chapter is the abnormal anatomy and physiology of the corpus callosum (CC) in ADHD. Further, this chapter aims to discuss the neurobiological causes of behavioral changes in children and adults diagnosed with ADHD; the alterations in brain development, structure, and executive function of the brain; and the neurophysiological differences in children diagnosed with ADHD versus a control group, as these topics pertain to the CC; lastly, the chapter aims to discuss changes in the CC following current treatments with dopamine transporter blockers and amphetamines.

G. Dupont (✉)
Department of Cellular and Structural Biology, Tulane School of Medicine, New Orleans, LA, USA
e-mail: gdupont4@tulane.edu

R. S. Tubbs
Department of Neurosurgery, Tulane School of Medicine, New Orleans, LA, USA
e-mail: rtubbs@tulane.edu

39.2 ADHD-Related Changes in Corpus Callosum Neurophysiology

Historically ADHD has been a controversial diagnosis due to its epidemic rise in rates of diagnosis in the last two decades in the United States, a hot topic in the media, to the point where some may argue it is an American disease that has led to the "medicalization of American childhood"—the cost of ADHD in the United States per year has been estimated to be $42.5 billion per year [2–4]. Certainly, there are genetic and sociocultural elements of ADHD, notably being that boys are diagnosed and medicated for ADHD in higher proportion and display more of the inattentive, peer-directed aggressive-type features of the disorder, compared to girls [5]. Neuroimaging studies have shown consistently that there is abnormal brain structure, function, functional co-processing and connectivity, and white matter growth [6]. Advanced MRI techniques have pointed to the dysfunction of prefrontal and striatal regions of the brain [7]. Of these brain regions, abnormal formation of the CC seems to have a strong effect on social behavior and cohesion amongst peers, as CC malformation can result in a wide array of neurodevelopmental issues such as bipolar disorder, dyslexia, Tourette syndrome, and autism spectrum disorders [8–11]. ADHD is highly heritable (70%), and there is current research that aims to understand the genetic factors of ADHD, via susceptibility genes [2].

M. Turgut et al. (eds.), *The Corpus Callosum*, https://doi.org/10.1007/978-3-031-38114-0_39

The textbook neuroanatomy of the CC has it split into wedges based on the parent brain region it serves. From rostral to caudal, there are the prefrontal, premotor and supplementary motor, motor, sensory, parietal, temporal, and occipital regions. The rostral-most region of the CC contains the highest population of unmyelinated axons and smaller-diameter fibers, which will increase in diameter as the CC continues caudally. Functionally these diameter changes allow integration of higher-order prefrontal cortical functions in the rostral CC and higher conduction velocities in the mid-CC [12]. There is a varied consensus in ADHD-related CC abnormalities, as several studies report reduction in CC mass rostrally, with others reporting caudal reduction [8, 10, 12, 13]. The degree of brain lateralization, brain asymmetry, and CC size has been implicated in the development of ADHD.

Neurophysiological changes have been demonstrated by various authors; in particular a study on developmental trajectories of the subregions of the CC (I, prefrontal; II, premotor/supplementary motor; III, motor; IV, sensory; V, parietal, temporal, and occipital) by Gilliam et al. focused on the growth rates of these regions in right-handed children diagnosed with ADHD versus typically developing participants [12]. In right-handed children diagnosed with ADHD, region I (prefrontal) demonstrated the highest percent change per year (0.92%) compared to typical development (0.32%) [12]. There were no reported differences in CC surface area. Interestingly, there were no growth velocity differences found for growth rates in left-handed nor ambidextrous participants. These atypical CC growth patterns in ADHD can result in atypical cerebral lateralization in the prefrontal region of the brain, as evidenced by changes in EEG coherence, cerebral blood flow, and activation during cognitive tasks [14, 15]. Growth rates in regions II–V, however, were lower in the ADHD group, with a large difference in region IV compared with typical CC development [12]. In children, increases in the degree of impulsivity and hyperactivity are correlated with loss of rostral size in the CC [13].

39.3 ADHD, Lateralization, and Executive Function

Cerebral asymmetry and abnormal lateralization play an important role in the severity of ADHD, as mixed-handed individuals have a higher risk for developing ADHD and dyslexia. Further, non-right-handed individuals are far more likely to develop ADHD and to be considered "weaker students" [16]. Dysfunction in the right hemisphere and right striatal-parietal regions have been suggested to affect left motor function and spatial attention to the left visual field. In addition, children diagnosed with ADHD do poorly on tests involving maintaining attention and executive function, as well as tests designed to target interhemispheric processing [16]. Neuroimaging studies have suggested dysfunction in the anterior cingulate, dorsolateral prefrontal, and ventrolateral prefrontal cortices, in addition to the striatal-parietal and cerebellar regions of the brain [17].

39.4 Brain Mapping Patients with ADHD Versus Control

Dickstein et al. summarized activation patterns obtained from various neuroimaging techniques (PET, fMRI, EEG) in the brains of children, adolescents, and adults diagnosed with ADHD and control groups using an analytical technique called activation likelihood estimation (ALE), which detects the above-chance convergence of activation probabilities across multiple neuroimaging studies, of which a total of 134 foci of activation were studied in the ADHD group and 180 for the control group [17]. As a baseline, the control group demonstrated elevated probabilities of activation for frontal and posterior (anterior cingulate cortex, left dorsolateral prefrontal cortices, bilateral inferior prefrontal cortices) regions of the brain, as well as the thalamus, claustrum, insula, striatum, and parietal lobe. Patients with ADHD demonstrated increased probability of activation patterns in the middle frontal gyrus, frontal lobe, and left ventral prefrontal cortex (bilateral in controls), bilateral thalamus, and right putamen and globus pallidus [17]. When comparing the ADHD and control groups, the results were consistent with the consensus on ADHD, that being hypofrontality; control groups had higher probabilities of activation in the frontal and parietal lobes compared to those diagnosed with ADHD. Patients diagnosed with ADHD demonstrated higher probability of activation in the insular cortex, middle frontal gyrus, and left thalamus [17].

Current research is consistent with functional hypoactivation of the frontal and cingulate areas, as well as morphological and structural differences in the CC. In addition to hypoactivation, patients with ADHD display reduction in total cerebral volume, as well as the parietal lobe, basal nuclei in right caudate, putamen, and CC, respectively [18]. Fractional anisotropy (loss of white matter density, diameter, myelination) reduction in the CC is also seen in patients diagnosed with ADHD [19]. The CC is certainly implicated in ADHD; yet, it is one small piece of the puzzle, and currently researchers are adapting a multi-neurophysiological network model of ADHD that encompasses the dopaminergic and fronto-striato-cerebellar system which corresponds to cognitive tasks and executive function, the limbic system which corresponds to emotionality associated with ADHD, and the cortico-striato-thalamic-cortical circuit that regulates sensorimotor, cognitive, and limbic processes [18]. Reduction in inhibitory control is a hallmark of ADHD, and due to abnormal growth in striato-frontal-cingulate areas [20, 21].

39.5 Default Mode Network and the Corpus Callosum

The default mode network (DMN) in the brain is a resting process that governs basic consciousness, cognition, and behavior and is contained within the medial prefrontal cortex, posterior cingulate cortex/precuneus, and the medial, lateral, and inferior parietal regions [22]. The DMN is more active during rest, and activity decreases during focused cognition [23]. There is not yet a direct connection between the DMN, the CC, and the ADHD; however dysfunction of the DMN and diagnosis of ADHD are highly intertangled and have had a wide body of support grow in the last two decades [24–26]. DMN dysfunction in children with ADHD can result in brain maturational delay and result in difficulties in learning, completing tasks, and social cohesion. It has been suggested that children with ADHD may in fact not struggle with tasks due to DMN dysfunction during rest but that the DMN fails to attenuate when changing from rest- to task-oriented brain programs [27]. Interestingly, Mitaki et al. reported a case of a 29-year-old man with a reversible lesion spanning the entire CC that demonstrated disturbances in consciousness during the resting state and reduced functional connectivity between the posterior cingulate cortex/precuneus within the DMN, without resolution after disappearance of the lesion [28]. The CC appears to be a principal connecting point for the DMN, as demonstrated by Van Den Heuvel et al. and Greicius et al. [29, 30].

39.6 Methylphenidate Treatment and CC Reorganization

Methylphenidate (brand name: Ritalin) is a brain stimulant and a dopamine reuptake inhibitor, which increases the value of reward in children and adults diagnosed with ADHD and was found to enhance the neurophysiological value of reward, increase the levels of extra-cellular dopamine, and cause attenuation in the DMN [24]. Methylphenidate has an extreme therapeutic effect, as typically developing children are indistinguishable from those diagnosed with ADHD and being treated with methylphenidate [24]. Further, methylphenidate has been shown to increase fractional anisotropy in the CC, restoring and restructuring white matter in the brain [31]. A clinical study by Bouziane et al. demonstrated an age-based modulation of white matter in the CC, showing that children are more neuroplastic than adults receiving the same treatment with methylphenidate [31]. One of the hallmark studies by Castellanos et al. demonstrated an 8.9% increase in white matter volume in children on a medication regiment for ADHD [32]. Across the board, studies have demonstrated that children without treatment have smaller cerebral and cerebellar volumes; therefore treatment with methylphenidate or other first-line brain stimulants is indicated in order to reclaim the white matter loss and disorganization [12–14, 24, 32]. Methylphenidate works by initiating a dopamine transporter blockade, which in turn increases synaptic and extracellular dopamine, and allows for higher stored dopamine release and dopamine signal amplification in a dose-dependent fashion [33]. This effect differs slightly from current lines of treatment, amphetamines, which block dopamine reuptake but promote synthesis of dopamine [34]. Both methylphenidate and amphetamines are highly therapeutic for patients diagnosed with ADHD, as the pathways that have been demonstrated to have reduced activity (cortico-subcortical) are mediated by dopamine [34].

39.7 Conclusion

Connections between ADHD and subregions of the brain are still being researched, and while there is an undeniable therapeutic effect from first-line treatment, further therapies and research may be called for in order to develop more comprehensive models of treatment for medication-resistant types of ADHD. There are sociocultural predictors for ADHD diagnosis that can be alleviated with perspective and educational shifts on how children and young adults ought to conduct themselves in Western society. There are certainly more ways to learn than decades ago, when many of the epidemic-like ADHD diagnoses took place, and coincidentally, education was not as capable of adapting to children with different styles of learning or with learning deficits. The interactions between ADHD/neurobehavioral disorders and Western society are a convergence of psychiatry and sociocultural, and it is helpful for the clinician to be aware of these topics when developing treatment goals for the patient.

References

1. Taylor E, Chadwick O, Heptinstall E, Danckaerts M. Hyperactivity and conduct problems as risk factors for adolescent development. J Am Acad Child Adolesc Psychiatry. 1996;35:1213–26.
2. Arcos-Burgos M, Muenke M. Toward a better understanding of ADHD: LPHN3 gene variants and the susceptibility to develop ADHD. Atten Defic Hyperact Disord. 2010;2:139–47.
3. Kazda L, Bell K, Thomas R, McGeechan K, Sims R, Barratt A. Overdiagnosis of attention-deficit/hyperactivity disorder in children and adolescents: a systematic scoping review. JAMA Netw Open. 2021;4:e215335.
4. Wedge M. A disease called childhood: why ADHD became an American epidemic. New York: Avery; 2016.
5. Gaub M, Carlson CL. Gender differences in ADHD: a meta-analysis and critical review. J Am Acad Child Adolesc Psychiatry. 1997;36:1036–45.
6. Connaughton M, Whelan R, O'Hanlon E, McGrath J. White matter microstructure in children and adolescents with ADHD. Neuroimage Clin. 2022;33:102957.
7. Wu W, McAnulty G, Hamoda HM, Sarill K, Karmacharya S, Gagoski B, Ning L, Grant PE, Shenton ME, Waber DP, Makris N. Detecting microstructural white matter abnormalities of frontal pathways in children with ADHD using advanced diffusion models. Brain Imaging Behav. 2020;14:981–97.
8. Baumgardner TL, Singer HS, Denckla MB, Rubin MA, Abrams MT, Colli MJ, Reiss AL. Corpus callosum morphology in children with Tourette syndrome and attention deficit hyperactivity disorder. Neurology. 1996;47:477–82.
9. Caetano SC, Silveira CM, Kaur S, Nicoletti M, Hatch JP, Brambilla P, Sassi R, Axelson D, Keshavan MS, Ryan ND, Birmaher B, Soares JC. Abnormal corpus callosum myelination in pediatric bipolar patients. J Affect Disord. 2008;108:297–301.
10. Hynd GW, Hall J, Novey ES, Eliopulos D, Black K, Gonzalez JJ, Edmonds JE, Riccio C, Cohen M. Dyslexia and corpus callosum morphology. Arch Neurol. 1995;52:32–8.
11. Mulas F, Roca P. Concordancias entre lostrastornos del espectro del autismo y eltrastornopor deficit de atencion/hiperactividad [Concordances between autism spectrum disorders and attention deficit hyperactivity disorder]. Rev Neurol. 2018;66:S91–6.
12. Gilliam M, Stockman M, Malek M, Sharp W, Greenstein D, Lalonde F, Clasen L, Giedd J, Rapoport J, Shaw P. Developmental trajectories of the corpus callosum in attention-deficit/hyperactivity disorder. Biol Psychiatry. 2011;69:839–46.
13. Giedd JN, Castellanos FX, Casey BJ, Kozuch P, King AC, Hamburger SD, et al. Quantitative morphology of the corpus callosum in attention deficit hyperactivity disorder. Am J Psychiatr. 1994;151:665–9.

14. Chabot RJ, Serfontein G. Quantitative electroencephalographic profiles of children with attention deficit disorder. Biol Psychiatry. 1996;40:951–63.
15. Langleben DD, Austin G, Krikorian G, Ridlehuber HW, Goris ML, Strauss HW. Interhemispheric asymmetry of regional cerebral blood flow in prepubescent boys with attention deficit hyperactivity disorder. Nucl Med Commun. 2001;22:1333–40.
16. Rodriguez A, Kaakinen M, Moilanen I, Taanila A, McGough JJ, Loo S, Järvelin MR. Mixed-handedness is linked to mental health problems in children and adolescents. Pediatrics. 2010;125:e340–8.
17. Dickstein SG, Bannon K, Xavier Castellanos F, Milham MP. The neural correlates of attention deficit hyperactivity disorder: an ALE meta-analysis. J Child Psychol Psychiatry. 2006;47:1051–62.
18. Vieira de Melo B, Trigueiro MJ, Rodrigues P. Systematic overview of neuroanatomical differences in ADHD: definitive evidence. Am J Occup Ther. 2019;73:7311505175p1.
19. Chen L, Hu X, Ouyang L, He N, Liao Y, Liu Q, Zhou M, Wu M, Huang X, Gong Q. A systematic review and meta-analysis of tract-based spatial statistics studies regarding attention-deficit/hyperactivity disorder. Neurosci Biobehav Rev. 2016;68:838–47.
20. Schachar R, Mota VL, Logan GD, Tannock R, Klim P. Confirmation of an inhibitory control deficit in attention-deficit/hyperactivity disorder. J Abnorm Child Psychol. 2000;28:227–35.
21. Schachar RJ, Tannock R, Logan G. Inhibitory control, impulsiveness, and attention deficit hyperactivity disorder. Clin Psychol Rev. 1993;13:721–39.
22. Raichle ME, MacLeod AM, Snyder AZ, Powers WJ, Gusnard DA, Shulman GL. A default mode of brain function. Proc Natl Acad Sci. 2001;98:676–82.
23. Janssen TW, Hillebrand A, Gouw A, Geladé K, Van Mourik R, Maras A, Oosterlaan J. Neural network topology in ADHD; evidence for maturational delay and default-mode network alterations. Clin Neurophysiol. 2017;128:2258–67.
24. Liddle EB, Hollis C, Batty MJ, Groom MJ, Totman JJ, Liotti M, Scerif G, Liddle PF. Task-related default mode network modulation and inhibitory control in ADHD: effects of motivation and methylphenidate. J Child Psychol Psychiatry. 2011;52:761–71.
25. Uddin LQ, Kelly AC, Biswal BB, Margulies DS, Shehzad Z, Shaw D, Ghaffari M, Rotrosen J, Adler LA, Castellanos FX, Milham MP. Network homogeneity reveals decreased integrity of default-mode network in ADHD. J Neurosci Methods. 2008;169:249–54.
26. Wilson TW, Franzen JD, Heinrichs-Graham E, White ML, Knott NL, Wetzel MW. Broadband neurophysiological abnormalities in the medial prefrontal region of the default-mode network in adults with ADHD. Hum Brain Mapp. 2013;34:566–74.
27. Sonuga-Barke EJ, Castellanos FX. Spontaneous attentional fluctuations in impaired states and pathological conditions: a neurobiological hypothesis. Neurosci Biobehav Rev. 2007;31:977–86.
28. Mitaki S, Onoda K, Ishihara M, Nabika Y, Yamaguchi S. Dysfunction of default-mode network in encephalopathy with a reversible corpus callosum lesion. J Neurol Sci. 2012;317:154–6.
29. Greicius MD, Supekar K, Menon V, Dougherty RF. Resting-state functional connectivity reflects structural connectivity in the default mode network. Cereb Cortex. 2009;19:72–8.
30. Van Den Heuvel M, Mandl R, Luigjes J, Pol HH. Microstructural organization of the cingulum tract and the level of default mode functional connectivity. J Neurosci. 2008;28:10844–51.
31. Bouziane C, Filatova OG, Schrantee A, Caan MW, Vos FM, Reneman L. White matter by diffusion MRI following methylphenidate treatment: a randomized control trial in males with attention-deficit/hyperactivity disorder. Radiology. 2019;293:186–92.
32. Castellanos FX, Lee PP, Sharp W, Jeffries NO, Greenstein DK, Clasen LS, Blumenthal JD, James RS, Ebens CL, Walter JM, Zijdenbos A. Developmental trajectories of brain volume abnormalities in children and adolescents with attention-deficit/hyperactivity disorder. JAMA. 2002;288:1740–8.
33. Volkow ND, Fowler JS, Wang G, Ding Y, Gatley SJ. Mechanism of action of methylphenidate: insights from PET imaging studies. J Atten Disord. 2002;6(Suppl 1):31–43.
34. Scahill L, Carroll D, Burke K. Methylphenidate: mechanism of action and clinical update. J Child Adolesc Psychiatr Nurs. 2004;17:85.

Part VI

Other Features of the Corpus Callosum

Editor's Summary

The first chapter in the last part of this book will review medicolegal aspects of the corpus callosum (CC) describing the lesions that can cause neurological symptoms and morbidity with a delay in diagnosis and treatment. As exemplified through interesting case scenarios in Chap. 40, a delay can lead to medicolegal consequences. Afterwards, statistical shape analysis of the CC, investigating the morphological changes in the CC induced by neurodegenerative and neurocognitive disorders, will be reviewed in Chap. 41 for readers of the book.

nance (MRI), which are the modalities of choice for evaluating cases with this rare disorder of the CNS or at autopsy [4, 8]. Therefore, it is apparent that these lesions should be carefully evaluated by a team comprising genetic counselors, obstetricians, radiologists, neurologists, and neurosurgeons.

Because the disorder is rare, there are few relevant data in the current literature; case reports and small case series have given variable conclusions regarding the optimal management of patients with various lesions involving the CC. This chapter will focus on these fascinating but complex issues, with a case scenario of malpractice claim including a delayed or inaccurate diagnosis and/or inappropriate management. As described below in the case scenario, the usual cases of negligence include a delay in diagnosis and treatment of agenesis of the CC, a delay in obtaining diagnostic fetal US or MRI studies, or improper management of such CC lesions.

40.2 Medicolegal Issues Regarding Lesions Involving the Corpus Callosum

There is still controversy about some aspects of the management of lesions involving the CC. Two important medicolegal issues related to this management are as follows:

40.2.1 Wrongful Birth

"Wrongful birth" is an allegation of malpractice by parents of a child born with various birth defects, who claim that inadvertent treatment or suggestion deprived them of the opportunity to prevent or terminate the pregnancy [9–11]. The mother takes the stance that if a genetic counselor, obstetrician, radiologist, neurologist, or neurosurgeon had reported the genetic defect early in testing, she would not have borne the child with the deformity to term [9–11]. The reason why the child was born with a genetic disorder was negligence by the physician or other health-care professionals, and the mother can claim the defendant to have been negligent in testing or counseling. From a legal point of view, the cause of action for the wrongful birth claim stems from the tortious act's interference with the mother's legal right to terminate a pregnancy involving a genetic defect.

40.2.2 Informed Consent and Parents' Right to Information

From a legal perspective, in informed consent cases, the plaintiff should prove that a reasonably prudent patient in the plaintiff's situation would refuse treatment if he/she was aware of the risks that the defendant had failed to disclose through negligence [9]. In a wrongful birth lawsuit, the plaintiff does not have to prove that negligence by a physician or other health-care professionals was the medical cause of their child's birth defect, but she would have terminated her pregnancy had she known of this risk [9]. The emotional distress and economic loss resulting from the absence of opportunity to decide whether to terminate the pregnancy constitutes the plaintiff's compensation claim. The task of a physician or other health-care professionals is to convey to the patient's sufficient material information to allow him/her to make an informed choice regarding continuation of the pregnancy.

Medicolegally, there is a difference between causing a birth defect and causing the parents to lose the option to terminate the pregnancy [9]. The critical question for a decision on this topic is whether inadequate explanations by a physician or other health-care professionals deprived the parents of their deep personal right to decide whether to give birth to a child who could possibly suffer a physical abnormality. In any case, plaintiffs must demonstrate that the resulting birth defect was fairly foreseeable and that the pregnancy would have been terminated had it not been for the defendant's negligence.

40.3 Case Scenario

A case of malpractice due to negligent management of agenesis of the CC, described below, provides useful guidelines for correct diagnosis and appropriate management by physicians from various disciplines including genetic counselors, obstetricians, radiologists, neurologists, and neurosurgeons in their future professional lives.

A claim was settled against the sonologist for "misdiagnosis" of agenesis of the CC of the patient as an interhemispheric arachnoid cyst (AC) leading to severe brain abnormalities. The brief case scenario is as follows: *In her 20th week of gestation, a woman underwent routine diagnostic US examination. Because a radiologist interpreted the study as disclosing possible dilatation of the third ventricle in the brain of the fetus, the patient's obstetrician performed an amniocentesis and referred her to a well-known expert in obstetric US. After reexamining the patient, the sonologist concluded that the cranial structures of the fetus were normal, but he nevertheless requested that the patient return in 4 weeks for a follow-up examination.*

By the time the patient reached her 24th week of gestation, the material obtained from amniocentesis had been evaluated and found to be normal. On this follow-up US, the sonologist noted that the third ventricle fluctuated in size during the examination but nonetheless appeared to be normal. The patient was asked to return in 3 weeks.

On the patient's third US examination, the sonologist noted a small oblong cyst measuring 1.5 × 0.8 cm just cephalad to the third ventricle that he believed was most likely an interhemispheric AC. He considered the possibility that the findings could indicate agenesis of the CC, but because he believed that the third ventricle itself was not dilated, he told the parents that he believed the fetus was free of any significant abnormality. Eight weeks later the patient gave birth to a baby girl who had agenesis of the CC and multiple CNS and congenital cardiac anomalies. Afterwards, the findings were clearly confirmed by MRI scans of the newborn.

At the age of 2 years the child, who remained markedly mentally retarded and physically disabled, and her parents, filed a wrongful life medical malpractice lawsuit against the sonologist. The suit alleged that the sonologist's failure to advise the parents of the substantial probability that their fetus suffered from severe brain abnormalities prevented them from terminating the pregnancy.

In this case scenario, there was a "misdiagnosis" of agenesis of the CC of the patient as an interhemispheric AC leading to severe brain abnormalities. Interestingly, the claim was settled against a radiologist (sonologist) nationally/internationally well known for expertise in obstetric US. A brain MRI scan subsequently revealed complete agenesis of the CC in this patient. As anticipated, the report from an expert consultant radiologist confirmed negligence by the defendant sonologist regarding recognition of the atypical orientation of the lateral ventricles and their teardrop shape. Radiologically, the abnormalities were not gross, but the failure to diagnose agenesis of the CC constituted negligence. Therefore, the statement by the sonologist that the US studies were reasonable and within the standard of care was not acceptable. Experts will confirm that the vast majority of babies born with isolated agenesis of the CC have normal intelligence; when the amniocentesis findings are normal, they would not recommend termination of the pregnancy solely on the basis of US findings reporting an isolated agenesis of the CC. In such a case, selection of termination of pregnancy by the child's parents is very important if they are informed about the likelihood of agenesis of the CC. In clinical practice, the parents will proceed with termination of pregnancy only if the amniocentesis (amniotic fluid test) has abnormal findings.

In accordance with the standard management algorithm for congenital lesions involving the CC, including agenesis, it can be crucial to terminate the pregnancy rather than observing the fetus because such a lesion with neurological deficits can lead to an undesirable postnatal outcome. Medicolegally, it is evident that these errors caused a failure to terminate the pregnancy to avoid child's birth defects and malformations of the CNS with severe disabilities.

40.4 Comment upon Case Scenario

Today it is well known that patients with lesions involving the CC can present with variable clinical signs and symptoms and neuroimaging appearances that depend principally on their anatomical locations, cyst sizes, and etiological forms and on the patient's age. Unfortunately, neuroimaging features of lesions involving the CC are neither uniform nor pathognomonic; therefore, these cysts can be confused with many normal anatomical variants, as described above in a case scenario with agenesis of the CC. As a result, the differential diagnosis is very important because all of these diagnoses can influence the management of the patient.

As is clear in the case scenario presented above, most malpractice related to the negligent management of lesions involving the CC includes misdiagnosis or delayed diagnosis and/or mistakes in treating patients by physicians including genetic counselors, obstetricians, radiologists, neurologists, and neurosurgeons. In fact, the case scenario highlights the importance of proper management of agenesis of the CC and that termination of the pregnancy can be indicated on the basis of prenatal imaging findings and/or results of genetic tests.

In principle, there is medical malpractice if a child is born with birth defects as a result of negligent management, depriving parents of the opportunity to end the pregnancy, by the physician or other health-care professionals failing to perform standard, up-to-date practice during diagnosis or treatment of a child with agenesis of the CC [12–17]. Unfortunately, many examples of negligence that can result in a lawsuit can be encountered nowadays, as exemplified by the above case scenario: absence or inadequacy of informed consent by the parents, failure to order the proper prenatal genetic testing for making a correct diagnosis, failure to diagnose, misdiagnosis or improper management, etc.

As explained earlier, malpractice lawsuits with allegations of wrongful life are filed by a disabled child against a physician or other health-care professionals for depriving the child's parents of the medical information necessary to make an informed decision about termination of the pregnancy. Parents of a disabled child often file their own lawsuit alleging "wrongful birth," claiming they were injured because they had to live with and care for a child who could not live a normal healthy life.

Legally, there is no doubt that physicians and other health-care professionals are responsible for the damages arising from lack of care in their efforts, actions, and behaviors to reach this result [13, 15–17]. It is well known that physicians and/or other health-care professionals are responsible for their smallest errors [16]. As a legitimate request, the patients and their relatives have the right to expect their physicians and other health-care professionals to be meticulous and alert at every stage of management of patients with agenesis of the CC.

It is critical that patients and their relatives be informed by physicians or other health-care professionals about the potential risks and benefits of surgical intervention. As a general rule, two main questions must be considered in resolving any medical malpractice case involving alleged negligence: (1) Did the behavior of physician or other health-care professionals violate the standard of medical care? If the answer is believed to be positive, then (2) did the negligence of the physician or other health-care professionals harm the plaintiff patient [18]? In most medical malpractice cases, as in the case scenario, the main debate centers on whether the physician or other health-care professionals was negligent, although the issue of patient's injury is almost certainly accepted. However, the type of harm claimed by the plaintiff requires much more attention because it is unusual when put simply. To begin with, the party claiming the damage was an unborn fetus at the time of the alleged injury, and the injury suffered was the actual birth of the claimant baby, but the nature of the alleged injury further obscures the matter.

In a wrongful life or wrongful birth lawsuit, it must be shown that the defendant's negligent behavior prevented a child's parents from making an informed decision regarding the continuation or termination of the pregnancy. If the child's parents had been properly informed by the physician about the possibility of abnormalities, they would have chosen to terminate the pregnancy, resulting in absence of the baby. If the defendant is a genetic counselor or obstetrician, the allegation could be that the parent was not informed about a test to identify a genetic defect or that the test was poorly performed or interpreted. If the defendant is a radiologist, the claim could be that the prenatal US was obtained or interpreted improperly; this could result in the radiologist, the referring obstetrician, or both, negligently withholding appropriate information from the parents.

Nowadays, most courts conclude that some lives are not worth living thanks to the development of technology for detecting genetic and fetal abnormalities. Therefore, if a physician or other health-care professionals fails to disclose the findings of imaging studies or results of other genetic tests to inform a couple's decision to continue or terminate a pregnancy, he/she could have violated a duty owed to both the couple and the prospective child. If the pregnancy ends when the child is exposed to congenital abnormalities, the child or their parents, or both, can have a valid basis for filing a malpractice action against the physician or other health-care professionals.

40.5 Conclusion

Nowadays, a lesion involving the CC can be mistreated because of the high frequency of misdiagnoses, leading to delayed or wrong management. In clinical practice, therefore, cases of malpractice regarding the management of agenesis of the CC are common. In terms of forensic medicine, it is important that physicians and other health-care professionals realize that any negligence and carelessness at any stage will lead to serious medical malpractice in the management of cases affected by agenesis of the CC. In the future, large multicenter prospective randomized controlled trials and evidence-based medicine are required to yield strong conclusions regarding this topic.

References

1. Blaauw J, Meiners LC. The splenium of the corpus callosum: embryology, anatomy, function and imaging with pathophysiological hypothesis. Neuroradiology. 2020;62:563–85.
2. Shah A, Jhawar S, Goel A, Goel A. Corpus callosum and its connections: a fiber dissection study. World Neurosurg. 2021;151:e1024–35.
3. Unterberger I, Bauer R, Walser G, Bauer G. Corpus callosum and epilepsies. Seizure. 2016;37:55–60.
4. Edwards TJ, Sherr EH, Barkovich AJ, Richards LJ. Clinical, genetic and imaging findings identify new causes for corpus callosum development syndromes. Brain. 2014;137:1579–613.
5. Alby C, Malan V, Boutaud L, Marangoni MA, Bessières B, Bonniere M, Ichkou A, Elkhartoufi N, Bahi-Buisson N, Sonigo P, Millischer AE, Thomas S, Ville Y, Vekemans M, Encha-Razavi F, Attié-Bitach T. Clinical, genetic and neuropathological findings in a series of 138 fetuses with a corpus callosum malformation. Birth Defects Res A Clin Mol Teratol. 2016;106:36–46.
6. Kidron D, Shapira D, Ben Sira L, Malinger G, Lev D, Cioca A, Sharony R, Lerman ST. Agenesis of the corpus callosum. An autopsy study in fetuses. Virchows Arch. 2016;468:219–30.
7. Aldinger Kimberly A, Doherty D. The genetics of cerebellar malformations. Semin Fetal Neonatal Med. 2016;21:321–32.
8. Li X, Wang Q. Magnetic resonance imaging (MRI) diagnosis of fetal corpus callosum abnormalities and follow-up analysis. J Child Neurol. 2021;36:1017–26.
9. Berlin L. Wrongful life. AJR Am J Roentgenol. 2003;181:1181–8.
10. Schmidt SM. Wrongful life. JAMA. 1983;250:2209–10.
11. Weil E. A wrongful birth? N Y Times Mag. 2006;12:48–53.
12. Arikan M, Kalkan E, Erdi F, Deniz M, İzci E. The distinction between cases and malpractice-complications in medical law: the perspectives of the senior students of the faculty of medicine and the faculty of law, prob-

lems and suggestions for solution (in Turkish). Türk Nöroşir. 2016;26:40–8.
13. Bal BS. An introduction to medical malpractice in the United States. Clin Orthop Relat Res. 2009;467(2):339–47.
14. Birtek F. Complication-malpractice distinction in terms of medical interventions (in Turkish). İstanbul Barosu Dergisi. 2007;81:1997–2007.
15. Hakeri H. Distiction between malpractice and complication in medical law (in Turkish). Toraks Cerrahisi Bülteni. 2014;1:23–8.
16. Kök N. Malpractice and complication distinction. In: Kalkan E, Serel TA, Yilmaz EN, editors. Legal guide for physicians (in Turkish). Ankara: Sage Publishing; 2018. p. 109–39. Turkish Neurosurgery Academy Publications No:1.
17. Marcus P. Book Review of medical malpractice law: a comparative law study of civil responsibility arising from medical care. Hastings Int Comp Law Rev. 1981;5:235–43.
18. Berlin L. Proximate cause. AJR Am J Roentgenol. 2002;179:569–73.

41 Statistical Shape Analysis of Corpus Callosum

Yaşar Türk

41.1 Introduction

Exactly 30 years ago, one of the most fundamental concepts in image analysis was presented [1]. Shape is the geometrical information that remains when position, scale, and rotation are removed. A landmark is a point of matching between and within populations. An outline can be described by a limited number of points, so a landmark is adopted. Dryden and Mardia divided landmarks into three subgroups: (1) anatomical landmarks, points assigned by an expert that correlate between species in a scientifically relevant way; (2) mathematical landmarks, points on an object with mathematical or geometrical significance such as strong curvature or an extremum; and (3) pseudo landmarks, constructed points on an item, either on the outline or between landmarks [2].

Procrustes analysis and large deformation diffeomorphic metric mapping (LDDMM) are two methods for analyzing the shape of the corpus callosum (CC) statistically. This chapter discusses both methods in relation to a variety of neurodegenerative diseases and neurocognitive disorders that affect the brain.

Y. Türk (✉)
Radiology Department, Şişli Kolan International Hospital, İstanbul, Turkey

41.2 Statistical Shape Analysis

Statistical shape analysis is the statistical assessment of a set of geometrical properties of a structure. It includes determining the distance between morphologies, calculating mean shapes from (possibly random) samples, estimating shape variability within samples, clustering, and testing for differences between shapes [2, 3]. One of the most widely used approaches is principal component analysis (PCA). Statistical shape analysis has a range of applications including medical imaging [4], computer vision, computational anatomy, sensor measurement, and geographical profiling [5].

41.2.1 Landmark-Based Techniques

A shape in the point distribution model is defined by a small set of coordinate points known as landmark points. These landmark locations are often linked to important distinguishing features such as the corners of the eyes. There is some kind of registration once the points have been established. This is one of Bookstein's typical anthropological geometric morphometric techniques. Another technique is Procrustes analysis, which calculates the shape's median. Kendall investigated the statistical distribution of triangular forms and represented each as a point on a sphere. He explored ley lines and whether three stones were more likely to be colinear than

M. Turgut et al. (eds.), *The Corpus Callosum*, https://doi.org/10.1007/978-3-031-38114-0_41

expected from this distribution on the sphere [6]. Statistical methods such as the Kent distribution can be used to analyze the distribution of such spaces. Forms can also be depicted as curves or surfaces expressing their contours, depending on the amount of space they occupy [7–10].

Geometric morphometrics is a statistical shape analysis concept that involves the use of landmarks [8]. It is a relatively new paradigm for quantitatively analyzing the variability and covariance of the shapes of biological entities [9, 11]. Coordinate-based techniques keep all geometric information intact throughout the data gathering, processing, and visualization phases. Statistical shape or image analysis, which is used to evaluate data in two-dimensional or three-dimensional form, is rapidly expanding in bioscience. The increasing application of statistical shape analysis in medicine is due to developments in neuroimaging and a desire to elucidate the influence of disorders and environment on organ or organism structure [12].

41.2.2 Shape Deformations

Deformations that transform one shape into another can be used to examine shape disparities. Smoothness is preserved in the deformation of a diffeomorphism in particular. This was pioneered before computers in D'Arcy Thompson's "On Growth and Form" [13]. A deformation can be regarded as the result of a force acting on a shape and is described mathematically as a transformation function (ϕ) mapping from shape x to shape y (i.e., $y = \phi.x$) [2]. Given a concept of deformation size, the distance between two forms can be defined as the size of their smallest distortion.

Large deformation diffeomorphic metric mapping (LDDMM) is a cutting-edge anatomical manifold matching approach that has been successfully used to characterize the local morphometrics of discrete brain regions of interest in diverse neurodegenerative diseases [14–17]. This global registration framework can be used to register landmarks, curves, surfaces, and dense images of different dimensions.

LDDMM for curve (LDDMM-curve) has been used to analyze the statistical forms of numerous two-dimensional brain regions of interest (ROIs), including the two-dimensional mid-sagittal CC [18–20]. The CC is the largest white matter structure in the human brain and the largest commissural fiber bundle connecting the two cerebral hemispheres, making it a focus for research into Alzheimer's disease (AD) [21], Huntington's disease [22], and schizophrenia [23].

LDDMM-curve registrations between a template curve and all target population curves could take hours or days, particularly when the sample size of the population of interest is enormous. Lowering the computational cost of the LDDMM-curve registration technique is therefore vital both technically and practically. Previously, GPU-based parallel computing was employed to accelerate LDDMM [24–27]. In addition to GPU-based acceleration, Wu et al. proposed a unique optimization strategy to speed up the LDDMM registration process, but only for registering photographs [28]. Other activities have also been launched in this area. Most recently, Hernandez recast the problem in the space of band-limited vector fields and connected it with semi-Lagrangian integration for three types of band-limited PDE-constrained LDDMM to boost computation performance [29].

Another approach to characterizing anatomical manifolds of interest is to decrease the discretized sample points. Downsampling the images of interest to have fewer voxels in the LDDMM-image, downsampling the surfaces of interest to have fewer vertices in the LDDMM-surface, and downsampling the curves of interest to have fewer points in the LDDMM-curve would all reduce calculation time. However, difficulties can arise. First, registration accuracy can be compromised. Second, study findings can be affected, especially in identifying and quantifying disease-related anomalies.

LDDMM employs dual matching. First, three-dimensional volume pictures, multi-valued vectors, and DTI tensor matrices are used [30, 31]. Second, landmark matching requires human or semiautomated selection [32]. Sulcal or gyral curves derived from segmented cortical surfaces are common brain imaging landmarks. Such examples use landmark matching [33, 34], which is unsatisfactory because it requires the same number of points to be picked on the source and destination curves and defines arbitrary correspondences between them, confining registration. Level-set curve representations solved this problem [35]. In Glaunès et al., forms are unlabeled landmark groupings [36]. Higher-order information (tangent vectors or curvature) is lost when one-dimensional or two-dimensional manifolds are converted to zero-dimensional point sets. Vaillant and Glaunès devised a surface representation using vector measurements. Location and surface information can be input [37]. Glaunès et al. and Vaillant and Glaunès use the same paradigm to compute metrics on m-dimensional R^d submanifolds with $m \leq d$ [36, 37].

41.3 Magnetic Resonance Imaging

Magnetic resonance imaging (MRI) sequences are chosen to give strong contrast between gray and white matter, and the exams are performed using standard sagittal T1- or T2-weighted spin echo sequence images (FOV, 230,250 mm; section thickness, 5 mm; TR/TE, 5800-6700/95-115 ms; flip angle, 90°) from patients who had undergone conventional MRI.

41.4 Landmark Acquisition

In Procrustes analysis or landmark-based techniques, an experienced radiologist carefully picks the mid-sagittal sections from the sagittal planes, ensuring that the cerebral aqueduct, CC, and superior colliculus are clearly visible. The anterior–posterior commissure line and the interhemispheric fissure are located and used to align the brains of all patients in the same place. Using the existing Witelson framework, which divides the CC into 7 subdivisions on the basis of certain geometric components [38], 16 homologous markers are applied, and CC shape variations are studied for each segment. To highlight the specified landmarks on the digital images, the TPSDIG 2.17 program is used [39]. To measure landmark reliability, the intra-rater reliability coefficient for a two-facet crossed design (landmark pairs-by-rater-by-subject) based on generalizability theory is calculated [40, 41]. An example of patient-to-control distortion follows (Fig. 41.1).

Fig. 41.1 Sample graph of patient-to-control deformation. If we deduce meaning from the color scale, a hue with a value of 1 indicates no distortion. Red areas in the CC show considerable deformation. Likewise, dark blue toward the bottom indicates significant distortion

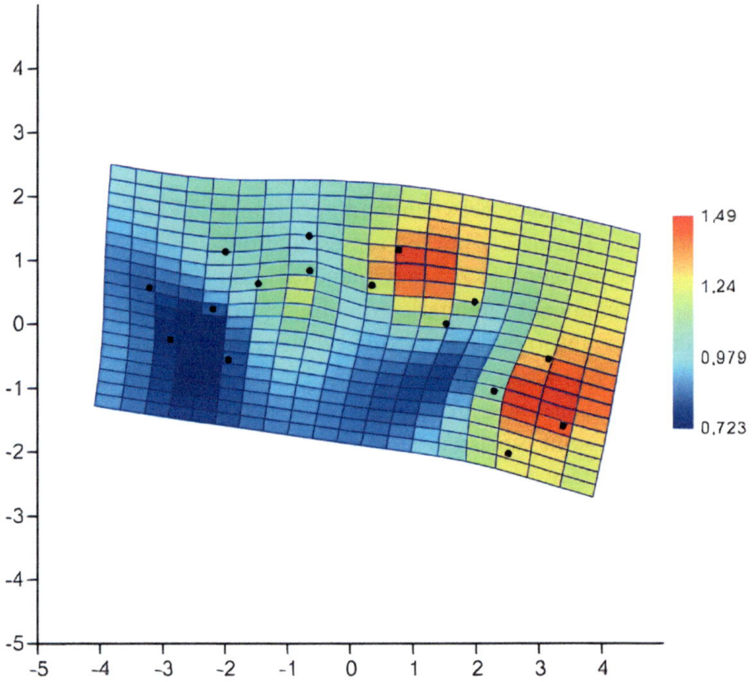

41.5 Discussion

The CC links the hemispheres of the brain. It is composed of nerve fibers of various thicknesses that gather in the midline and spread to topographically equivalent places in either hemisphere. Its involvement in hemispheric communication, lateralized brain activity, and cognition has been extensively studied.

MRI investigations of CC morphometry have focused mostly on its three-dimensional global volume [42–44] or its two-dimensional area measurement [23, 45, 46] as opposed to its local shape properties. In this section, we examine a few studies of mid-sagittal CC shapes using Procrustes analysis, followed by studies of such shapes using curve-based shape deformation or large deformation diffeomorphic metric mapping.

Our research team studied the CC in patients with deficit and non-deficit schizophrenia. They discovered deformations in the genu, along the superior border between the posterior body and the isthmus and between the isthmus and the splenium, indicating that schizophrenia affects the prefrontal and parietotemporal cortices [47]. The superior zone of the CC between the isthmus, splenium, and rostrum is impaired in the DS group, possibly because DS affects the superior temporal cortex and cingulum. This investigation of CC form was coordinated with Catani and Thiebaut de Schotten's parcellations of the CC along the midline and their study of the connectional division of the cortex [48].

Guneş et al. [49] found abnormalities in the genu area of the CC linking the prefrontal cortex in mild to moderate OSA patients and in the genu and rostral body regions in severe OSA cases [49]. Zhang et al. used DTI-based fiber tractography to show that the affected CC subregions in OSA were in the front rather than the back. They also pointed out that the earlier myelination and different fiber composition of the front CC could make it more vulnerable to OSA [50], similar to the findings of Guneş et al. [49]. The prefrontal cortex regulates attention, distinguishes conflicting ideas and plans, and predicts future behaviors and impulses. The premotor cortex regulates spatial attention, working memory, movement planning, and execution [51, 52]. Attention, vigilance, and prospective memory are multistage processes regulated by the prefrontal cortex and other cortices. Sensory and prefrontal cortex cells focus attention and form working memory for movement. The premotor and supplementary motor cortexes receive input from those locations [53].

Sigirli et al. analyzed the shape of the CC statistically in restless leg syndrome (RLS) [54]. They found that distortion was most evident in the posterior midbody and somewhat less in the anterior area, which they interpreted as evidence that the development of RLS affects the somatosensory cortex and, to a lesser extent, the motor region. In addition, longer illness duration and greater disease severity were related to substantial shrinkage of the CC, indicating that demyelination is connected to disease severity in RLS patients. The results of this statistical shape analysis were consistent with published neuroimaging studies [55, 56].

Huang et al. studied the deformation-based shape of the CC in first-episode schizophrenia [57]. The totally automated nature of the shape analysis pipeline in that study made it superior to previous CC analyses, which mostly used manual feature extraction. Additionally, rather than focusing just on each individual point, the authors' point-wise shape deformation markers from the LDDMM-curve took the surrounding morphometry into account. In their investigation, HC and FES were compared in both global and local CC morphometrics. The area was studied to obtain an overall measurement. A well-known deformation-based statistical shape analysis pipeline, validated in earlier research, was used to assess shape for local measurements [14, 17, 20]. According to their global area study findings, there was a substantial difference between the HC and FES groups in the gCC subregion. There was significant FES-related inward deformation of shape in the superior section of the genu and the anterosuperior area of the CC body. These shape findings were reasonably compatible with previous investigations [58, 59] that focused on regional thickness but not deformation. The Pearson correlation coefficient (PCC) between the average thickness and average deformation marker of the full two-dimensional CC showed a substantial linear correlation. The two metrics are connected but not comparable. In two dimensional CC, the deformation marker indicates area change but not thickness. The deformation marker value should be substantially connected to a region's area when averaged. To validate this, the authors computed the PCC between the area and the average deformation marker of the full two-dimensional CC curve, revealing a substantial linear association. The statistical association between these two metrics was considerably stronger than that between deformation marker and thickness. The authors also compared the area of each CC curve to the result of multiplying the template area by the average target-specific deformation marker. No significant group differences were detected. Their gender results were notable. In both HC and FES, females had bigger areas than men in the total CC and each of its three subregions, consistent with earlier results [23, 59, 60], and female patients had more CC shrinkage than male patients compared to healthy control peers, notably in the genu. Morphometric analysis revealed statistically significant global and local inward genu deformations only in the female group. The authors hypothesized that female CC morphology, specifically genu morphology, was more sensitive to FES neuropathology. Therefore, the global area and localized form of the genu could be useful indicators for screening female FES patients. The genu is the most curved region of the anterior CC, which projects the forceps minor fibers. It links the frontal lobe's medial and lateral surfaces and functions with it [61]. It is involved in emotional regulation and its ventral side is connected to attention and mood processing, which could explain the FES-associated genu abnormalities identified in this research. The gender difference in schizophrenia could be driven by social environment, genetic and hormonal variations, brain development and morphology, and age- and gender-specific behavior patterns. Häfner found that the only significant difference between men and women was the age at which the illness began and at all subsequent milestones through the early course of the illness [62]. As estrogen production falls in preclimacterium, women suffer a second peak of onsets with more severe first-episode symptoms and a more severe illness course than males. This could explain why this investigation identified morphometric abnormalities in female but not male FES cases.

41.6 Conclusion

The results of several studies linking callosal morphology to neurodegenerative and neurocognitive disorders suggest that individual variations in callosum anatomy are not random. This raises the possibility that functional consequences can follow variations in callosum anatomy or vice versa. Gross architectural variations in the CC can provide information about individual differences in cortical and white matter histology. The nature of such putative histological differences, such as neuronal cell or axonal packing density, can elucidate the mechanisms behind neuroanatomical variety and the associated psychological variation among individuals.

References

1. Kass M, Witkin A. Snakes: active contour models. Int J Comput Vis. 1988;1:321–31.
2. Dryden IL, Mardia KV. Statistical shape analysis, vol. 347. New York: Wiley; 1998.
3. Ziezold H. Mean figures and mean shapes applied to biological figure and shape distributions in the plane. Biom J. 1994;36:491–510. https://doi.org/10.1002/BIMJ.4710360409.
4. Zheng G, Li S, Székely G. Statistical shape and deformation analysis: methods, implementation and applications. New York: Academic; 2017.
5. Giebel S. Zur Anwendung der statistischen Formanalyse Einstieg in die drei- und vierdimensionale Formanalyse; 1976.
6. Professor David Kendall | Ghostarchive [www.document]. n.d. https://ghostarchive.org/archive/2BGNM. Accessed 15 Jul 2022.
7. Bauer M, Bruveris M, Michor PW. Overview of the geometries of shape spaces and diffeomorphism groups. J Math Imaging Vision. 2014;50:60–97. https://doi.org/10.1007/S10851-013-0490-Z.
8. Ercan I, Ocakoglu G, Sigirli D, Ozkaya G. Statistical shape analysis and usage in medical sciences. Türkiye Klinikleri J Biostat. 2012;4(1):27–35.
9. Rohlf FJ. On the use of shape spaces to compare morphometric methods. Hystrix Ital J Mammal. 2000;11:27. https://doi.org/10.4404/hystrix-11.1-4134.
10. Zhang D, Lu G. Review of shape representation and description techniques. Pattern Recogn. 2004;37:1–19. https://doi.org/10.1016/J.PATCOG.2003.07.008.
11. Rohlf FJ. Shape statistics: procrustes superimpositions and tangent spaces. J Classif. 1999;16:197–223. https://doi.org/10.1007/s003579900054.
12. Slice DE. Geometric morphometrics. Annu Rev Anthropol. 2007;36:261–81. https://doi.org/10.1146/annurev.anthro.34.081804.120613.
13. Thompson DW. On growth and form: the complete revised edition. Mineola: Dover Publications; 1992. p. 1116.
14. Garg A, Appel-Cresswell S, Popuri K, McKeown MJ, Beg MF. Morphological alterations in the caudate, putamen, pallidum, and thalamus in Parkinson's disease. Front Neurosci. 2015;9:101. https://doi.org/10.3389/FNINS.2015.00101.
15. Qiu A, Miller MI. Cortical hemisphere registration via large deformation diffeomorphic metric curve mapping. In: Lecture notes in computer science, Lecture notes in artificial intelligence and lecture notes in bioinformatics, vol. 4791. Berlin: Springer; 2007. p. 186–93. https://doi.org/10.1007/978-3-540-75757-3_23/COVER/.
16. Qiu A, Vaillant M, Barta P, Ratnanather JT, Miller MI. Region-of-interest-based analysis with application of cortical thickness variation of left planum temporale in schizophrenia and psychotic bipolar disorder. Hum Brain Mapp. 2008;29:973–85. https://doi.org/10.1002/HBM.20444.
17. Tang X, Holland D, Dale AM, Younes L, Miller MI. Shape abnormalities of subcortical and ventricular structures in mild cognitive impairment and Alzheimer's disease: detecting, quantifying, and predicting. Hum Brain Mapp. 2014;35:3701. https://doi.org/10.1002/HBM.22431.
18. Glaunès J, Qiu A, Miller MI, Younes L. Large deformation diffeomorphic metric curve mapping. Int J Comput Vis. 2008;80:317. https://doi.org/10.1007/S11263-008-0141-9.
19. Jiang Z, Yang H, Tang X. Deformation-based statistical shape analysis of the corpus callosum in mild cognitive impairment and Alzheimer's disease. Curr Alzheimer Res. 2018;15:1151–60. https://doi.org/10.2174/1567205015666180813145935.
20. Qiu A, Tuan TA, Woon PS, Abdul-Rahman MF, Graham S, Sim K. Hippocampal-cortical structural connectivity disruptions in schizophrenia: an integrated perspective from hippocampal shape, cortical thickness, and integrity of white matter bundles. NeuroImage. 2010;52:1181–9. https://doi.org/10.1016/J.NEUROIMAGE.2010.05.046.
21. Hampel H, Teipel SJ, Alexander GE, Horwitz B, Teichberg D, Schapiro MB, Rapoport SI. Corpus callosum atrophy is a possible indicator of region- and cell type-specific neuronal degeneration in Alzheimer

disease: a magnetic resonance imaging analysis. Arch Neurol. 1998;55:193–8. https://doi.org/10.1001/ARCHNEUR.55.2.193.
22. Rosas HD, Lee SY, Bender AC, Zaleta AK, Vangel M, Yu P, Fischl B, Pappu V, Onorato C, Cha JH, Salat DH, Hersch SM. Altered white matter microstructure in the corpus callosum in Huntington's disease: implications for cortical "disconnection". NeuroImage. 2010;49:2995–3004. https://doi.org/10.1016/J.NEUROIMAGE.2009.10.015.
23. Arnone D, McIntosh AM, Tan GMY, Ebmeier KP. Meta-analysis of magnetic resonance imaging studies of the corpus callosum in schizophrenia. Schizophr Res. 2008;101:124. https://doi.org/10.1016/j.schres.2008.01.005.
24. Brunn M, Himthani N, Biros G, Mehl M, Mang A. Fast GPU 3D diffeomorphic image registration. J Parallel Distrib Comput. 2021;149:149–62. https://doi.org/10.1016/J.JPDC.2020.11.006.
25. Sommer S. Accelerating multi-scale flows for LDDKBM diffeomorphic registration. In: Proceedings of the IEEE international conference on computer vision, 2011. p. 499–505. https://doi.org/10.1109/ICCVW.2011.6130284.
26. Tward DJ, Kolasny A, Sicat CS, Brown T, Miller MI. Tools for studying populations and time series of neuroanatomy enabled through GPU acceleration in the computational anatomy gateway. In: ACM international conference proceeding series 17-21 July 2016. https://doi.org/10.1145/2949550.2949574.
27. Wu J, Tang X. Fast diffeomorphic image registration via GPU-based parallel computing: an investigation of the matching cost function. Proc SPIE 97. https://doi.org/10.1117/12.2293082.
28. Wu J, Tang X. A large deformation diffeomorphic framework for fast brain image registration via parallel computing and optimization. Neuroinformatics. 2020;18:251–66. https://doi.org/10.1007/S12021-019-09438-7.
29. Hernandez M. Combining the band-limited parameterization and semi-Lagrangian Runge-Kutta integration for efficient PDE-constrained LDDMM. J Math Imaging Vision. 2021;63:555. https://doi.org/10.1007/s10851-021-01016-4.
30. Beg MF, Miller MI, Trouvé A, Younes L. Computing large deformation metric mappings via geodesic flows of diffeomorphisms. Int J Comput Vision. 2005;61(2):139–57. https://doi.org/10.1023/B:VISI.0000043755.93987.AA.
31. Cao Y, Miller MI, Winslow RL, Younes L. Large deformation diffeomorphic metric mapping of vector fields. IEEE Trans Med Imaging. 2005;24:1216. https://doi.org/10.1109/TMI.2005.853923.
32. Joshi SC, Miller MI. Landmark matching via large deformation diffeomorphisms. IEEE Trans Image Process. 2000;9:1357–70. https://doi.org/10.1109/83.855431.
33. Bakircioglu MM, Joshi SC, Miller MI. Landmark matching on brain surfaces via large deformation diffeomorphisms on the sphere. Proc SPIE. 1999;3661:710–5. https://doi.org/10.1117/12.348628.
34. Joshi AA, Shattuck DW, Thompson PM, Leahy RM. Registration of cortical surfaces using sulcal landmarks for group analysis of MEG data. Int Cong Ser. 2007;1300:229. https://doi.org/10.1016/J.ICS.2006.12.052.
35. Leow A, Leow A, Thompson P, Brain warping with implicit representations. Presented at ISBI; n.d.
36. Glaunes J, Trouvé A, Younes L. Diffeomorphic matching of distributions: a new approach for unlabelled point-sets and sub-manifolds matching. In: Proceedings of the IEEE Computer Society conference on computer vision and pattern recognition, vol. 2, 2004. https://doi.org/10.1109/CVPR.2004.1315234.
37. Vaillant M, Glaunès J. Surface matching via currents. Lect Notes Comput Sci. 2005;3565:381–92. https://doi.org/10.1007/11505730_32/COVER/.
38. Witelson SF. Hand and sex differences in the isthmus and genu of the human corpus callosum. A postmortem morphological study. Brain. 1989;112(Pt 3):799–835. https://doi.org/10.1093/BRAIN/112.3.799.
39. Rohlf FJ. Digitize landmarks and outlines, version 2.17. New York: Department of Ecology and Evolution, State University of at Stony Brook; 2004.
40. Ercan I, Ocakoglu G, Guney I, Yazici B. Adaptation of generalizability theory for inter-rater reliability for landmark localization. Int J Tomogr Simul. 2008;9:51–8.
41. Erford BT. Assessment for counselors. London: Pearson; 2007. p. 574.
42. Collinson SL, Gan SC, Woon PS, Kuswanto C, Sum MY, Yang GL, Lui JM, Sitoh YY, Nowinski WL, Sim K. Corpus callosum morphology in first-episode and chronic schizophrenia: combined magnetic resonance and diffusion tensor imaging study of Chinese Singaporean patients. Br J Psychiatry. 2014;204:55–60. https://doi.org/10.1192/bjp.bp.113.127886.
43. de Moura MTM, Zanetti MV, Duran FLS, Schaufelberger MS, Menezes PR, Scazufca M, Busatto GF, Serpa MH. Corpus callosum volumes in the 5 years following the first-episode of schizophrenia: effects of antipsychotics, chronicity and maturation. Neuroimage Clin. 2018;18:932–42. https://doi.org/10.1016/J.NICL.2018.03.015.
44. Evans AC, Collins DL, Mills SR, Brown ED, Kelly RL, Peters TM. 3D statistical neuroanatomical models from 305 MRI volumes. In: IEEE nuclear science symposium & medical imaging conference, 1994. p. 1813–17. https://doi.org/10.1109/NSSMIC.1993.373602.
45. Bachmann S, Pantel J, Flender A, Bottmer C, Essig M, Schröder J. Corpus callosum in first-episode patients with schizophrenia--a magnetic resonance imaging study. Psychol Med. 2003;33:1019–27. https://doi.org/10.1017/S0033291703008043.
46. Keshavan MS, Diwadkar VA, Harenski K, Rosenberg DR, Sweeney JA, Pettegrew JW. Abnormalities of the corpus callosum in first episode, treatment naive schizophrenia. J Neurol Neurosurg

Psychiatry. 2002;72:757–60. https://doi.org/10.1136/jnnp.72.6.757.
47. Türk Y, Ercan I, Sahin I, Erdemli Gursel B, Uzunoglu A, Öge C, Beyazyüz E, Albayrak Y. Corpus callosum in schizophrenia with deficit and non-deficit syndrome: a statistical shape analysis. Gen Psychiatry. 2021;34:e100635. https://doi.org/10.1136/GPSYCH-2021-100635.
48. Catani M, Thiebaut de Schotten M. Atlas of human brain connections. Oxford: Oxford University Press; 2013. https://doi.org/10.1093/med/9780199541164.001.0001.
49. Gunes A, Sigirli D, Ercan I, Turan Ozdemir S, Durmus Y, Yildiz T. Evaluation of the corpus callosum shape in patients with obstructive sleep apnea. Sleep Breath. 2021;1:1–7. https://doi.org/10.1007/S11325-021-02502-0/FIGURES/3.
50. Zhang B, Zhu DM, Zhao W, Zhang Y, Yang Y, Zhang C, Zhu J, Yu Y. Selective microstructural integrity impairments of the anterior corpus callosum are associated with cognitive deficits in obstructive sleep apnea. Brain Behav. 2019;9:e01482. https://doi.org/10.1002/BRB3.1482.
51. Baleydier C, Achache P, Froment JC. Neurofilament architecture of superior and mesial premotor cortex in the human brain. Neuroreport. 1997;8:1691–6. https://doi.org/10.1097/00001756-199705060-00026.
52. Simon SR, Meunier M, Piettre L, Berardi AM, Segebarth CM, Boussaoud D. Spatial attention and memory versus motor preparation: premotor cortex involvement as revealed by fMRI. J Neurophysiol. 2002;88:2047–57. https://doi.org/10.1152/JN.2002.88.4.2047.
53. Feher J. Quantitative human physiology: an introduction. Am J Med Sci. 2012;243:386–400.
54. Sigirli D, Gunes A, Turan Ozdemir S, Ercan I, Durmus Y, Erdemli Gursel B. Statistical shape analysis of corpus callosum in restless leg syndrome. Neurol Res. 2020;42:760–6. https://doi.org/10.1080/01616412.2020.1773631.
55. Chang Y, Paik JS, Lee HJ, Chang HW, Moon HJ, Allen RP, Earley CJ, Cho YW. Altered white matter integrity in primary restless legs syndrome patients: diffusion tensor imaging study. Neurol Res. 2014;36:769–74. https://doi.org/10.1179/1743132814Y.0000000336.
56. Connor JR, Ponnuru P, Lee BY, Podskalny GD, Alam S, Allen RP, Earley CJ, Yang QX. Postmortem and imaging based analyses reveal CNS decreased myelination in restless legs syndrome. Sleep Med. 2011;12:614. https://doi.org/10.1016/J.SLEEP.2010.10.009.
57. Huang W, Chen M, Lyu G, Tang X. A deformation-based shape study of the corpus callosum in first episode schizophrenia. Front. Psychiatry. 2021;12:621515. Published 2021 Jun 4. https://doi.org/10.3389/fpsyt.2021.621515.
58. Walterfang M, Wood AG, Reutens DC, Wood SJ, Chen J, Velakoulis D, McGorry PD, Pantelis C. Corpus callosum size and shape in first-episode affective and schizophrenia-spectrum psychosis. Psychiatry Res Neuroimaging. 2009;173:77. https://doi.org/10.1016/j.pscychresns.2008.09.007.
59. Walterfang M, Wood AG, Reutens DC, Wood SJ, Chen J, Velakoulis D, McGorry PD, Pantelis C. Morphology of the corpus callosum at different stages of schizophrenia: cross-sectional study in first-episode and chronic illness. Br J Psychiatry. 2008;192:429–34. https://doi.org/10.1192/BJP.BP.107.041251.
60. Ardekani BA, Figarsky K, Sidtis JJ. Sexual dimorphism in the human corpus callosum: an MRI study using the OASIS brain database. Cereb Cortex. 2013;23:2514–20. https://doi.org/10.1093/CERCOR/BHS253.
61. Grey H. Gray's anatomy: the anatomical basis of clinical practice. 40th ed. London: Churchill Livingstone/Elsevier; 2008.
62. Häfner H. Gender differences in schizophrenia. Psychoneuroendocrinology. 2003;28(Suppl 2):17–54. https://doi.org/10.1016/S0306-4530(02)00125-7.

Conclusion

Our understanding of the corpus callosum (CC) has come a long way since its earliest descriptions. However, there is still much to learn and decipher about this fascinating component of the central nervous system. With an improved understanding of the CC, clinicians will be able to provide their patients with better treatments. Additionally, improved imaging modalities such as tractography and functional MRI will allow the surgeon to target pathologies more accurately in or around the CC. Future imaging technologies will assuredly show us more details about this structure. The current offering on this topic will provide the reader with the most up-to-date information and will be a go-to resource for those interested in this commissural tract of the brain for years to come. The expert authors who have contributed their knowledge to this book are to be commended.

M. Turgut et al. (eds.), *The Corpus Callosum*, https://doi.org/10.1007/978-3-031-38114-0

Index

M. Turgut et al. (eds.), *The Corpus Callosum*, https://doi.org/10.1007/978-3-031-38114-0

GPSR Compliance

The European Union's (EU) General Product Safety Regulation (GPSR) is a set of rules that requires consumer products to be safe and our obligations to ensure this.

If you have any concerns about our products, you can contact us on ProductSafety@springernature.com

In case Publisher is established outside the EU, the EU authorized representative is:

Springer Nature Customer Service Center GmbH
Europaplatz 3
69115 Heidelberg, Germany

Batch number: 10731383

Printed by Printforce, the Netherlands